SIGNALS and SYSTEMS PRIMER with MATLAB®

THE ELECTRICAL ENGINEERING AND APPLIED SIGNAL PROCESSING SERIES
Edited by Alexander Poularikas

The Advanced Signal Processing Handbook: Theory and Implementation for Radar, Sonar, and Medical Imaging Real-Time Systems
Stergios Stergiopoulos

The Transform and Data Compression Handbook
K.R. Rao and P.C. Yip

Handbook of Multisensor Data Fusion
David Hall and James Llinas

Handbook of Neural Network Signal Processing
Yu Hen Hu and Jenq-Neng Hwang

Handbook of Antennas in Wireless Communications
Lal Chand Godara

Noise Reduction in Speech Applications
Gillian M. Davis

Signal Processing Noise
Vyacheslav P. Tuzlukov

Digital Signal Processing with Examples in MATLAB®
Samuel Stearns

Applications in Time-Frequency Signal Processing
Antonia Papandreou-Suppappola

The Digital Color Imaging Handbook
Gaurav Sharma

Pattern Recognition in Speech and Language Processing
Wu Chou and Biing-Hwang Juang

Propagation Handbook for Wireless Communication System Design
Robert K. Crane

Nonlinear Signal and Image Processing: Theory, Methods, and Applications
Kenneth E. Barner and Gonzalo R. Arce

Smart Antennas
Lal Chand Godara

Mobile Internet: Enabling Technologies and Services
Apostolis K. Salkintzis and Alexander Poularikas

Soft Computing with MATLAB®
Ali Zilouchian

Wireless Internet: Technologies and Applications
Apostolis K. Salkintzis and Alexander Poularikas

Signal and Image Processing in Navigational Systems
Vyacheslav P. Tuzlukov

Medical Image Analysis Methods
Lena Costaridou

MIMO System Technology for Wireless Communications
George Tsoulos

Signals and Systems Primer with MATLAB®
Alexander Poularikas

SIGNALS and SYSTEMS PRIMER with MATLAB®

Alexander D. Poularikas

CRC Press
Taylor & Francis Group
Boca Raton   London   New York

CRC Press is an imprint of the
Taylor & Francis Group, an informa business

CRC Press
Taylor & Francis Group
6000 Broken Sound Parkway NW, Suite 300
Boca Raton, FL 33487-2742

© 2007 by Taylor & Francis Group, LLC
CRC Press is an imprint of Taylor & Francis Group, an Informa business

No claim to original U.S. Government works
Printed in the United States of America on acid-free paper
10 9 8 7 6 5 4 3 2 1

International Standard Book Number-10: 0-8493-7267-4 (Hardcover)
International Standard Book Number-13: 978-0-8493-7267-4 (Hardcover)

This book contains information obtained from authentic and highly regarded sources. Reprinted material is quoted with permission, and sources are indicated. A wide variety of references are listed. Reasonable efforts have been made to publish reliable data and information, but the author and the publisher cannot assume responsibility for the validity of all materials or for the consequences of their use.

No part of this book may be reprinted, reproduced, transmitted, or utilized in any form by any electronic, mechanical, or other means, now known or hereafter invented, including photocopying, microfilming, and recording, or in any information storage or retrieval system, without written permission from the publishers.

For permission to photocopy or use material electronically from this work, please access www. copyright.com (http://www.copyright.com/) or contact the Copyright Clearance Center, Inc. (CCC) 222 Rosewood Drive, Danvers, MA 01923, 978-750-8400. CCC is a not-for-profit organization that provides licenses and registration for a variety of users. For organizations that have been granted a photocopy license by the CCC, a separate system of payment has been arranged.

Trademark Notice: Product or corporate names may be trademarks or registered trademarks, and are used only for identification and explanation without intent to infringe.

Library of Congress Cataloging-in-Publication Data

Poularikas, Alexander D., 1933-
   Signals and systems primer with MATLAB / Alexander D. Poularikas.
      p. cm. -- (The electrical engineering and applied signal processing series ; 20)
   Includes bibliographical references and index.
   ISBN-13: 978-0-8493-7267-4 (0-8493-7267-4 : alk. paper)
   1. Signal processing--Mathematics. 2. System analysis. 3. MATLAB. I. Title.

TK5102.9.P683 2006
621.382'2--dc22   2006019962

Visit the Taylor & Francis Web site at
http://www.taylorandfrancis.com

and the CRC Press Web site at
http://www.crcpress.com

To our grandchildren Colton-Alexander and Thatcher-James who have given us so much pleasure and happiness and have shown us how difficult it is to teach a time-varying, nonlinear, noncausal, and a multiple-input–multiple-output (MIMO) feedback system.

Preface

Signals and Systems is designed for use as a one-semester analog track, a one-semester digital track, a one-semester analog–digital track, or a two-semester analog–digital course.

We have included several carefully chosen examples from many fields to show the wide applicability of the material and methods presented in this text. The selection of examples from diverse engineering fields, as well as from nonengineering fields, helps to motivate and excite the student to delve into this course with enthusiasm. It is important that the student studies and learns the material presented, since this course is one of the most fundamental in the electrical engineering curriculum. To illustrate this diversity, the examples we include are drawn from chemical engineering, mechanical engineering, biomedical engineering, process control, economics, heat transfer, and other areas. They show that systems in general possess two features: the input signals to the system are functions of one or more independent variables, and the systems produce signals when excited by input signals. It is well known that engineers in their long careers will face problems that are not purely electrical engineering. They will be involved in heating problems of their electronic chips, the use of electromechanical elements driven by small motors in the development of disc drives, the modeling of parts of the body when helping biomedical engineers, the development of edge detectors when involved in pattern recognition of targets, and numerous other applications.

It is the author's belief that a good background, first of analog signals and systems, is necessary for the study of discrete signals and systems. It is known that discrete signals do not exist in nature for the simple fact that their detection needs the use of a physical transducer. Therefore, introducing arbitrarily the signal $x(n)$, which is derived from the continuous signal $x(t)$ by taking its values at equal time distances T, it is difficult for the student to understand why $x(n)$ and $x(t)$ are two different signals. Furthermore, unless the student is well versed and has a good knowledge of the spectrum of analog signals, he or she will have difficulties understanding the spectrum of sampled signals. Some methods to build digital filters are based on analog filter considerations and use.

In this text we have strived to balance the modeling and presentation of systems and their interaction with signals, as well as the study of signals.

We have tried to balance these two entities and avoid the presentation of a couple of systems, based on electrical circuits alone, for example, and then proceed to deal only with signals and their interaction with system models arbitrarily given without deriving them. It is important that the student be able to create the mathematical representation of a system from the physical laws that govern it, its physical presentation, and its underline constants. Engineers, if they want to be creative, must be able to model accurately the systems they investigate and find their responses under many different excitations. This type of work is at the heart of the engineering field.

The electrical engineering field, as it was taught for a long time based on electrical notion, has been changed considerably. For example, microelectromechanical systems (MEMS) are part of the electrical engineering study today and include mechanical systems, signal processing, mathematics, solid-state physics, vibration analysis (deterministic and random), etc. The above discussion points to the fact that electrical engineering students must possess a wide variety of knowledge in many fields. The engineers must be able to build and study systems and predict results accurately. This book tries to do that by introducing systems from many diverse disciplines.

In the beginning, we introduce the block diagram presentation of systems, starting from the most elementary systems represented by one element, such as resistors, capacitors, mass, etc. This approach starts the student in the right direction so that he or she builds the right foundation. During their career, engineers will deal with block diagram representations of complicated and interrelated systems and will be expected to produce the right solution.

In this text, by expounding and covering continuous and discrete systems, we try to emphasize that some operations are done by computer software, and therefore we are talking about software systems. It is important for the student to realize that circuits, mechanical control systems, and physical media and computers are the everyday compound systems found in most of the devices in the market.

Because coverage of the different fields starts from the fundamental laws and devices, there is an additive flexibility, and the book can be used in other fields of engineering with equal ease. The prerequisites for this course are the standard mathematics and engineering courses up to the junior year.

In this text, we have also introduced advanced concepts in their elementary form such that students require the fundamentals of those principles that will become the basis for their future studies. Concepts such as correlation, match filtering, least squares estimation, adaptation, edge detection, etc., are skillfully introduced so that the student builds his or her knowledge on these important signal processing procedures. Furthermore, although new situations and systems are studied, we require their presentations and solutions to use methods already introduced, such as spectra, convolution, impulse response, etc. This repetition is one of the most basic pedagogical methods of learning available in the field of education.

The modeling of signals and systems is inherently mathematical and involves mathematical forms not ordinarily studied by undergraduate engineering majors. Therefore, to improve the student's skills in different fields of needed mathematics, we have included enough mathematical material to serve in understanding the classical solution of differential and difference equations, convolutions and correlations, Fourier transforms, z-transforms, and other important topics. For the same purpose, we have also included as appendices mathematical formulas, definitions, algebraic summations, etc.

We have introduced MATLAB functions and m-files to produce the desired results. We have not introduced SIMULINK or any other canned software program because we feel, at this early learning stage, that the student should program the steps needed to find his or her answer. With this form of programming, the student must first understand the problem, understand the necessary steps in the digital format, find the requested results, and, finally, compare them with the corresponding mathematical results whenever they exist. Educators should refrain from canned programs. Very often, even in my graduate classes, when we learn about Fourier transforms and ask the students to obtain the transform of a finite signal using computers, they come with plots in the frequency domain from 0 to ½, although we have emphasized in class that every finite signal has infinite spectrum. When asked what 0 to ½ means in the frequency axis, the answer is that MATLAB gave it to them in this form. In my undergraduate classes, I explain to the students the history and importance of the fast Fourier transform (FFT), but I require that they find the discrete Fourier transform (DFT) from the definition.

Pedagogical features

1. The book introduces all the needed mathematical background within the chapters; some additional material is included in the appendices.
2. There is a balance between signals and systems, and both are emphasized equally.
3. Examples are derived from many diverse fields to emphasize that this course is applicable to any field as long as the systems are characterized by the same form of mathematical equations.
4. Key equations are boxed to emphasize their importance.
5. There are examples in the text and problems. In addition, appendices are also added in some chapters to elucidate the mathematics of the chapter.
6. We use paragraphs with the indication "**Note**" to emphasize the meaning of a concept or to improve the understanding of an operation.
7. Analog and digital signal processing and systems are presented in separate chapters. From experience, we have found that students get confused easily by jumping back and forth between analog and digital. It may seem appropriate and even novel for an experienced

person, but for a student that is meeting many of these concepts and mathematical areas for the first time, it is rather confusing.
8. We strive to repeat introduced concepts, such as convolution, correlation, spectra, etc., throughout the book. Repetition is the most fundamental pedagogical learning process, and in this book we try to use it.
9. The author believes that at an early stage of an engineering student's career, exposure to random signals and their interaction with linear time-invariant (LTI) systems is essential. There is nothing deterministic in the physical world. Most students in their professional life will deal with and study random phenomena, such as earthquake signals, electroencephalograms, thermal noise in electrical elements, noise introduced during the transportation of signals through different communication channels, radar returns, target recognition, acoustic signals in the ocean, etc. For this reason, Chapter 13 was added.
10. It is very important that our undergraduate students come in contact with and understand filtering processes that are adaptive and learn where such devices are used in practice. For this reason, we have added Chapter 14, where Wiener and least mean square (LMS) filtering are introduced. We provide numerous examples of how these filters are used, and we expect that this will stimulate the student's interest to continuously be involved in the digital signal processing discipline.
11. We introduce the fundamentals of digital sampling and the effect it has on the spectrum.
12. At the end of each chapter we summarize the new definitions and concepts that were introduced in the chapter.
13. The material is introduced in a concise format but is elucidated by a large number of examples.
14. The book has been arranged in such a way that it can be used for a one- or two-semester course. Furthermore, the track can be analog, digital, or analog–digital.

The text consists of 14 chapters. Chapter 1 introduces both analog and digital signal presentation. It also introduces the presentation of signals by a complete set of orthogonal signals.

Chapter 2 introduces linear and time-invariant continuous-time systems. In addition, the chapter covers convolution and impulse response of such systems.

Chapter 3 introduces discrete-type systems and their solution. It also discusses how to simulate analog systems using discrete methods.

Chapter 4 presents the analysis of periodic signals in Fourier series, the amplitude and phase spectra of these signals, as well as the effect that linear time-invariant systems have on them.

Chapter 5 develops the Fourier transform and its great utility in identifying systems, studying the spectra of signals as well as the influence that systems have on these spectra, known as filtering of signals. The Gibbs' phenomenon is also introduced, along with its significance in the study of signals.

Chapter 6 introduces the sampling theorem and the importance of signal bandwidth in the aliasing problem.

Chapter 7 introduces the discrete-time Fourier transform and the discrete Fourier transform. With examples, this chapter gives a clear understanding of the effects of decreasing the sampling time and padding with zeros the sequences on the spectra of discrete signals.

Chapter 8 presents in some detail the Laplace transform and its usefulness in solving differential equations, identifying analog systems, and producing transfer functions of systems, as well as its use in feedback control of systems. The Bode plots are also introduced.

Chapter 9 presents the z-transform and its applications in solving difference equations, developing the transfer functions of discrete systems, and finding the frequency response of discrete systems.

Chapter 10 brings forth the analog filter design. Specifically, it introduces the Butterworth and Chebyshev filter, and finally, it introduces the design of analog filters with the use of MATLAB functions.

Chapter 11 introduces the digital filter, known as the finite impulse response or the nonrecursive filter.

Chapter 12 introduces the infinite impulse response digital filter and its use for filtering discrete signals.

Chapter 13 develops the fundamentals of random variables and their use for finding the spectrums of stationary processes' random sequences. This chapter introduces the Wiener–Kinthcin relation and the nonparametric spectral estimation.

Chapter 14 characterizes different types of filtering and their uses. Particularly, it introduces the use of Wiener filtering and least mean squares filtering. It presents several examples with applications to noise elimination, system identification, and channel equalization.

The following are some suggestions in using the text.

One-semester course: Analog and digital signal processing
Chapter 1: 1.1 → 1.3
Chapter 2: 2.1 → 2.7
Chapter 3: 3.1 → 3.4
Chapter 5: 5.1 → 5.3
Chapter 6: 6.1 → 6.2
Chapter 7: 7.5 → 7.6
Chapter 8: 8.1 → 8.3
Chapter 10: 10.1 → 10.5
Chapter 11: 11.1 → 11.3

One-semester course: Analog signal processing
Chapter 1: 1.1 → 1.3
Chapter 2: 2.1 → 2.7
Chapter 4: 4.1 → 4.4
Chapter 5: 5.1 → 5.2

Chapter 6: 6.1 → 6.2
Chapter 8: 8.1 → 8.7
Chapter 10: 10.1 → 10.6

One-semester course: Discrete signal processing
Chapter 1: 1.1 → 1.3
Chapter 3: 3.1 → 3.3
Chapter 5: 5.1 → 5.2
Chapter 6
Chapter 7: 7.5 → 7.6
Chapter 11: 11.1 → 11.3
Chapter 13

Two-semester course
All the chapters could be covered, including the starred chapters and sections. Some sections may be skipped at the discretion of the instructor. The starred sections may be skipped without loss of continuity.

The author acknowledges the valuable and helpful comments of Professor Nicolaos Karayiannis of the University of Houston. The author is indebted to Dr. Haluk Ogmen, department chair, and Dr. Raymond Flumerfelt, dean of the College of Engineering, who provided an excellent academic and research environment at the University of Houston.

MATLAB® is a registered trademark of The MathWorks, Inc. For product information, please contact:

The MathWorks, Inc.
3 Apple Hill Drive
Natick, MA 01760-2098 USA
Tel: 508-647-7000
Fax: 508-647-7001
E-mail: info@mathworks.com
Web: www.mathworks.com

The Author

Alexander D. Poularikas received his Ph.D. from the University of Arkansas and became professor at the University of Rhode Island. He became chairman of the Engineering Department at the University of Denver and then became chairman of the Electrical and Computer Engineering Department at the University of Alabama in Huntsville. He has published six books and has edited two. Dr. Poularikas served as editor-in-chief of the Signal Processing series (1993–1997) with Artech House and is now editor-in-chief of the Electrical Engineering and Applied Signal Processing series, as well as the Engineering and Science Primers series (1998–present) with Taylor & Francis. He was a Fulbright scholar, is a lifelong senior member of IEEE, and is a member of Tau Beta Pi, Sigma Nu, and Sigma Pi. In 1990 and 1996, he received the Outstanding Educator Award of IEEE, Huntsville Section.

Abbreviations

A/D	Analog-to-digital conversion
AM	Amplitude modulation
AR	Autoregressive
ARMA	Autoregressive moving average
cdf	Cumulative density function
dB	Decibel
DFT	Discrete Fourier transform
DSBSC	Double-sideband suppressed carrier
DTFT	Discrete-time Fourier transform
FFT	Fast Fourier transform
FIR	Finite impulse response
FM	Frequency modulation
GHz	Gigahertz
IDTFT	Inverse discrete-time Fourier transform
IFFT	Inverse fast Fourier transform
iid	Independent and identically distributed
IIR	Infinite impulse response
ILT	Inverse Laplace transform
KCL	Kirchhoff current law
KVL	Kirchhoff voltage law
LMS	Least mean squares
LT	Laplace transform
LTI	Linear time invariant
MMSE	Minimum mean square error
MSE	Mean square error
NLMS	Normalized least mean squares
PAM	Pulse amplitude modulation
pdf	Probability random function
PID	Proportional integral differential controller
ROC	Region of convergence
rv	Random variable
SFG	Signal flow graph
SSB	Single sideband
WGN	White Gaussian
WN	White noise
WSS	Wide-sense stationary

Contents

Chapter 1 Signals and their functional representation 1
1.1 Some applications involving signals 1
1.2 Fundamental representation of simple time signals 3
    Periodic discrete-time signals 6
    Nonperiodic continuous signals 7
        Unit step function 7
        Rectangular pulse function 8
        Sinc function 9
    Nonperiodic special discrete signals 10
        Delta function 10
    Comb function 13
    Arbitrary sampled function 13
1.3 Signal conditioning and manipulation 14
    Modulation 14
    Shifting and flipping 14
    Time scaling 15
    Windowing of signals 17
*1.4 Representation of signals 18
Important definitions and concepts 25
Chapter 1 Problems 26
Appendix 1.1: Elementary matrix algebra 33
Appendix 1.2: Complex numbers 36
Appendix 1.1 Problems 38
Appendix 1.2 Problems 39

Chapter 2 Linear continuous-time systems 41
2.1 Properties of systems 41
2.2 Modeling simple continuous systems 43
    Electrical elements 43
        Capacitor 43
        Inductor 45
        Resistor 47
    Mechanical translation elements 48
        Ideal mass element 48
        Spring 48
        Damper 49

	Mechanical rotational elements	50
	Inertial elements	50
	Spring	51
	Damper	52
2.3	Solutions of first-order systems	52
	Zero-input and zero-state solution	54
	Standard solution techniques of differential equations	56
2.4	Evaluation of integration constants: initial conditions	63
	Switching of sources	64
	Conservation of charge	64
	Conservation of flux linkages	64
	Circuit behavior of L and C	65
	General switching	65
2.5	Block diagram representation	68
2.6	Convolution and correlation of continuous-time signals	71
	Matched filters	83
	Correlation	84
2.7	Impulse response	86
Important definitions and concepts		96
Chapter 2 Problems		97

Chapter 3 Discrete systems 111
3.1 Discrete systems and equations 111
3.2 Digital simulation of analog systems 118
*3.3 Digital simulation of higher-order differential equations 131
3.4 Convolution of discrete-time signals 135
Important definitions and concepts 141
Chapter 3 Problems 142
Appendix 3.1: Method of variation of parameters 149
Appendix 3.2: Euler's approximation for differential equations 152

Chapter 4 Periodic continuous signals and their spectrums 157
4.1 Complex functions 157
    Continuous-time signals 158
    Discrete-time signals 160
4.2 Fourier series of continuous functions 161
    Fourier series in complex exponential form 162
    Fourier series in trigonometric form 165
4.3 Features of periodic continuous functions 169
    Parseval's formula 169
    Symmetric functions 170
        Even function 172
        Odd function 172
    *Finite signals 172
    *Convolution 173

4.4 Linear systems with periodic inputs 174
Important definitions and concepts 180
Chapter 4 Problems 181

Chapter 5 Nonperiodic signals and their Fourier transform **189**
5.1 Direct and inverse Fourier transform 190
    Real functions 194
    Real and even functions 195
    Real and odd functions 196
5.2 Properties of Fourier transforms 196
    Linearity 196
    Symmetry 196
    Time shifting 199
    Scaling 200
    Central ordinate 200
    Frequency shifting 202
    Modulation 202
    Derivatives 208
    Parseval's theorem 213
    Time convolution 216
    Frequency convolution 217
    Summary of continuous-time Fourier properties 218
*5.3 Some special Fourier transform pairs 220
*5.4 Effects of truncation and Gibbs' phenomenon 225
*5.5 Linear time-invariant filters 226
    Distortionless filter 228
    Ideal low-pass filter 229
    Ideal high-pass filter 230
Important definitions and concepts 231
Chapter 5 Problems 231
Appendix 5.1 237

Chapter 6 Sampling of continuous signals **239**
6.1 Fundamentals of sampling 239
6.2 The sampling theorem 244
    Construction of analog signal from its sampled values 252
Important definitions and concepts 253
Chapter 6 Problems 254

Chapter 7 Discrete-time transforms **257**
7.1 Discrete-time Fourier transform (DTFT) 257
    Approximating the Fourier transform 257
7.2 Summary of DTFT properties 260
7.3 DTFT of finite time sequences 262
    Windowing 266

7.4 Frequency response of linear time-invariant (LTI) discrete systems...268
7.5 The discrete Fourier transform (DFT)..269
7.6 Summary of the DFT properties...271
*7.7 Multirate digital signal processing..284
  Down sampling (or decimation)..284
  Frequency domain of down-sampled signals...................................286
  Interpolation (up sampling) by a factor U...291
  Frequency domain characterization of up-sampled signals...........292
Important definitions and concepts..295
Chapter 7 Problems...295
Appendix 7.1: Proofs of the DTFT properties..298
Appendix 7.2: Proofs of DFT properties..300
Appendix 7.3: Fast Fourier transform (FFT)...304

Chapter 8   Laplace transform 309
8.1 One-sided Laplace transform 309
8.2 Summary of the Laplace transform properties 312
8.3 Systems analysis: transfer functions of LTI systems 316
8.4 Inverse Laplace transform (ILT)...328
  MATLAB function residue 329
8.5 Problem solving with Laplace transform 336
8.6 Frequency response of LTI systems..352
8.7 Pole location and the stability of LTI systems...................................361
  Simple-order poles..361
  Multiple-order poles...363
*8.8 Feedback for linear systems..365
  Cascade stabilization of systems..365
  Parallel composition...366
  Feedback stabilization..367
  Sensitivity in feedback...368
  Rejection of disturbance using feedback 369
  Step response...370
  Proportional controllers..371
  Proportional integral differential (PID) controllers.......................375
*8.9 Bode plots 377
  Bode plots of constants..377
  Bode diagram for differentiator 378
  Bode diagram for an integrator 379
  Bode diagram for a real pole...379
Important definitions and concepts..383
Chapter 8 Problems...383
Appendix 8.1: Proofs of Laplace transform properties.............................397

Chapter 9   The z-transform, difference equations, and discrete systems 401
9.1   The z-transform..401
9.2   Convergence of the z-transform..405
9.3   Properties of the z-transform...412
        Summary of z-transform properties...413
9.4   z-Transform pairs...423
9.5   Inverse z-transform 423
9.6   Transfer function..431
        *Higher-order transfer functions...437
9.7   Frequency response of first-order discrete systems...............438
        Phase shift in discrete systems..443
*9.8 Frequency response of higher-order digital systems.............443
9.9   z-Transform solution of first-order difference equations.....447
*9.10 Higher-order difference equations...450
        Method of undetermined coefficients.......................................454
Important definitions and concepts..459
Chapter 9 Problems..459
*Appendix 9.1: Proofs of the z-transform properties 473

Chapter 10   Analog filter design...477
10.1 General aspects of filters 477
10.2 Butterworth filter 479
10.3 Chebyshev low-pass filter 486
10.4 Phase characteristics...494
10.5 Frequency transformations 494
         Low-pass-to-low-pass transformation....................................494
         Low-pass-to-high-pass transformation 495
         Low-pass-to-band-pass transformation.................................495
         Low-pass-to-band-stop transformation..................................498
10.6 Analog filter design using MATLAB functions....................500
         Butterworth filter design...500
Important definitions and concepts..501
Chapter 10 Problems..501

Chapter 11   Finite Impulse Response (FIR) filters...............................505
11.1 Properties of FIR filters..505
         Causality...505
         Frequency normalization..505
         Phase consideration...506
         Scaling the digital transfer function 507
         Symmetric FIR low-pass filters 508
11.2 FIR filters using the Fourier series approach.......................509

11.3 FIR filters using windows 513
    Windows 513
    High-pass FIR filters 515
    Band-pass FIR filters 516
    Band-stop FIR filters 516
*11.4 Prescribed filter specifications using a Kaiser window 519
11.5 MATLAB FIR filter design 522
    Low-pass FIR filter 522
    High-pass FIR filter 522
    Band-pass FIR filter 522
    Band-stop FIR filter 523
    Window use 523
Important definitions and concepts 523
Chapter 11 Problems 523

Chapter 12 Infinite Impulse Response (IIR) filters 525
12.1 The impulse-invariant method approximation in the time domain .. 525
12.2 Bilinear transformation 532
12.3 Frequency transformation for digital filters 538
    Low-pass-to-low-pass transformation 538
    Low-pass-to-high-pass transformation 539
    Low-pass-to-band-pass transformation 540
    Low-pass-to-band-pass transformation 542
12.4 Recursive versus non-recursive design 542
    Important defintions and concepts 543
Chapter 12 Problems 543

Chapter 13 Random variables, sequences, and power spectra densities 545
13.1 Random signals and distributions 545
    Stationary and ergodic processes 548
13.2 Averages 548
    Mean value 548
    Correlation 549
    Covariance 553
    Independent and uncorrelated rv's 553
13.3 Stationary processes 554
    Autocorrelation matrix 554
    Purely random process (white noise, WN) 557
    Random walk (RW) 557
13.4 Special random signals and probability density functions 557
    White noise 557
    Gaussian processes 558
        Algorithm to produce normalized Gaussian distribution 558

       Lognormal distribution..559
            Algorithm to produce lognormal distribution.........................559
       Chi-square distribution..560
13.5 Wiener–Kintchin relations...560
13.6 Filtering random processes...562
       Spectral factorization...565
       Autoregressive process (AR) 565
13.7 Nonparametric spectra estimation..568
       Periodogram 568
       Correlogram...568
       Computation of $\hat{S}_p(ej\omega)$ and $\hat{S}_c(ej\omega)$ using FFT 568
       General remarks on the periodogram..570
       Blackman–Tukey (BT) method 570
       Bartlett method 572
       Welch method..573
       Modified Welch method 575
       The Blackman–Tukey periodogram with the Bartlett window 577
Important definitions and concepts..578
Chapter 13 Problems...579

*Chapter 14   Least square system design, Wiener filter, and the LMS filter 583

14.1 The least-squares technique 583
       Linear least squares...587
14.2 The mean square error..590
       The Wiener filter 591
       The Wiener solution 594
       Orthogonality condition 597
14.3 Wiener filtering examples 597
       Minimum mean square error (MMSE) 602
       Optimum filter (w^o) 603
14.4 The least mean square (LMS) algorithm 608
       The LMS algorithm 609
14.5 Examples using the LMS algorithm...614
Important definitions and concepts..623
Chapter 14 Problems...623

Appendix A   Mathematical formulas 627

A.1 Trigonometric identities..627
A.2 Orthigonality 629
A.3 Summation of trigonometric forms..629
A.4 Summation formulas...630
      Finite summation formulas..630
      Infinite summation formulas 630
A.5 Series expansions...631

A.6	Logarithms	631
A.7	Some definite integrals	632

Appendix B Suggestions and explanations for MATLAB use 633

B.1	Creating a directory	633
B.2	Help	633
B.3	Save and load	634
B.4	MATLAB as calculator	634
B.5	Variable names	634
B.6	Complex numbers	634
B.7	Array indexing	635
B.8	Extracting and inserting numbers in arrays	635
B.9	Vectorization	635
B.10	Matrices	636
B.11	Produce a periodic function	636
B.12	Script files	636
	Script file pexp.m	637
B.13	Functions	637
B.14	Subplots	638
B.15	Figures	638
B.16	Changing the scales of the axes of a figure	639
B.17	Writing Greek letters	639
B.18	Subscripts and superscripts	640
B.19	Lines in plots	640

Index 641

chapter 1

Signals and their functional representation

In this chapter we will learn about signals, how to represent them in their mathematical form, how to manipulate them to create new ones, and, finally, why it is necessary to study them.

1.1 Some applications involving signals

Electrical engineers, as well as engineers from other disciplines, are concerned with the detection of signals, their analysis, their processing, and their manipulation. They are interested in the signal **amplitude**, **time duration**, and **shape**. Electrical engineers are concerned with detecting radar pulses returned from targets that are hidden in noise. Mechanical engineers are interested in signals that engines emit to detect any change that may suggest engine trouble. Bioengineers are interested in biological signals such as electrocardiograms (EKGs), brain signals, etc. Economists are interested in how stock market signals, import–export signals, and the gross national product (GDP) are changing, as well as many other economic signals, so that they can predict how the economy will change in the future.

The most fundamental instrument that is associated with the detection of signals is the **transducer**. All of us are aware of our eye–brain combination transducer that detects two-dimensional signals (images). Our ear is another transducer that detects air pressure variations, and the result is an acoustic signal, such as a song. Since all the transducers found in nature and those we build, such as voltage and current transducers, which are made up of resistors, inductors, and capacitors, cannot respond instantaneously to the abrupt excitation, the signals the transducers produce are continuous. We are basically living in an analog world.

Since the early 1950s, when the computer appeared, it became obvious that by describing the signals in their discrete form, it is possible to process the signals quickly and accurately, without the need to build elaborate and large analog systems. To create **discrete-time signals**, we use an electronic instrument called **analog-to-digital** (A/D) converter. Since the discrete signals take all the values within a range, we further process them and assign only one value if their values fall within a prespecified range; for example, if we split the range between 0 and 2 and assign the value 0 to any discrete signal of values between 0 and 0.5, the value 1 if its value falls between 0.5 and 1.5, and the value 2 if the value falls between 1.5 and 2.0. This process will produce the **digital** signals, and in this particular case, any discrete signal with values between 0 and 2.0 will be represented by a digital signal having three levels only. In the present text, we will primarily study the discrete-type signals.

Ordinarily, signals cannot be transmitted over long distances directly as produced. They must be modified or conditioned, usually "riding" on another signal for their transmission. This type of modification is known as **modulation** of signals. For example, if we want to transmit speech, whose mean frequency is 3000 Hz (cycles per second), using an antenna, we know from our course in electromagnetics that $c = f\lambda$, where c is the speed of light, 3×10^8 m/s, and f = 3000 Hz in this example. Hence, the wavelength of the 3000-Hz sine signal is $\lambda = 3 \times 10^8/3000 = 10^5$ m, or about 60 miles. Communication engineers tell us that an efficient antenna must have a length in the range of ½ or ¼ of the wavelength. It is the practical considerations that force us to modify the signals to be sent. At other times, we may translate a particular signal into a completely different form. The use of smoke signals by Indian tribes is an example of **coding** signals for transmission through space. Most of us are familiar with the Morse code as a means of transforming letters and numbers into dots and dashes.

Present-day engineers have gone further in studying signals and have been able to establish a **quantitative** description and measure of the information contained in a signal. To develop their theory, they created a conceptual model communication system, as shown in Figure 1.1.1. To simplify the discussion, let us consider a person who is transmitting a message as an **information source**. The output of the information source is called a

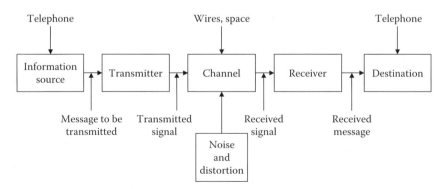

Figure 1.1.1 General communication system.

Chapter 1: Signals and their functional representation 3

massage — it is what the person is saying. The transmitter is an apparatus that transforms the message into a form that allows it to be transmitted. A telephone handset that converts the acoustic message into a varying electrical current in the telephone line is such a transmitter. The communication **channel** is the medium over which the message is transmitted. The wires between two telephone receivers and the space between two microwave towers constitute two different types of channels.

Because of the properties of the elements making up the communication system, distortion often occurs. **Distortion** is an operation effect of the system on the signal, whereas noise involves unpredictable perturbations (variations) of the signal. Clearly, to be useful, the output of the channel, which is the **receiving signal**, must resemble to some extent the original signal. For example, the received message is what we hear in the telephone receiver. The destination is the final element of a communication system. It can be a person or an apparatus, depending on the original intent of the system.

1.2 *Fundamental representation of simple time signals*

The most fundamental periodic continuous-time signal that we will encounter in our studies is the trigonometric sine function shown in Figure 1.2.1.

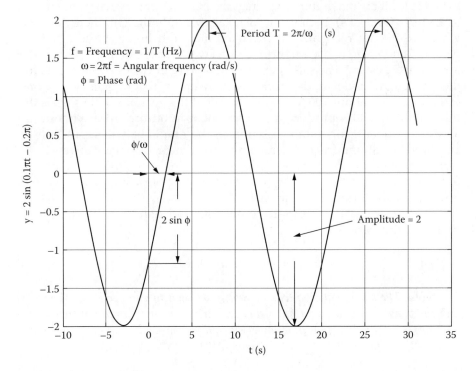

Figure 1.2.1 The sine function.

For plotting the sine signal we used the specific function $s(t) = 2\sin(0.1\pi t - 0.2\pi)$. The reason for the importance of this waveform is that any periodic signal can be approximated by the sum of sine and cosine functions. The idea of adding sine and cosine functions in describing general periodic functions dates back to the time of the Babylonians, who used ideas of this type in their prediction of astronomical events. In the 18th century, Euler's observation that the vibrating strings produce sinusoidal motion was followed half a century later by the work of Jean Baptiste Joseph Fourier. Fourier claimed that any periodic signal could be represented by an infinite sum of sine and cosine functions, each one with different amplitude and each function having a different, although integrally related, frequency. This representation feature is also true for nonperiodic functions, but with the difference that the frequencies constitute a continuum. The amplitude and phase of these sinusoidal waves are known as the **amplitude** and **phase spectra**. This particular matter will be studied in detail in later chapters.

Sometimes, eliminating some of the frequencies that make up a complicated signal, known as **filtering**, causes signal distortion and, sometimes, produces undesirable effects. An example familiar to us occurs in our telephone conversations. Telephone companies transmit many different voices over the same telephone line in the interest of making optimum use of the transmission lines. To optimize this number for the cable wire used, they are forced to limit the maximum frequency, the **bandwidth**, of each voice channel to 4000 Hz (cycles/s). Because of this bandwidth limitation, there are times when we do not recognize the speaker, although he or she is a familiar person. Bandwidth is also important when we want to purchase an amplifier for our hi-fi system. The sales person will defend the price by telling us that the particular model under consideration has a flat frequency response extending from about 100 to 14,000 Hz. What the person is telling us is that this amplifier will reproduce any instrument in an orchestra without any loss in the quality of sound. Essentially, the amplifier will reproduce the amplitude of all the frequencies within the indicated range. However, distortion may occur due to phase changes. Filtering is the most fundamental process that engineers, most importantly electrical engineers, will be involved with during their career.

The sine wave is a **periodic continuous** function. Any function $f(t)$ is periodic with period T if

$$f(t \pm nT) = f(t) \qquad n = 0, \pm 1, \pm 2, \cdots \qquad (1.1)$$

Note: *The above equation tells us that if we identify any particular time t, we find the value of the function to be equal to f(t). If we next go to the time values $t + T$ or $t - T$, the value of the function is the same as that which we found previously. By repeating this process for all t, we obtain the periodic signal.*

Figure 1.2.2 shows two periodic functions. If we had plotted a longer range of time, we would have observed the periodicity in Figure 1.2.2b.

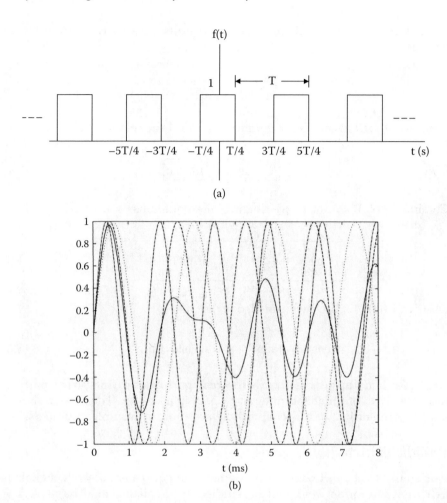

Figure 1.2.2 (a) A general periodic signal. (b) $f(t) = \sin(2\pi 440t)$ (A note ···); $f(t) = \sin(2\pi 494t)$ (B note –·–); $f(t) = \sin(2\pi 659t)$ (E note ---); (average sum ——).

In general, signals are real functions of time since they denote the results of physical phenomena. However, for mathematical convenience, it is very useful to present sinusoidal functions as components of complex-valued functions of time called **complex signals**. An important complex signal is the exponential function $e^{j\omega t}$. If we use the Maclaurin series expansion $\exp(x) = 1 + (x/1!) + (x^2/2!) + (x^3/3!) + \ldots$ and substitute $j\omega$ for x, we obtain

$$e^{j\omega t} = 1 + j\omega t + \frac{(j\omega t)^2}{2!} + \cdots + \frac{(j\omega t)^n}{n!}$$
$$= \left[1 - \frac{(\omega t)^2}{2!} + \frac{(\omega t)^4}{4!} - \cdots\right] + j\left[\omega t - \frac{(\omega t)^3}{3!} + \frac{(\omega t)^5}{5!} - \cdots\right]$$

(1.2)

The expansions in (1.2) are cos(ωt) and sin(ωt), respectively. Hence, (1.2) becomes

$$\boxed{e^{j\omega t} = \cos \omega t + j \sin \omega t} \qquad (1.3)$$

If we set $-j$ for j in the above equation, we find the relation

$$\boxed{e^{-j\omega t} = \cos \omega t - j \sin \omega t} \qquad (1.4)$$

Combining (1.3) and (1.4), we obtain Euler equations

$$\boxed{\cos \omega t = \frac{e^{j\omega t} + e^{-j\omega t}}{2}, \qquad \sin \omega t = \frac{e^{j\omega t} - e^{-j\omega t}}{2j}} \qquad (1.5)$$

Observe from (1.3) that we can also write

$$\cos \omega t = \operatorname{Re}\{e^{j\omega t}\}, \qquad \sin \omega t = \operatorname{Im}\{e^{j\omega t}\} \qquad (1.6)$$

where Re and Im denote the words **real part of** and **imaginary part of**, respectively. Note that the imaginary part is also real. The reader should refer to Appendix 2 of this chapter to learn more on complex numbers.

Periodic discrete-time signals

Our ability to convert continuous signals into equivalent discrete signals is extremely important and will be studied in detail in later chapters. The conversion is extremely important since we are obligated to use computers to process every signal today. Here, we will examine certain features of typical discrete signals. To create a discrete sine wave, we keep the values of the function at distances nT_s, where T_s is the **sampling time**, $1/T_s$ is the **sampling frequency** in Hz, and $2\pi/T_s = \omega_s$ is the **sampling frequency** in rad/s. Therefore, if we set $t = nT_s$, where n is a set of integers, we obtain the sampled sine function

$$f(nT_s) = \sin \omega T_s n \qquad (1.7)$$

If we substitute ω with ω + $(2\pi/T_s)$ in (1.7), we find that

$$f(nT_s) = \sin\left(\left(\omega + \frac{2\pi}{T_s}\right)nT_s\right) = \sin(\omega T_s n + 2\pi n) = \sin \omega T_s n \qquad (1.8)$$

Chapter 1: Signals and their functional representation

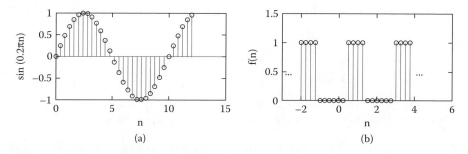

Figure 1.2.3 Typical discrete periodic signal.

Note: *The above equation shows that a sampled periodic signal is periodic with period $2\pi/T_s$.*

The frequency π/T_s is called the **fold-over** frequency. The frequency ωT_s has units of radians and is known as the **discrete frequency**. If this distinction is not obvious from the context in the development, we will explicitly state the units of the frequency present. Figure 1.2.3 presents two typical discrete signals.

Nonperiodic continuous signals

This section introduces the reader to a number of special functions for describing particular classes of signals. Many of these functions have features that make them particularly useful directly or indirectly in describing other functions in the solution of engineering problems. At this point, little more than a catalog of these signals and their mathematical description is presented.

Unit step function

The unit step function $u(t)$ is an important signal for analytic studies; moreover, it has many practical applications. When we switch on a constant voltage source (battery) across a circuit, the source is modeled by a unit step function. Likewise, when a constant force is applied to a body at a particular time, the mathematical modeling of such a force is accomplished by using a unit step function.

The unit step function is defined by the relation

$$\boxed{u(t) = \begin{cases} 1 & t \geq 0 \\ 0 & t < 0 \end{cases} \quad \text{or} \quad u(t-a) = \begin{cases} 1 & t \geq a \\ 0 & t < a \end{cases}} \quad (1.9)$$

Note: *To plot a unit step function, we introduce values for t, and if the number in parentheses is 0 or positive, we set the value of the function equal to 1, and if the value is negative, we set the value of the function equal to 0.*

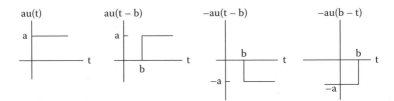

Figure 1.2.4 The step function and some of its shifted and reflected forms.

The step function of height a and its shifted forms are shown in Figure 1.2.4.

Rectangular pulse function

The pulse function is the result of an on–off switching operation of a constant voltage source to an electric circuit. This waveform has considerable practical utility. For example, its narrow pulses are used to modulate the transmitters in radar equipment, and it is the waveshape of the recurring timing signals (clock pulses) that controls the total operation of any digital computer.

A rectangular pulse function of height b and its shifted form are shown in Figure 1.2.5. Its mathematical form is developed from the appropriate choice of step functions. Hence, the just-mentioned pulse functions are specified by

$$p_a(t) = [u(t+a) - u(t-a)] = \begin{cases} 0 & |t| > a \\ 1 & |t| < a \end{cases}$$

$$p_a(t-t_0) = [u(t-t_0+a) - u(t-t_0-a)] = \begin{cases} 0 & |t-t_0| > a \\ 1 & |t-t_0| < a \end{cases}$$

(1.10)

Note: *The representation of the pulse function by $p_a(t)$ is not a mathematical function. In calculations we must use the representation with the appropriate step functions. The definition $p_a(t)$ means the following: when t takes a value such that the factor inside the parentheses becomes zero, the pulse is centered at that time value and the width of the pulse is equal to a in either side of its center.*

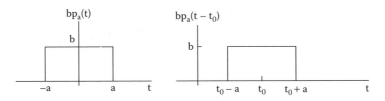

Figure 1.2.5 Rectangular pulse function.

Chapter 1: Signals and their functional representation

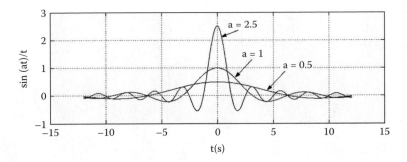

Figure 1.2.6 Sinc function for three different values of *a*.

Sinc function

This function plays an important role in the reconstruction of special signals. We will talk more about this function in the section discussing the sampling of continuous functions and their reconstruction. As a function of ω, it also represents the frequency content of the pulse function, as we will learn in a later chapter. The sinc function is defined by the expression

$$\boxed{\sin c_a(t) = \frac{\sin at}{t} \qquad -\infty < t < \infty} \qquad (1.11)$$

This function is shown in Figure 1.2.6 for three different values of *a*.

Note: *If we try to plot the sinc function, we find that at point t = 0 the fraction becomes 0/0, which is an undefined number. To obtain the limit, we use the L'Hopital rule, which says: Take the derivative of the numerator with respect to t and set t = 0 in the new function. Next, take the derivative of the denominator with respect to t and set t = 0 in the new function.*

The following MATLAB program was used to plot Figure 1.2.6:

```
t=-12:0.05:12;%this creates a vector with values of t from -12
         %to 12 in steps of 0.05

s1=sin(0.5*t+eps)./(t+eps);%eps is a small number and is used
         %to avoid warnings from
         %MATLAB that a division with zero is present; the period
         %in the numerator tells MATLAB to divide element by
         %element the two vectors (numerator and denominator);

s2=sin(t+eps)./(t+eps);
s3=sin(2.5*t+eps)./(t+eps);
s1(241)=0.5;%MATLAB gives always 1 for the form
         %of any sinc function, hence we
         %must substitute the 0/0
```

```
s3(241)=2.5;%undifined number by its exact values;
subplot(2,1,1);plot(t,s1,'k')%'k' makes the curves black;
hold on;plot(t,s2,'k')
hold on;plot(1,s3,'k')
xlabel('t (s)');ylabel('sin(at)/t');
```

Nonperiodic special discrete signals

Delta function

The **delta function** $\delta(t)$, often called the **impulse** or **Dirac delta** function, occupies a central place in signal analysis. Many physical entities, such as point sources, point charges, concentrated loads on structures, and voltages or current sources acting for very short times, can be modeled as delta functions. The delta function, not a regular function, has very peculiar properties, and it is defined as follows:

$$\delta(t - t_0) = 0 \qquad t \neq t_0$$

$$\int_{t_1}^{t_2} f(t)\delta(t - t_0)dt = f(t_0) \qquad t_1 < t_0 < t_2 \qquad (1.12)$$

$$\int_{t_1}^{t_2} \delta(t - t_0)dt = 1 \qquad f(t) = 1, \quad t_1 < t_0 < t_2$$

Note: *Equation (1.12) indicates that the delta function exists only at one point, in this case at t_0. At that point the argument is 0, and hence, the value of t that makes the argument 0 is the position of the delta function. If $f(t) = 1$ for all t, then the integral of the delta function is equal to 1. This indicates that the area under the delta function is 1. Furthermore, (1.12) indicates that the result of integration of the product of a regular function with a delta function is equal to the value of the function at the point the delta function is located. Hence, we do not integrate when a delta function is involved, but we simply insert in the rest of the functions of the integrand the value of the independent variable t at which the delta is located.*

Several delta functions are shown in Figure 1.2.7a. The height of the spikes shown denotes the area of the delta function. Figure 1.2.7b illustrates graphically the second equation of (1.12). Note that the delta function is not presented in a functional form, but is presented by its behavior under the integral sign. Despite the fact that the delta function is not an analytic one (regular function), a rigorous mathematical formulation was given by Laurent Schwartz in his work on the theory of **generalized functions**.

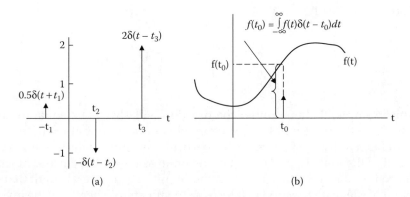

Figure 1.2.7 (a) Representation of several delta functions. (b) Illustration of the effect of multiplying a well-behaved function and a delta function.

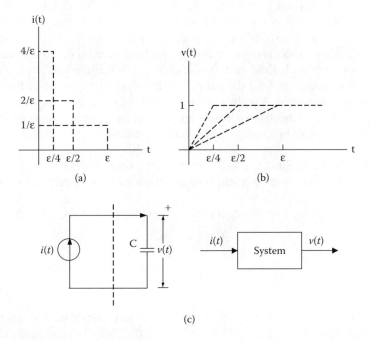

Figure 1.2.8 Voltage on a capacitor due to an impulse current source.

To see how the delta function can be created, consider that a current of very short duration ε and amplitude $1/\varepsilon$ is applied to the circuit shown in Figure 1.2.8c. From the well-known expression for the voltage across the capacitor,

$$v(t) = \frac{1}{C}\int_0^t i(\tau)d\tau$$

we obtain for C = 1 F,

$$v(t) = \int_0^{\varepsilon} \frac{1}{\varepsilon} dt = \int_0^{\varepsilon/2} \frac{2}{\varepsilon} dt = \int_0^{\varepsilon/4} \frac{4}{\varepsilon} dt = 1$$

In this example, the area under the function will always be equal to unity for any $\varepsilon \to 0$, and it will be zero everywhere else except at $t = 0$, where it is undefined. Observe further in this particular example, which involves the integral of the delta function, that in the limit, as $\varepsilon \to 0$, the current function will produce a step voltage of 1 V in the circuit. It is a general result that the integral of the delta function produces a step function. Therefore, the inverse mathematical operation to integration, the derivative, will produce a delta function at the point the unit step function changes from zero to one.

Despite its special properties, the importance of the delta function will become apparent when we study the response of linear systems (those with linear elements) to an arbitrary input. We will learn that if we know the output of a linear and time-invariant system (a system composed of elements that do not vary with time and are linear) to a delta function input, we can use this knowledge to find the output of the system to an input that is any well-behaved function. A well-behaved function is one that may have a finite number of discontinuities of finite amplitude on a finite interval and has finite derivatives at both sides of each discontinuity.

Example 1.2.1: Find the velocity of a free body with mass M on a frictionless surface when a unit impulse function force is applied at $t = 0$.

Solution: Using the Newton force law $f(t) = Ma(t) = M\, dv(t)/dt$, we find

$$v(t) = \frac{1}{M} \int_{0-}^{t} f(t)dt = \frac{1}{M} \int_{0-}^{t} \delta(t)dt = \frac{1}{M} \qquad t > 0 \qquad (1.13)$$

This result indicates that the velocity is a step function: it is zero at $t = 0-$ and assumes the value $1/M$ for times $t \geq 0+$. We also conclude that the derivative of the unit amplitude step function is equal to the delta function $\delta(t)$. This conclusion is the converse property made above. We also know that the derivative of a definite integral is equal to the integrand. We may write, in general,

$$\frac{du(t-t_0)}{dt} = \delta(t-t_0) \qquad (1.14)$$

■

Chapter 1: Signals and their functional representation

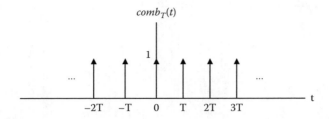

Figure 1.2.9 The comb function.

Comb function

The **comb function** is an array of delta functions that are spaced T units apart and extend in the range $-\infty < t < \infty$. The function is illustrated in Figure 1.2.9; its mathematical description is

$$\text{comb}_T(t) = \sum_{n=-\infty}^{\infty} \delta(t - nT) \tag{1.15}$$

Arbitrary sampled function

The comb function can be used in the presentation of any continuous function in its sampled or discrete form. The property of the delta function permits us to write (integrate each side of the expressions)

$$f(t)\delta(t) = f(0)\delta(t)$$
$$f(t)\delta(t-a) = f(a)\delta(t-a) \tag{1.16}$$

Therefore,

$$f_s(t) = f(t)\text{comb}_T(t) = \sum_{n=-\infty}^{\infty} f(nT)\delta(t-nT) \tag{1.17}$$

where $f_s(t)$ denotes the sampled version of the function $f(t)$. Figure 1.2.10a shows graphically the function $f(t)$, $\text{comb}_T(t)$, and $f_s(t)$. Figure 1.2.10b displays operationally the process of combining $f(t)$ and $\text{comb}_T(t)$. We shall show in a later chapter that the sampled function has no relation to the function it was produced from by the process of sampling. However, we shall present specific conditions under which we shall be able to recover the original signal from its sampled version.

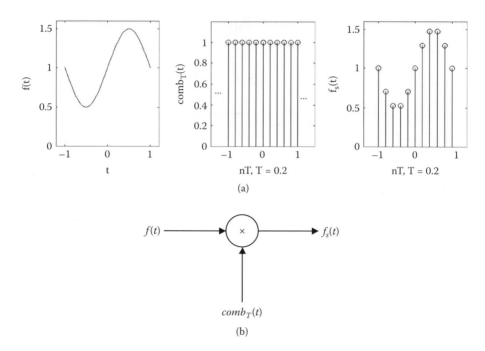

Figure 1.2.10 (a) Representation of the discrete form of an arbitrary function. (b) Representation of the operational form.

1.3 Signal conditioning and manipulation

Modulation

In communications it is desirable to transmit slow varying signals over long distances as it is done the amplitude modulation (AM) radio stations. If we assume that the slowest frequency is 1000 Hz, the wavelength of this radio signal is $\lambda = 3 \times 10^8 / 1000 = 300$ km. Since an efficient antenna must have a length of about 1/2 of the wavelength of the transmitting signal, it is impractical to build a 150-km antenna. To circumvent this problem, one often resorts to **amplitude modulation**, which requires the multiplication of the signal with a high-frequency sinusoidal signal, e.g., $\cos(\omega_c t)$, known as the **carrier**. The frequency used for the AM carrier ranges from 80 to 120 kHz. Figure 1.3.1 shows the signal to be transmitted and the modulated signal that is transmitted from the antenna.

Shifting and flipping

Figure 1.3.2 shows the exponential function and two shifted positions. Figure 1.3.3 shows the same function in its sampled form. Figure 1.3.4 shows a finite ramp function, its shifted, and its flipped positions.

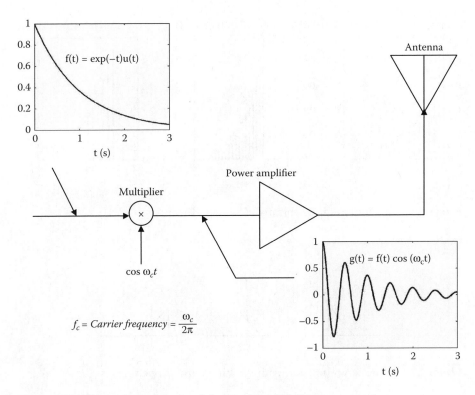

Figure 1.3.1 Illustration of the usefulness of modulation.

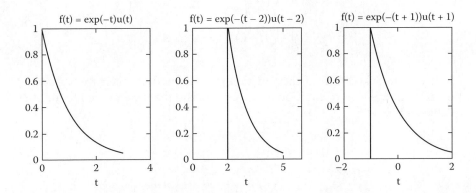

Figure 1.3.2 Illustration of the shifting of functions.

Time scaling

The compression or expansion of a signal is known as **time scaling**. Let us assume that the original signal is

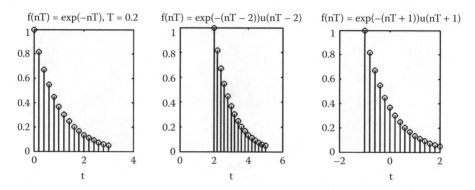

Figure 1.3.3 Illustration of the shifting of discrete functions.

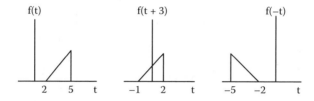

Figure 1.3.4 Finite ramp function with its shifted and flipped forms.

$$f(t) = \begin{cases} t+1 & -1 \leq t \leq 0 \\ -\frac{1}{2}(t-2) & 0 \leq t \leq 2 \end{cases} \quad (1.18)$$

If we are asked to find the function $f(2.5t)$, we first insert the value $2.5t$ for each t in the above equation and then rewrite the time range inequalities as shown below

$$f(2.5t) = \begin{cases} 2.5t+1 & -1 \leq 2.5t \leq 0 \\ -\frac{1}{2}(2.5t-2) & 0 \leq 2.5t \leq 2 \end{cases} \quad (1.19)$$

$$f(2.5t) = \begin{cases} 2.5t+1 & -\frac{1}{2.5} \leq t \leq 0 \\ -\frac{1}{2}(2.5t-2) & 0 \leq t \leq \frac{2}{2.5} \end{cases} \quad (1.20)$$

Chapter 1: Signals and their functional representation

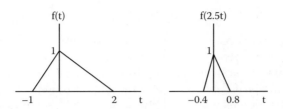

Figure 1.3.5 Illustration of the scaling property.

The functions $f(t)$ and $f(2.5t)$ are shown in Figure 1.3.5. Note that using the multiplier $2.5 > 1$ the function $f(2.5t)$ is compressed. If, on the other hand, we used a multiplier less than 1, the function would have been expanded.

Windowing of signals

One important signal processing operation is the **windowing** of signals. This operation is necessary in practical cases when computations are required. For example, when the Fourier transform (the spectrum of the signal) must be found using computers (we will discuss this integration in a later chapter), it is not possible to use the infinite length of a signal, for example. To accomplish the integration, we must terminate the function at some time length. This is equivalent to multiplying the original signal with a rectangular function, known as **window**. It turns out that the rectangular window is not satisfactory for accurately finding the Fourier transform of a signal. Therefore, many other types of windows were suggested, and the most important ones are given below:

Windows $w(t)$		First Sidelobe Level (dB)
1. Rectangular window: $w_{rect} = \begin{cases} 1 & -\frac{T}{2} \leq t \leq \frac{T}{2} \\ 0 & \text{otherwise} \end{cases}$		−13.3
2. Triangular (Bartlett): $w_{tr}(t) = 1 - \frac{|t|}{T}$	$-T \leq t \leq T$	−26.5
3. Hanning: $w_{hn}(t) = 0.5\left[1 + \cos\left(\frac{2\pi t}{T}\right)\right]$	$-\frac{T}{2} \leq t \leq \frac{T}{2}$	−31.5
4. Hamming: $w_{hm}(t) = 0.54 + 0.46\cos\left(\frac{2\pi t}{T}\right)$	$-\frac{T}{2} \leq t \leq \frac{T}{2}$	−42.7
5. Blackman: $w_{bl}(t) = 0.42 + 0.5\cos\left(\frac{2\pi t}{T}\right) + 0.08\cos\left(\frac{2\pi t}{T}\right)$	$-\frac{T}{2} \leq t \leq \frac{T}{2}$	−58.1

*1.4 Representation of signals

In the previous sections, it was noted that we presented the signals using mathematical forms. Another approach, and sometimes very important for signal manipulation and processing, is to represent the signal by a combination of elementary functions called **basis functions**. For example, the exponential function $\exp(j\omega t)$ and the cosine and sine functions are basis functions. The principal reason in employing basis functions is that mathematical operations may be performed easier; also, a better understanding of the system behavior is possible using such presentation.

Because systems are usually described by mathematical operators, it is evident that the output of the system is the result of the operator operating on the input signal $f(t)$. However, many mathematical operations, such as integration, are difficult to perform unless the function $f(t)$ possesses certain special properties. One approach is to find a suitable set of basis functions $\{\varphi_i(t)\}$ such that the function $f(t)$ is constructed by using a linear combination of these functions. Since we are dealing with linear systems, it is relatively easy to determine the system output because it is the result of the system operator operating on each basis function and adding these effects, an application of the superposition principle.

In order to understand better the use of basis functions, we shall recall the use of vectors in vector spaces. Let $\mathbf{a_i}$ (i = 1, 2, 3), known as the **basis set**, and denote the three orthogonal unit vectors along the Cartesian coordinate axis (see Figure 1.4.1). An arbitrary vector $\mathbf{F}$ can be expressed in the form

$$\mathbf{F} = F_1\mathbf{a_1} + F_2\mathbf{a_2} + F_3\mathbf{a_3} = \sum_{i=1}^{3} F_i\mathbf{a_i} \qquad (1.21)$$

where F_i's are projections of the vector $\mathbf{F}$ on each coordinate axis. Taking into consideration the orthogonality of the basis vector set $\mathbf{a_i}$, we have the relation

$$\mathbf{a_i} \cdot \mathbf{a_j} = a_i a_j \cos\theta_{i,j} = \begin{cases} 1 & i = j \\ 0 & i \neq j \end{cases} \qquad (1.22)$$

where $\theta_{i,j}$ is the angle between the two vectors. This mathematical operation is known as the **dot product** of vectors. Dot multiply (1.21) by $\mathbf{a_i}$ and take into consideration the orthogonality property of the basis vectors to find the ith component of the vector $\mathbf{F}$

$$F_i = \frac{\mathbf{F} \cdot \mathbf{a_i}}{\mathbf{a_i} \cdot \mathbf{a_i}} \qquad (1.23)$$

Chapter 1: Signals and their functional representation

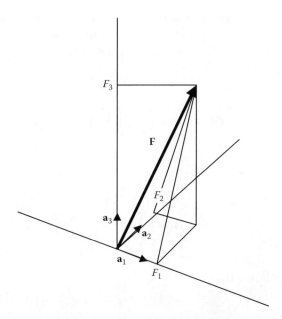

Figure 1.4.1 Three-dimensional vector space with its basis set.

where F_i is the projection of the vector **F** on the $\mathbf{a_i}$ axis. Since **any** three-dimensional vector **F** can be represented as a linear combination of the three basis orthogonal unit vectors, the system is **complete** — that is, **F** is contained in the space **spanned** by the basis vectors $\{\mathbf{a_i}\}$. On the other hand, a three-dimensional vector **F** cannot be represented as a linear combination of only two orthogonal vectors $\mathbf{a_1}$ and $\mathbf{a_2}$. Therefore, the two-dimensional orthogonal set is **incomplete** to represent three-dimensional vectors, and the basis set does not span the space containing **F**.

Our next step is to project the above ideas to functions. We first need functions that are orthogonal. However, functions are not vectors, and therefore, we must find another analogous way. It is known that when we multiply two cosine functions, e.g., $\cos(\pi t)$ and $\cos(2\pi t)$, and integrate from 0 to 1, we obtain zero value. However, by integrating over the same range the function $\cos^2(\pi t)$ we obtain $1/2$. Mathematicians tell us that there are several other functions that may become basis functions over some interval $t = a$ to $t = b$, and thus, we can express an arbitrary function $f(t)$ as a linear sum of these basis functions. If we proceed analogously from the three-dimensional vector space, we may set the equivalent identities: $\varphi_i(t) \triangleq \mathbf{a_i}$ and $c_i \triangleq F_i$. Hence, (1.21) assumes the equivalent form

$$f(t) = \sum_{i=1}^{\infty} c_i \varphi_i(t) \qquad (1.24)$$

Multiply (1.24) by $\varphi_j(t)dt$ or by $\varphi_j^*(t)$ if the φ_i's are complex functions (the asterisk means complex conjugate function — the functions with all complex components replaced by their negative values) and integrate over the range of definition $[a, b]$ of $f(t)$. The result is

$$\int_a^b f(t)\varphi_j(t)dt = \sum_{i=1}^{\infty} c_i \int_a^b \varphi_i(t)\varphi_j(t)dt \qquad (1.25)$$

or

$$\int_a^b f(t)\varphi_i(t)dt = c_i \int_a^b \varphi_i^2(t)dt = c_i \|\varphi_i\|^2$$

from which it follows that

$$c_i = \frac{\int_a^b f(t)\varphi_i(t)dt}{\int_a^b \varphi_i^2(t)dt} = \frac{\int_a^b f(t)\varphi_i(t)dt}{\|\varphi_i\|^2} \qquad i = 1, 2, 3, \cdots \qquad (1.26)$$

where

$$\|\varphi_i\| = \left[\int_a^b \varphi_i^2(t)dt\right]^{1/2} \qquad (1.27)$$

is called the **norm** of the function φ_i and is a real number. Equations (1.25) to (1.27) are the result of assumed orthogonality property over the interval $t = a$ to $t = b$:

$$\int_a^b \varphi_i(t)\varphi_j(t)dt = \begin{cases} \|\varphi_i\|^2 & i = j \\ 0 & i \neq j \end{cases} \qquad (1.28)$$

If the norm is equal to 1, the basis functions are called **orthonormal**. One of the conditions to analyze functions as a linear combination of basis functions is that both must be **absolutely** and square **integrable**, that is, they must obey the relations

Chapter 1: Signals and their functional representation 21

$$\int_a^b |f(t)| \, dt < \infty \quad \text{and} \quad \int_a^b |f(t)|^2 \, dt < \infty \qquad (1.29)$$

including the case when the interval $[a, b] = (-\infty, \infty)$.

Example 1.4.1: Show that the set of functions $f(t) = \{1, \cos t, \cos 2t, \cdots, \cos nt, \cdots\}$ constitutes orthogonal basis functions over the interval $[0, \pi]$.

Solution: From the relations

$$\int_0^\pi 1 \cos nt \, dt = 0; \quad \int_0^\pi \cos nt \cos mt \, dt = 0 \quad m \neq n; \quad \int_0^\pi \cos^2 nt \, dt = \frac{\pi}{2}$$

we observe that this set constitutes basis functions on the interval $0 \leq t \leq \pi$. The basis functions will become orthonormal if each one is multiplied by $\sqrt{2/\pi}$. ∎

The trigonometric function $\cos n\omega_0 t$ (similarly, $\sin n\omega_0 t$) is one of the most important basis functions and one that we will use frequently in our study of linear systems. The complex representation of these functions is

$$\varphi_n(t) = e^{jn\omega_0 t} \qquad n = 0, \pm 1, \pm 2, \cdots \qquad (1.30)$$

since $\text{Re}\{\exp(jn\omega_0 t)\} = \text{Re}\{\cos n\omega_0 t + j \sin n\omega_0 t\} = \cos n\omega_0 t$; $\text{Im}\{\exp(jn\omega_0 t)\} = \sin n\omega_0 t$. These expressions form the basis for the Fourier series expansion of periodic functions. We will study this matter in more detail in a later chapter. The main reason for learning about basis functions is that they are the eigenfunctions of linear time-invariant (LTI) systems.

A **linear time-invariant** (LTI) **system** is one that when its input is shifted, its output is also shifted. In addition, superposition applies to LTI systems, which means that the output of a system to multiple inputs is the sum of the outputs due to each component of the input. Hence, an eigenfunction input to LTI system results in an output that is equal to the input multiplied by a complex constant. If the input is a sinusoidal, the output is a sinusoidal having the same frequency, with the only difference that its amplitude may be changed and an additional phase is added. These ideas and results will be discussed in more detail in the next chapter.

Orthogonal functions are also important in detection theory. Suppose, for example, that we send four messages corresponding to four orthogonal functions φ_1, φ_2, φ_3, and φ_4, respectively. If we construct a receiver having the properties shown in Figure 1.4.2, we observe that each channel will respond only when the incoming signal is equal to the input signal of the

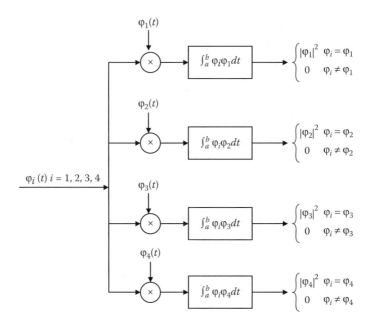

Figure 1.4.2 Detection scheme using orthogonal functions.

channel. Because of the orthogonality property of the functions, the remaining channels will not produce an output.

Example 1.4.2: Given below is some of the complete set of orthonormal functions in the range $-1 \leq t \leq 1$, known as **Legendre polynomial functions**.

$$\varphi_0(t) = 1/\sqrt{2}, \varphi_1(t) = \sqrt{3/2}\, t, \varphi_2(t) = (1/2)\sqrt{5/2}(3t^2 - 1), \varphi_3(t)$$
$$= (1/2)\sqrt{7/2}(5t^3 - 3t)$$

Find the value of the integral

$$I = \text{mean square error} = \frac{1}{2}\int_{-1}^{1}[t^2 - f_a(t)]^2\, dt$$

using the two different approximating functions given below. These functions try to approximate the parabolic function $f(t) = t^2$ in the range from -1 to 1.

a. $f_a(t) = c_0\varphi_0(t) + c_1\varphi_1(t)$

b. $f_a(t) = c_0\varphi_0(t) + c_1\varphi_1(t) + c_2\varphi_2(t)$

Chapter 1: Signals and their functional representation

Solution: To begin with, we must first find the coefficients c_i so that the function $f_a(t)$ approximates $f(t) = t^2$ in the range $-1 \leq t \leq 1$. Using (1.26), we find

$$c_0 = \frac{\int_{-1}^{1} t^2 \left(1/\sqrt{2}\right) dt}{\int_{-1}^{1} \left(1/\sqrt{2}\right)^2 dt} = \frac{\sqrt{2}}{3}, \quad c_1 = \frac{\int_{-1}^{1} t^2 \sqrt{3/2}\, t\, dt}{\int_{-1}^{1} [\sqrt{3/2}\, t]^2 dt} = 0$$

Therefore, the first order approximating function is $f_a(t) = \sqrt{2}/3$. Using this function, the mean square error I is 0.1080. For the case b, following the same procedure we find that $c_2 = 2\sqrt{10}/15$, and so $f_a(t) = (\sqrt{2}/3) + (1/3)(3t^2 - 1)$. For this case, the mean square error I is 0.0191.

From these results, we observe that the **mean square error** (MSE), which is an indicator of how close the approximation function is to the given function, decreases as we add more and more members from the basis set. It can be shown that if we use a complete set of orthogonal functions, the MSE will approach zero as the number of terms approaches infinity. ∎

Another approach is to approximate the function by polynomials. The use of a polynomial for curve fitting can never be mathematically exact unless the function $f(t)$ is a polynomial. However, in suitable circumstances, it can be as accurate as the tabulated values that are available.

Suppose that the function $f(t)$ is approximated by the polynomial

$$f_a(t) = \sum_{n=0}^{N} a_n t^n \tag{1.31}$$

It is assumed that the function is known (collocates) at $N + 1$ points with values $f(t_0), f(t_1), \cdots, f(t_N)$, where $t_0 < t_1 < t_2 \cdots < t_N$. Equation (1.31) leads to the set of equations

$$\begin{aligned} a_0 + a_1 t_0 + a_2 t_0^2 + \cdots + a_N t_0^N &= f(t_0) \\ a_0 + a_1 t_1 + a_2 t_1^2 + \cdots + a_N t_1^N &= f(t_1) \\ &\vdots \\ a_0 + a_1 t_N + a_2 t_N^2 + \cdots + a_N t_N^N &= f(t_N) \end{aligned} \tag{1.32}$$

This set of equations can be solved for the unknown coefficients a_i. This is conveniently done by writing this set of equations in matrix form:

$$\begin{bmatrix} 1 & t_0 & t_0^2 & \cdots & t_0^N \\ 1 & t_1 & t_1^2 & \cdots & t_0^N \\ & & \vdots & & \\ 1 & t_N & t_N^2 & \cdots & t_N^N \end{bmatrix} \begin{bmatrix} a_0 \\ a_1 \\ \vdots \\ a_N \end{bmatrix} = \begin{bmatrix} f(t_0) \\ f(t_1) \\ \vdots \\ f(t_N) \end{bmatrix} \qquad (1.33)$$

or, more compactly,

$$\mathbf{TA = F} \qquad (1.34)$$

where **T**, **A**, and **F** denote matrices in (1.33). Using matrix methods as given below, we find for **A** the expression

$$\mathbf{A = T^{-1}F} \qquad (1.35)$$

where $\mathbf{T}^{-1}$ is the matrix inverse of **T**, which involves a procedure discussed in Appendix 1.1.

Example 1.4.3: Find a three-term polynomial approximation for the function

$$f(t) = \cos\frac{\pi}{2}t$$

over the interval $-1 \le t \le 1$.

Solution: If we arbitrarily select the three points at $t_0 = -1$, $t_1 = 0$, and $t_2 = 1$ in a three-term expansion, as given by (1.31), the resulting set of equations is

$$f_a(-1) = a_0(-1)^0 + a_1(-1)^1 + a_2(-1)^2 = a_0 - a_1 + a_2 = 0$$
$$f_a(0) = a_0(0)^0 + a_1(0)^1 + a_2(0)^2 = a_0 + 0a_1 + 0a_2 = 1$$
$$f_a(1) = a_0(1)^0 + a_1(1)^1 + a_2(1)^1 = a_0 + a_1 + a_2 = 0$$

Solving this set of equations, we obtain the coefficients $a_0 = 1$, $a_1 = 0$, and $a_2 = -1$. The approximating function is $f_a(t) = 1 - t^2$, which is a circle. A better approximation is possible by using more terms in the polynomial expansion. However, we can also use MATLAB in the form:

```
a=inv(T)*F;%inv(.) is a MATLAB function which finds the inverse
           %of a matrix;
           %a is a vector containing the values 1,0 and -1.
```

$$T = \begin{bmatrix} 1 & -1 & 1 \\ 1 & 0 & 0 \\ 1 & 1 & 1 \end{bmatrix}; \quad F = \begin{bmatrix} f(-1) \\ f(0) \\ f(1) \end{bmatrix} = \begin{bmatrix} 0 \\ 1 \\ 0 \end{bmatrix}$$

■

The reader should refer to Appendix 1 of this chapter to read more about matrices and to find out about MATLAB functions to manipulate matrices.

Important definitions and concepts

1. Basic signal characteristics: amplitude, time duration, and shape
2. The transducer
3. Analog and discrete-time signals
4. Modulation of a signal
5. The communication channel
6. The amplitude and phase spectra of signals
7. The bandwidth of a signal
8. How to verify that a signal is periodic
9. The Euler equations
10. The sampling frequency
11. Nonperiodic continuous signals: pulse, unit step function, sinc function
12. Nonperiodic special discrete signals: delta function, comb function, sampled function
13. Signal conditioning: modulation, shifting and flipping, time scaling, windowing of signals
14. Windows: rectangular, triangular, Hanning, Hamming, Blackman
15. Representation of signals using basis functions
16. Basis set functions
17. Norm of a function
18. Orthonormal functions
19. Linear time-invariant systems
20. Mean square error
21. Even and odd functions (see Problem 1.2.11)
22. Energy and power signals (see Problem 1.3.7 and Problem 1.3.8)
23. Projection theorem (see Problem 1.4.5)
24. Gram–Schmidt orthogonalization process (see Problem 1.4.2)
25. Walsh functions (see Problem 1.4.7)

Chapter 1 Problems
Section 1.2

1. Plot e^{j2t} for $t = 0$ to $t = \pi$ in the complex plane and define the shape of the curve. Also use MATLAB to verify your conclusion.
2. Using MATLAB, plot $\sin\omega t$ for $\omega = 0, \pi/8, \pi/4, \pi/2, \pi, 2\pi, 4\pi, 8\pi$. Similarly, plot $\sin\omega nT = \sin\omega n$ ($T = 1$) vs. n for $\omega = 0, \pi/8, 2\pi/8, 3\pi/8, 4\pi/8, 5\pi/8, 6\pi/8, 7\pi/8, 8\pi/8, 9\pi/8, 10\pi/8, 11\pi/8, 12\pi/8$. State an important conclusion from the plots.
3. Find the amplitude r and phase φ of the cosine function $f(t) = r\cos(\omega t + \varphi)$ if $f(t)$ is given by the expression $f(t) = a\cos\omega t + b\sin\omega t$.
4. Draw the following functions:

    a. $f_1(t) = u(t) - u(t - 2)$
    b. $f_2(t) = u(t - 2) - u(t - 4)$
    c. $f_3(t) = u(-t + 2) - u(-t - 2)$
    d. $f_4(t) = 2u(t + 1) - 2u(t - 1)$
    e. $f_5(t) = p(t + 2)$
    f. $f_6(t) = -p(t - 2)$

5. Find the outputs of the systems shown in Figures P1.2.5a and b.

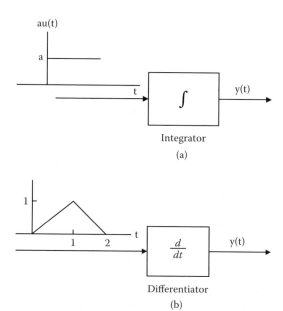

Figure P1.2.5

Chapter 1:   Signals and their functional representation

6. Sketch the functions:

   a. $f_1(t) = 2\dfrac{\sin \dfrac{1}{2}t}{t}$

   b. $f_2(t) = \dfrac{\sin 2(t-1)}{t-1}$

7. If $f(t) = u(-t - 1) + \exp(-t)u(t)$, sketch the following functions and state their changes:

   a. $f(-t)$
   b. $-f(t)$
   c. $f(t-1)$
   d. $f(-t-2)$

8. Sketch the functions:

   a. $f_1(t) = \dfrac{\sin(t-1)}{t-1} u(-t+1)$

   b. $f_2(t) = \cos 2t\, u(t-1.5)$

   c. $f_3(t) = e^{-t} \cos 2t\, u(t)$

   d. $f_4(t) = \cos t + \cos 2t$

9. Show that the following sequences lead to delta functions as a changes:

   a. $\dfrac{1}{a}[u(t) - u(t-a)]\quad a \to 0,\ a \geq 0$

   b. $a e^{-at} u(t)\quad a \to \infty,\ a > 0$

   c. $\dfrac{1}{2\sqrt{\pi a}} e^{-t^2/4a}\quad a \to 0,\ a > 0$

   Hint: Show that the peak increases and the duration decreases as a varies; also show that the area under each curve is 1.

10. Evaluate the integrals:

    a. $\displaystyle\int_{-2}^{2} (3t^2 + 1)[\delta(t) + 2\delta(t+1)]dt$

    b. $\displaystyle\int_{-2}^{2} e^{-t}[\delta(t+1) + 2\delta(t-1)]dt$

    c. $\displaystyle\int_{-2}^{2} e^{-2t} \cos 3t [\delta(t) + \delta(t-1)]dt$

11. If $f(t) = f(-t)$, the function is known as an **even** function. If $f(t) = -f(-t)$, the function is known as an **odd** function. The even and odd parts of a general function are given respectively by

$$f_e(t) = \frac{f(t) + f(-t)}{2}, \quad f_o(t) = \frac{f(t) - f(-t)}{2}$$

For the function shown in Figure P1.2.11, sketch its even and odd parts.

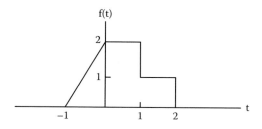

Figure P1.2.11

12. Evaluate the integrals:

   a. $\int_{-2}^{2} e^{-t^2} \dot{\delta}(t) dt$

   b. $\int_{-2}^{2} e^{-t^2} \dot{\delta}(t-1) dt$

   Because $\exp(-t^2)$ is a symmetrical function (even function), and in the light of the value of the integrals, what can we tell about the type function $\dot{\delta}(t)$ (the time derivative of $\delta(t)$)?
   Hint: In the evaluation, integrate by parts.

13. If $f(t) = \exp(-0.5t)u(t)$, plot the functions given below and compare their values with the exact ones for $n = 0, 1, 2, 3, \ldots$:

   a. $f_1(n) = e^{-0.5n} u(n)$
   b. $f_2(n0.5) = e^{-0.5n0.5} u(0.5n)$

14. Use the Euler relations to derive the expressions:

   a. $\cos^2 t = \frac{1}{2}(1 + \cos 2t)$
   b. $\sin x \sin y = \frac{1}{2}\cos(x-y) - \frac{1}{2}\cos(x+y)$
   c. $\sin(x+y) = \sin x \cos y + \cos x \sin y$

Chapter 1: Signals and their functional representation 29

15. Sketch the following discrete signals for $n = 0, 1, 2, 3, \ldots$:

    a. $f_1(n) = 2(0.9)^n$
    b. $f_2(n) = \delta(n+2) - 2\delta(n-2)$
    c. $f_3(n) = u(n) - u(n-4)$
    d. $f_4(n) = -u(3-n)$
    e. $f_5(n0.5) = 2(0.9)^{0.5n}$

16. If $f(n) = 2(0.9)^n u(n)$, sketch the signals:

    a. $f_1(n) = f(n-2)$
    b. $f_2(n) = f(2-n)$
    c. $f_3(n) = -f(-2-n)$

17. Plot the following discrete signals for $n = 0, 1, 2, 3, \ldots$:

    a. $f_1(n) = (-1)^n \, 2(0.8)^n$
    b. $f_2(n) = 0.2(1.1)^n$
    c. $f_3(n) = 0.9^n \sin(\pi n/8)$
    d. $f_4(n) = (1.1)^n \sin(\pi n/8)$

18. Plot the following discrete functions, $t = 0.2n$:

    a. $f_1(n0.2) = e^{-|t|} \mathrm{comb}_{0.2}(t)$
    b. $f_2(n0.2) = \dfrac{2\sin t}{t} \mathrm{comb}_{0.2}(t)$

Section 1.3

1. If $f(t) = e^{-(t-1)} u(t-1)$, plot the functions:

    a. $f(-t)$
    b. $f(-t+2)$
    c. $f(-t-2)$

2. If the carrier signal is $v_c(t) = \cos \omega_c t$ and the modulating function is $f(t) = \cos \omega_m t$ with $\omega_c \gg \omega_m$, sketch the form of the amplitude-modulated function $g(t) = v_c(t) f(t)$.

3. Determine the frequency spectrum (frequency present) in the modulated signal $g(t) = \cos \omega_m t \cos \omega_c t$, where the modulation frequency is $\omega_m = 10^3$ rad/s and the carrier frequency is $\omega_c = 1.2 \times 10^6$ rad/s.

4. Determine the spectrum of the modulated signal, $g(t) = (1 + 0.1 \cos \omega_m t) \cos \omega_c t$, where the modulating frequency is $\omega_m = 10^3$ rad/s and the carrier frequency is $\omega_c = 1.2 \times 10^6$ rad/s. What observation is evident when comparing these results with those of Problem 1.3.3?

5. If

$$f(t) = \begin{cases} t & 0 \le t < 2 \\ 2 & 2 \le t < 3 \\ 1 & 3 \le t < 4 \\ 0 & \text{otherwise} \end{cases},$$

   find and sketch the functions:
   a. $f_1(t) = f(2t - 1)$
   b. $f_2(t) = f(0.5t + 1)$

6. If

$$f(t) = \begin{cases} 1 - |t - 1| & 0 \le t < 1 \\ 0 & \text{otherwise} \end{cases},$$

   find and sketch the functions:
   a. $f_1(t) = f(2t)$
   b. $f_2(t) = f(0.5t)$

7. A signal that satisfies the relation

$$E = \int_{-\infty}^{\infty} |f(t)|^2 \, dt < \infty$$

   is known as the **energy signal**. Indicate whether or not the following are energy signals:
   a. $f(t) = \exp(-t)u(t)$
   b. $f(t) = u(t)$

8. A signal that satisfies the relation

$$P = \lim_{T \to \infty} \frac{1}{T} \int_{-T/2}^{T/2} |f(t)|^2 \, dt < \infty$$

   is known as the **power signal**. Indicate whether or not the following are power signals:
   a. $f(t) = \exp(-t)u(t)$
   b. $f(t) = u(t)$

Section 1.4

1. If the basis functions are complex, show that

$$\int_a^b |f(t)|^2 \, dt = \int_a^b f(t)f^*(t)\,dt = \sum_{n=0}^{\infty} |c_n|^2 |\varphi_n|^2$$

2. Given a linear independent system of functions $\varphi_1, \varphi_2, \ldots, \varphi_n, \ldots$ defined on the interval $[a, b]$, define a new system $\psi_1, \psi_2, \ldots, \psi_n, \ldots$ as follows: $\psi_1 = \varphi_1$,

$$\psi_2 = \varphi_2 - \frac{\int_a^b \psi_1(t)\varphi_2(t)\,dt}{\int_a^b \psi_1^2(t)\,dt} \psi_1(t), \quad \psi_3 = \varphi_3 - \frac{\int_a^b \psi_1(t)\varphi_3(t)\,dt}{\int_a^b \psi_1^2(t)\,dt} \psi_1(t) - \frac{\int_a^b \psi_2(t)\varphi_3(t)\,dt}{\int_a^b \psi_2^2(t)\,dt} \psi_2(t), \ldots$$

    This procedure is called the **Gram–Schmidt orthogonalization process**. Apply this process to the set of functions $1, t, t^2, t^3, \ldots$ in the interval $-1 \le t \le 1$ and generate the Legendre polynomials within a constant.

3. Are the basis functions $\{\varphi_n(t)\} = \{\exp(jn\omega_0 t)\}$ orthonormal in the range $0 \le t \le 2\pi/\omega_0$ ($\omega_0 = 2\pi/T$)? If not, find the appropriate constant that will make them orthonormal.

4. Show that the functions

$$\varphi_0(t) = \sqrt{\frac{1}{2}}, \quad \varphi_1(t) = \sqrt{\frac{3}{2}} t$$

    are orthonormal in the interval $-1 \le t \le 1$, in the approximate form $f_a(t) = c_0 \varphi_0(t) + c_1 \varphi_1(t)$.

5. (**Projection theorem**) One of the most celebrated theorems in signal processing is the projection theorem, which states: *If $f(t)$ is an element of a linear (vector) space S spanned by the orthogonal basis $\{\varphi_i\}$, and $f_a(t)$ is an element in a lower-dimensional space $S_a \subset S$, the error $\varepsilon(t) = f(t) - f_a(t)$ is a minimum in the mean square sense if, and only if, $f_a(t)$ is the orthogonal projection of $f(t)$ into $f_a(t)$. The minimum error and the projection are orthogonal.* Figure P1.4.5 illustrates the projection theorem. Observe that for vectors the orthogonality is defined by their dot product,

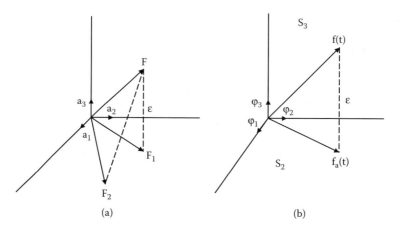

Figure P1.4.5

whereas in functions it is defined by their integrals. Let $f(t) = t^3$ $-1 \le t \le 1$ and let $f_a(t) = c_0\varphi_0(t) + c_1\varphi_1(t)$, where

$$\varphi_0(t) = \sqrt{\frac{1}{2}}, \quad \varphi_1(t) = \sqrt{\frac{3}{2}}t, \quad \varphi_2(t) = \sqrt{\frac{5}{8}}(3t^2 - 1), \cdots$$

is the orthonormal set of Lengendre functions. Use the minimum mean square error approach to find c_0 and c_1. The error then is equal to $\varepsilon(t) = f(t) - f_a(t)$. Show that the error is orthogonal to $f_a(t)$.

Hint: Find the value of the integral,

$$\int_{-1}^{1} f_a(t)\varepsilon(t)dt$$

6. Determine the MSE if we approximate $f(t)$ = sint over the range $0 \le t \le \pi$ by the functions:

   a. $f_a(t) = (1/\pi)t$
   b. $f_a(t) = 0.5$
   c. $f_a(t) = \sin^2(t)$

7. Find an approximation to the function shown in Figure P1.4.7a using the first three **Walsh functions** shown in Figure P1.4.7b. Show also that the Walsh functions are orthonormal.

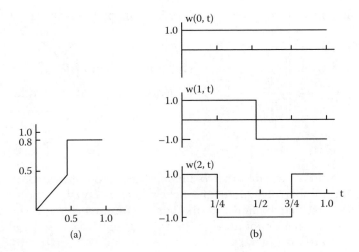

Figure P1.4.7

Appendix 1.1: Elementary matrix algebra

Some basic matrix algebra

We will present the algebraic methods for matrices with constant elements using a 2 × 2 and a 2 × 3 matrix as examples. The generalization to larger matrices is done in a similar way.

Addition and subtraction of matrices

We can add and subtract two matrices if they have the same number of rows and columns. The element a_{ij} of the matrix **A** is the one located at the intersection of the ith row and the jth column. Hence, the ij^{th} element of the matrix $\mathbf{A} \pm \mathbf{B}$ is $a_{ij} \pm b_{ij}$. For example,

$$\mathbf{A} \pm \mathbf{B} = \begin{bmatrix} a_{11} & a_{12} & a_{13} \\ a_{21} & a_{22} & a_{23} \end{bmatrix} \pm \begin{bmatrix} b_{11} & b_{12} & b_{13} \\ b_{21} & b_{22} & b_{23} \end{bmatrix} = \begin{bmatrix} a_{11} \pm b_{11} & a_{12} \pm b_{12} & a_{13} \pm b_{13} \\ a_{21} \pm b_{21} & a_{22} \pm b_{22} & a_{23} \pm b_{23} \end{bmatrix} \quad (1.1)$$

Multiplication of matrices

The **multiplication** of a matrix with any **number** is a matrix with each element been the product of the element of the initial matrix times the number.

$$c_{ij} = b a_{ij} \quad (1.2)$$

The **multiplication** of two **matrices** is given as follows:

$$\mathbf{AB} = \begin{bmatrix} a_{11} & a_{12} & a_{13} \\ a_{21} & a_{22} & a_{23} \end{bmatrix} \begin{bmatrix} b_{11} & b_{12} \\ b_{21} & b_{22} \\ b_{31} & b_{32} \end{bmatrix} = \begin{bmatrix} a_{11}b_{11} + a_{12}b_{21} + a_{13}b_{31} & a_{11}b_{12} + a_{12}b_{22} + a_{13}b_{32} \\ a_{21}b_{11} + a_{22}b_{21} + a_{23}b_{31} & a_{21}b_{12} + a_{22}b_{22} + a_{23}b_{32} \end{bmatrix}$$

$$2 \times 3 \qquad 3 \times 2 \qquad\qquad 2 \times 2$$

(1.3)

The columns of the first matrix must be equal to the rows of the second matrix. The number of rows of the new matrix is equal to the number of rows of the first matrix, and the number of columns of the new matrix is equal to the number of columns of the second matrix. The general element of the new matrix is given by

$$c_{ij} = \sum_{k=1}^{p} a_{ik} b_{kj} \tag{1.4}$$

where **A** is any $m \times p$ matrix and **B** is any $p \times n$ matrix. The new matrix **C** is a $m \times n$ matrix.

Determinant of a matrix

The **determinant** of a 3×3 matrix is given as follows:

$$\det \left\{ \begin{bmatrix} a_{11} & a_{12} & a_{13} \\ a_{21} & a_{22} & a_{23} \\ a_{31} & a_{32} & a_{33} \end{bmatrix} \right\} = (-1)^{1+1} a_{11} \det \left\{ \begin{bmatrix} a_{22} & a_{23} \\ a_{32} & a_{33} \end{bmatrix} \right\} + (-1)^{1+2} a_{12} \det \left\{ \begin{bmatrix} a_{21} & a_{23} \\ a_{31} & a_{33} \end{bmatrix} \right\}$$

$$+ (-1)^{1+3} a_{13} \det \left\{ \begin{bmatrix} a_{21} & a_{22} \\ a_{31} & a_{32} \end{bmatrix} \right\} = a_{11}(a_{22}a_{33} - a_{23}a_{32}) - a_{12}(a_{21}a_{33} - a_{23}a_{31}) \tag{1.5}$$

$$+ a_{13}(a_{21}a_{32} - a_{22}a_{31})$$

The inverse of a matrix

A matrix **A** multiplied with its inverse $\mathbf{A}^{-1}$, if it exists, produces the identity matrix **I** (**I** is a square matrix that has all the off-diagonal elements equal to 0 and the element along the main diagonal equal to 1).

The **adjoint** of a matrix **A** is defined below using a 3×3 matrix **A** as an example:

Chapter 1: Signals and their functional representation

$$adj\{A\} = adj\left\{\begin{bmatrix} a & b & c \\ d & e & f \\ g & h & i \end{bmatrix}\right\} = [c_{ij}] = \begin{bmatrix} ei-fh & -(di-fg) & dh-eg \\ -(bi-ch) & ai-cg & -(ah-bg) \\ bf-ce & -(af-cd) & ae-bd \end{bmatrix}^T$$

$$= \begin{bmatrix} ei-fh & -(bi-ch) & bf-ce \\ -(di-fg) & ai-cg & -(af-cd) \\ dh-eg & -(ah-bg) & ae-bd \end{bmatrix}$$

(1.6)

The **inverse** of a matrix **A** is given by

$$\mathbf{A}^{-1} = \frac{adj\{\mathbf{A}\}}{det\{\mathbf{A}\}}$$

(1.7)

Example 1.1: Find the inverse of the following 2 × 2 matrix:

$$\mathbf{A} = \begin{bmatrix} 2 & 3 \\ 1 & 6 \end{bmatrix}$$

Solution:

$$adj\{A\} = \begin{bmatrix} 6 & -1 \\ -3 & 2 \end{bmatrix}^T = \begin{bmatrix} 6 & -3 \\ -1 & 2 \end{bmatrix}; \quad det\{A\} = 6 \times 2 - 1 \times 3 = 9;$$

$$A^{-1} = \frac{1}{9}\begin{bmatrix} 6 & -3 \\ -1 & 2 \end{bmatrix} = \begin{bmatrix} \frac{6}{9} & \frac{-3}{9} \\ \frac{-1}{9} & \frac{2}{9} \end{bmatrix}$$

The reader should verify the product $\mathbf{AA}^{-1} = \mathbf{I}$.

MATLAB operations and functions to manipulate matrices

Writing a matrix

```
A = [1 2 5; 2 -10 0.2]    a 2 x 3 matrix;
B = [0.1 1 -2 0.5 5]      a 1 x 5 matrix, a vector.
```

Multiplication of two matrices

```
C=A*B;%matrices are multiplied if they have dimensions of the
    %form mxn and nxp
```

Addition and subtraction of two matrices

$$\mathbf{C} = \mathbf{A} \pm \mathbf{B}$$

The dimension of the matrices to be added must be the same.

Inverse of a matrix

$$\mathbf{A}^{-1} = inv(\mathbf{A});$$

Transpose of a matrix

$$\mathbf{B} = \mathbf{A}';$$

Matrix **B** is the transpose of **A**.

Appendix 1.2: Complex numbers

Addition and subtraction of complex numbers

If $z_1 = 2 + j$ and $z_2 = 1 - 3j$, then $z = z_1 + z_2 = 3 - 2j$. The Cartesian representation of these complex numbers and their sum is shown in Figure A1.2.1. The plane on which the complex numbers are presented is known as the complex plane. Observe that complex numbers add in the complex plane as forces add. This is also true for subtraction of complex numbers. Figure A1.2.2 presents the following forms of a complex number: $z = 2 + j$; $z^* =$ conjugate of $z = 2 - j$; $-z =$ reflected of z; $-z^*$ reflected of its conjugate form; $z = (2^2 + 1^2)^{1/2} \exp(j\tan^{-1}(1/2)) =$ polar form.

Multiplication of complex numbers

Cartesian format

If $z_1 = 2 + j$ and $z_2 = 1 - 3j$, their product is $z = z_1 \times z_2 = (2 + j)(1 - 3j) = 2 \times 1 + j \times 1 - 2 \times 3j - 3j^2 = 5 - j5$.

Polar form

$$z_1 z_2 = \sqrt{2^2 + 1^2} e^{j\tan^{-1}(1/2)} \sqrt{1^2 + (-3)^2} e^{j\tan^{-1}(-3/1)} = \sqrt{5 \times 10} e^{j(\tan^{-1}(1/2) - \tan^{-1}(3))}$$

$$= \sqrt{50} e^{j(-0.7854)} = 7.0711 e^{-j\frac{\pi}{4}}$$

Chapter 1: Signals and their functional representation

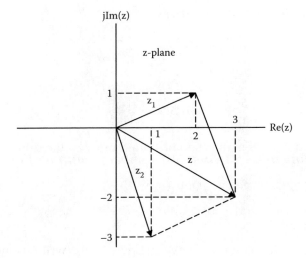

Figure A1.2.1

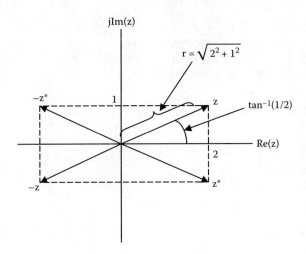

Figure A1.2.2

Division of complex numbers

Cartesian form

$$\frac{z_1}{z_2} = \frac{2+j}{1-3j} = \frac{(2+j)(1+j3)}{(1-3j)(1+3j)} = \frac{2+j+j6-3}{1-3j+3j+9} = -\frac{1}{10} + j\frac{7}{10}$$

Polar form

$$\frac{z_1}{z_2} = \frac{\sqrt{2^2+1^2}\,e^{j\tan^{-1}(1/2)}}{\sqrt{1^2+(-3)^2}\,e^{j\tan^{-1}(-3/1)}} = \sqrt{\frac{5}{10}}\,e^{j(\tan^{-1}(1/2)+\tan^{-1}(3))}$$

$$= \sqrt{\frac{5}{10}}\,e^{j1.7127} = 0.7071 e^{j0.5452\pi}$$

MATLAB functions for complex numbers manipulation

changing Cartesian to polar form

```
[angle_in_rad, magn]=cart2pol(2,3);%will change the complex
        %number 2+j3 to 3.6056e^j0.9828
```

To change the radians found above to degrees, we write in the command window: angle_in_rad*(180/pi). For this example, we find that 0.9828 radians corresponds to 56.31 degrees.

Changing polar to Cartesian form

```
The MATLAB function pol2cart(θ,r) converts the complex number
        rexp(jθ) to Cartesian form.
[real,imag]=pol2cart(0.9828,3.6056);%for this case we find the
        %original values: 2+j3;
```

Appendix 1.1 Problems

1. Add the following matrices:

$$A = \begin{bmatrix} -1 & 2 \\ 3 & 4 \end{bmatrix}, \quad B = \begin{bmatrix} 2 & -1 \\ -1 & -1 \end{bmatrix}$$

2. Add the matrices:

$$A = \begin{bmatrix} -1 & 2 & 3 \\ -1 & -2 & -3 \end{bmatrix}, \quad B = \begin{bmatrix} j & 2+j & 1 \\ -4 & 2 & 1 \end{bmatrix}$$

3. Multiply the matrices:

$$A = \begin{bmatrix} 1 & 2 & 3 \\ 4 & 5 & 6 \end{bmatrix}, \quad B = \begin{bmatrix} 2 & 3 \\ 4 & 6 \\ 1 & 1 \end{bmatrix}$$

Chapter 1: Signals and their functional representation

4. Multiply the matrices:

$$A = \begin{bmatrix} 1 & 2 \\ 3 & 4 \end{bmatrix}, \quad B = \begin{bmatrix} 1 & 2 & 3 \\ -1 & -1 & -2 \end{bmatrix}, \quad C = \begin{bmatrix} 1 \\ 2 \\ 4 \end{bmatrix}$$

5. Find the determinant of the following matrices:

$$A = \begin{bmatrix} 1 & 2 \\ 2 & 1 \end{bmatrix}, \quad B = \begin{bmatrix} 2 & 1 \\ 2 & 2 \end{bmatrix}, \quad C = \begin{bmatrix} 1 & 2 & 3 \\ 4 & 5 & 6 \\ 7 & 8 & 9 \end{bmatrix}$$

6. Find the inverse of the matrix:

$$A = \begin{bmatrix} 1 & 2 \\ 3 & 4 \end{bmatrix}$$

7. Find the inverse of the following matrix:

$$A = \begin{bmatrix} 1 & 2 & 0 \\ -1 & 3 & 1 \\ 0 & 2 & 3 \end{bmatrix}$$

Appendix 1.2 Problems

1. Add the following complex numbers and plot them: $z_1 = 2 + j$, $z_2 = -1 - 2j$, $z_3 = 2j$.
2. Multiply the following complex numbers and plot them: $z_1 = 2 + j$, $z_2 = z_1^*$, $z_3 = 2$.
3. Represent the following complex numbers in their Cartesian and polar form and plot them: $z_1 = -2 + j$, $z_2 = -z_1^*$, $z_3 = z_1 + 2 - j4$, $z_4 = |z_3| + z_1$.

chapter 2

Linear continuous-time systems

In this chapter we will learn about continuous systems. We will learn how to model the systems and find their outputs if they are excited by different types of inputs. Furthermore, we will learn the mathematical operation of convolution and its importance to system operations. We will focus our attention on how to find the impulse response of a system that is the cornerstone of system identification and system responses.

2.1 Properties of systems

Every **physical system** is broadly characterized by its ability to accept an **input** (voltage, current, pressure, magnetic fields, electric fields, etc.) and to produce an **output response** to this input. To study the behavior of a system, the procedure is to model mathematically each element that the system comprises and then consider the interconnected array of elements. The analysis of most systems can be reduced to the study of the relationship among certain inputs and resulting outputs. The interconnected system is described mathematically, the form of the description being dictated by the domain of description — time or frequency domain.

Critical to this process are:

1. A mathematical modeling of the elements that the system comprises; these elements might be complicated structures.
2. A mathematical description of their interconnection.
3. A solution to the mathematical equations for the specified input and prescribed initial state of the system.
4. Desirably, as a criterion of reasonableness, a careful examination of the final response equation.

This chapter addresses these aspects of system description and system analysis. System description is discussed both mathematically and graphically.

Three different methods for carrying out the solution process for continuous-type systems are studied initially. A somewhat parallel approach is considered for discrete-time systems, although the details differ from those for continuous-type systems. Our study will show the applicability of the developed methods to disciplines outside of the engineering domain.

As already noted, every physical system is able to accept an input and to produce an output response to this input. For example, a modern video-disc player is an optoelectronic system whose input is the reflection (or transmission) of laser from the grooves of the record and whose output is a video signal that is viewed on a television system. A telescope is an optical system that accepts the irradiance of the stars as its input and produces their image on a film or CCD camera as its output. A seismograph is an apparatus that registers the shocks and motions of the earth due to earthquakes. An elementary seismograph consists of a mass and a delicately mounted spring. Its input is the force caused by the earth's movement, and its output is the subsequent oscillations, which are registered on a paper as it moves under the stylus.

When the input–output properties of a system's elements can be described as proportional, time derivative, or integral functions, the resulting description is given by an integrodifferential equation. The form is evident when we apply the Kirchhoff voltage law to a series RLC circuit that has a voltage source as input and the circuit current as its output. Such systems can be shown in a circuit diagram or block diagram form. We will restrict our attention to systems made up of **lumped parameters**, that is, systems with a finite number of discrete elements, each of which is able to store or dissipate energy or, if it is a source, deliver energy. Such systems are described by ordinary differential equations. We can also represent the same systems in a **block diagram** form that describes the **terminal properties** of the networks, that is, the relationship between its input and its output. Therefore, the output $y(t) = \mathcal{O}\{f(t)\}$, where $f(t)$ is the input to the system and $\mathcal{O}$ is the operator of the system that operates (differentiates, integrates, multiplies, etc.) on the input to produce the output. Figure 2.1.1 shows the above definition in a block diagram form.

We can also have systems with **distributed parameters** instead of distinct discrete elements, although each infinitesimal part of these systems can be modeled in lumped parameter form. Their description is accomplished using partial differential equations. One such system is the telephone line. These systems are not dealt with in this text.

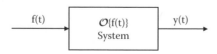

Figure 2.1.1 Block diagram representation of a system.

Chapter 2: Linear continuous-time systems

Systems with random parameters or randomly varying parameters are called **stochastic** systems, and those with parameters that are not varying randomly are called **deterministic** systems. Both system categories may use time in a continuous form and may be described by differential equations; in such cases, both are continuous-time systems. If excitations are introduced to systems at distinct instants of time and the systems are described by difference equations, these systems are termed **discrete-time systems**.

Systems can be linear or nonlinear. A **linear system** is one for which a linear relation exists between cause and effect, or between excitation and response. If the relation between cause and effect is nonlinear, the system is termed **nonlinear**. We will find that systems analysis can often be carried out in closed mathematical form for systems composed of linear elements. This analysis is rarely possible for nonlinear systems. This book is confined to linear systems only.

Systems for which an output at any time t_0 is a function only of those inputs that have occurred for $t \leq t_0$ — that is, the system response never precedes the system excitation — are **nonanticipatory**, or **causal**, systems. Ordinary physical systems are causal systems. Signals that do not start at minus infinity are called causal. Although noncausal systems exist, in this text we will study only causal systems.

Systems whose parameters vary with time are called **time-varying systems**, and they are not considered in this book. The focus here is on systems with constants, called **linear time-invariant** (LTI) **systems**. It is important to note that within these limitations, the number of specific systems is almost limitless. Thus, while the book emphasizes the study of some relatively simple electrical and mechanical systems, the procedures and methods can be applied to many other types of systems, such as economic, chemical, biological, etc.

2.2 *Modeling simple continuous systems*

As already noted, an essential requirement in continuous-time, or analog, systems analysis is a mathematical input–output description of the elements that make up the interconnected system. The requirement exists because we wish to perform mathematical studies on systems comprising hardware components or carry out designs mathematically that will ultimately be realized by hardware. It must be stressed that the analysis or design can be no better than the quality of the models used. The essential ideas in the modeling process are reviewed here.

Electrical elements

Capacitor

The linear capacitor is an idealized circuit element in which energy is stored in electric form. In its most elementary form, the capacitor consists of two closely spaced metallic plates that are separated by a single or multiple layers

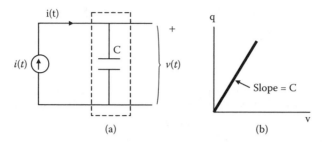

Figure 2.2.1 (a) Capacitor element and its circuit representation. (b) Charge–voltage characteristics of a linear capacitor element.

of nonconducting (insulating) dielectric material, such as air, glass, or paper. A schematic representation of the capacitor is shown in Figure 2.2.1a. The terminal properties of a linear time-invariant capacitor are described graphically by a charge–voltage relationship of the form shown in Figure 2.2.1b. The capacitors used in most electronic circuits are essentially time invariant, although the capacitor microphone used in radio studios is an example of a linear time-varying capacitor.

By definition, the capacitance C is

$$C = \frac{q}{v} \qquad \frac{\text{coulomb}}{\text{volt}} = \text{farad (F)} \qquad (2.1)$$

Since the current is given by

$$i = \frac{dq}{dt} \qquad \frac{\text{coulomb}}{\text{second}} = \text{ampere (A)} \qquad (2.2)$$

we obtain the relations

$$\boxed{\begin{aligned} v(t) &= \frac{1}{C} \int_{-\infty}^{t} i(x)dx \quad \text{a)} \\ i(t) &= C \frac{dv(t)}{dt} \quad \text{b)} \end{aligned}} \qquad (2.3)$$

As descriptive of their inherent properties, the current $i(t)$ is termed a **through variable** and the voltage $v(t)$ is called **across variable**.

Example 2.2.1: Find the voltage across a 1-μF capacitor at $t = 2$ s if the current through it has the waveshape shown in Figure 2.2.2.

Chapter 2: Linear continuous-time systems

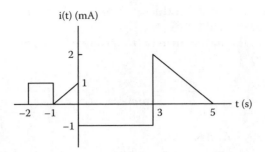

Figure 2.2.2 Current delivered to 1 µF capacitor.

Solution: Apply (2.3) part a) to obtain

$$v(2) = \frac{1}{1\times 10^{-6}} \int_{-\infty}^{2} i(x)dx = 10^{6}\left[\int_{-2}^{-1} 10^{-3} dx + \int_{-1}^{0} 10^{-3}(x+1)dx + \int_{0}^{2}(-10^{-3})dx\right]$$

$$= -0.5\times 10^{3} \ V$$

The negative sign indicates that the voltage polarity at $t = 2$ s is opposite of the polarity initially assumed as the reference condition shown in Figure 2.2.1a.

Inductor

Another important electrical element that stores energy in the form of magnetic field is the **inductor**, sometimes called coil or solenoid. The terminal properties of linear inductors are described graphically by the flux linkage–current relationship ψ, I shown in Figure 2.2.3b. Figure 2.2.3a shows the

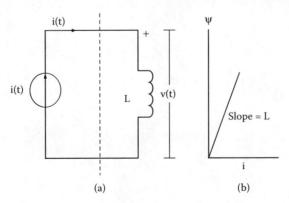

Figure 2.2.3 (a) Inductor element and its circuit representation. (b) Linear flux–current relationship.

circuit representation of the inductor. The simple telephone receiver is an example of a time-varying inductor.

By definition, the inductance of an inductor is written

$$L = \frac{\psi}{i} \quad \frac{\text{weber}}{\text{ampere}} = \text{henry (H)} \tag{2.4}$$

Combine this with Faraday's law,

$$v(t) = \frac{d\psi}{dt} \quad \text{volt (V)} \tag{2.5}$$

to obtain the relation (L is assumed to be time invariant; see Figure 2.2.3b)

$$\boxed{\begin{aligned} i(t) &= \frac{1}{L} \int_{-\infty}^{t} v(x)dx \quad \text{a)} \\ v(t) &= L \frac{di(t)}{dt} \quad \text{b)} \end{aligned}} \tag{2.6}$$

Example 2.2.2: Find the voltage across a 3-H inductor at $t = 2$ s if the current through it is as shown in Figure 2.2.4.

Solution: From (2.6) part b) we obtain

$$v(2) = 3 \left. \frac{di(t)}{dt} \right|_{t=2} = 3 \left. \frac{d[e^{-(t-1)}u(t-1)]}{dt} \right|_{t=2} = -3e^{-1} \qquad \blacksquare$$

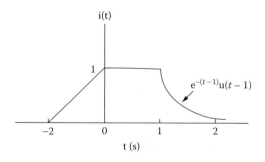

Figure 2.2.4 Current delivered to a 3-H inductor.

Chapter 2: Linear continuous-time systems

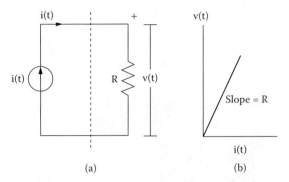

Figure 2.2.5 Resistor. (a) Network representation of the resistor. (b) Linear voltage–current relationship.

Resistor

Unlike the capacitor and inductor, each of which stores energy, the resistor shown in Figure 2.2.5a dissipates energy. The v, i characteristic for a linear resistor, as shown in Figure 2.2.5b, is a straight line with proportionality factor

$$\boxed{\begin{aligned} R &= \frac{v(t)}{i(t)} \qquad \frac{\text{volt}}{\text{ampere}} = \text{ohm}\,(\Omega) \qquad \text{a)} \\ G &= \frac{i(t)}{v(t)} \qquad \text{mho}\,(\mho) \qquad \text{b)} \end{aligned}} \qquad (2.7)$$

where R = **resistance** (ohm) and $G = 1/R$ = **conductance** (mho). Ordinary resistors are usually assumed to be time invariant, but carbon microphone in the ordinary telephone set is a time-varying resistor.

The above three simple systems can also be represented in their block diagram form (Figure 2.2.6). The input is manipulated (mathematically) by the system **operator** to form the output. It is desired that the reader becomes familiar with this type of presentation since it helps engineers and scientists to study systems and their behavior. Observe in Figure 2.2.6 that the system operator is different for each system, although the input and output for all the systems are the same. From this we can conclude that operator form

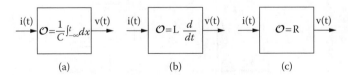

Figure 2.2.6 $\mathcal{O}$ = operator. (a) Capacitor block diagram–operator representation. (b) Inductor block diagram–operator representation. (c) Resistor block diagram–operator representation.

Mechanical translation elements

Ideal mass element

Bulk matter as a single unit is defined as a **mass element**. The dynamics of a mass element are described by Newton's second law of motion:

$$f(t) = M\frac{dv(t)}{dt} = M\frac{d^2x(t)}{dt^2} \qquad \text{newton} = \text{kg} \cdot \text{m} \cdot \text{s}^{-2} \text{ (N)} \qquad (2.8)$$

which relates the force $f(t)$ to the acceleration $dv/dt = d^2x/dt^2$. The motion variable v enters in a relative form: it is the velocity of the mass relative to the velocity of the ground, which is zero; hence, v is an **across** variable. The force $f(t)$ is transmitted through an element; hence, it is a **through** variable. The integral form of (2.8) is given by

$$v(t) = \frac{1}{M}\int_{-\infty}^{t} f(\tau)d\tau \qquad (2.9)$$

Note: *It is interesting to compare (2.9) and (2.3) part a). This comparison shows that an analogy exists between the mass in a mechanical system and the capacitor in an electrical system. A schematic representation of the mass element is shown in Figure 2.2.7.*

Spring

A spring element is one that stores energy due to the elastic deformation that results from the application of a force. Over its linear region, the spring satisfies Hook's law that relates the force to the displacement by the expression

$$f(t) = Kx(t) \qquad \text{newton (N)} \qquad (2.10)$$

where K is the spring constant, with units newton/m.

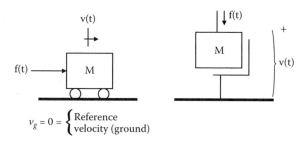

Figure 2.2.7 Schematic representation of the mass element.

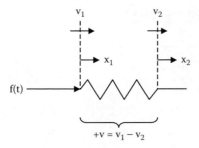

Figure 2.2.8 Schematic representation of the spring.

By differentiating (2.10) with respect to time, we obtain the relation

$$v(t) = \frac{1}{K}\frac{df(t)}{dt} \qquad (2.11)$$

Refer to the schematic representation of the spring in Figure 2.2.8 and use (2.10) and (2.11) to find

$$f(t) = K[x_1(t) - x_2(t)] = K\int_{-\infty}^{t}[v_1(x) - v_2(x)]dx = K\int_{-\infty}^{t}v(x)dx \qquad (2.12)$$

If $x_1 > x_2$, there is compressive force and $f > 0$. If $x_1 < x_2$, there is a negative, or extensive, force.

Note: *Observe the analogy between the spring and the inductor.*

Damper

Consideration of this element is limited here to viscous friction, which, for a linear dependence between force and velocity, is given by

$$f(t) = Dv(t) \qquad (2.13)$$

from which

$$v(t) = \frac{1}{D}f(t) \qquad (2.14)$$

where D is the damping constant (newton-second/m). A mechanical damper and its schematic representation are shown in Figure 2.2.9.

Note: *Observe the analogy between the damper and the resistor.*

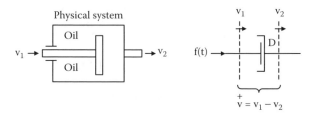

Figure 2.2.9 Physical and diagrammatic representation of a dash pot (damper).

By maintaining the parallelism between through and across variables, which also leads to a parallelism in the resulting network topology, we found that mass and capacitor, spring and inductor, and damper and resistor are analogous quantities. From a purely mathematical point of view, a parallelism between velocity and current and force and voltage could be chosen, in which case, we could obtain analogy between mass and inductor, spring and capacitor, and damper and resistor. In this case, the topology would be dually related — that is, parallel connection of mechanical elements would lead to an analogous series connection of electrical elements. Some of the older writings, particularly in acoustics, employed this dual relationship; this text is confined only to the through and across variable parallelisms, thereby also maintaining the parallel topological structure.

Mechanical rotational elements

Inertial elements

A set of rotational mechanical elements and rotational variables exists and bears a one-to-one correspondence to the translational mechanical elements and the translational variables that have been discussed above. In the rotational system, **torque** ($\mathcal{T}$) is the through variable and **angular velocity** ($\omega = d\theta/dt$) is the motional or across variable. The corresponding fundamental quantities are

J = polar moment of inertia	corresponds to M in translation
K = rotational spring constant	corresponds to K in translation
D = rotational damping constant	corresponds to D in translation
$\mathcal{T}$ = torque	corresponds to f in translation
$\omega = d\theta/dt$, angular velocity	corresponds to $v = dx/dt$ in translation

In this rotational set, J is a rotational parameter and the proportionality factor between torque and angular acceleration. When the motion is considered on one axis only,

Chapter 2: Linear continuous-time systems

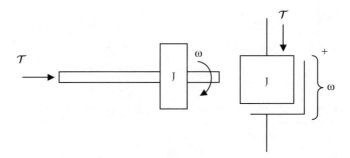

Figure 2.2.10 Schematic representation of rotational inertial systems.

$$\mathcal{T} = \frac{d(J\omega)}{dt} \qquad (2.15)$$

and for systems with constant J,

$$\boxed{\mathcal{T} = J\frac{d\omega(t)}{dt} = J\frac{d^2\theta}{dt^2}} \quad \text{newton-meter} \qquad (2.16)$$

J, the moment of inertia of a rotational body, depends on the mass and the square of a characteristic distance of the body, called the **radius of gyration**, k. This is given by

$$J = Mk^2 \qquad \text{kg} \cdot \text{m}^2 \qquad (2.17)$$

For a simple point mass rotating about an axis at a distance r from the center of mass, $k = r$ and

$$J = Mr^2 \qquad (2.18)$$

For a simple disc of radius r rotating about its center, $k = r/\sqrt{2}$, with

$$J = (1/2)Mr^2 \qquad (2.19)$$

Figure 2.2.10 shows the linear rotational inertial elements.

Spring
A rotational spring is one that will twist under the action of torque. A linear spring element is described by the pair of equations

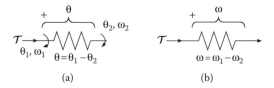

Figure 2.2.11 Schematic representation of rotational spring: (a) physical system and (b) circuit representation.

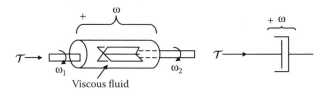

Figure 2.2.12 Schematic representation of a rotational damper.

$$\mathcal{T}(t) = K\theta(t) = K\int_{-\infty}^{t} \omega(\tau)d\tau$$
$$\omega(t) = \frac{1}{K}\frac{d\mathcal{T}(t)}{dt}$$
(2.20)

A schematic representation of the rotational spring is shown in Figure 2.2.11.

Damper

A rotational damper differs from the translational damper principally in the character of the motion. A schematic representation is given in Figure 2.2.12, and the equations that describe a rotational damper are

$$\mathcal{T}(t) = D\omega(t)$$
$$\omega(t) = \frac{1}{D}\mathcal{T}(t)$$
(2.21)

2.3 Solutions of first-order systems

First, we wish to study the characteristic behavior of systems of interconnected elements through their mathematical formulations. Second, we will study other than electrical and mechanical systems. It will be evident from our studies that, knowing the fundamental approach and the basic principles for engineering systems, we will be able to solve any other system that is described by the same differential equations. We develop the mathematical description by the use of Kirchhoff's current and voltage laws for electrical systems and by the use of D'Alembert's principle for mechanical systems. The resulting system equations that are given as one or more differential equations require, as the next step, the solution of the mathematical problem in a form that will yield the trajectory (output value) that satisfies the **initial condition** that applies to the system at some specified time, known as the **initial time**.

First, we shall examine **first-order systems**, that is, those systems described by first-order differential equations. It is very important to understand the principles and fundamental properties of the first-order differential equations because the basic principles can be extrapolated to any order of differential equations such as initial conditions, input, output, homogeneous solution, etc.

Consider the simple electric system shown in Figure 2.3.1a. An application of Kirchhoff's voltage law (the algebraic sum of the voltages around a closed loop is equal to zero) yields

$$L\frac{di(t)}{dt} + Ri(t) = v(t)$$

or

$$\frac{L}{R}\frac{dv_o(t)}{dt} + v_o(t) = v(t) \tag{2.22}$$

where we set $v_o(t) = Ri(t)$. Observe that (2.22) is a first-order, ordinary differential equation since it involves only the first derivative of the **dependent variable**, $v_o(t)$. In the present system we have only one **energy-storing** element (the inductor in the form of magnetic field) and only one forcing function (source), $v(t)$.

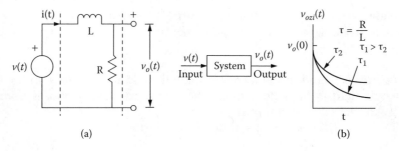

Figure 2.3.1 (a) Simple electrical system. (b) Zero-input response.

Note: *Every system described by an ordinary differential equation is characterized by its input function (in this case by v(t)) and its output function (in this case, $v_o(t)$). A system is also characterized by its operator, which operates on its input to produce the output. Although in simple systems finding the operator may be easy, in most cases, it is very difficult, if not impossible.*

Finding the unknown function $v_o(t)$ requires that we find the solution to (2.22). A fundamental requirement in the solution of differential equations is that the complete solution must contain as many arbitrary constants as the order of the differential equation being solved. Therefore, we must have additional information if we are to obtain a solution that meets all the conditions of the problem. This added information is specified by the knowledge of the **state**; the condition of the system at a particular time is known as the **initial condition**. Without loss of generality we use the particular time $t = 0$.

There are several approaches in solving ordinary differential equations. In this chapter we will present the method of undetermined coefficients and convolution. The transform methods for both continuous and discrete systems will be presented in later chapters.

Zero-input and zero-state solution

To apply the method of **undetermined coefficients**, we use a two-step process. First, we write the differential equation in its **homogeneous** form, which is equivalent to writing the equation without its forcing function, or equivalently the input to the system. Hence, we write

$$\frac{L}{R}\frac{dv_o(t)}{dt} + v_o(t) = 0 \qquad (2.23)$$

The solution of this homogeneous equation is known as the **zero-input response**.

Note: *The zero-input response is due **entirely** to the energy stored in the system. If there is no energy stored in the system, the zero-input response of the system is zero.*

Energy is stored in the form of magnetic field in inductors, in the form of electric field in capacitors, as kinetic energy in masses, and as potential energy in the deformation of springs.

To solve (2.23), we try a solution of the form

$$v_o(t) = Ae^{st} \qquad (2.24)$$

Chapter 2: Linear continuous-time systems

where A is an unknown constant and s is a constant having the units of frequency s^{-1}. The basis for the suggested form of the solution is found in (2.23), which requires that the sum of its derivative and its solution must be equal to zero (hence the exponential function). Substituting this trial solution in (2.23), we obtain

$$\left(\frac{L}{R}s+1\right)Ae^{st} = 0 \quad \text{or} \quad \left(s+\frac{R}{L}\right)Ae^{st} = 0 \tag{2.25}$$

If we were to set $A = 0$, we would obtain the trivial solution. Therefore, we impose the requirement that $(s + R/L) = 0$ or $s = -(R/L)$. Hence, the homogeneous solution takes the form

$$v_o(t) = Ae^{-t(R/L)} \tag{2.26}$$

If we choose $v_o(t)|_{t=0} = v_o(0)$, then, for this **initial state**, $v_o(0) = A \exp(0R/L) = A$, and hence the **zero-input** solution (response) is

$$v_{ozi}(t) = v_o(0)e^{-(R/L)t} \tag{2.27}$$

Note: *Observe that it was necessary to evaluate only a single arbitrary constant for this first-order differential equation.*

The zero-input solution is shown in Figure 2.3.1b for two values of the ratio R/L, known as the **time constant** of the system.

Our next step is to solve (2.22) with the forcing function present (input). Thus, we are to find a solution to the **nonhomogeneous** differential equation

$$\frac{dv_o(t)}{dt} + \frac{R}{L}v_o(t) = \frac{R}{L}v(t) \tag{2.28}$$

To proceed, we multiply both sides of this equation by $\exp(tR/L)$, the form suggested by (2.27). We recognize that we can write the result as follows:

$$\frac{d}{dt}[e^{t(R/L)}v_o(t)] = \frac{R}{L}e^{t(R/L)}v(t) \tag{2.29}$$

Integrate both sides of the above equation from 0 to t to obtain

$$e^{t(R/L)}v_o(t)\Big|_0^t = \frac{R}{L}\int_0^t e^{x(R/L)}v(x)dx \quad \text{or} \quad e^{t(R/L)}v_o(t) - v_o(0) = \frac{R}{L}\int_0^t e^{x(R/L)}v(x)dx$$

This result is written in the form

$$v_o(t) = \underbrace{v_o(0)e^{-t(R/L)}}_{\text{zero-input response } v_{ozi}(t)} + \underbrace{\frac{R}{L}e^{-t(R/L)}\int_0^t e^{x(R/L)}v(x)dx}_{\text{zero-state response } v_{ozr}(t)} \qquad (2.30)$$

$$= v_o(0)e^{-t(R/L)} + \frac{R}{L}\int_0^t v(x)e^{-(R/L)(t-x)}dx$$

where the integral indicates a **convolution** of the two functions $v(t)$ and $\exp(-(R/L)t)$. More will be said about convluition in Section 2.6.

Let us assume that at $t = 0$, the initial voltage $v_o(0) = 2$ V and the ratio $R/L = 1$. Under these conditions, the output voltage for an assumed unit step input voltage $v(t) = u(t)$ is

$$v_o(t) = \underbrace{2e^{-t}}_{v_{ozi}} + \underbrace{e^{-t}\int_0^t e^x dx = 2e^{-t} + (1-e^{-t})}_{v_{ozs}} = \underbrace{e^{-t}}_{v_{ot}} + \underbrace{1}_{v_{oss}} \qquad (2.31)$$

Here v_{ot} identifies the **transient** component of the solution and v_{oss} identifies the **steady-state** component of the solution.

Standard solution techniques of differential equations

In this subsection we carry out the solution of (2.22) using the standard approach that provides directly the transient and steady-state components of the solution. The steps are as follows:

Step 1: First solve the homogeneous equation

$$\frac{L}{R}\frac{dv_o(t)}{dt} + v_o(t) = 0 \qquad (2.32)$$

Following the procedure discussed above, we found (see (2.26)) that the homogeneous solution is ($L/R = 1$)

$$v_{oh}(t) = Ae^{-t} \qquad (2.33)$$

Step 2: Next, we solve the nonhomogeneous equation, with input voltage $u(t) = 1$ ($t > 0$), which is

$$\frac{dv_o(t)}{dt} + v_o(t) = 1 \qquad (2.34)$$

Since the input voltage is constant, the solution, known as the **particular solution**, is also a constant. Let $v_{op}(t) = B$ = constant, where B

Chapter 2: Linear continuous-time systems

is unknown. We next substitute the trial solution into this equation and force both sides to be equal. Hence, we find that $B = 1$. Therefore, the total solution is

$$v_o(t) = v_{oh}(t) + v_{op}(t) = Ae^{-t} + 1 \qquad t > 0 \qquad (2.35)$$

Step 3: Finally, we apply the initial conditions in the total solution. Hence, we find

$$v_o(0) = 2 = A + e^{-0} + 1 \quad \text{or} \quad A = 1$$

Therefore, the solution is

$$v_o(t) = e^{-t} + 1 \qquad t > 0$$

which is identical to (2.31), as it should be.

Example 2.3.1: Computer system. Find the zero-input, zero-state, transient, and steady-state responses of the magnetic head of the computer disc drive modeled in Figure 2.3.2a. We consider only the mass of the head structure and the friction due to the air and arm of the assembly. The mass is equal to 10^{-4} kg, and the air friction constant $D = 10^{-4}$ N s/m.

Solution: First, we present the system in its physical representation, as shown in Figure 2.3.2b. Next, we create a diagram that indicates the two

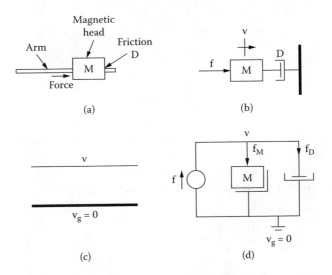

Figure 2.3.2 Computer disc drive: (a) physical representation, (b) velocity diagram, and (c) velocity response.

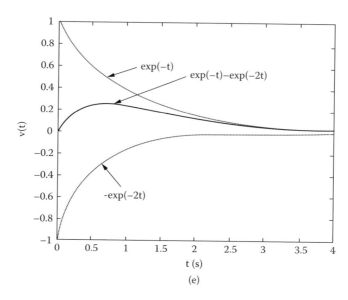

(e)

Figure 2.3.2 (continued).

velocities: the ground velocity equal to zero and the velocity of the mass, as shown in Figure 2.3.2c. From the ground node to velocity, we connect the force source, the mass, and the damping effect, as shown in Figure 2.3.2d. This figure is the equivalent circuit representation of the system.

At the node, D'Alembert's principle requires that the algebraic sum of the forces must be equal to zero. Hence, we write

$$f_M + f_D = f \qquad (2.36)$$

or

$$M\frac{dv(t)}{dt} + Dv(t) = f(t) \qquad (2.37)$$

It has been assumed that the frictional force is proportional to velocity v, with a frictional coefficient D.

We assume that the force f is an exponential function, $f(t) = 10^{-4}e^{-2t}u(t)$. The differential equation under consideration is

$$\frac{dv(t)}{dt} + v(t) = e^{-2t} \qquad t > 0 \qquad (2.38)$$

We first employ the method of variation of parameters. The solution to the homogeneous equation is easily found to be

Chapter 2: Linear continuous-time systems

$$v_{zi}(t) = v(0)e^{-t} \tag{2.39}$$

where $v(0)$ is the initial velocity (state) of the system. To find the solution of the nonhomogeneous equation, multiply both sides of (2.38) by exp(t) and rearrange the result to the form

$$\frac{d}{dt}[e^{t}v(t)] = e^{t}f(t)$$

Integrate both sides of this equation from 0 to t, combine the result with $v_{zi}(t)$, and rearrange the resulting equation to

$$v(t) = \underbrace{v(0)e^{-t}}_{v_{zi}} + \underbrace{e^{-t}\int_{0}^{t} e^{x}e^{-2x}dx}_{v_{zs}} = v(0)e^{-t} + (e^{-t} - e^{-2t}) \tag{2.40}$$

To proceed, we employ the standard methods of differential equations as it was discussed above. The particular solution of (2.38) is obtained if we assume a trial solution of the form $v_p(t) = Be^{-2t}$. Again, the trial solution was selected to be proportional to the input to the system. Combine this solution with (2.38) to find the relation

$$B(-2)e^{-2t} + Be^{-2t} = e^{-2t} \quad \text{or} \quad B = -1$$

The complete general solution is given by

$$v(t) = Ae^{-t} - e^{-2t} \tag{2.41}$$

where A, the unknown constant of the homogeneous solution, is deduced by the specified initial condition. At $t = 0$, the initial velocity is $v(0)$, so that

$$v(0) = A - 1 \quad \text{or} \quad A = v(0) + 1$$

Combine this value of A with (2.41) to find

$$v(t) = v(0)e^{-t} + (e^{-t} - e^{-2t}) \tag{2.42}$$

which is the same result given by (2.40). If we further assume that the initial velocity $v(0)$ is zero, the velocity of the head becomes

$$v(t) = (e^{-t} - e^{-2t}) \tag{2.43}$$

This response is a pulse-type function and does not show the discontinuity at $t = 0$ of the exponential input function. Figure 2.3.2e graphically shows, (2.43). ∎

Example 2.3.2: Electro-optics. In many applications, such as reading product codes in supermarkets, printing, and so forth, an optical scanner, as shown in Figure 2.3.3a, is used. If the applied torque is given by $\mathcal{T}(t) = 0.1\cos\omega_o(t)u(t)$, find the zero-input and zero-state responses of the angular velocity of the mirror. As the mirror rotates, a friction force is developed that is proportional to its angular speed. The friction constant D is equal to 0.05 N m s/rad, and the polar moment of inertial J is equal to 0.1 kg-m². Also determine the transient and steady terms using the standard differential equation techniques.

Solution: The physical system is modeled as shown in Figure 2.3.3b; the circuit diagram is developed in Figures 2.3.3c and d. Apply D'Alembert's principle for rotational systems, which requires that the algebraic sum of the torques be zero. We thus have

$$\mathcal{T}(t) = \mathcal{T}_J(t) + \mathcal{T}_D(t) \equiv \text{inertial torque} + \text{frictional torque}$$

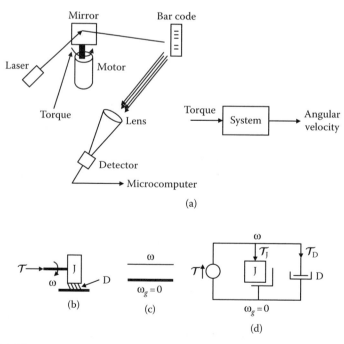

Figure 2.3.3 Electro-optic scanner: (a) physical representation, (b) velocity diagram, and (c) circuit representation.

Chapter 2: Linear continuous-time systems

Using the known form for each term, we write

$$0.1\frac{d\omega(t)}{dt} + 0.05\omega(t) = 0.1\cos\omega_0 t\, u(t)$$

or equivalently,

$$\frac{d\omega}{dt} + 0.5\omega = \cos\omega_0 t\, u(t) \qquad (2.44)$$

The zero-input solution is readily found from the homogeneous equation, and the result is

$$\omega_{zi}(t) = \omega(0)e^{-0.5t} \qquad t > 0 \qquad (2.45)$$

Also, following the same procedure as in the previous example, we obtain

$$\omega(t) = \underbrace{\omega(0)e^{-0.5t}}_{\text{zero input}} + \underbrace{e^{-0.5t}\int_0^t e^{0.5x}\cos\omega_0 x\, dx}_{\text{zero state}} \qquad (2.46)$$

$$= \omega(0)e^{-0.5t} + \frac{0.5\cos\omega_0 t - 0.5e^{-0.5t} + \omega_0\sin\omega_0 t}{0.25 + \omega_0^2}$$

Now, let us use standard techniques of differential equations to find the transient and steady-state solutions, assuming $\omega(0) = 1$. The homogeneous solution is

$$\omega_h(t) = Ae^{-0.5t} \qquad (2.47)$$

Since the input to the system is a cosine function, the particular solution is found by assuming a combination of a sine and a cosine function:

$$\omega_p(t) = B\cos\omega_0 t + C\sin\omega_0 t$$

Note that even though the input is a cosine function, we must use both the sine and cosine terms in the trial solution. Combine the particular solution with (2.44) to find

$$(C\omega_0 + 0.5B)\cos\omega_0 t + (0.5C - B\omega_0)\sin\omega_0 t = \cos\omega_0 t$$

Equating the coefficients of like terms, we find the system

$$0.5B + \omega_0 C = 1$$
$$-\omega_0 B + 0.5C = 0$$

Solving the system, we obtain the values of B and C to be

$$B = \frac{0.5}{0.25 + \omega_0^2}, \quad C = \frac{\omega_0}{0.25 + \omega_0^2}$$

The complete solution is

$$\omega(t) = Ae^{-0.5t} + \frac{0.5}{0.25 + \omega_0^2} \cos \omega_0 t + \frac{\omega_0}{0.25 + \omega_0^2} \sin \omega_0 t$$

For the initial condition $\omega(0) = 1$, the above equation becomes

$$1 = A + \frac{0.5}{0.25 + \omega_0^2} \quad \text{or} \quad A = 1 - \frac{0.5}{0.25 + \omega_0^2}$$

Introducing A in the above equation, we find that the angular velocity is identical to that given by (2.46) with $\omega(0) = 1$. ∎

It is important for the reader to recognize the applicability of the methods we have presented up to here; they can also be applied to any other discipline as long as they are characterized by the same form of differential equations. Let us consider an example from engineering economics.

Example 2.3.3: Economics. Suppose that the rate of change in the price $p(t)$ of a commodity, e.g., electronic chips, is proportional to the difference between the demand $d(t)$ and the supply $s(t)$ in the market at time t. Determine and study the equation that describes the economic system for the case where $d(t)$ and $s(t)$ are linear and proportional to the price, respectively: $d(t) = a + bp(t)$ and $s(t) = cp(t)$, with a, b, and c constants.

Solution: The controlling equation, as expressed by the problem, is

$$\frac{dp(t)}{dt} = h[d(t) - s(t)] \quad \text{or} \quad \frac{dp(t)}{dt} + h(c-b)p(t) = ah \quad (2.48)$$

where h is an adjustment constant. The solution of this differential equation is found (following the procedures developed above) to be

Chapter 2: Linear continuous-time systems

$$p(t) = \left[p(0) - \frac{a}{c-b} \right] e^{-h(c-b)t} + \frac{a}{c-b} \qquad (2.49)$$

The reader is easily able to see that c cannot be equal to b and that c > b must be the case, since the price will continuously be increasing. In economics, the price of a commodity when supply is equal to demand is known as the **equilibrium price**, p_e. Then,

$$d(t) = s(t) \quad \text{or} \quad a + p_e = cp_e$$

Hence,

$$p_e = \frac{a}{c-b}$$

and (2.49) becomes

$$p(t) = [p(0) - p_e] e^{-h(c-b)t} + p_e \qquad (2.50)$$

From this we observe that as the time increases to infinity, the final price approaches the equilibrium price. It is also obvious that the speed of convergence depends on the constants in the exponential as well as on the difference between the initial price and the equilibrium price. ∎

2.4 Evaluation of integration constants: initial conditions

Once the complementary solution and the particular solution have been deduced for a given differential equation, the next step is to determine the constants of integration that arise in the complementary solution, if the general solution is to be converted to a solution that satisfies all conditions of a particular problem. The data necessary for this determination are provided by initial conditions that exist in the system. For electrical systems, these conditions are the known charges on all capacitors (usually given as initial voltages across the capacitors) and known currents through all inductors at some specified initial time, usually taken as $t = 0$. In the more general case, the initial conditions would be specified by initial through and across variables pertinent to the system elements. If these initial voltages and currents are zero, the system is said to be **initially relaxed**. In all cases, the specified initial conditions must be applied to the complete solution to determine the constants of integration. In this way, the character of the system disturbance is taken into account. If the system excitation involves more than the switching of input sources — for example, the switching of system elements — such changes must be taken into consideration.

Table 2.3.1 Particular Solutions

Excitation Function (Input)	Trial Solution
1. Polynomial $f(t) = a_0 + a_1 t + a_2 t^2 + \cdots + a_n t^n$ (Some a_i's may be zero)	Polynomial $f(t) = c_0 + c_1 t + c_2 t^2 + \cdots + c_n t^n$ (No c_i's are set to zero)
2. Exponential function $f(t) = A e^{at}$	$f(t) = B e^{at}$
3. Sine or cosine function $f(t) = A \cos t + B \sin t$ (A or B may be zero)	$f(t) = C \cos t + D \sin t$ (Both C and D are different from zero)
4. Combination of above cases	Split excitation into its separate parts, solve the individual equations, then add the results

Switching of sources

The most common switching operation is that of introducing an excitation source into the circuit. The most common input waveforms are the step function, sinusoids, pulses, and decaying exponentials. As shown in Table 2.3.1, the excitation function establishes the form of the particular solution.

Conservation of charge

If the initial voltages across capacitors are not zero and if, during the switching operation, the total system capacitance remains unchanged, then the voltage across each capacitor will be the same before and after the switching instant. This result assumes the absence of switching impulse to one or more of the capacitors. It follows from the fact that the terminal relation for the capacitor is $i(t) = C dv(t)/dt$, and in the switching operation during the interval from $t = 0-$ to $t = 0+$, when C is constant,

$$\int_{v(0-)}^{v(0+)} C\, dv = \int_{0-}^{0+} i\, di \qquad (2.51)$$

The value of the integral on the right side is zero unless i is an impulse function (delta function). Hence, for a constant C, (2.51) becomes

$$Cv(0+) - Cv(0-) = 0 \qquad (2.52)$$

Equation (2.52) is a statement of conservation of charge (conservation of momentum in mechanics). If no current impulse is applied during the switching operation, the voltage across the capacitor remains constant.

Conservation of flux linkages

If initial currents exist in inductors and if, during the switching operation, the total inductance of the system remains unchanged, then the current through each inductor will remain unchanged over the switching instant. This result follows directly from the fact that the terminal relation for the inductor is $v(t) = L\,di(t)/dt$ and, with L constant over the switching interval,

$$\int_{i(0-)}^{i(0+)} L\,di = \int_{0-}^{0+} v\,dt \tag{2.53}$$

The right-hand side will be zero in the absence of voltage impulses. If such a situation exists, then

$$Li(0+) - Li(0-) = 0 \tag{2.54}$$

and the current through the inductor remains constant during the switching interval. This is a statement of conservation of flux linkages. This condition is the dual complement of that of the voltage across the capacitor.

Circuit behavior of L and C

In the light of the foregoing discussion, for circuits with **constant excitation**:

1. Capacitors behave as open circuits in the steady state.
2. Inductors behave as short circuits in the steady state.

General switching

In those cases where L or C or both are changed instantaneously during a switching operation, and in the absence of switching impulses, (2.52) and (2.54) must be modified to the following forms:

For capacitors: conservation of charge:

$$\boxed{\begin{array}{c} q(0+) = q(0-) \\ C(0+)v(0+) = C(0-)v(0-) \end{array}} \tag{2.55}$$

For inductors: conservation of flux linkages:

$$\boxed{\begin{array}{c} \psi(0+) = \psi(0-) \\ L(0+)i(0+) = L(0-)i(0-) \end{array}} \tag{2.56}$$

As a practical matter, circuits with switched L or C are easily accomplished by placing switches across all or part of the L or C of a circuit.

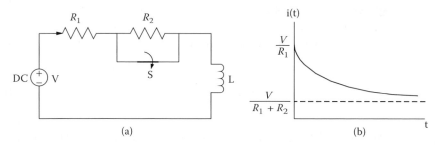

Figure 2.4.1 Resistor switching. (a) Switching of R in a circuit with an initial current. (b) Transient response following switching.

Example 2.4.1: Consider the circuit in Figure 2.4.1a, which shows the switching of the circuit resistance. Observe that the total circuit resistance is R_1 before opening the switch and $R_1 + R_2$ after the switch operation at $t = 0$. Find $i(t)$ for $t > 0$.

Solution: The initial current prior to the switching instant is constant since the source is assumed to be on for a long time and all the transients have died out. The initial current has the value

$$i_0 = i(0-) = \frac{V}{R_1} \qquad (2.57)$$

After opening the switch at time $t(0+)$, the controlling differential equation is

$$L\frac{di(t)}{dt} + (R_1 + R_2)i(t) = V \qquad (2.58)$$

The differential equation has the homogeneous solution $i_h(t) = Ae^{-(R_1+R_2)t/L}$ and a particular solution $i_p(t) = V/(R_1 + R_2)$. The total solution is

$$i(t) = i_h + i_p = \frac{V}{R_1 + R_2} + Ae^{-(R_1+R_2)t/L} \qquad (2.59)$$

To evaluate the constant A, we impose the initial condition, which is $i(0-) = i(0+)$, since no energy-storing elements were altered. Then by (2.54), no instantaneous change in current will occur, and $i(0-) = i(0+)$. Thus, from (2.57) and (2.59), with $t = 0$, we find

$$\frac{V}{R_1} = \frac{V}{R_1 + R_2} + A \quad \text{or} \quad A = \frac{VR_2}{R_1(R_1 + R_2)}$$

Therefore, the final solution is

$$i(t) = \frac{V}{R_1 + R_2}\left(1 + \frac{R_2}{R_1} e^{-(R_1+R_2)t/L}\right) \qquad (2.60)$$

This is sketched in Figure 2.4.1b. ∎

Example 2.4.2: Refer to Figure 2.4.2a, which shows the switching of L in a circuit with an initial current. Prior to switching, the inductance is L_1, and after the switching, the total circuit inductance is $L_1 + L_2$. Switching occurs at $t = 0$. Find the current in the circuit after switching.

Solution: The current in the circuit prior to switching for assumed steady-state conditions is

$$i(0-) = \frac{V}{R_1} \qquad (2.61)$$

To find the current after switching, we employ the law of conservation of flux linkages given by (2.56). Thus, over the switching period

$$L_1 i(0-) = (L_1 + L_2) i(0+) \qquad (2.62)$$

from which

$$i(0+) = \frac{L_1}{L_1 + L_2} i(0-) = \frac{L_1}{L_1 + L_2} \frac{V}{R_1} \qquad (2.63)$$

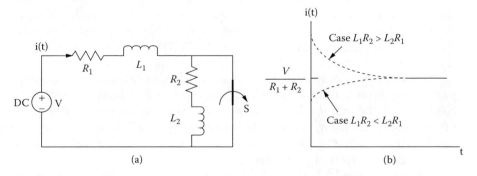

Figure 2.4.2 Illustration of Example 2.4.2.

The differential equation for $t > 0$ is now of the form

$$(L_1 + L_2)\frac{di(t)}{dt} + (R_1 + R_2)i(t) = V \tag{2.64}$$

Following the above discussion, the total solution of this equation is

$$i(t) = Ae^{-(R_1+R_2)t/(L_1+L_2)} + \frac{V}{R_1 + R_2} \tag{2.65}$$

At the switching instant $t = 0+$,

$$i(0+) = A + \frac{V}{R_1 + R_2} = \frac{L_1}{L_1 + L_2}\frac{V}{R_1} \quad \text{or} \quad A = \frac{(L_1 R_2 - L_2 R_1)V}{R_1(R_1 + R_2)(L_1 + L_2)}$$

Therefore, the final solution is given by the expression

$$i(t) = \frac{V}{R_1 + R_2}\left[1 + \frac{L_1 R_2 - L_2 R_1}{R_1(L_1 + L_2)}e^{-(R_1+R_2)t/(L_1+L_2)}\right] \tag{2.66}$$

The form of this function is illustrated in Figure 2.4.2b. Observe that the current may rise or fall at the switching instant, depending on the relative values $L_1 R_2$ and $L_2 R_1$. ∎

2.5 Block diagram representation

The description of system interconnections to this point has been a mathematical one that employed the fundamental laws of Kirchhoff, D'Alembert, and others to specify the interconnection of the components of a system by differential equations. A second procedure is one that graphically displays the interconnected models and then employs techniques of graphical reductions to write the circuit equations. Two important graphical methods exist: the block diagram and signal flow graph (SFG). These techniques are closely related, and consideration will only be given to the **block diagrams**.

Block diagram portrayals possess the important feature that the signal paths from input to output are placed in sharp focus without displaying the hardware of the system. Furthermore, they provide tools of analysis that often possess advantages over other methods. We will usually proceed in block diagram developments through a series of mathematical equations that are given graphical portrayal.

Refer to Figure 2.5.1. Figure 2.5.1a shows a **pick-off point** that transmits the variable to different branches. Note that the variable is not divided but

Chapter 2: Linear continuous-time systems

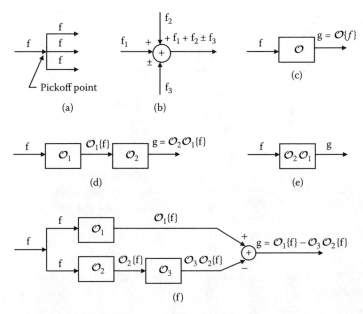

Figure 2.5.1 Elementary rules of block diagram operations.

transmitted to all branches at its original strength. Figure 2.5.1b shows a **summation point** and the resulting output, Figure 2.5.1c shows a system that is characterized by an **operator** $\mathcal{O}$, for example, an integrator, a differentiator, a multiplier, or the like.

To understand the block diagram representation of systems, it is convenient to think of the block as a multiplier, where the operator operates on the input. Hence, if $\mathcal{O} = a$ in Figure 2.5.1c, the output is $g = af$. Figure 2.5.1d shows two blocks in cascade. Clearly, the output is just the successive operation of the operators on the input. If $\mathcal{O}_1 = a$ and $\mathcal{O}_2 = b$, the output is abf. If each of the operators is a derivative, then the output is $d^2f(t)/dt^2$. Figure 2.5.1e shows the equivalent configuration of Figure 2.5.1d. Figure 2.5.1.f shows the details of a more complicated block diagram.

Figure 2.5.2 shows the most celebrated block diagram in systems and electronics and, of course, in many other fields, known as the **feedback system**. The terminology is extensively used in control systems engineering. It is one of the most fundamental configurations of physical and man-made systems. Without this feedback mechanism, no living creature would have survived.

To understand and learn how to build block diagrams for different systems, let us consider the *RL* circuit shown in Figure 2.3.1. The differential equation that represents the system is

$$\frac{L}{R}\frac{dv_o}{dt} + v_o = v(t) \tag{2.67}$$

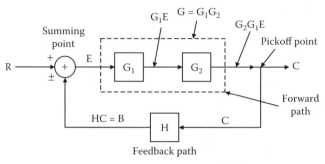

R = Input signal
G = C/E = Forward transfer function or open-loop transfer function
C = G_2G_1E = GE = Output signal
T = C/R = System transfer function or closed-loop transfer function

E = R ± CH = R ± B = Error signal
H = B/C = Feedback transfer function
GH = B/E = Loop transfer function

Figure 2.5.2 Block diagram of a system with feedback. All operators are assumed constant multipliers.

To produce a summation point, we write the equation in the form

$$v_o(t) = v(t) - \frac{L}{R}\frac{dv_o(t)}{dt} \qquad (2.68)$$

This expression shows that the output is equal to the input minus (L/R) times the derivative of the output. To build the block diagram, we start with two arrows at opposite sides of the page. The left-side one represents the input and the right-side one represents the output. Taking into consideration the properties as given in Figure 2.5.1, we complete the block diagram. The representation of the above equation is shown in Figure 2.5.3a.

Suppose that we write (2.67) in the form

$$\frac{L}{R}\frac{dv_o(t)}{dt} = v(t) - v_o(t) \qquad (2.69)$$

This expression shows that the output of the summation point $(L/R)dv_o(t)/dt$ is the difference between the input $v(t)$ and the output $v_o(t)$. The block diagram representation is shown in Figure 2.5.3b.

A block diagram representation of a given system can often be reduced by block diagram reduction techniques to achieve a block diagram with fewer blocks than the original diagram. A number of block reduction rules are contained in Figure 2.5.4.

Example 2.5.1: Use Figure 2.5.4 to deduce the complex systems shown in Figure 2.5.5a and b to a single-block diagram. The operators are constant multipliers.

Chapter 2: Linear continuous-time systems 71

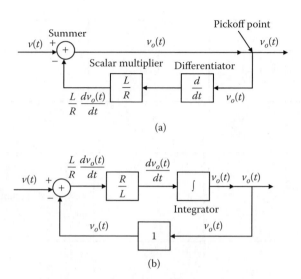

Figure 2.5.3 Block diagram representation of the system shown in Figure 2.3.1.

Solution: We observe from Figure 2.5.5a that the output of the summer is $H_1 f - H_2 f$. The summer output is the input to the next block, the output of which is $g = H_3(H_1 f - H_2 f) = H_3 H_1 f - H_3 H_2 f$. This result is shown in Figure 2.5.5c.

Refer next to the Figure 2.5.5b. By entry 4 of Figure 2.5.4 the equivalent representation of the first feedback loop operator is equal to $1/(1 - H_1)$. By entry 6 of Figure 2.5.4 the equivalent representation of the second feedback loop operator is $H_2/(1 + H_2 H_3)$. These two blocks are in cascade and the final operator is the product of these operators, thus forming the final block. The final reduction is sown in Figure 2.5.5d. ∎

2.6 Convolution and correlation of continuous-time signals

The convolution operation on functions is one of the most useful operations encountered in the study of signals and systems. The importance of the convolution integral in systems studies stems from the fact that a knowledge of the output of the system to an impulse (delta) function excitation allows us to find its output to any input function (subject to some mild restrictions).

To help us develop the convolution integral, let us begin with the properties of the delta function (see Chapter 1). Based on the delta properties, we write

$$f(t) = \int_{-\infty}^{\infty} f(\tau)\delta(t - \tau)d\tau \qquad (2.70)$$

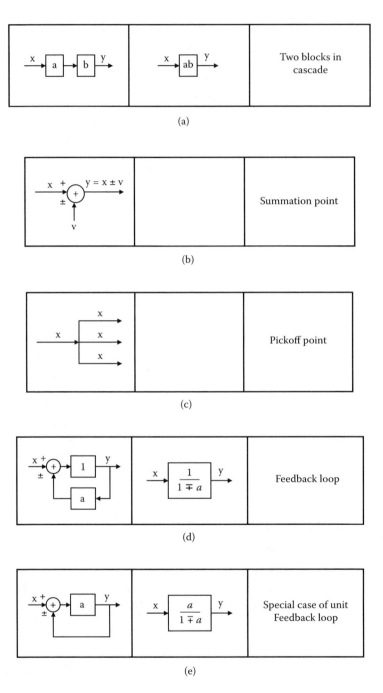

Figure 2.5.4 Properties of block diagrams.

Chapter 2: Linear continuous-time systems

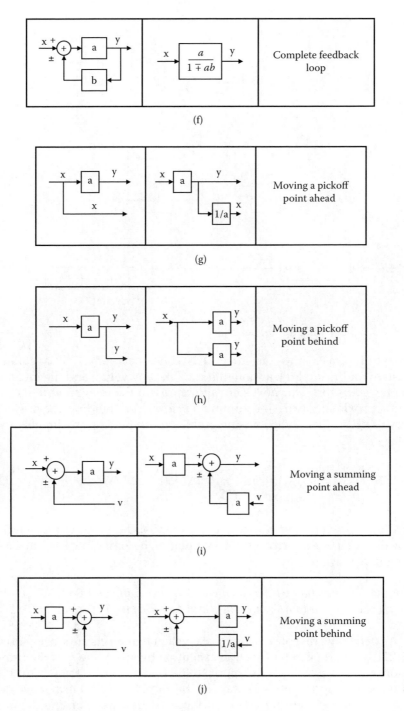

Figure 2.5.4 (continued).

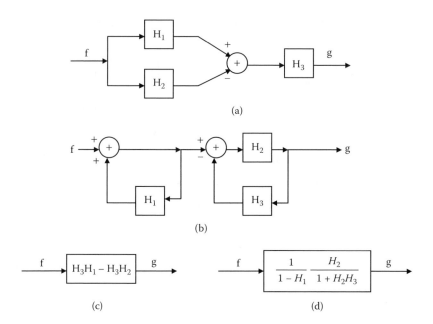

Figure 2.5.5 Block diagram reduction.

Observe that, as far as the integral is concerned, time *t* is a parameter (constant for the integral, although it can take any value) and the integration is with respect to τ. Our next step is to represent the integral with its equivalent approximate form, the summation form, by dividing the τ axis into intervals of ΔT; then the above integral is represented approximately by the sum

$$f_a(t) = \lim_{\Delta T \to 0} \sum_{n=-\infty}^{\infty} f(n\Delta T)\delta(t - n\Delta T)\Delta T \qquad (2.71)$$

As ΔT goes to zero and n increases to infinity, the product nΔT takes the value of τ, ΔT becomes dτ, and the summation becomes integral, thus recapturing (2.70).

Note: *The function f(t) has been approximated with an infinite sum of shifted delta functions equal to nΔT, and their area (see Chapter 1) is equal to f(nΔT)ΔT.*

We define the response of a causal (system that reacts after being excited) and LTI system to a delta function excitation by $h(t)$, known as the **impulse response** of the system. If the input to the system is $\delta(t)$, the output is $h(t)$, and when the input is $\delta(t - t_0)$, then the output is $h(t - t_0)$. Further, we define the output of a system by $g(t)$ if its input is $f(t)$. Based on the definitions discussed so far, it is obvious that if the input to the system is $f_a(t)$, the output

Chapter 2: Linear continuous-time systems

is a sum of impulse functions shifted identically to the shifts of the input delta functions of the summation, and therefore, the output is equal to

$$g(t) = \lim_{\Delta T \to 0} \sum_{n=-\infty}^{\infty} f(n\Delta T) h(t - n\Delta T) \Delta T$$

In the limit as ΔT approaches zero, the summation becomes an integral of the form

$$g(t) = \int_{-\infty}^{\infty} f(\tau) h(t - \tau) d\tau \qquad (2.72)$$

This expression is the **convolution integral** for any two functions $f(t)$ and $h(t)$ (see also (2.30)).

Let us represent an LTI system by its operator $\mathcal{O}$. Clearly, the operator $\mathcal{O}$ denotes a linear process in the variable t that operates on the input function to yield the system response. In actual systems, the operator $\mathcal{O}\{.\}$ can be a differentiator d/dt, an integrator $\int dt$, a multiplier, or a combination of all. We write this operation as

$$g(t) = \mathcal{O}\{f(t)\} \qquad (2.73)$$

Further, we make use of (2.70), which specifies $f(t)$ as an infinite sum of weighted impulses. Thus,

$$g(t) = \mathcal{O}\left\{ \int_{-\infty}^{\infty} f(\tau) \delta(t - \tau) d\tau \right\} = \int_{-\infty}^{\infty} f(\tau) \mathcal{O}\{\delta(t - \tau)\} d\tau = \int_{-\infty}^{\infty} f(\tau) h(t - \tau) d\tau$$

which is the convolution integral given by (2.72). This expression employed the definition $\mathcal{O}\{\delta(t)\} = h(t)$, the system response to a delta function (impulse function). Because the system is time invariant, $\mathcal{O}\{\delta(t - \tau)\} = h(t - \tau)$ for a shifted delta input, and the response to $f(\tau)\delta(t - \tau)$ is $f(\tau)h(t - \tau)$. By superposition, the response for all t is that given by (2.72).

Convolution is a general mathematical operation, and for any two real-valued functions, their convolution, indicated mathematically by the asterisk between the functions, is given by

$$\boxed{g(t) \triangleq f(t) * h(t) = \int_{-\infty}^{\infty} f(\tau) h(t - \tau) d\tau = \int_{-\infty}^{\infty} f(t - \tau) h(\tau) d\tau} \qquad (2.74)$$

Observe from the definition that $f(t)*h(t) = h(t)*f(t)$, which indicates that convolution obeys the commutation property.

Note: *Equation (2.74) tells us the following: given two functions in the time domain t, we find their convolution g(t) by doing the following steps:*

1. *Rewrite one of the functions in the τ domain by substituting the variable τ wherever there is t; the shape of the function is identical to that in the t domain.*
2. *To the second function we substitute t – τ wherever we see t; this produces a function in the τ domain that is flipped (the minus sign in front of the τ) and shifted by t (positive values of t shift the function to the right and negative values shift the function to the left).*
3. *Multiply these two functions and find another function of τ, since t is a parameter and constant as far as the integration is concerned.*
4. *Next find the area under the product function whose value is equal to the output of the convolution at t (in our case, g(t)). By introducing the infinite values of t's, from minus infinity to infinity, we obtain the output function g(t).*

Example 2.6.1: Deduce the convolution of the following two functions: $f(t) = \exp(-t)u(t)$ and $h(t) = \exp(-0.5t)u(t)$.

Solution: First, we observe from (2.74) that one of the functions is unchanged when it is mapped from the *t* domain to the τ domain. The second function is reversed or folded over (mirrored with respect to the vertical axis) in the τ domain, and it is shifted by an amount *t*, which is just a parameter in the integrand. Figures 2.6.1a and b shows two functions in the *t* and τ domains, respectively. We now write

$$g(t) = f(t) * h(t) = \int_{-\infty}^{\infty} e^{-\tau}u(\tau)e^{-0.5(t-\tau)}u(t-\tau)d\tau = \int_{0}^{t} e^{-\tau}e^{-0.5(t-\tau)}d\tau$$

$$= e^{-0.5t}\int_{0}^{t} e^{-0.5\tau}d\tau = 2(e^{-0.5t} - e^{-t})$$

Figure 2.6.1c shows the result of the convolution. ∎

Note: *Observe that the resulting function from the convolution is smoother than either of them. This smoothing effect is a fundamental property of the convolution process.*

Example 2.6.2: Discuss the convolution of the pulse functions $p_a(t)$ and $p_{2a}(t)$.

Solution: Refer to Figure 2.6.2a, which shows the overlapping of the two pulse functions for different values of *t*. Figure 2.6.2b shows the resulting function *g(t)*. The points on the curve represent the values of the integrals

Chapter 2: Linear continuous-time systems 77

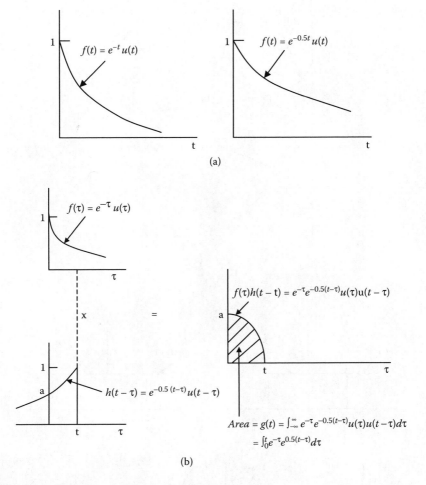

Figure 2.6.1 Illustration of Example 2.6.1.

at the values of *t* shown in Figure 2.6.2a. Since the two rectangular pulses have unit amplitudes, the value of the function is smoother than either of the convolving functions, a feature already noted. Figure 2.6.2b shows the result of the convolution. ∎

Example 2.6.3: Determine the convolution $g(t) = p_a(t)*\delta(t - 2a)$.

Solution: This convolution, in integral form, is

$$g(t) = \int_{-\infty}^{\infty} p_a(\tau)\delta(t - \tau - 2a)d\tau$$

By the use of the properties of the delta function, we proceed as follows: Set $t - \tau - 2a = 0$, from which $\tau = t - 2a$. Next, introduce this value into the $p_a(\tau)$

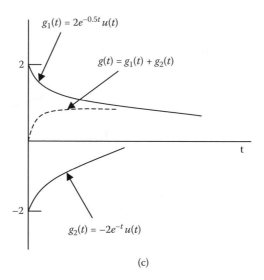

(c)

Figure 2.6.1 (continued).

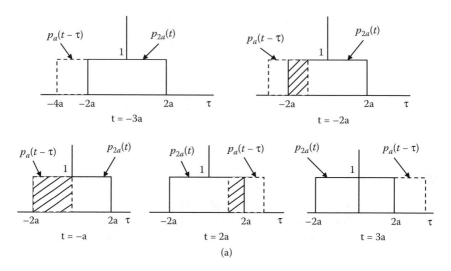

Figure 2.6.2 Graphical representation of the convolution of two-unit pulse functions.

function. The resulting function is given by $g(t) = p_a(t - 2a)$. This shows that we have recaptured the function $p_a(t)$, although it is shifted to the point where the delta function was located. ∎

Note: *Any time we convolve a function with the delta function, there is no need to do the integration. Simply, we shift the function by the same amount the delta function has been shifted.*

Chapter 2: Linear continuous-time systems 79

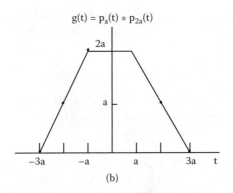

Figure 2.6.2 (continued).

Example 2.6.4: Determine the convolution of the functions $f(t) = p_{1/2}[t - (1/2)]$ and $h(t) = 2p_{1/2}[t - (1/4)]$.

Solution: Figure 2.6.3 includes the details of the solution. ∎

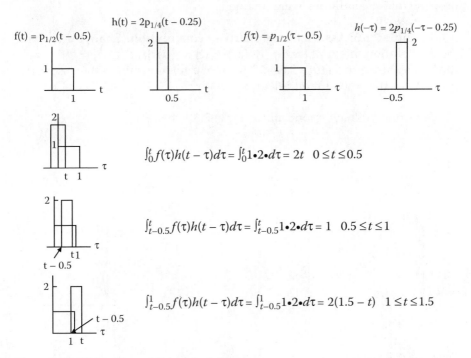

$\int_0^t f(\tau)h(t - \tau)d\tau = \int_0^t 1 \cdot 2 \cdot d\tau = 2t \quad 0 \leq t \leq 0.5$

$\int_{t-0.5}^t f(\tau)h(t - \tau)d\tau = \int_{t-0.5}^t 1 \cdot 2 \cdot d\tau = 1 \quad 0.5 \leq t \leq 1$

$\int_{t-0.5}^1 f(\tau)h(t - \tau)d\tau = \int_{t-0.5}^1 1 \cdot 2 \cdot d\tau = 2(1.5 - t) \quad 1 \leq t \leq 1.5$

Figure 2.6.3 Illustration of Example 2.6.4.

When a signal is applied to a causal system, its output (response) can never precede the input. We can split the convolution integral (see (2.74)) into two parts:

$$g(t) \triangleq f(t) * h(t) = \underbrace{\int_{-\infty}^{t_{0-}} f(\tau)h(t-\tau)d\tau}_{g_{zi}(t)} + \underbrace{\int_{t_{0+}}^{\infty} f(\tau)h(t-\tau)d\tau}_{g_{zs}(t)} = g_{zi}(t) + g_{zs}(t) \quad (2.75)$$

The function $g_{zi}(t)$ is the **zero-input** response of the system, and $g_{zs}(t)$ is the **zero-state** response of the system, where t_0 is arbitrarily taken as the initial time. The symbol t_{0-} indicates that functions with finite discontinuities are included in the definition of the zero-state response.

The second convolution integral represents the zero-state response of the system. This integral specifies that at time $t = t_0$, the system is relaxed; that is, there were not charges on capacitors and currents through the inductors, no deformation of springs, no velocities of masses, etc. However, if the initial state of the system is not zero, we must add a zero-input response produced by the initial state at $t = t_0$. The integral $g_{zi}(t)$ specifies that the state at $t = t_0$ depends on the inputs prior to that time. The solution after t_0 depends on the state of the system at that time (initial conditions), and it is immaterial how the initial conditions were attained.

Example 2.6.5: Associative property of convolution. Shaw, by applying the convolution integral, that $[f_1(t)*f_2(t)]*f_3(t) = f_1(t)*[f_2(t)*f_3(t)] = g(t)$ for the functions shown in Figure 2.6.4a. The above relationship is the associative property of the convolution integral.

Solution: We have found in Example 2.6.1 that

$$f(t) = f_1(t) * f_2(t) = 2(e^{-0.5t} - e^{-t})u(t)$$

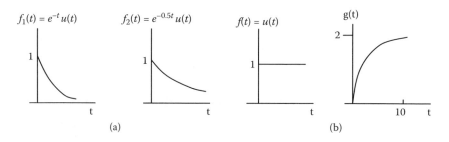

Figure 2.6.4 Illustration of Example 2.6.5.

Chapter 2: Linear continuous-time systems 81

Therefore, $g(t) = f(t)*f_3(t)$ is given by

$$g(t) = \int_{-\infty}^{\infty} f(\tau)f_3(t-\tau)d\tau = 2\int_{-\infty}^{\infty} (e^{-0.5\tau} - e^{-\tau})u(\tau)u(t-\tau)d\tau$$

However, $u(t-\tau) = 0$ for $t < \tau$ and $u(\tau) = 0$ for $\tau < 0$, so that the integral becomes

$$g(t) = 2\int_{0}^{t} (e^{-0.5\tau} - e^{-\tau})d\tau = (2 + 2e^{-t} - 4e^{-0.5t})u(t)$$

Next, we consider the alternate form. We begin with the convolution

$$f_2(t)*f_3(t) = \int_{-\infty}^{\infty} e^{-0.5\tau}u(\tau)u(t-\tau)d\tau = \int_{0}^{t} e^{-0.5\tau}d\tau = (2 - 2e^{-0.5t})u(t)$$

Then,

$$g(t) = \int_{-\infty}^{\infty} (2 - 2e^{-0.5\tau})u(\tau)e^{-(t-\tau)}u(t-\tau)d\tau = e^{-t}\int_{0}^{t}(2e^{\tau} - 2e^{0.5\tau})d\tau$$

$$= (2 - 4e^{-0.5t} + 2e^{-t})u(t)$$

The two relationships for these particular functions are identical. It can be proved that this associative property is always true for all functions that can be used in the convolution integral. ∎

An important input function in system studies is $f(t) = e^{j\omega t}$ (see also Section 1.2). For any specific value of omega, $f(t)$ is a complex function having a real part $\cos\omega t$ and imaginary part $\sin\omega t$. In polar form representation, the function has magnitude 1 and phase ωt. If $t = 0$, the point in the complex plane is at the real axis and at a distance 1. As t increases, the values of the function describe a unit circle. After the value $2\pi/\omega$ of t, the circle is repeated. Because $f(t)$ is the input to an LTI system, its output is given by

$$g(t) = \int_{-\infty}^{\infty} h(\tau)f(t-\tau)d\tau = \int_{-\infty}^{\infty} h(\tau)e^{j\omega(t-\tau)}d\tau$$

$$= e^{j\omega t}\int_{-\infty}^{\infty} h(\tau)e^{-j\omega\tau}d\tau = e^{j\omega t}H(\omega)$$

(2.76)

where

$$H(\omega) = \int_{-\infty}^{\infty} h(\tau)e^{-j\omega\tau}d\tau \qquad (2.77)$$

is the **system function** or **transfer function** of the system at frequency ω. The transfer function $H(\omega)$ is called the Fourier transform (more will be said about the Fourier transform of signals in a later chapter) of the system's impulse function $h(t)$. Also in this context, $e^{j\omega t}$ is the **eigenfuction** of the system and $H(\omega)$ is its **eigenvalue**. As we know from our circuit analysis definition, the transfer function is equal to the ratio of the output to the input of a relaxed system when the excitation is the sinusoidal signal.

The transfer function is complex and can be separated into its real and imaginary parts, Cartesian form. It can also be represented in its polar form. To see how to find the transfer function using the circuit analysis approach, let us find the transfer function for the system shown in Figure 2.3.1a.

The differential equation describing the system is given by

$$L\frac{di(t)}{dt} + Ri(t) = v(t) \quad \text{or} \quad \frac{dv_o(t)}{dt} + \frac{R}{L}v_o(t) = \frac{R}{L}v(t) \quad v_o(t) = Ri(t) \quad (2.78)$$

If the input to this relaxed system (zero initial conditions) is the complex exponential function $\exp(j\omega t)$, the output is $A\exp(j\omega t)$, where A is an arbitrary constant. Introducing this function in the above equation, we obtain

$$\frac{d(Ae^{j\omega t})}{dt} + \frac{R}{L}Ae^{j\omega t} = \frac{R}{L}e^{j\omega t} \quad \text{or} \quad j\omega Ae^{j\omega t} + \frac{R}{L}Ae^{j\omega t} = \frac{R}{L}e^{j\omega t} \quad \text{or}$$

$$\frac{Ae^{j\omega t}}{e^{j\omega t}}\left(\frac{R}{L} + j\omega\right) = \frac{R}{L} \quad \text{or} \quad H(\omega) = \frac{voltage\ out}{voltage\ in} = \frac{V_o(\omega)}{V(\omega)} = \frac{R}{R + j\omega} \qquad (2.79)$$

The Cartesian form of the above transfer function is

$$H(\omega) = \frac{R(R - j\omega)}{(R + j\omega)(R - j\omega)} = \frac{R^2}{R^2 + \omega^2} - j\frac{\omega}{R^2 + \omega^2} \triangleq H_r(\omega) + j(-H_i(\omega)) \quad (2.80)$$

where $H_r(.)$ and $H_i(.)$ are real functions of omega. The corresponding polar form of this transfer function for any value of omega is

$$H(\omega) = [H_r^2(\omega) + H_i^2(\omega)]^{1/2}e^{j\theta(\omega)} = |H(\omega)|e^{j\theta(\omega)};$$

$$\theta(\omega) = \tan^{-1}[-H_i(\omega)/H_r(\omega)] \qquad (2.81)$$

Chapter 2: Linear continuous-time systems

In case the input function was $\cos\omega t = \text{Re}\{e^{j\omega t}\}$, the output would have been

$$g(t) = \text{Re}\{e^{j\omega t} H(\omega)\} = \text{Re}\{|H(\omega)| e^{j(\omega t + \theta(\omega))}\} = |H(\omega)| \cos(\omega t + \theta(\omega))$$

The above equation indicates that the system for every sinusoidal signal input will create a sinusoidal with the same frequency, but phase shifted and amplitude modified. This shows the importance of transfer function. In purchasing an audio amplifier, we are interested in knowing if the amplitude of the transfer function $H(\omega)$ vs. frequency is flat for all audio frequencies. This observation indicates that all frequencies of speech or melodies will pass undisturbed. The phase also plays a role in passing undisturbed inputs.

Note: *For any linear operation, such as integration, we can find the real (imaginary) part of the resulting quantity by taking first the real (imaginary) part of the integrand and then integrate, or integrate first and then taking the real (imaginary) part of the integration results.*

Note: *LTI systems that have real-valued impulse responses, the eigenvalues $H(\omega)$, are Hermitian. A transfer function is Hermitian if its real part is even and its imaginary part is odd. Equivalently, the amplitude is an even function and the phase is an odd function of frequency.*

A reasonable question is whether, given the output $g(t)$ and the input $f(t)$, it is possible to deduce $h(t)$. The process, known as **deconvolution**, is often encountered in communications while trying to measure the response of the communication channel (space, wires, fibers). However, deconvolution has no direct mathematical definition in the time domain and merely connotes the inverse of the convolution operation. Owing to errors and noise associated with practical measurements, exact knowledge of the deconvolved time signals is not possible, and approximations must be made. The deconvolution process lends itself to better interpretation in the frequency domain, where it becomes an approximation and a filtering problem.

Matched filters

One important procedure for signal detection is the **matched filtering** operation. If we receive a signal $s(t)$ and want to maximize at t_0 in some sense, we pass the signal through a system whose impulse response is the shifted and flipped form of the received signal, $h(t) = s(t_0 - t)$. Let the signal be that shown in Figure 2.6.5a. Let the impulse response be that shown in Figure 2.6.5b, $h(t) = s(-t)$, where it is assumed that $t_0 = 0$. Therefore, the output of the system is the convolution of these two signals, and the result is shown in Figure 2.6.5c. Observe that the maximum point of the output is at $t = 0$. Next, we set $t_0 = 1$. Then, the two corresponding functions are shown in Figures 2.6.5d and e. The output of the system is the convolution of these

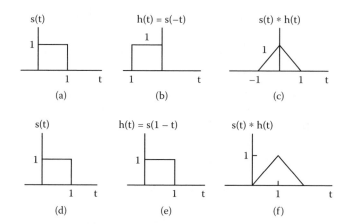

Figure 2.6.5 Illustration of matched filtering operation.

two signals, and the result is shown in Figure 2.6.5f. Since in this case $t_0 = 1$, the maximum point is located at $t = 1$.

Correlation

Correlation is an important concept in the study of random signals and serves to relate the average intensity (or power), an observable quantity, with specified time-average values of members of a family of functions defined on a probability space. Correlation techniques are used in many fields, including medicine when acoustic (ultrasonic) waves are used as probes to detect anomalies inside the human body. This technique also finds important applications in many areas of physics and technology. It is routinely used in radar signal detection when correlation is performed between the emitted signal and the signal returned from a target. A large peak indicates a resemblance between the returned signal and the emitted signal, from which it is inferred that a target is present.

Mathematically, correlation is a process somewhat similar to convolution. The mathematical procedures are identical to those used in a convolution operation, with the only difference that the shifted signal is not flipped, and the commutative property of cross-correlation (correlation between two different signals) is not obeyed. The cross-correlation of two different functions is defined by the relation

$$R_{fh}(t) \triangleq f(t) \odot h(t) = \int_{-\infty}^{\infty} f(\tau)h(\tau-t)d\tau = \int_{-\infty}^{\infty} f(\tau+t)h(\tau)d\tau \quad (2.82)$$

When $f(t) = h(t)$, the correlation operation is called **autocorrelation**. It can be shown, using Schwartz inequality, that the following relation is true:

$$\left|R_{ff}(t)\right| \triangleq \left|f(t) \odot f(t)\right| \leq R_{ff}(0) \tag{2.83}$$

This equation indicates that there is a time $t = 0$ at which the absolute value of the autocorrelation function is equal to or greater than at any other time. This fact is of great importance in signal detection.

Example 2.6.6: Find the correlation between a pulse transmitted by radar equipment, as shown in Figure 2.6.6a, and each of the two possible received pulses shown in Figures 2.6.6b and c.

Solution: The correlation between the transmitted signal and the identical received one can be easily accomplished by using the graphical approach. At zero shifting, $t = 0$, the two pulses do not overlap, and the output is zero. This result is the same until $t = 1$. Then the two pulses start overlapping, and since the overlap is linear, the resulting output is that shown in Figure 2.6.6d. Observe that the correlation signal is displaced by one unit of time. This indicates the time it took the signal to reach the target and return to radar. This is how air traffic controllers know exactly how far airplanes are from the airport. For the case of $s_2(t)$, we proceed to perform the calculations for each pulse independently and then add the two results. Figure 2.6.6e is the final correlation output. We observe that if the returned pulse is highly distorted, we may not be able to detect for sure the existence

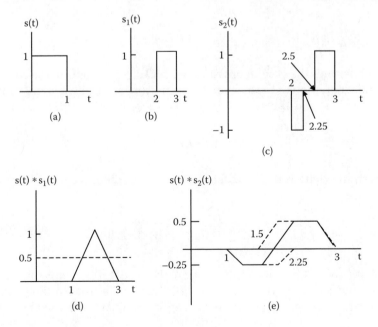

Figure 2.6.6 Illustration of the correlation principle.

of a target. If, for example, we had set a threshold level at 0.5, to avoid small variations of the returned signal, then the second returned signal would not have indicated a target. ∎

Now that we have learned about convolution, let us return to (2.30) and assume the initial condition $v_o(0) = 0$. Hence, the equation takes the form

$$v_o(t) = \frac{R}{L} e^{-t(R/L)} \int_0^t e^{x(R/L)} v(x) dx = \frac{R}{L} \int_0^t e^{-(R/L)(t-x)} v(x) dx \qquad (2.84)$$

The above equation indicates that the output of the system is the convolution of the input $v(t)$ and the function $h(t) = (R/L)e^{-(R/L)t}$, which is the impulse response of the system. This fact will be shown in the next section.

Note: *The output of an LTI system is equal to the convolution of its impulse response (the output of a relaxed system when its input is a delta function) and its input.*

2.7 Impulse response

As discussed above, we require a knowledge of the impulse response $h(t)$ if we are to apply convolution techniques to the solution of system problems with general excitation functions. The following sequence arises from the system description:

$$g(t) = \mathcal{O}\{f(t)\} \qquad (2.85)$$

where $\mathcal{O}$ denotes the system operator, with $\mathcal{O}^{-1}\{g(t)\}$ denoting the controlling differential equation of the system. By definition, $h(t)$ is the response function when $f(t) = \delta(t)$ and the system is relaxed (zero initial conditions), that is

$$h(t) = \mathcal{O}\{\delta(t)\} \qquad (2.86)$$

However, the delta function exists only at $t = 0$; then, we split the problem into two parts:

$$h(t)\big|_{t=0} = \mathcal{O}\{\delta(t)\} \; a); \quad h(t)\big|_{t>0} = \mathcal{O}\{0\} \; b) \qquad (2.87)$$

Note that (2.87) a) provides information about features of $h(t)$ only at $t = 0$, and so provides initial conditions on $h(t)$ that can be used in the solution of the resulting homogeneous differential equation in $h(t)$ specified by (2.87) b).

Chapter 2:   Linear continuous-time systems

A second approach is to use the fact that the impulse response $h(t)$ of a system is the time derivative of the response of the system, $y_u(t)$, to a unit step function, $u(t)$. This response function is given the name **indicial response**. Symbolically,

$$y_u(t) = \mathcal{O}\{u(t)\} \tag{2.88}$$

$$h(t) = \frac{dy_u(t)}{dt} \tag{2.89}$$

In the chapter dealing with the Laplace transform we will show how to find the impulse response of a system. Here, however, we will use the approach provided by (2.87) and (2.89).

Example 2.7.1: Find the zero-input and zero-state response of the system described in Example 2.3.1 if the initial condition for the velocity is $v(0) = 2$, the input force is $f(t) = e^{-2t}u(t)$, $M = 1$ kg, and $D = 0.1$ N s/m.

Solution: The differential equation that describes the system is given by

$$\frac{dv(t)}{dt} + 0.1v(t) = f(t) \tag{2.90}$$

If $f(t)$ is the delta function, then the above equation becomes

$$\frac{dh(t)}{dt} + 0.1h(t) = \delta(t) \tag{2.91}$$

We have changed the v to h to indicate that the input to the system is the delta function. However, the impulse response $h(t)$ has the same units as velocity, that is, m/s. Next, we multiply the above equation by dt and integrate from 0− to 0+. Hence,

$$\int_{0-}^{0+} \frac{dh(t)}{dt} dt + 0.1 \int_{0-}^{0+} h(t) dt = \int_{0-}^{0+} \delta(t) dt \tag{2.92}$$

from which we obtain

$$h(0+) - h(0-) + 0.1 \int_{0-}^{0+} h(t) dt = \int_{0-}^{0+} \delta(t) dt = 1 \tag{2.93}$$

where the property of the delta function was used on the right side of the equation. The second integral vanishes since $h(t)$ does not have an infinite discontinuity at $t = 0$, and it is assumed to be a well-behaved function. This indicates that the area under the $h(t)$ curve of zero distance is zero. Furthermore, the system is relaxed at $t = 0$, and therefore $h(0-) = 0$. Thus, we have that

$$h(0+) = 1$$

For $t > 0$, (2.91) becomes a homogeneous differential equation:

$$\frac{dh(t)}{dt} + 0.1h(t) = 0$$

This equation has the solution

$$h(t) = Ae^{-0.1t} \qquad t > 0$$

Using the known value for $h(0+)$, it follows that $A = 1$. Therefore, the impulse response of the system is

$$h(t) = e^{-0.1t} \qquad t > 0 \quad (m/s) \tag{2.94}$$

The zero-state response is equal to the convolution of the input and the impulse response of the system. Hence, we obtain

$$v_{zs}(t) = h(t) * f(t) = \int_0^t e^{-0.1x} e^{-2(t-x)} dx = \frac{1}{1.9}(e^{-0.1t} - e^{-2t}) \qquad t > 0 \tag{2.95}$$

From the homogeneous equation of (2.90) we find its solution to be

$$v_h(t) = Be^{-0.1t}$$

Applying next the initial condition $v(0) = 2$, we find the value of B to be equal to 2. Hence, the total solution is

$$v(t) = \underbrace{2e^{-0.1t}}_{\text{zero-input solution}} + \underbrace{\frac{1}{1.9}(e^{-0.1t} - e^{-2t})}_{\text{zero-state solution}} \tag{2.96}$$

∎

Chapter 2: Linear continuous-time systems

Example 2.7.2: Determine the output of the system shown in Figure 2.7.1a if the input is a unit pulse function.

Solution: To determine the impulse response function of the system, we first write the controlling differential equation that describes the system:

$$\frac{di(t)}{dt} + i(t) = v(t) \qquad v_o(t) = 1 \times i(t) = i(t)$$

Therefore, the solution of the above equation gives the voltage output. For a delta function input we obtain

$$\frac{dh(t)}{dt} + h(t) = \delta(t)$$

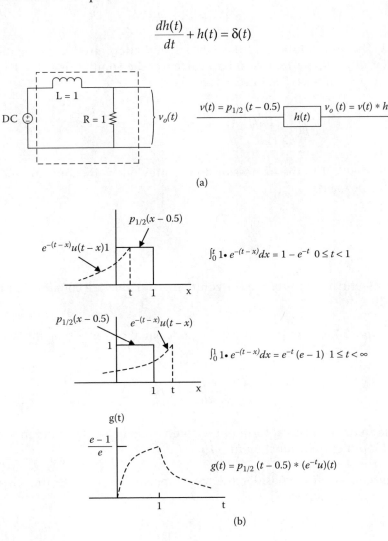

Figure 2.7.1 Illustration of Example 2.7.2.

Following the procedure developed in the previous example, the impulse response is

$$h(t) = e^{-t} \qquad t > 0$$

Let us also find the impulse response using the second method, which is the solution to the relaxed system when the input is a step function. We first find the homogeneous solution, which is $i_h(t) = Ae^{-t}$. Next we find the particular solution by introducing the constant B as the solution. Hence, we obtain

$$0 + B = 1$$

or $B = 1$. The total solution is $i(t) = Ae^{-t} + 1$, and after introducing the initial condition $i(0) = 0$, we find A to be -1. Hence, the solution is $i(t) = 1 - e^{-t}$. Therefore, the impulse response is (see (2.89))

$$h(t) = \frac{di(t)}{dt} = 0 - (-e^{-t}) = e^{-t} \qquad t > 0$$

which is identical to the expression found previously. By convolution, the system response is

$$v_o(t) = \begin{cases} 1 - e^{-t} & 0 \leq t < 1 \\ e^{-t}(e - 1) & 1 \leq t < \infty \end{cases}$$

The essential details of the convolution evaluation are contained in Figure 2.7.1b. ∎

Example 2.7.3: Low-pass filter: The **node** equation describing the system shown in Figure 2.7.2a is

$$\frac{dv(t)}{dt} + \frac{v(t)}{RC} = \frac{1}{C} i(t) \qquad (2.97)$$

Determine the general form of the solution and also the solution to the delayed input current function $i(t - t_0)$.

Solution: We first consider the step function response to this circuit, which is the solution to

$$\frac{dv(t)}{dt} + \frac{v(t)}{RC} = \frac{1}{C} u(t) \qquad (2.98)$$

Chapter 2: Linear continuous-time systems

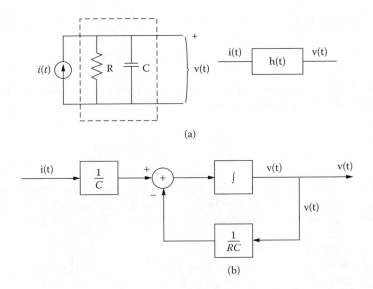

Figure 2.7.2 Low-pass filter.

The solution to the homogeneous equation is found to be

$$v_h(t) = Ae^{-t/RC} \tag{2.99}$$

where A is an unknown constant to be determined from the initial condition.

Now, we consider the particular solution to a step function excitation $u(t)$. This requires the solution of

$$\frac{dv(t)}{dt} + \frac{v(t)}{RC} = \frac{1}{C} \qquad t > 0 \tag{2.100}$$

We observe that the right-hand side is a constant. Hence, we assume that the particular solution is equal to some constant B, that is, $v_p(t) = B$. Introducing this value into (2.100), we obtain $B/(RC) = 1/C$. From this, the particular solution is $v_p(t) = R$. Attention is called to the fact that R is multiplied by a unit current; hence, this result is dimensionally correct. Therefore, the total solution is

$$v(t) = v_h(t) + v_p(t) = Ae^{-t/RC} + R \tag{2.101}$$

The value of A is deduced by applying the initial condition $v(0) = 0$ for the relaxed system. As a consequence, we have

$$v(0) = 0 = Ae^{-0/RC} + R \quad \text{or} \quad A = -R$$

The complete solution is

$$v(t) = R(1 - e^{-t/RC}) \qquad t > 0 \qquad (2.102)$$

Now, making use of the fact that the impulse response is the time derivative of the unit step function response, we have

$$h(t) = \frac{dv(t)}{dt} = \frac{1}{C} e^{-t/RC} \qquad t > 0 \qquad (2.103)$$

With the knowledge of the impulse response of the system, we can determine the response to an arbitrary input $i(t - t_0)$ using the convolution theorem. For the output, we find

$$\int_{-\infty}^{\infty} \frac{1}{C} e^{-(t-x)/RC} u(t - x) i(x - t_0) u(x - t_0) dx = \int_{t_0}^{t} \frac{1}{C} e^{-(t-x)/RC} u(t - x) i(x - t_0) dx$$

The limits in the second integral were defined from the definition of the unit step function and the condition imposed on the current. The step function $u(t - x)$ is zero for $x > t$ and $i(x - t_0)$ is zero for $x < t_0$. To carry out the integration, set $x - t_0 = y$, then $dx = dy$. In addition, at $x = t_0$, $y = 0$, and at $x = t$, $y = t - t_0$, and the step function is unity within these limits. We thus obtain

$$\frac{1}{C} \int_0^{t-t_0} e^{-[(t-t_0)-y]/RC} i(y) dy \triangleq v(t - t_0)$$

This equation indicates that this system is time invariant, that is, the quantity $t - t_0$ substitutes for t in (2.97). In other words, a shifted input of t_0 produces a shifted output of t_0. Figure 2.7.2b is a block diagram representation of (2.97). The block diagram was obtained after we multiplied the equation with dt and integrated both sides. This procedure produced the following expression:

$$v(t) = -\frac{1}{RC} \int v(t) dt + \frac{1}{C} \int i(t) dt$$

∎

Example 2.7.4: Determine the impulse response of an initially relaxed RLC series circuit with $L = 1$ H, $R = 5$, and $C = 0.25$ F.

Solution: The integrodifferential equation describing the system is

$$L \frac{di(t)}{dt} + Ri(t) + \frac{1}{C} \int i(t) dt = v(t) \qquad (2.104)$$

Chapter 2:   Linear continuous-time systems

We write this equation in differential equation form by writing $i(t) = dq(t)/dt$, with the result

$$\frac{d^2q(t)}{dt^2} + 5\frac{dq(t)}{dt} + 4q(t) = v(t) \tag{2.105}$$

For a delta function input, this equation becomes

$$\frac{d^2h(t)}{dt^2} + 5\frac{dh(t)}{dt} + 4h(t) = \delta(t) \tag{2.106}$$

First, we establish the initial conditions that permit a solution to the above equation. The system is initially relaxed so that

$$\frac{dh(0-)}{dt} = h(0-) = 0 \tag{2.107}$$

Now, let us integrate (2.106) between the limits $t = 0-$ and $t = 0+$. We thus write

$$\int_{0-}^{0+} \frac{d^2h(t)}{dt^2} dt + 5\int_{0-}^{0+} \frac{dh(t)}{dt} dt + 4\int_{0-}^{0+} h(t)dt = \int_{0-}^{0+} \delta(t)dt \tag{2.108}$$

The third integral is zero because the area under a continuous function from 0– to 0+ is zero. Also, the area under the delta function is unity. The result is

$$\left[\frac{dh(0+)}{dt} - \frac{dh(0-)}{dt}\right] + 5[h(0+) - h(0-)] = 1 \tag{2.109}$$

which becomes, by (2.107),

$$\frac{dh(0+)}{dt} + 5h(0+) = 1 \tag{2.110}$$

Now, we integrate (2.106) twice, which yields

$$\int_{0-}^{0+} \frac{dh(t)}{dt} dt + 5\int_{0-}^{0+} h(t)dt + 4\int_{0-}^{0+}\int_{0-}^{0+} h(t)dtdt = \int_{0-}^{0+}\int_{0-}^{0+} \delta(t)dtdt$$

Assuming $h(t)$ to be a well-behaved function, the second term is zero because the area under $h(t)$ within the limits from 0– to 0+ is zero. The first integration of the third term will produce a well-behaved function, and hence the second integration will give a value of zero. The first integration of the fourth term will produce the value of 1, a well-behaved function, and thus the second integration will produce zero. Hence, the above equation becomes

$$h(0+) - h(0-) = 0 \quad \text{or} \quad h(0+) = 0 \tag{2.111}$$

Combining (2.111) with (2.110) yields

$$\frac{dh(0+)}{dt} = 1 \tag{2.112}$$

This result indicates that the impulse source forces the current $dq(t)/dt \triangleq dh(t)/dt$ to jump instantaneously, in this case from 0 to 1, while the charge remains at zero.

We next proceed with the solution (2.106), noting as before that $\delta(t) = 0$ for $t > 0$. Thus, the equation under consideration is

$$\frac{d^2h(t)}{dt^2} + 5\frac{dh(t)}{dt} + 4h(t) = 0 \quad t > 0$$

$$\text{Initial conditions:} \quad h(0+) = 0 \quad \frac{dh(0+)}{dt} = 1 \tag{2.113}$$

For an assumed solution of the form $h(t) = A\exp(st)$, the characteristic equation becomes

$$s^2 + 5s + 4 = 0 \quad \text{with roots} \quad s_1 = -4, \; s_2 = -1$$

The solution of the above homogeneous equation is

$$h(t) = Be^{-4t} + Ce^{-t}$$

Introducing the initial conditions in the above equation and solving the system for the two unknown constants B and C, we obtain

$$h(0+) = 0 = B + C, \quad \frac{dh(0+)}{dt} = 1 = -4B - C$$

or $B = -1/3$ and $C = 1/3$. Therefore, the solution is

$$h(t) \triangleq q(t) = -\frac{1}{3}e^{-4t} + \frac{1}{3}e^{-t} \quad t > 0 \tag{2.114}$$

Chapter 2: Linear continuous-time systems

The current through the circuit is found by taking the derivative of the above equation. ∎

Example 2.7.5: Repeat Example 2.7.4, but now proceeding from (2.104).

Solution: We proceed now by writing the equation $[h(t) \triangleq i(t)]$

$$\frac{dh(t)}{dt} + 5h(t) + 4\int h(t)dt = \delta(t) \tag{2.115}$$

We consider this equation at two specific instants of time, $t = 0-$ and $t = 0+$, and subtract the two. The result is

$$\left[\frac{dh(0+)}{dt} - \frac{dh(0-)}{dt}\right] + [5h(0+) - 5h(0-)] + 4\int_{0-}^{0+} h(t)dt = \delta(t)\Big|_{0-}^{0+}$$

We recall that the delta function is zero at any value of t besides $t = 0$; thus, $\delta(0-) = \delta(0+) = 0$. Further, the area under the $h(t)$ curve within those limits is zero. In addition, $h(0-) = dh(0-)/dt = 0$, and thus the above equation becomes

$$\frac{dh(0+)}{dt} + 5h(0+) = 0 \tag{2.116}$$

Now we consider the integral of (2.115) over the limits $t = 0-$ and $t = 0+$. This gives

$$\int_{0-}^{0+} \frac{dh(t)}{dt}dt + 5\int_{0-}^{0+} h(t)dt + 4\int_{0-}^{0+}\left(\int h(t)dt\right)dt = \int_{0-}^{0+} \delta(t)dt$$

Remembering that the area under the delta function is unity and that $h(t)$ is a smooth function, we then have that

$$h(0+) - h(0-) = 1 \quad \text{or} \quad h(0+) = 1 \tag{2.117}$$

Combine this result with (2.116) to get

$$\frac{dh(0+)}{dt} = -5 \tag{2.118}$$

To find the solution for $h(t)$ and for $t > 0$, differentiate (2.115), which yields the differential equation

$$\frac{d^2h(t)}{dt} + 5\frac{dh(t)}{dt} + 4h(t) = \frac{d\delta(t)}{dt} = 0 \qquad t > 0$$

Since $\delta(t) = 0$ for $t > 0$, the derivative of $\delta(t)$ is also equal to zero for $t > 0$. The resulting differential equation is precisely that given by (2.113), and hence its final solution is

$$h(t) \triangleq i(t) = Be^{-4t} + Ce^{-t} \qquad t > 0$$

Next, apply the initial conditions given by (2.117) and (2.118) to obtain the final solution:

$$h(t) \triangleq i(t) = \frac{4}{3}e^{-4t} - \frac{1}{3}e^{-t} \qquad t > 0 \qquad (2.119)$$

Observe that the derivative of (2.114) yields (2.119), as it should, since $dq(t)/dt = i(t)$. ∎

Important definitions and concepts

1. Mathematical modeling of analog electrical systems
2. Causality
3. Mathematical modeling of linear and rotating mechanical systems
4. Zero-input and zero-state solutions
5. Transient and steady-state solution
6. Initial conditions
7. Block diagram representation of systems
8. Block diagram transformations
9. The feedback concept
10. Convolution of signals
11. Correlation of signals
12. Impulse response of systems
13. Matched filtering
14. Edge detection (see Problem 2.6.8)
15. Monte Carlo calculations (see Problem 2.2.7)
16. Energy associated with electrical circuits
17. Environmental engineering (see Problem 2.3.14)
18. Heat transfer (see Problem 2.3.8)

Chapter 2 Problems

Section 2.2

1. Develop the block diagram representation of the system shown in Figure P2.2.1 and explicitly identify the operator of the system, $\mathcal{O}$.

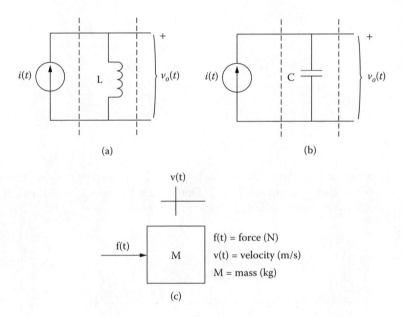

Figure P2.2.1

2. a. Find the current through an inductor of 1 H at $t = 0$ s and at $t = 2$ s if the voltage waveform across the inductor is that shown in Figure P2.2.2a.

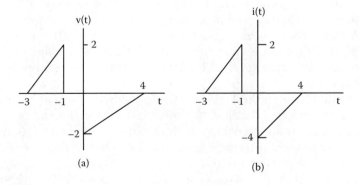

Figure P2.2.2

b. Find the voltage across an inductor of 1 H at $t = 0$ s and at $t = 2$ s, if the current waveform through the inductor is that shown in Figure P2.2.2b.
3. Find the energy dissipated by a resistor of 5 Ω for currents through it that have the waveforms shown in Figure P2.2.3.

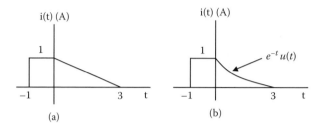

Figure P2.2.3

4. Forces of the form shown in Figure P2.2.4 are applied to a spring with a spring constant $K = 2$. Determine the functional form of the velocity corresponding to these two forces.

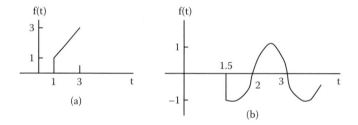

Figure P2.2.4

5. A time-varying torque is applied to a rotating body that has a polar moment of inertia of 2 kg-m². Plot the angular velocity of the rotating body if the torque is that shown in Figure P2.2.5.
6. A current source having the fundamental form $i(t) = e^{-|t|}u(t + 2)$ is applied through a capacitor of 0.5 F. Find the voltage and charge across the capacitor if the current source remains applied for an infinite time.
7. **Monte Carlo method**. (Monte Carlo method is one of the most important practical methods for solving differential equations, finding values of multidimensional integrals, performing simulations, filtering signals using particle filters, etc.) A force having the functional form $i(t) = e^{-t}p_{1/2}(t - 0.5)$ is applied to a mass of a 1 kg. Determine the velocity of the mass. Compare your results with the results obtained by using the following numerical procedure, known as the

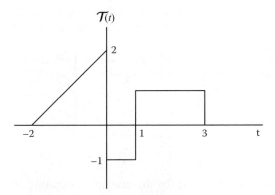

Figure P2.2.5

Monte Carlo method. Here, because the curve lies within a unit square, use a random number generator supplied by MATLAB, for example, to produce randomly values from 0 to 1. For any pair of generated numbers (t_{i1}, t_{i2}) corresponds a point inside the square. If the total number of trials is N (use $N = 100$, 1000, 10,000) and n denotes the number of times that $f(t_{i1}) > t_{i2}$ is satisfied, then the value of the integral is approximately equal to n/N.

Book MATLAB function

```
function[v]=ssvelocitymc(N)
n1=0;
for n=1:N
   t1=rand;
   t2=rand;
   if exp(-t1)>=t2;
   n1=n1+1;
   end;
end;
v=n1/N
```

(It is recommended that the reader develops his or her own MATLAB function and compare results.)

8. The energy delivered to each electrical element — resistor, inductor, and capacitor — between $t_1 \le t \le t_2$ is given by $\int_{t_1}^{t_1} v(t)i(t)dt$, where $v(t)$ is the voltage across the element and $i(t)$ is through it. Deduce expressions for the energies E_R, E_L, and E_C.

9. The switch to the relaxed circuit (zero initial conditions) shown in Figure P2.2.9 is closed at $t = 0$. The voltage across the RC parallel combination is $v(t) = 0.5(1 - e^{-10t})$. Determine (a) the energy delivered

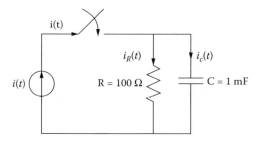

Figure P2.2.9

by the source at time t, (b) the energy dissipated in the resistor, and (c) the energy stored in the capacitor. From your answer, draw a fundamental conclusion that is the property of any type of RLC circuits.

Section 2.3

1. Represent the following differential equations in block form:

   a. $0.5(dy/dt) = 4y + \exp(-t)u(t)$
   b. $(dy/dt) - y - 2tu(t) = \sin t\, u(t)$

2. Find the equilibrium equations for each network shown in Figure P2.3.2. Specify the initial conditions, assuming that the situation shown existed for a time before the switching operation was initiated. The solutions are not required. What conclusions can be drawn from these results?

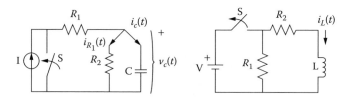

Figure P2.3.2

3. Find the transient and steady-state solutions for the relaxed system given in Figure P2.3.3.

   Hint: Assume a trial solution of the form $v(t) = a_0 + a_1 t + a_2 t^2$ in determining the particular solution.

4. Find the zero-input, zero-state, transient, and the steady-state solutions for the system shown in Figure 2.3.1. The input voltage is $v(t) = \exp(-0.5t)u(t)$. The initial current is $i(0) = 1.5$ A.

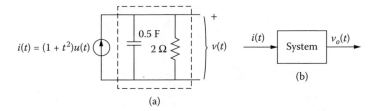

Figure P2.3.3

5. Find the zero-input, zero-state, transient, and steady-state solutions of the system shown in Figure 2.3.3a. The input force is $f(t) = 10^{-4} \exp(-t)u(t) + u(t)$. The initial velocity is $v(0) = 2$ m/s.
6. A current is applied to the circuit shown in the block diagram in Figure P2.3.3. Determine the zero-input and zero-state responses. Set $v(0) = 1$.
7. Repeat Problem 2.3.3 for the transient and steady-state responses, but now by determining the zero-input and zero-state responses of the initially relaxed system. Compare these results with those obtained in Problem 2.3.3.
8. **Heat transfer.** The temperature outside a digital chip is constant and equal to $T_0 = 22°C$. Because of currents in the chip, heat is generated uniformly at the rate Q. The temperature inside the chip is considered to be uniform throughout the chip at temperature $T°C$. The total rate of change of heat within the chip is equal to $mc_p(dT/dt)$, where m is the mass of the chip and c_p is its heat capacity. This heat is equal to the rate at which heat is supplied by the source Q plus the rate at which the heat is transferred to the chip from the surrounding air. Thus, we write $mc_p(dT/dt) = -K_o(T - T_o) + Q$, where $K_o(T - T_o)$ is the rate of heat transfer into the chip from the surrounding air (or heat sink). Given that $m = 50$ g, $c_p = 0.15$ cal/g-C°, $K_o = 3$ cal/C°, and $Q = e^{-t} + \cos t$ cal for $t > 0$, determine the transient and steady-state solutions for the chip temperature.

    Hint: Find the particular solution separately for $\exp(-t)$ and $\cos t$ and add the results. This superposition is valid because differentiation is a linear process.

9. The current in the circuit shown in Figure P2.3.9 is $i(t) = 2(1 - \exp(-2t))u(t)$. Determine the input voltage.
10. Determine the zero-input, zero-state, transient, and steady-state solutions for the current in the circuit shown in Figure P2.3.9. The input voltage is $v(t) = 4 + \exp(-4t)$ for $t > 0$. The circuit is initially relaxed.

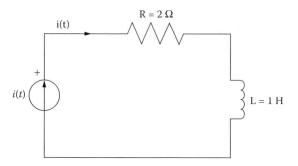

Figure P2.3.9

11. **Vibration isolator.** A force $f(t) = 2\exp(-t)$ for $t > 0$ is applied to the system shown in Figure P2.3.11. This type of system is often used as a vibration isolator. Determine the distance x traveled by the top level of the initially relaxed system.

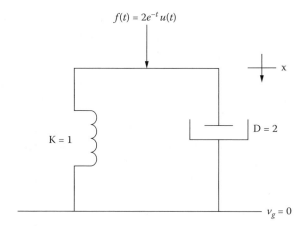

Figure P2.3.11

12. The voltage applied to the circuit of Figure P2.3.12 is $v(t) = \cos\omega t$ for $t > 0$. The initial charge on the capacitor is zero.

    a. Find an expression for the current $i_C(t)$.
    b. Is there some value of $v_C(0-)$ for which the transient term is zero?

13. **Biomedical engineering (biology).** Assume that a colony of bacteria increases at a rate proportional to the numbers present. If the number of bacteria doubles in 5 h, how long will it take for the number of bacteria to triple? Draw a block diagram that represents the system.

14. **Environmental engineering.** A tank contains 400 l of water. By accident, 120 kg of chlorine is poured into the tank instead of 60 kg. To

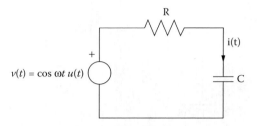

Figure P2.3.12

correct the mixture, a stopper is removed from the bottom of the tank allowing 3 l of the mixture to flow out each minute. At the same time, 3 l/min of pure water is poured into the tank. If the mixture is kept uniform by stirring, how long will it take for the mixture to attain the desired amount of clorine?

15. **Finance**. The sum of $6000.00 is invested at the rate of 6% per year, compounded continuously. What will be the amount after 50 years?

Section 2.4

1. The circuit shown in Figure P2.4.1 is initially in its quiescent state when the switch is opened.
   a. Deduce an expression for the inductor currents for $t > 0$.
   b. Determine the conditions for the current to be zero.

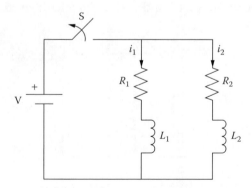

Figure P2.4.1

2. A pulse of height V and duration $T = RC$ is applied to the series RC circuit shown in Figure P2.4.2. The circuit is initially relaxed. Sketch the output voltage $v_o(t)$ as a function of time. Label all features of the curve.

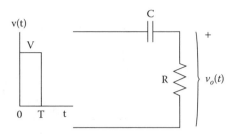

Figure P2.4.2

3. A two-pulse wave-train is applied to the initially relaxed RC circuit shown in Figure P2.4.3. From knowledge of the step function response of the RC circuit, deduce the voltage waveform $v_o(t)$.

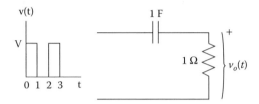

Figure P2.4.3

Hint: Use a step-by-step procedure beginning at $t = 0$, and then determine the values of $v_o(t)$ at the successive switching times.

4. In the circuit of Figure P2.4.4, $v_C(0-) = 0$. The switch is closed at time $t = 1$ s. Determine $v_C(t)$ and $i_C(t)$ and sketch their time variations.

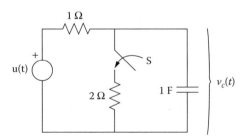

Figure P2.4.4

Section 2.5

1. Prove the block diagram transformations shown in Figure 2.5.4 and specify (d) and (j).
2. Draw block diagrams for the systems shown in Figure P2.5.2.

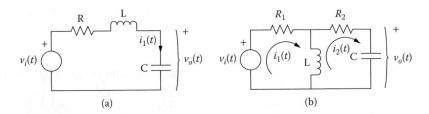

Figure P2.5.2

3. Draw block diagrams for the following sets of differential equations, where $p = d/dt$, $p^2 = d^2/dt^2$, and $p^3 = d^3/dt^3$.

   a. $(p^3 + 3p + 2)y = t + 7$
   b. $(p^2 + 5)y = 3t$

4. Deduce the input–output (transmittance $H(p)$) of the systems represented by the block diagrams shown in Figure P2.5.4.

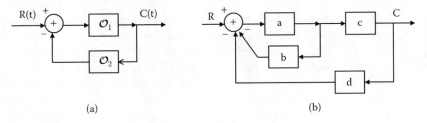

Figure P2.5.4

Section 2.6

1. Determine the convolution of the following pairs of functions:

   a. $p_1(t)$, $p_1(t - 2)$
   b. $p_1(t)$, $\delta(t)$
   c. $\exp(-(t - 2))u(t - 2)$, $\delta(t)$
   d. $u(t)$, $\exp(-t)u(t)$

2. Determine the convolution of the following pairs of functions:

   a. $p_1(t) - p_1(t - 2)$, $\delta(t)$
   b. $2p_1(t - 2)$, $\exp(-t)u(t)$

3. Using the functions $f(t) = p_1(t - 1)$ and $h(t) = \exp(-t)u(t)$, verify the commutatitive property of the convolution $f(t)*h(t) = h(t)*f(t)$.

4. Determine the convolution of the following pairs of functions:
   a. $tu(t)$, $t^2u(t)$
   b. $t^2u(t)$, $t^3u(t)$

5. Show that if $h(t)$ is a real function and we define $[f_1(t) + jf_2(t)]*h(t) = g_1(t) + jg_2(t)$, then $f_1(t)*h(t) = g_1(t)$ and $f_2(t)*h(t) = g_2(t)$.

6. If $g(t) = f_1(t)*f_2(t)$, show that $dg(t)/dt = [df_1(t)/dt]*f_2(t) = f_1(t)*[df_2(t)/dt]$.

7. Find the convolution of the functions shown in Figure P2.6.7.

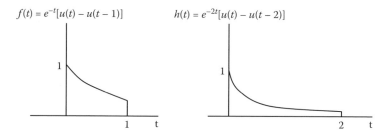

Figure P2.6.7

8. **Edge detection.** One of the most important operations in signal processing, especially in image processing for recognizing patterns (targets in military terms) embedded in an image, is the exaggeration of a signal transition. An operator (signal) that will exaggerate a transition (from white to gray to black) is known as the **edge detector**. For verifying the above assertion, find the convolution of the signals shown in Figure P2.6.8 for $a = 1$ and $a = 1/2$ and state your observations.

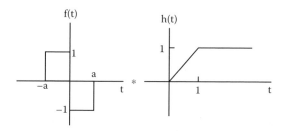

Figure P2.6.8

9. Deduce the correlation between the functions given below:
   a. $p_1(t)$, $\exp(-t)u(t)$
   b. $p_1(t)$, $\delta(t-1) + p_1(t-3)$
   c. $\exp(-t)u(t)$, $\exp(-t)u(t)$
   d. $tp_1(t-1)$, $\exp(-t)u(t)$

10. Verify that $f(t) \odot h(t) = f(t) * h(-t)$, where the circle with the dot means correlation.

Section 2.7

1. Deduce the zero-input and zero-state responses of the system shown in Figure P2.7.1. The initial condition for the charge is $q(0) = 1$ C and the input voltage is $v(t) = tu(t)$. Use the convolution approach.

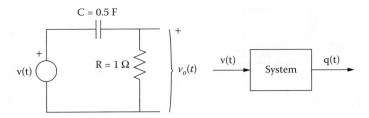

Figure P2.7.1

2. Determine the output of the initially relaxed systems shown in Figure P2.7.2 if the input is $v(t) = \exp(-t)u(t)$. Use the convolution approach.

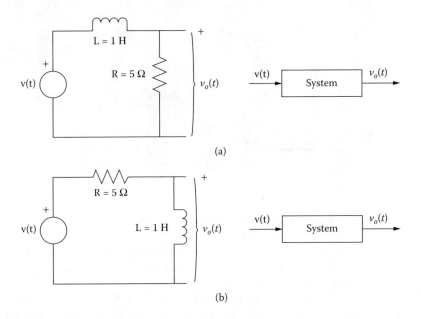

Figure P2.7.2

3. Determine the output of the initially relaxed systems shown in Figure P2.7.3 if the inputs are, respectively, $f(t) = p_1(t-1)$ and $\mathcal{T} = \exp(-t)u(t)$. Use the convolution method.

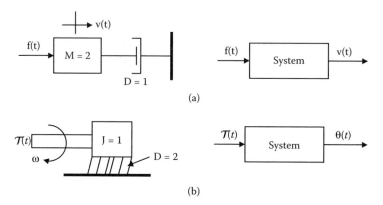

Figure P2.7.3

4. Find the impulse response of the initially relaxed systems shown in Figure P2.7.4. The charge is $q(t)$ for initial charge $q(0+) = 0$.

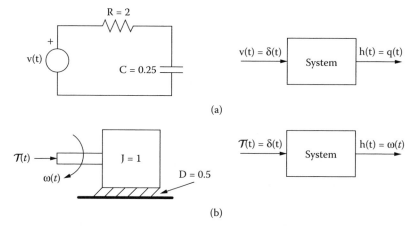

Figure P2.7.4

5. Find the output of the system described in Example 2.7.2 if the input is $v(t) = u(t-1) + \delta(t-3)$.
6. The input to the system discussed in Example 2.7.3 is $i(t) = p(t-4)$. Determine the output using convolution methods. Choose $C = 1$ F and $R = 2$ ohms.
7. The input to the system discussed in Example 2.7.3 is $p(t-1) - p(t-5)$. Use the convolution approach to determine the output. Choose $C = 1$ F and $R = 2$ ohms.
8. Determine the output of the initially relaxed system shown in Figure P2.7.8 for the following inputs:
   a. $i_1(t) = u(t-2)$
   b. $i_2(t) = \exp(-(t-4))u(t-4)$

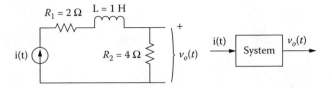

Figure P2.7.8

9. Determine the output to the system shown in Figure P2.7.9 if the input is $f(t) = \cos t \, u(t)$. Use convolution methods.

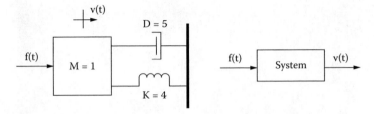

Figure P2.7.9

10. Determine the output of the system discussed in Example 2.7.5 if the input is the unit step function $v(t) = u(t)$.

chapter 3

Discrete systems

In this chapter we will learn how to describe and model discrete systems. The systems that we will study will be primarily derived from continuous systems appropriately digitized. As we learn the mathematical operations of convolution, we shall proceed in this chapter to learn similar operations for discrete systems. Furthermore, we shall develop approaches such that the discrete systems approximate the continuous ones to a degree that is acceptable for practical applications.

3.1 Discrete systems and equations

Note: The reader should have in mind that, in this chapter, we will arbitrarily define discrete signals and will also introduce discrete signals created from analog ones, without presenting their similarities and differences. This will be done when we study the sampling theorem in a later chapter. The effect of sampling will be brought to the attention of the reader when applicable.

When we dealt with continuous systems, our interest was in the relationship between an input signal $v(t)$ and output signal $g(t)$. Hence, we were interested in finding the appropriate operator $\mathcal{O}$ that, when operated on the input, would yield the relation between the input and output. Similarly, a **discrete**, or **digital**, **system** establishes a relationship between two discrete signals: an input $v(n)$ and output $g(n)$. The values of n are integers. However, we will also deal with discrete signals whose values will not be separated by a unit time, but by a fraction of time, e.g., nT, where T is equal to a value in the range, $0 < T < 1$. An actual discrete system is a combination of many different electronic components, such as amplifiers, shift registers, and gates. However, it is represented here in simple block diagram form.

In discrete-time systems, a **discrete source** produces discrete-time signals that are pulses of assumed zero width and finite height. A discrete signal sequence may arise by pulse sampling a continuous-time excitation function, usually at uniform time intervals. It might represent a sequence of narrow pulses generated in a pulse-generating source. As we mentioned above, the

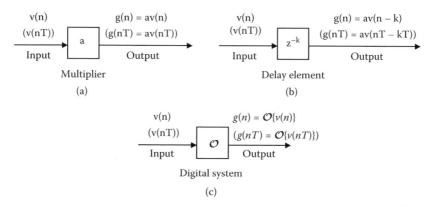

Figure 3.1.1 Digital system entities: (a) multiplier, (b) delay element, and (c) digital system.

generation of discrete signals having zero width and finite height is not possible using any physical system. Therefore, there is no physical source that is able to produce discrete signals.

Another element of a discrete system is the **scalar multiplier**, a component that produces pulses at the output that are proportional to input pulses. There is also a **delay element** that produces an output identical to its input, but delayed by a predetermined number of time units or a multiple fraction of a time unit. The symbol z^{-1} will be used to denote a delay of a unit time or a fraction of unit time (T); two delay units will be denoted by z^{-2} and k delay units by z^{-k}. These delay elements are shown in Figure 3.1.1.

A group of such connected discrete elements and adders (summing points or summers) with appropriate pick-off points constitute a digital system. Discrete sources are its inputs, and the resulting signals in its various parts are the outputs, or responses, of the system. The analysis of the behavior of such systems will parallel, to some extent, the systems analysis procedures in continuous-time systems.

Example 3.1.1: A discrete system and its input is shown in Figure 3.1.2. Find its output.

Solution: By inspection, it is seen that the output from the summer is a signal consisting of two pulses: a pulse of height 1 at $n = 0$ and a pulse of height 2 at $n = 2$. These pulses are each delayed by two time units, giving the output shown. ∎

Example 3.1.2: Find the output of the system shown in Figure 3.1.3.

Solution: Clearly, the output signal is the sum of the input signal plus a replica of this signal, but delayed by two time units. The result is shown in the same figure. ∎

Chapter 3: Discrete systems

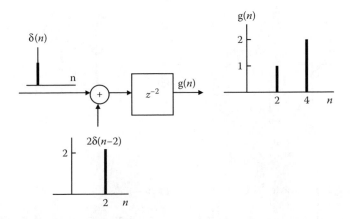

Figure 3.1.2 Simple discrete system with delay element.

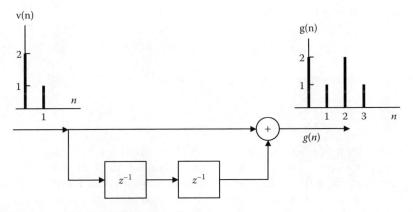

Figure 3.1.3 Single-time discrete repeater.

A system consisting of discrete-time elements with a discrete-time input signal sequence is described by a **difference equation**. This description is to be contrasted with the description of linear, lumped, time-invariant systems that are described by ordinary differential equations. Depending on the number and character of the elements in the discrete system, the system is described by difference equation of different order. The following examples will help us understand how difference equations are developed. The structure layout of the discrete elements will produce specific type of difference equations, which will be identified with specific names. Furthermore, the iterative method of solution will be adopted at present and other methods will be considered in later chapters.

Example 3.1.3: Consider the system shown in Figure 3.1.4a. Find the difference equation that describes the system and deduce $g(n)$ if the input is

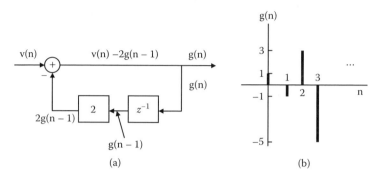

Figure 3.1.4 A first-order IIR discrete system.

$v(n) = u(n)$, the unit step function. The system is relaxed at $n = 0$, which implies that $g(-1) = 0$.

Solution: From the figure we observe that

$$g(n) = v(n) - 2g(n-1) \quad \text{or} \quad g(n) + 2g(n-1) = v(n) \tag{3.1}$$

The above equation may be put in the general form

$$\boxed{y(n) + a_1 y(n-1) = b_0 x(n)} \tag{3.2}$$

Since the difference between the independent variables of the output of the system is $(n - (n - 1)) = 1$, the difference equation is of the **first order**. Furthermore, since the output is given in its unshifted and shifted formats, but the input is only given by its zero shift position, the system is of the **infinite impulse response** (**IIR**) system. Observe that this type of discrete system produces a feedback-type block diagram form.

Note: *The infinite impulse response (IIR) discrete system consists of the output plus additional delayed outputs multiplied by constants. This type of system has only one input and not delayed ones. The block diagram structure is a feedback type.*

To find a numerical solution, we proceed by successively introducing values of n into the difference equation, starting with $n = 0$ since the input function starts at time $n = 0$. We proceed iteratively thereafter. Hence,

$$g(0) = v(0) - 2g(-1) = 1 - 2 \times 0 = 1$$

$$g(1) = v(1) - 2g(0) = 1 - 2 \times 1 = -1$$

$$g(2) = v(2) - 2g(1) = 1 - 2 \times (-1) = 3$$

$$\vdots$$

Chapter 3: Discrete systems 115

The output has been plotted in Figure 3.1.4b. The following MATLAB function can solve any first-order IIR discrete system. ∎

Book MATLAB function for an IIR system

```
function[y]=ssfirstorderiir(a1,b0,N,x)
y(1)=0;%if the initial condition is different than 0
       %the new value must be introduced here;
for n=0:N
    y(n+2)=-a1*y(n+1)+b0*x(n+1);
end;
%since matlab starts from 1, everything must be shifted by
%two; the answer will also contain the value of the initial
%condition at n=1; to have the exact plot we must set: k=-1:N;
%stem(k,y);
```

Example 3.1.4: Find the difference equation that describes the system shown in Figure 3.1.5a. Determine its output $g(n)$ if its input is $v(n) = u(n)$. The system is relaxed at $n = 0$, which implies that $g(-1) = 0$.

Solution: From the figure we deduce that

$$g(n) = v(n) + 2v(n-1) - 4v(n-2) \tag{3.3}$$

The above equation may be put in the general form

$$\boxed{y(n) = b_0 x(n) + b_1 x(n-1) + b_2 x(n-2)} \tag{3.4}$$

Note that the output of the system is expressed only in terms of the input function. Such a system is called a **nonrecursive, transversal,** or **finite**

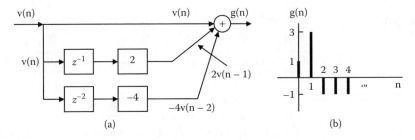

Figure 3.1.5 A three-term FIR system.

duration impulse response (**FIR**). To find the solution, we solve the equation recursively. Hence,

$$g(0) = v(0) + 2v(-1) - 4v(-2) = 1 + 2 \times 0 - 4 \times 0 = 1$$

$$g(1) = v(1) + 2v(0) - 4v(-1) = 1 + 2 \times 1 - 4 \times 0 = 3$$

$$g(2) = v(2) + 2v(1) - 4v(0) = 1 + 2 \times 1 - 4 \times 1 = -1$$

$$\vdots$$

The MATLAB function that will solve a FIR system is given below. ∎

Book MATLAB function for four-term FIR system

```
function[y]=ssfir4term(N,x,b)
for n=0:N
    y(n+1)=[x(n+4)  x(n+3)  x(n+2)  x(n+1)]*b';
end;
%the input vector x must have its first three
%elements zero for a relaxed system; to plot
%the output must write:k=0:N;stem(k,y);
```

Note: *The finite impulse response system has only one output, an input, and additional delayed inputs multiplied by constants. The output of the system is the sum of both the input and the delayed inputs. The block diagram form is of the forward format.*

Example 3.1.5: Find the difference equation that describes the system shown in Figure 3.1.6a. Determine the output $g(n)$ for an input $v(n) = u(n)$. The system is relaxed at $n = 0$, which implies that $g(-1) = g(-2) = 0$.

Solution: From the figure we see that

$$g(n) = v(n-1) - 2v(n-2) - 2g(n-2) \qquad (3.5)$$

The general second-order mixed (IIR and FIR) discrete system is given by

$$\boxed{y(n) + a_1 y(n-1) + a_2 y(n-2) = b_0 x(n) + b_1 x(n-1) + b_2 x(n-2)} \qquad (3.6)$$

Because the difference between n and $n-2$ belonging to the dependent variables $g(n)$ and $g(n-2)$ is equal to 2, the difference equation is of the **second order**. We note that the output is the sum of the inputs and the delayed outputs with their proportionality factors. The solution of (3.5) is

Chapter 3: Discrete systems

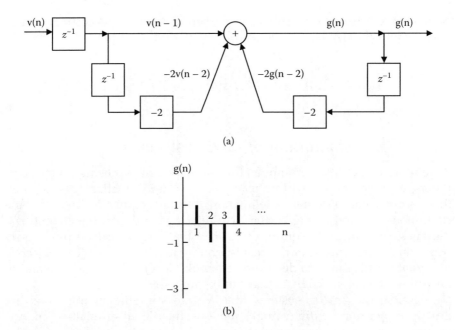

Figure 3.1.6 Mixed IIR and FIR discrete systems.

$$g(0) = v(-1) - 2v(-2) - 2g(-2) = 0 - 2 \times 0 - 2 \times 0 = 0$$

$$g(1) = v(0) - 2v(-1) - 2g(-1) = 1 - 2 \times 0 - 2 \times 0 = 1$$

$$g(2) = v(1) - 2v(0) - 2g(0) = 1 - 2 \times 1 - 2 \times 0 = -1$$

$$g(3) = v(2) - 2v(1) - 2g(1) = 1 - 2 \times 1 - 2 \times 1 = -3$$

$$\vdots$$

The output is plotted in Figure 3.1.6b. ∎

Example 3.1.6: Engineering economics (banking). The interest rate of a savings account is r% per year compounded k times per year ($k = 4$ would correspond to quarterly compounding). Determine what the total bank account balance, $y(n)$, would be at the end of the nth compounded period if the total deposits during the nth compounding period are $x(n)$. Assume that deposits at any period do not earn interest until the next compounding period.

Solution: The interest is compounded at the rate r/k% for each compounding period. Therefore, at the end of any compounding period, the total bank account balance is equal to the sum of the following: the bank account balance at the start of the compounding period, the interest accrued on this balance, and the deposits made during the period. We, thus write

$$y(n) = y(n-1) + \frac{r}{k}y(n-1) + x(n) = \left(1 + \frac{r}{k}\right)y(n-1) + x(n)$$

∎

3.2 Digital simulation of analog systems

If we wish to carry out continuous-time systems analysis using digital computers, a procedure we will be using very often in practice, it is required that we determine a digital system whose output is equivalent to that of the continuous-time or analog system. We refer to the digital equivalent as a **digital simulator** and require, of course, that the digital output closely approximate the output of the corresponding analog system. The objective of digital simulation is to determine a simulator of an anlog system and to determine the class of signals that can be processed.

Because the analog systems that we are studying in this book are described by ordinary differential equations, the digital simulators will correspond to a recursive equation that can be represented by delay elements and scalar multipliers. To accomplish these objectives, we must study how to create the digital simulators from their analog counterparts.

To understand the process, let us study the analog system shown in Figure 3.2.1a. We first replace the differential relationship $di(t)/dt$ by an approximately equivalent difference relationship. This transformation is accomplished by replacing the derivative by the approximate form

$$\boxed{\frac{di(t)}{dt} \cong \frac{i(nT) - i(nT - T)}{T} = \frac{i(nT) - i[(n-1)T]}{T}} \qquad (3.7)$$

or

$$\frac{di(t)}{dt} \cong i(n) - i(n-1), \qquad T = 1 \qquad (3.8)$$

The approximation (3.7) is shown graphically in Figure 3.2.1b. Observe that as T approaches zero and n increases, so that nT remains the same, the inclination of the line AB approaches the inclination of the exact tangent at nT.

For our case and for $T = 1$, the approximation to the analog expression (the output voltage) is

$$v(n) = L[i(n) - i(n-1)] \qquad (3.9)$$

Chapter 3: Discrete systems

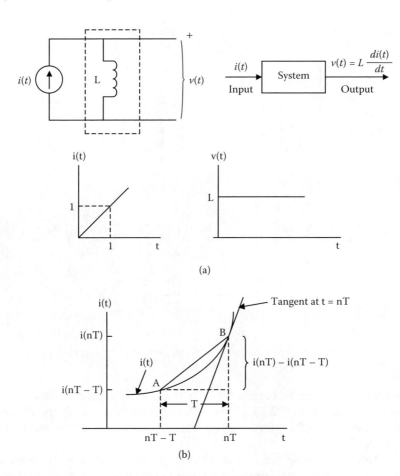

Figure 3.2.1 Analog-to-digital simulation of electrical circuits. (a) Simple analog circuit, input and output. (b) Approximation to derivative.

Assume an initially relaxed system $v(-1) = 0$; for the discrete or sampled version of the analog input $i(t)$ shown in Figure 3.2.1a, we obtain $i(n) = n$. The solution is found based on the same procedure as that done in the previous example, as follows:

$$v(0) = L[i(0) - i(-1)] = L[0 - 0] = 0$$

$$v(1) = L[i(1) - i(0)] = L[1 - 0] = L$$

$$v(2) = L[i(2) - i(1)] = L[2 - 1] = L$$

$$v(3) = L[i(3) - i(2)] = L[3 - 2] = L$$

$$\vdots$$

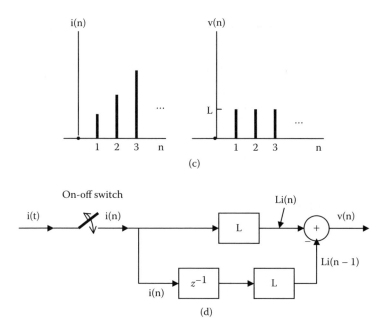

Figure 3.2.1 (continued). (c) Sampled input and output from digital equivalent system. (d) Block diagram of digital simulation.

These data are shown in Figure 3.2.1c. The numerical process described by (3.9) is shown in Figure 3.2.1d. Observe that the instantaneous on–off switch (ideal) accomplishes the production of discrete signals from the analog with infinite accuracy.

We now investigate the system shown in Figure 3.2.2a with a unit step input. In this example, we approximate an integral with its simulated discrete form. Hence, we write

$$v(nT) = \frac{1}{M}\int_0^{nT} f(t)dt = \frac{1}{M}\int_0^{nT-T} f(t)dt + \frac{1}{M}\int_{nT-T}^{nT} f(t)dt \qquad (3.10)$$

which becomes

$$\boxed{v(nT) = v(nT-T) + \frac{1}{M}\int_{nT-T}^{nT} f(t)dt} \qquad (3.11)$$

However, the integral represents the area under the curve of $f(t)$ in the interval $nT - T \le t \le nT$, and this area is approximately equal to $Tf(nT)$. Hence, (3.11) becomes

$$v(nT) \cong v(nT-T) + \frac{1}{M}Tf(nT) \qquad n = 0, 1, 2, \cdots \qquad (3.12)$$

Chapter 3: Discrete systems 121

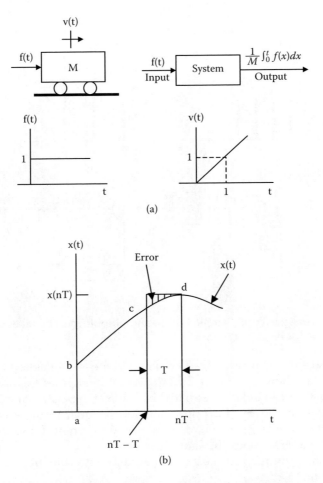

Figure 3.2.2 Analog-to-digital simulation of mechanical system. (a) Mechanical system, step force input and output velocity. (b) Simulation of integration.

In general, the integral

$$y(t) = \int_0^{t=nT} x(t)dt \qquad (3.13)$$

is approximated as follows

$$y(t=nT) = y(nT-T) + \int_{nT-T}^{nT} x(t)dt \cong y(nT-T) + Tx(nT) \qquad (3.14)$$

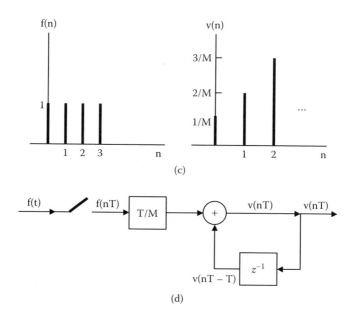

Figure 3.2.2 (continued). (c) Sampled input and output from digital equivalent system. (d) Simulator block diagram.

Observe that the quantity $y(nT - T)$ is equal to the area enclosed by the lines $abc(nT - T)$, as shown in Figure 3.2.2b. The integral gives the area $(nT - T)$ $cd(nT)$. This area can be approximated by the rectangle shown in the figure. The error committed by this approximation is equal to the area enclosed between the curve $x(t)$ and the top side of the rectangle. Observe that as T diminishes, the error does the same.

For an initially relaxed mechanical system with unit step function and $T = 1$, we find

$$v(0) = v(-1) + (1/M)f(0) = 0 + (1/M) = 1/M$$
$$v(1) = v(0) + (1/M)f(1) = (1/M) + (1/M) = 2/M$$
$$v(2) = v(1) + (1/M)f(2) = (2/M) + (1/M) = 3/M$$
$$\vdots$$

These values are shown in Figure 3.2.2c. The numerical process described by (3.12) is shown in Figure 3.2.2d, which is the digital simulator.

Example 3.2.1: Find the output in analog and discrete form of the system shown in Figure 3.2.3a for a unit step function $v(t) = u(t)$. The system is initially relaxed at $t = 0$.

Chapter 3: Discrete systems

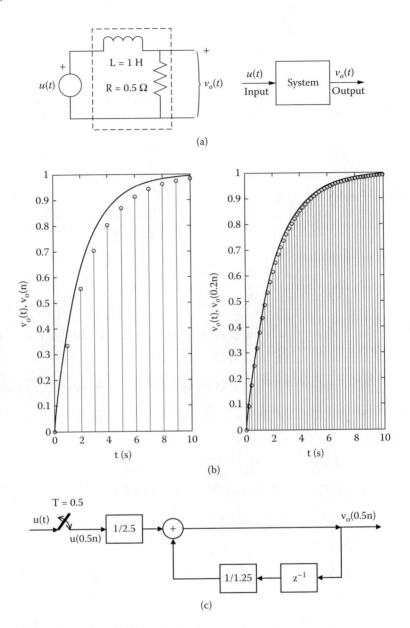

Figure 3.2.3 Illustration of Example 3.2.1.

Solution: The differential equation that describes the system is

$$\frac{di(t)}{dt} + 0.5i(t) = u(t) \quad \text{and} \quad v_o(t) = 0.5i(t)$$

or

$$\frac{dv_o(t)}{dt} + 0.5v_o(t) = 0.5u(t) \qquad (3.15)$$

By an application of the methods previously discussed for solving first-order differential equations, we obtain the following solution:

$$v_o(t) = (1 - e^{-0.5t})u(t) \qquad (3.16)$$

Next, we deduce the equivalent digital format of (3.15) using (3.7). Equation (3.15) is approximated by

$$\frac{v_o(nT) - v_o(nT-T)}{T} + 0.5v_o(nT) = 0.5u(nT)$$

or

$$v_o(nT) - a_1 v(nT-T) = a_1 T 0.5 u(nT) \qquad a_1 = \frac{1}{1+0.5T} \qquad (3.17)$$

To have zero value at $n = 0$, we introduce $n = 0$ in the above equation to find the relation

$$v_o(0T) - a_1 v_o(-T) = 0 - a_1 v_o(-T) = a_1 T 0.5 \quad \text{or} \quad v_o(-T) = -T 0.5 \qquad (3.18)$$

Next, we solve (3.17) by iteration to obtain the general form of solution, which is

$n = 0 \quad v_o(0T) = a_1 v_o(-T) + a_1 T 0.5 = -a_1 T 0.5 + a_1 T 0.5 = 0$

$n = 1 \quad v_o(1T) = a_1 v_o(0T) + a_1 T 0.5 = 0 + a_1 T = a_1 T 0.5$

$n = 2 \quad v_o(2T) = a_1 v_o(1T) + a_1 T 0.5 = a_1^2 T 0.5 + a_1 T 0.5$

$n = 3 \quad v_o(3T) = a_1 v_o(2T) + a_1 T 0.5 = a_1^3 T 0.5 + a_1^2 T 0.5 + a_1 T 0.5$

$n = 4 \quad v_o(4T) = a_1 v_o(3T) + a_1 T 0.5 = a_1^4 T 0.5 + a_1^3 T 0.5 + a_1^2 T 0.5 + a_1 T 0.5$

$\qquad = a_1 T 0.5 (1 + a_1 + a_1^2 + a_1^3)$

$\vdots$

Observe that if $n = N$, then the last set of parentheses will have the form $(1 + a_1 + a_1^2 + a_1^3 + \cdots + a_1^{N-1})$. This expression is a finite term geometric series, and thus, the output voltage is given for any n by

$$v_o(nT) = a_1 T 0.5 \frac{1 - a_1^n}{1 - a_1} \qquad a_1 = \frac{1}{1+0.5T} \qquad (3.19)$$

Chapter 3: Discrete systems

Book MATLAB m-file for example 3.2.1: ex3_2_1

```
%the name of this file is: Ex3_2_1;this is an m-file
%that produces outputs for the Ex. 3.2.1;the reader
%can change value of T and get different desired results;
T=1;      %T=sampling time;
t1=10;    %t1=time on the time axis we would like to plot
          %the functions;
N=t1/T;
a1=1/(1+0.5*T);
n=0:N;
v=a1*T*0.5*(1-a1.^n)/(1-a1);
```

Figure 3.2.3b gives plots of the analog system output and the output from its digital simulator. We observe that the smaller the sampling time introduced, the better approximation we obtain. Figure 3.2.3c shows the block diagram representation of the simulated continuous system for $T = 1$ and $T = 0.2$ sampling times. ∎

Example 3.2.2: Find the output of the analog and equivalent discrete form of the system shown in Figure 3.2.4a. The input to the system is the exponential function $v_i(t) = \exp(-t)u(t)$.

Solution: The integrodifferential equation describing the system is

$$4\int i(t)dt + 2i(t) = v_i(t) \tag{3.20}$$

We make use of the fact that the current through the capacitor is $C\, dv_c/dt$. This current is the same through the circuit. Substituting this expression in the above equation and taking into consideration the properties of the integration of a derivative, we obtain the equation

$$v_c(t) + 0.5\frac{dv_c}{dt} = v_i(t) \tag{3.21}$$

For an exponential decaying input function and relaxed initial conditions, the solution is easily found to be

$$v_c(t) = 2(e^{-t} - e^{-2t}) \quad t > 0 \tag{3.22}$$

Further, the output voltage is

$$v_o(t) = 2e^{-2t} - e^{-t} \quad t > 0 \tag{3.23}$$

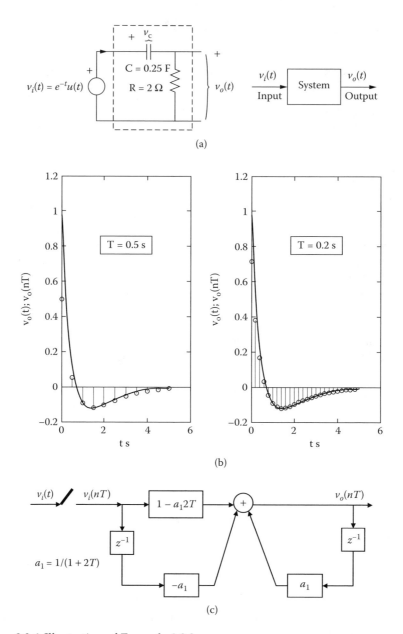

Figure 3.2.4 Illustration of Example 3.2.2.

A sketch of the function is given in Figure 3.2.4b by the solid line. For comparison, digital simulations with sampling times of $T = 0.5$ s and $T = 0.2$ s are also shown. As we have seen in the previous example, the smaller the sampling time, the better approximation we produce.

Next, we approximate (3.21) in its discrete form. The resulting equation is

Chapter 3: Discrete systems

$$v_c(nT) - a_1 v_c(nT-T) = a_1 2T v_i(nT) \qquad a_1 = 1/(1+2T) \qquad (3.24)$$

Also, by inspection of Figure 3.2.4a, we write Kirchhoff's voltage law,

$$v_i(nT) - v_c(nT) - v_o(nT) = 0 \qquad (3.25)$$

Combine these two equations by eliminating $v_c(nT)$ and $v_c(nT-T)$, the latter by appropriately changing the time variable nT to $nT-T$ in (3.25). The result is readily found to be

$$v_o(nT) = a_1 v_o(nT-T) + (1 - a_1 2T) v_i(nT) - a_1 v_i(nT-T) \quad a_1 = 1/(1+2T) \quad (3.26)$$

For the particular case of $T = 0.5$ s, the solution is calculated as follows:

$n = 0 \quad v_o(0) = 0.5 v_o(-0.5) + (1 - 0.5) v_i(0) - 0.5 v_i(-0.5) =$

$\qquad 0.5 \times 0 + 0.5 \times 1 - 0.5 \times 0 = 0.5$

$n = 1 \quad v_o(0.5) = 0.5 v_o(0) + (1 - 0.5) v_i(0.5) - 0.5 v_i(0) =$

$\qquad 0.25 + 0.5 \times 0.6065 - 0.5 \times 1 = 0.0533$

$n = 2 \quad v_o(1) = 0.5 v_o(0.5) + (1 - 0.5) v_i(1) - 0.5 v_i(0.5) =$

$\qquad 0.5 \times 0.0533 + 0.5 \times 0.3679 - 0.5 \times 0.6065 = -0.927$

$$\vdots$$

Figure 3.2.4b has been plotted using the Book MATLAB m-file given below.

Book MATLAB m-file for example 3.2.2: ex_3_2_2

```
%ex_3_2_2= name of book m-file to solve Ex. 3.2.2
T=0.2; %T and N can be changed to fit the desired accuracy;
N=26;
a1=1/(1+2*T);
v(1)=0;
n=2:N;
vi1=exp(-(n-2)*T);
vi=[0 vi1];
for m=2:N
    v(m)=a1*v(m-1)+(1-a1*2*T)*vi(m)-a1*vi(m-1);
end;
vo=v(1,2:N);
```

The reader will easily find that as T approaches zero, the accuracy will become better and better. Figure 3.2.4c gives the block diagram representation of the simulated continuous system. ∎

It is instructive to deduce the equivalent discrete form equation directly from (3.20). We proceed by writing (3.20) in the form

$$4\int_0^{nT-T} i(t)dt + 4\int_{nT-T}^{nT} i(t)dt + 2i(t) = v_i(t)$$

or

$$4q(nT-T) + 4Ti(nT) + 2i(nT) = v_i(nT)$$

But $2i(.)$ is equal to the voltage output $v_o(.)$, and the accumulated charge on the capacitor divided by the capacitance is equal to the voltage across the capacitor, $v_c(t)$. Hence, the above equation becomes

$$4 \times 0.25 \frac{q(nT-T)}{0.25} + (1+2T)[2i(nT)] = v_i(nT)$$

or

$$v_c(nT-T) + (1+2T)v_o(nT) = v_i(nT)$$

But from Kirchhoff's voltage law we have that

$$v_c(t) = v_i(t) - v_o(t)$$

and therefore by shifting the above equation we obtain

$$v_c(nT-T) = v_i(nT-T) - v_o(nT-T)$$

The combination of the last two discrete equations yields (3.26).

Example 3.2.3: Chemical engineering. A large tank contains 81 gal of brine that contains 20 lb of dissolved salt. Additional brine containing 3 lb of dissolved salt per gallon runs into the tank at the rate of 5 gal/min. The mixture, which is stirred to ensure homogeneity, runs out of the tank at the rate of 2 gal/min. How much salt is in the tank at the end of 37 min? Compare results using a continuous-time and discrete-time analysis.

Chapter 3: Discrete systems

Solution: The rate of change of salt is given by

$$\frac{dy(t)}{dt} = \text{rate in-rate out}$$

The rate in is given as

$$\text{rate in} = 3 \text{ (lb/gal)} \times 5 \text{ (gal/min)} = 15 \text{ lb/min}$$

To find the rate out, we must first find the concentration of salt at time *t*, that is, the amount of salt per gallon of the brine in the tank. Since

$$\text{concentration} = \frac{\text{pounds of salt in tank at time t}}{\text{gallons of brine in tank at time t}} = \frac{y(t)}{81+(5-2)t}$$

hence,

$$\text{rate out} = \left[\frac{y(t)}{81+3t}\text{ lbs/gal}\right]2\text{(gal/min)} = \frac{2y(t)}{81+3t}\text{ lbs/min}$$

Therefore, the differential equation that describes the mixture is

$$\frac{dy(t)}{dt} = 15 - \frac{2y(t)}{81+3t}$$

with the initial condition $y(0) = 20$.

To solve this differential equation, we employ the method of **variation of parameters** (see also Appendix 3.1). Based on the solution of the homogeneous equation, we use the multiplying factor (exponent with the positive sign)

$$\exp\left(\frac{2}{3}\int\frac{1}{27+t}dt\right) = \exp\left(\frac{2}{3}[\ln(27+t)]\right) = (27+t)^{2/3}$$

Next, by multiplying the differential equation with the multiplying factor $(27 + t)^{2/3}$, we obtain

$$\frac{dy(t)}{dt}(27+t)^{2/3} + \frac{2}{3}y(t)(27+t)^{-1/3} = 15(27+t)^{2/3}$$

This equation can be written in the form

$$\frac{d}{dt}[y(t)(27+t)^{2/3}] = 15(27+t)^{2/3}$$

Integrate both sides with respect to t to obtain

$$y(t)(27+t)^{2/3} = \frac{15(27+t)^{5/3}}{5/3} + C$$

where C is the constant of integration. This expression is written

$$y(t) = 9(27+t) + C(27+t)^{-2/3}$$

Applying the initial condition, we find that $C = -2007$. The final solution is

$$y(t) = 9(27+t) - 2007(27+t)^{-2/3}$$

For the special time $t = 37$, we find $y(37) = 450.56$ lb.
When this problem is converted into equivalent discrete form by writing $dy(t)/dt = [y(nT) - y(nT-T)]/T$, the discrete equivalent equation becomes

$$y(nT) = \frac{81+3nT}{81+2T+3nT}15T + \frac{81+3nT}{81+2T+3nT}y(nT-T)$$

For $n = 0$, $y(0T) = 20$, and hence, the above equation gives the following expression

$$y(-T) = \frac{20(81+2T)}{81} - 15T$$

for the value of $y(-T)$. The Book MATLAB m-file to solve the discrete simulation is given below.

Book MATLAB m-file for example 3.2.3: ex3_2_3

```
%ex3_2_3 is the name of the m-file to produce
%simulations for the Ex 3.2.3;to find desired results
%the reader can change sampling time T and appropriately
```

```
%N;
T=0.05;
N=500;
y(1)=(20*(81+2*T)/81)-15*T;
for n=0:N
    y(n+2)=((81+3*(n+2)*T)*15*T)/(81+2*T+3*(n+2)*T)...
        +((81+3*(n+2)*T)/(81+2*T+3*(n+2)*T))*y(n+1);
end;
```

For example, the exact value at t = 10 min is 152.2495 lb, and the corresponding values for T = 1, 0.2, and 0.05 are 139.8727, 147.3289, and 151.0224, respectively. ∎

*3.3 Digital simulation of higher-order differential equations

Because more complicated linear systems can be described by higher-order linear differential equations, it is desired to develop the equivalent digital representations of different-order derivatives. This development yields

$$\frac{dy(t)}{dt} \cong \frac{y(nT)-y(nT-T)}{T} \qquad (3.27)$$

$$\frac{d^2y(t)}{dt^2} = \frac{d}{dt}\frac{dy(t)}{dt} \cong \frac{y(nT)-2y(nT-T)+2y(nT-2T)}{T^2} \qquad (3.28)$$

The values of initial conditions are determined from those values specified for the continuous system. For example, given the values of $y(t)$ and $dy(t)/dt$ at t = 0, the required values of $y(-T)$ can be obtained approximately from the relationship

$$\left.\frac{dy(t)}{dt}\right|_{t=0} = \frac{dy(0)}{dt} \cong \frac{y(0T)-y(0T-T)}{T}$$

or

$$y(-T) = y(0) - T\frac{dy(0)}{dt} \qquad (3.29)$$

Following such a procedure, the value of $y(-nT)$ can be obtained from $d^n y(0)/dt^n$ by proceeding from the lower- to higher-order derivatives using those values already found for $y(0T)$, $y(-T)$, ..., $y(-nT + T)$.

Example 3.3.1: Bioengineering (linearization of systems). The system shown in Figure 3.3.1a represents an idealized model of a stiff human limb as a step in assessing the passive control process of locomotive action. It is required to find the movement of the system if the input torque is given by $T(t) = e^{-t}u(t)$. We also assume that friction during the movement is specified by the friction constant D. The initial conditions of the system are zero, that is, $\theta(0) = d\theta(0)/dt = 0$. Compare the analog solution with the corresponding digital simulation.

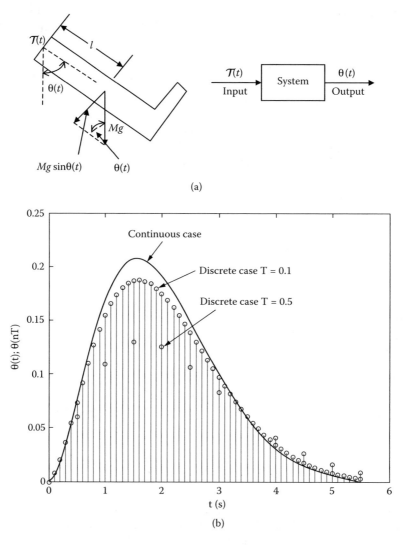

Figure 3.3.1 Model of human limb. (a) Modeling of the movement of a stiff human limb. (b) Continuous- and discrete-time simulation responses for $T = 0.5$ and $T = 0.1$.

Chapter 3:   Discrete systems

Solution: By an application of D'Alembert's principle, which requires that the algebraic sum of torques must be zero at a node, we write

$$\mathcal{T}(t) = \mathcal{T}_g(t) + \mathcal{T}_D(t) + \mathcal{T}_J(t) \qquad (3.30)$$

where

$\mathcal{T}(t) =$ input torque

$\mathcal{T}_g(t) =$ gravity torque $= Mgl\sin\theta(t)$

$\mathcal{T}_D(t) =$ frictional torque $= D\omega(t) = D\dfrac{d\theta(t)}{dt}$

$\mathcal{T}_J(t) =$ inertial torque $= J\dfrac{d\omega(t)}{dt} = J\dfrac{d^2\theta(t)}{dt^2}$

The equation that describes the system is

$$J\frac{d^2\theta(t)}{dt^2} + D\frac{d\theta(t)}{dt} + Mgl\sin\theta(t) = \mathcal{T}(t) \qquad (3.31)$$

This equation is nonlinear owing to the presence of the $\sin\theta(t)$ term in the expression of the gravity torque. To produce a linear equation, we must assume that the deflection is small enough ($\theta < 30°$) such that we can substitute the sine function with its approximate value equal to its angle. Hence, under these conditions (3.31) becomes

$$J\frac{d^2\theta(t)}{dt^2} + D\frac{d\theta(t)}{dt} + Mgl\theta(t) = \mathcal{T}(t) \qquad (3.32)$$

For the specific constants $J = 1$, $D = 2$, and $Mgl = 2$, the above equation becomes

$$\frac{d^2\theta(t)}{dt^2} + 2\frac{d\theta(t)}{dt} + 2\theta(t) = e^{-t}u(t) \qquad (3.33)$$

This is a second-order differential equation; hence, its solution must contain two arbitrary constants, the values of which will be determined from specified initial conditions.

We first find the homogeneous solution from the homogeneous equation. If we assume a solution of the form $\theta_h(t) = Ce^{st}$, the solution requirement is

$$s^2 + 2s + 2 = 0$$

from which we find the roots $s_1 = -1 + j$ and $s_2 = -1 - j$. The homogeneous solution is therefore

$$\theta_h(t) = C_1 e^{s_1 t} + C_2 e^{s_2 t}$$

where C_1 and C_2 are arbitrary constants to be found from the initial conditions.

To find the particular solution, we assume a trial solution of the form $\theta_p(t) = Ae^{-t}$ for $t \geq 0$. By inserting this value in (3.33), we obtain

$$Ae^{-t} - 2Ae^{-t} + 2Ae^{-t} = e^{-t} \quad \text{or} \quad A = 1$$

The total solution is

$$\theta(t) = \theta_h(t) + \theta_p(t) = C_1 e^{s_1 t} + C_2 e^{s_2 t} + e^{-t} \quad \text{for} \quad t \geq 0$$

We now apply the assumed zero initial conditions that require

$$\theta(0) = C_1 + C_2 + 1 = 0$$

$$\frac{d\theta(0)}{dt} = C_1 s_1 + C_2 s_2 - 1 = 0$$

Solving the system for the unknown constants, we obtain

$$C_1 = \frac{1 + s_2}{s_1 - s_2}; \quad C_2 = \frac{1 + s_1}{s_1 - s_2}$$

Introducing the above constants in the total solution, we find the final solution:

$$\theta(t) = -\frac{1}{2} e^{-t} e^{jt} - \frac{1}{2} e^{-t} e^{-jt} + e^{-t} = (1 - \cos t) e^{-t} \quad t \geq 0 \qquad (3.34)$$

The digital simulation of (3.33) is deduced by employing (3.27) and (3.28) in this expression. We obtain

$$\frac{\theta(nT) - 2\theta(nT - T) + \theta(nT - 2T)}{T^2} + 2 \frac{\theta(nT) - \theta(nT - T)}{T}$$

$$+ 2\theta(nT) = e^{-nT} \qquad n = 0, 1, 2, \ldots \qquad (3.35)$$

Chapter 3: Discrete systems

After rearrangement, this equation becomes

$$\theta(nT) = a(2 + 2T)\theta(nT - T) - a\theta(nT - 2T) + aT^2 e^{-nT}$$

$$a = \frac{1}{1 + 2T + 2T^2}, \quad n = 0, 1, 2, \ldots \tag{3.36}$$

Using (3.29), we find that $\theta(-T) = 0$. Next, introducing this value and the initial condition $\theta(0T) = 0$ in (3.36), we obtain $\theta(-2T) = T^2$. Hence, these initial conditions are used for the simulation of the discrete case. The following two Book MATLAB m-files solve the continuous and discrete cases, respectively.

Book MATLAB m-file for the continuous case: ex_3_3_1c

```
%ex_3_3_1c is the name of the m-file to produce
%and plot the continuous case presented in Ex 3.3.1
t=0:0.1:5.5;
th=(1-cos(t)).*exp(-t);
plot(t,th,'k');xlabel('t s');
```

Book MATLAB m-file for the discrete case: ex_3_3_1d

```
%ex_3_3_1d is the name of the m-file for
%the discrete case of the Ex 3.3.1
T=0.1;
N=5.5/T;
a=1/(1+2*T+2*T*T);
thd(2)=0;
thd(1)=T*T;
for n=0:N
   thd(n+3)=a*(2+2*T)*thd(n+2)-a*thd(n+1)+T*T*a*exp(-n*T);
end;
```

Figure 3.3.1b shows both the continuous case and two discrete simulations with $T = 0.5$ and $T = 0.1$. ∎

3.4 Convolution of discrete-time signals

The convolution of continuous signals is defined as follows (see also Chapter 2):

$$g(t) = \int_{-\infty}^{\infty} f(x)h(t-x)dx \tag{3.37}$$

The above equation is approximated as follows:

$$g(t) = \int_{-\infty}^{\infty} f(x)h(t-x)dx = \lim_{T \to 0} \sum_{m=-\infty}^{\infty} \int_{mT-T}^{mT} f(x)h(t-x)dx \cong \sum_{m=-\infty}^{\infty} Tf(mT)h(t-mT)$$

(3.38)

or

$$g(nT) = T \sum_{m=-\infty}^{\infty} f(mT)h(nT-mT) \qquad n = 0, \pm 1, \pm 2, \ldots \quad m = 0, \pm 1, \pm 2, \ldots$$

and for $T = 1$ the above convolution equation becomes

$$g(n) = \sum_{m=-\infty}^{\infty} f(m)h(n-m) \qquad n = 0, \pm 1, \pm 2, \ldots \quad m = 0, \pm 1, \pm 2, \ldots \quad (3.39)$$

Note: *To perform convolution of two sequences, we change n to m in one of the functions and in the other we substitute n – m in place of n. These substitutions mean that the first sequence is identical to the one in the n domain, and the second substitution performs a flipping of the sequence and a shifting by n in the m domain. For each n, we multiply first the corresponding elements of the two sequences, and then add the products. The result is equal to the output function at n.*

The impulse response of a discrete system is defined in the same way as for a continuous-time system. It specifies the output of a digital system $h(n)$ to a delta function excitation:

$$\delta(nT) = \begin{cases} 1 & n = 0 \\ 0 & n \neq 0 \end{cases} \qquad \delta(nT - mT) = \begin{cases} 1 & n = m \\ 0 & n \neq m \end{cases} \quad (3.40)$$

This process is represented by the input–output relationship of a discrete system in the form shown in Figure 3.4.1. Note that if $h(n)$ is the response of a system to $\delta(n)$, then $h(n - m)$ is the response of the system to a shifted delta function $\delta(n - m)$. Clearly, if the input is of the general form

$$\sum_{n=-\infty}^{\infty} \delta(n-m),$$

the output will be of the form

$$\sum_{n=-\infty}^{\infty} h(n-m).$$

Chapter 3: Discrete systems

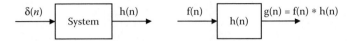

Figure 3.4.1 Diagrammatic representation of the input–output relationship of a discrete system.

For linear time-invariant (LTI) discrete systems, **causality, linearity,** and **stability** are defined in a manner analogous to their definition in continuous systems. Therefore, we expect that we will not observe a response ahead of a discrete input, that the resulting output due to many inputs will be equal to the sum of the outputs to each input alone, and that the infinite sum of impulse functions is produced by a physical source. As in continuous systems, the convolution of discrete sequences obeys the commutative and associative properties.

Suppose that the input function to a system is

$$f(n) = e^{-0.5n}u(n)$$

and the system impulse response is $h(n) = \delta(n)$. The output of this system is

$$g(n) = \sum_{m=-\infty}^{\infty} \delta(n-m)e^{-0.5m}u(m) = \sum_{m=0}^{\infty} \delta(n-m)e^{-0.5m}$$

$$= \delta(n-0)e^{-0.5 \times 0} + \delta(n-1)e^{-0.5 \times 1} + \delta(n-2)e^{-0.5 \times 2} + \cdots$$

The output is

$$g(0) = 1 + 0 + 0 + \cdots \qquad = 1$$
$$g(1) = 0 + e^{-0.5} + 0 + 0 + \cdots \qquad = e^{-0.5}$$
$$g(2) = 0 + 0 + e^{-1} + 0 + \cdots \qquad = e^{-1}$$
$$\vdots$$

We observe that the output is identical with the input, an anticipated result since the impulse response was a delta function.

Example 3.4.1: Determine the output of the system shown in Figure 3.4.2a. The input to the system is the unit step function $u(n)$. Use discrete convolution.

Solution: The application of Kirchhoff's voltage law leads to the differential equation

$$\frac{di(t)}{dt} + 0.5i(t) = v(t) \tag{3.41}$$

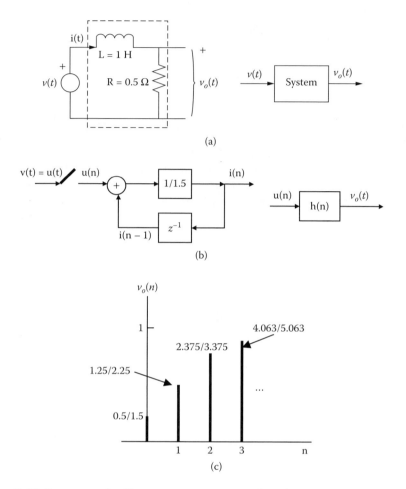

Figure 3.4.2 Response of a discrete system to a step function.

Since we have assumed from the beginning $T = 1$, the corresponding difference equation is

$$i(n) - i(n-1) + 0.5i(n) = u(n) \quad \text{or} \quad 1.5i(n) - i(n-1) = u(n) \qquad (3.42)$$

The equivalent discrete form of the system is shown in Figure 3.4.2b. To apply the convolution, we must first find the impulse response of the system. To do this, we write (3.42) in the form (remember that $h(n)$ has the units of current)

$$h(n) = \frac{1}{1.5}\delta(n) + \frac{1}{1.5}h(n-1)$$

We obtain (since the system is causal, that is, the system reacts after it is excited, $h(-1) = 0$)

Chapter 3: Discrete systems

$$h(0) = \frac{1}{1.5}, \quad h(1) = 0 + \frac{1}{1.5}h(1-1) = \frac{1}{1.5^2}, \quad h(2) = 0 + \frac{1}{1.5}h(2-1) = \frac{1}{1.5^3}, \ldots$$

or

$$h(n) = \frac{1}{1.5^{n+1}}$$

The output of the system is given by

$$v_o(n) = 0.5i(n) = 0.5u(n) * h(n) = 0.5 \sum_{m=-\infty}^{\infty} u(n-m)h(m) = 0.5 \sum_{m=0}^{n} u(n-m) \frac{1}{1.5^{m+1}}$$

From this expression, we deduce the successive values by iteration. Hence,

$$v_o(0) = 0.5 \sum_{m=0}^{0} u(0-m) \frac{1}{1.5^{m+1}} = 0.5 u(0) \frac{1}{1.5} = \frac{0.5}{1.5}$$

$$v_o(1) = 0.5 \sum_{m=0}^{1} u(1-m) \frac{1}{1.5^{m+1}} = 0.5 \left[u(1) \frac{1}{1.5^1} + u(0) \frac{1}{1.5^2} \right]$$

$$= 0.5 \frac{1}{1.5} \left(1 + \frac{1}{1.5} \right) = \frac{1.25}{2.25}$$

$$v_o(2) = 0.5 \sum_{m=0}^{2} u(2-m) \frac{1}{1.5^{m+1}} = 0.5 \left[u(2) \frac{1}{1.5^1} + u(1) \frac{1}{1.5^2} + u(0) \frac{1}{1.5^3} \right]$$

$$= \frac{0.5}{1.5} \left(1 + \frac{1}{1.5} + \frac{1}{1.5^2} \right)$$

$$= \frac{2.375}{3.375}$$

$$\vdots$$

The general expression is of the form

$$v_o(n) = 0.5 \frac{1}{1.5} \left[1 + \frac{1}{1.5} + \frac{1}{1.5^2} + \cdots + \frac{1}{1.5^n} \right] = 0.5 \frac{1}{1.5} \frac{1 - \left(\frac{1}{1.5}\right)^{n+1}}{1 - \frac{1}{1.5}}$$

The result is due to the fact that the expression in parentheses is a finite geometric series. ∎

Example 3.4.2: Repeat Example 3.4.1 with the difference that the sampling time T is included so that the solution can be found for any value of T.

Solution: The corresponding difference equation of (3.41) is

$$i(nT) - ai(nT - T) = aTv(nT), \qquad a = \frac{1}{1 + 0.5T} \qquad (3.43)$$

To apply the convolution, we must first find the impulse response function $h(nT)$ of the system. Hence, we write

$$h(nT) = ah(nT - T) + aT\delta(nT) \qquad a = \frac{1}{1 + 0.5T} \qquad (3.44)$$

In the above equation we have introduced the delta function property:

$$\delta(at) = \frac{1}{a}\delta(t).$$

Since the system is causal, $h(-T) = 0$. Following the procedure of the previous example, the impulse response is

$$h(nT) = Ta^{n+1} \qquad a = \frac{1}{1 + 0.5T} \qquad n = 0, 1, 2, \cdots \qquad (3.45)$$

If we compare the present impulse response with that found above, we see the correspondence between $1/1.5$ and a. Hence, the output voltage is

$$v_o(nT) = 0.5aT \frac{1 - a^{n+1}}{1 - a} \qquad (3.46)$$

To compare the output at $t = 1.5$ for $T = 0.5$ and $T = 0.01$, we introduce these values in the above equation. Hence, the values are 0.5904 and 0.5291, respectively. The exact value at $t = 1.5$ is 0.5276. ∎

Example 3.4.3: Determine the convolution of the two discrete functions shown in Figure 3.4.3a.

Solution: The two functions in Figure 3.4.3b are shown in the m domain for three different values of n. Now, apply the steps required in the convolution formula:

$$g(n) = \sum_{m=0}^{n} [u(m) - u(m - 4)] e^{-(n-m)} u(n - m)$$

Chapter 3: Discrete systems 141

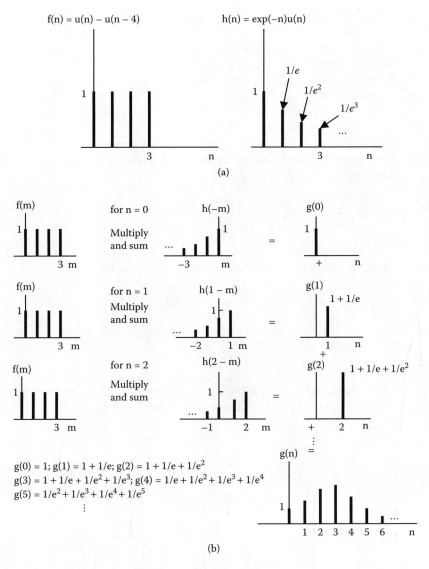

Figure 3.4.3 Illustration of Example 3.4.3.

This can be accomplished by graphical construction. The resulting $g(n)$ is shown in Figure 3.4.3b. ∎

Important definitions and concepts

1. Discrete system
2. Block diagram representation of discrete systems
3. Difference equations

4. Finite duration impulse response system (FIR), or nonrecursive or transversal
5. Infinite duration impulse response (IIR)
6. Digital simulation of analog systems
7. Linearization of nonlinear systems
8. Convolution of sequences
9. Smoothing systems
10. Correlation of sequences
11. Method of variation of parameters (see Appendix 3.1)
12. The approximate Euler's method for solving differential equations (see Appendix 3.2)

Chapter 3 Problems

Section 3.1

1. A signal generator produces a signal specified by

$$v(n) = \begin{cases} 0 & n < 0 \\ 1 + 2n & 0 \leq n \leq 2 \\ 0 & n > 2 \end{cases}$$

With signal source as a basis unit, draw the block diagram of the system that will produce the signal shown in Figure P3.1.1.

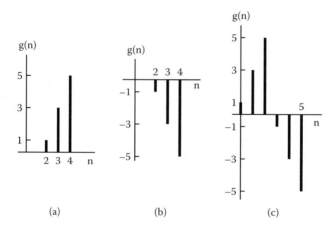

Figure P3.1.1

Chapter 3: Discrete systems

2. A discrete system is shown in Figure P3.1.2a. If the input function is that shown in Figure P3.1.2b, determine the output.

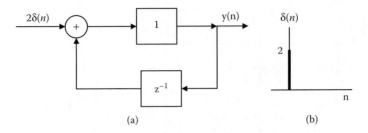

Figure P3.1.2

3. A discrete system is shown in Figure P 3.1.3a. If the input is that shown in Figure P3.1.3b, find the output.

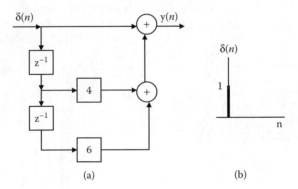

Figure P3.1.3

4. Refer to Figure P3.1.4. Deduce the difference equation and find its solution when the input voltage $v(n)$ is that shown in the same figure. Identify the system.

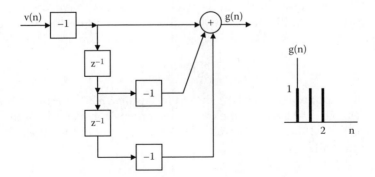

Figure P3.1.4

5. **Smoothing system.** Deduce the difference equation and its solution for the system shown in Figure P3.1.5. What is the effect of this system to the input?

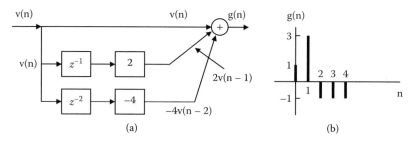

Figure P3.1.5

6. A unit step sequence is applied to the input of the system shown in Figure P3.1.6. Deduce the difference equation and its solution. Identify the nature of the output.

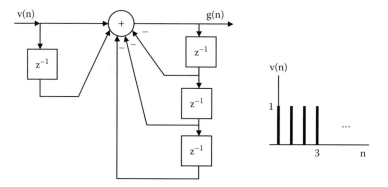

Figure P3.1.6

7. **Banking.** If a family invests $1000 per quarter at 12% interest for 20 years, how much money will the family have accumulated? Assume no initial money was in the account. Interest is compounded quarterly. Draw the block diagram of the system and identify it.

Section 3.2

1. The input to each system shown in Figure P3.2.1 is a ramp function $f(t) = 2r(t)$. Determine the outputs in analog form and in digital form for $T = 0.2$ s. Contrast results. Draw analog and digital block diagram representations.

Chapter 3: Discrete systems

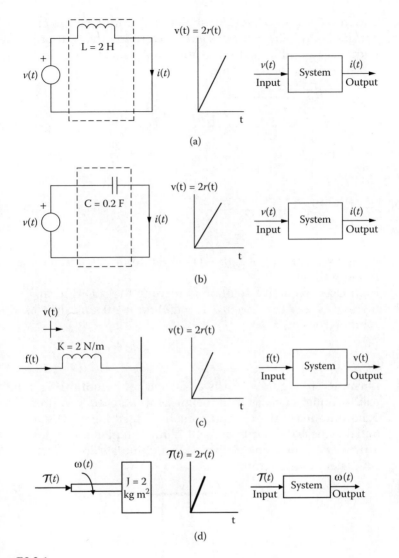

Figure P3.2.1

2. The initially relaxed system shown in Figure P3.2.2 is excited by a current source $i(t) = \exp(-2t)u(t)$. Determine the output voltage in both analog and digital forms for $T = 0.2$ s. Give the block diagram representation for each case.

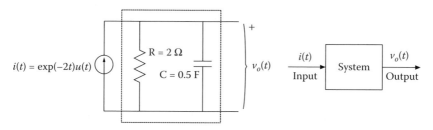

Figure P3.2.2

3. If $g(-1) = 0$ and $g(n) - 2g(n-1) - 4 = 0$ for $n \geq 0$, find a closed-form solution for $g(n)$.
   Hint: Use a recursive solution to identify the closed form.
4. If $g(-1) = 2$ and $g(n) - 2g(n-1) = \delta(n)$ for $n \geq 0$, find the closed-form solution for $g(n)$.

Section 3.3

1. Illustrated in Figure P3.3.1 is a ballistic pendulum that is initially at rest. A bullet of mass m traveling with a speed v is fired into the pendulum mass at $t = 0$ and remains lodged therein. Determine the initial value $\omega(0+)$. Write the controlling equation for angle $\theta(t)$. Draw a network equivalent of the system. Simulate the system with its equivalent discrete form.

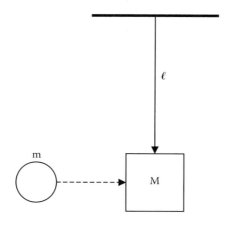

Figure P3.3.1

Chapter 3: Discrete systems

2. Determine the particular solution of the following differential equations:

   a. $(d^2y/dt^2) + y = 2t$ for $t > 0$

      Hint: Assume a solution of the form $y = A + Bt$.

   b. $(d^2y/dt^2) - 4y = \exp(2t)$ for $t > 0$

      Hint: Since $A\exp(2t)$ is a solution of the homogeneous equation, try $y = At\exp(2t)$.

   c. $(d^2y/dt^2) - y = \sin t$ for $t > 0$

      Hint: Assume the solution of the form $y = A\cos t + B\sin t$.

3. The capacitor of the RLC circuit shown in Figure P3.3.3 carries an initial charge of $Q_o = 2$ C. Determine the current in the circuit. Simulate the system with its discrete equivalent.

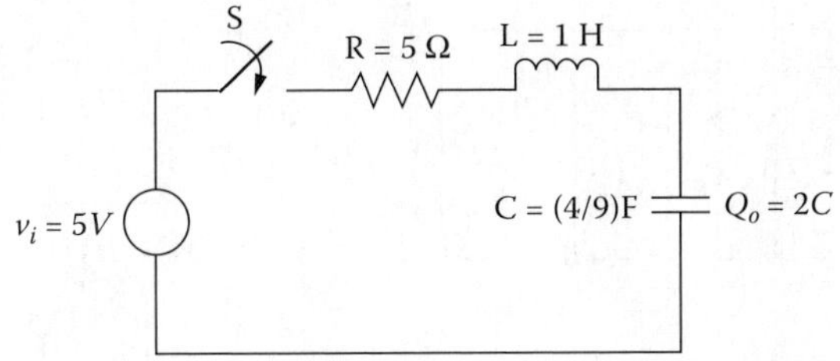

Figure P3.3.3

   Hint: Use the fact that $i = dq/dt$ to solve an equation for q, with $dq(0)/dt = 0$. Use this solution to deduce the equation for $i = dq/dt$. An alternative approach is to differentiate the KVL integrodifferential equation to obtain $L(d^2i/dt^2) + R(di/dt) + (1/C)i = dv_i/dt$. Initial conditions are $q(0) = Q_o =$ constant, then $i(0) = dq(0)/dt = 0$; from the integrodifferential equation, $(di(0)/dt) = (1/L)[v_i(0) - (1/C)q(0)]$.

4. Solve the equations given in Problem 3.3.2 for zero initial conditions: $y(0) = 0$ and $dy(0)/dt = 0$. Then, simulate the equations in digital form and solve for $T = 0.1$. Compare the results.

Section 3.4

1. Determine the convolution of the discrete signals shown in Figure P3.4.1.

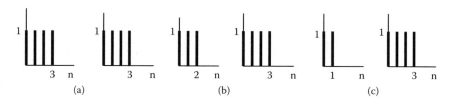

Figure P3.4.1

2. Deduce the output of the system shown in Figure P3.4.2 in discrete form using the convolution method. Assume a step function input $i(t) = u(t)$, a sampling time $T = 0.5$ s, $R = 0.5$ ohms, and $C = 1$ F.

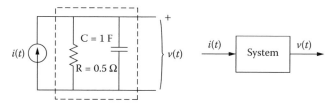

Figure P3.4.2

3. Deduce the convolution of the following pairs of functions:
   a. $f(n) = 3^n u(n)$, $h(n) = 4^n u(n)$
   b. $f(n) = 0.8^n u(n)$, $h(n) = u(n)$
   c. $f(n) = u(n)$, $h(n) = u(n)$

4. Determine the convolutions:
   a. $g(n) = [2^n u(n)]*[2^n u(n)]$
   b. $g(n) = [2^n u(n)]*[2^n u(n)]*[2^n u(n)]$

5. Determine the convolution of the functions shown in Figure P3.4.5.

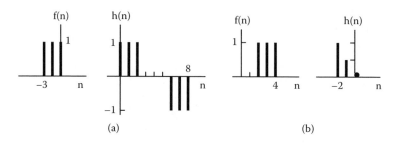

Figure P3.4.5

6. a. Show that

$$[f(n)u(n)]*[u(n-m)] = \sum_{n=0}^{n-m} f(n),$$

where m is a constant integer.

b. Determine the convolution of the functions shown in Figure P3.4.6.

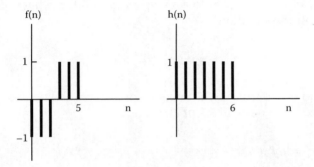

Figure P3.4.6

3. **Correlation of sequences.** The correlation of discrete signals is given by the relation

$$\boxed{g(nT) = f(nT) \odot h(nT) = \sum_{m=-\infty}^{\infty} f(mT)h(mT - nT)}$$

$$g(n) = f(n) \odot h(n) = \sum_{m=-\infty}^{\infty} f(m)h(m - n)$$

Discretize and deduce the correlation between the functions given.

a. $f(t) = p_1(t)$, $h(t) = \exp(-t)u(t)$
b. $f(t) = p_1(t)$, $h(t) = \delta(t - 1) + p_1(t - 3)$

Appendix 3.1: Method of variation of parameters

The method of variation of parameters

For convenience and clarity, we shall restrict our development on the first- and second-order differential equations with constant coefficients. Therefore, a second-order differential equation is of the form

$$a_2 y'' + a_1 y' + a_0 y = Q(t), \quad a_2 \neq 0 \tag{1.1}$$

In the method of undetermined coefficients, introduced in Chapter 2, the function Q(t) was assumed to have a finite number of linear independent derivatives. However, in this method we will assume that Q(t) is a continuous function on an interval I and it is different than zero on I.

The homogeneous equation of (1.1) is

$$a_2 y'' + a_1 y' + a_0 y = 0 \quad (1.2)$$

Let its two independent solutions be y_1 and y_2. With the help of these solutions we form the equation

$$y_p(t) = u_1(t) y_1(t) + u_2(t) y_2(t) \quad (1.3)$$

where u_1 and u_2 are **unknown** functions of t that are to be determined.

The successive derivatives of (1.3) are

$$y'_p = u_1 y'_1 + u'_1 y_1 + u'_2 y_2 + u_2 y'_2 = (u_1 y'_1 + u_2 y'_2) + (u'_1 y_1 + u'_2 y_2) \quad (1.4)$$

$$y''_p = (u_1 y''_1 + u_2 y''_2) + (u'_1 y'_1 + u'_2 y'_2) + (u'_1 y_1 + u'_2 y_2)' \quad (1.5)$$

Substituting the above values of y_p, y'_p, and y''_p in (1.1), we see that y_p will be a solution of (1.1) if, after some simple algebraic manipulation, the relation below is satisfied.

$$u_1(a_2 y''_1 + a_1 y'_1 + a_0 y_1) + u_2(a_2 y''_2 + a_1 y'_2 + a_0 y_2) + a_2(u'_1 y'_1 + u'_2 y'_2) \\ + a_2(u'_1 y_1 + u'_2 y_2)' + a_1(u'_1 y_1 + u'_2 y_2) = Q(t) \quad (1.6)$$

Since y_1 and y_2 are assumed to be solutions to homogeneous equation (1.2), the quantities in the first two parentheses in (1.6) are equal to zero. The remaining three terms will be equal to $Q(t)$ if we choose u_1 and u_2 such that

$$u'_1 y_1 + u'_2 y_2 = 0$$

$$u'_1 y'_1 + u'_2 y'_2 = \frac{Q(t)}{a_2} \quad (1.7)$$

The above equations can be solved for u'_1 and u'_2 in terms of the other functions by the ordinary algebraic methods. Or, using the determinant approach, the solutions of (1.7) are

Chapter 3: Discrete systems

$$u_1' = \frac{\begin{vmatrix} 0 & y_2 \\ \dfrac{Q(t)}{a_2} & y_2' \end{vmatrix}}{\begin{vmatrix} y_1 & y_2 \\ y_1' & y_2' \end{vmatrix}} = \frac{-y_2' \dfrac{Q(t)}{a_2}}{y_1 y_2' - y_1' y_2}, \quad u_2' = \frac{\begin{vmatrix} y_1 & 0 \\ y_1' & \dfrac{Q(t)}{a_2} \end{vmatrix}}{\begin{vmatrix} y_1 & y_2 \\ y_1' & y_2' \end{vmatrix}} = \frac{\dfrac{Q(t)}{a_2} y_1}{y_1 y_2' - y_1' y_2} \quad (1.8)$$

Equation (1.8) will always give solutions to u_1' and u_2', provided the denominator is different than zero.

Example A3.1.1: Find the general solution of

$$y'' - 3y' + 2y = \sin(e^{-t}) \quad (1.9)$$

Solution: We observe that this equation cannot be solved by the method of undetermined coefficients. We also see that $Q(t) = \sin(e^{-t})$ has an infinite number of linearly independent derivatives. The roots of the characteristic equation are $m = 1$, $m = 2$. Hence, the complementary function (or homogeneous solution) of (1.9) is

$$y_c = c_1 e^t + c_2 e^{2t} \quad (1.10)$$

The two independent solutions of the related homogeneous equation of (1.9) are

$$y_1 = e^t \quad \text{and} \quad y_2 = e^{2t} \quad (1.11)$$

Substituting the above values and their derivatives in (1.7), we obtain, with $a_2 = 1$ (a_2 is the coefficient of y''),

$$\begin{aligned} u_1' e^t + u_2' e^{2t} &= 0 \\ u_1' e^t + u_2'(2e^{2t}) &= \sin(e^{-t}) \end{aligned} \quad (1.12)$$

Solving (1.12) for u_1' and u_2', we find

$$u_1' = -e^{-t} \sin(e^{-t}), \quad u_2' = e^{-2t} \sin(e^{-t}) \quad (1.13)$$

Therefore, the values of u_1 and u_2 are

$$u_1 = \int \sin(e^{-t})(-e^{-t})dt = -\cos(e^{-t}),$$

$$u_2 = -\int e^{-t}\sin(e^{-t})(-e^{-t})dt = -\sin(e^{-t}) + e^{-t}\cos(e^{-t})$$

(1.14)

where we set $u = e^{-t}$ and $du = -e^{-t}dt$. Substituting (1.14) and (1.11) in (1.3), we obtain

$$y_p = -e^{2t}\sin(e^{-t})$$

(1.15)

Therefore, the total solution is

$$y = y_c + y_p = c_1 e^t + c_2 e^{2t} - e^{2t}\sin(e^{-t})$$

(1.16)

The unknown constants will be determined from the initial conditions.

■

Appendix 3.2: Euler's approximation for differential equations

Introduction

There are numerous numerical methods for solving differential equations. In this chapter we have introduced one of these methods. Since our treatment is introductory, we shall omit details on accuracy of results and how much error is involved in the approximation. In our case, we just decrease the sampling time and decide on it based on how much the values change from one T to the other.

In our present development, we assume that the differential equation in question is of the form $y' = f(t,y)$. We note that all differential equations can be written in this form since higher-order ones can be decomposed as a set of first-order differential equations. Therefore, to avoid cumbersome treatment of integration constants, we develop the **initial value problems** (IVPs). Thus, we investigate the following IVPs:

$$\frac{dy(t)}{dt} = f(t,y) \quad \text{or} \quad y' = f(t,y)$$

(2.1)

$$y(t_0) = y_0$$

(2.2)

Let assume that the solution to the IVP is the curve shown in Figure A3.2.1. Suppose that we want to know the coordinates of the point (t_1, y_1). In case

Chapter 3: Discrete systems

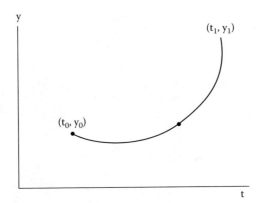

Figure A3.2.1

we know explicitly the equation of y as a function of t, it is a simple matter of knowing t_1. The following question arises: Is it possible in the absence of an explicit form of $y(t)$ to determine y_1 knowing t_1? The answer to this question is the goal of any numerical method. We will say that we have a numerical solution of our differential equation in the interval $[t_0, t_b]$ when we have obtained a set of values $[(t_0, y_0), (t_1, y_1), \ldots, (t_n, y_n)]$, where $t_n = b$ and y_i is an approximation to the solution at each point of t.

Euler method

Referring to Figure A3.2.2, we can consider that in transferring from the initial point (t_0, y_0) to nearby point (t_1, y_1) the following changes occur: $\Delta t = t_1 - t_0$, $\Delta y = y_1 - y_0$. The Euler method is to approximate Δy by its differential dy. Hence, we write

$$dy = \frac{dy}{dt} dt = f(t,y)dt \cong f(t,y)\Delta t$$

To obtain an accurate approximation, it is reasonable to make t as small as possible. Given the IVPs (2.1) and (2.2) and the specific value t_1, y_1 can be found using the Euler method:

$$y_1 = y_0 + f(t_0, y_0)\Delta t = y_0 + f(t_0, y_1)(t_1 - t_0)$$

Thus, one can calculate y_1 directly from the information given. Next, we find y_2 using the relationship

$$y_2 = y_1 + f(t_1, y_1)\Delta t = y_1 + f(t_1, y_1)(t_2 - t_1) \qquad t_2 > t_1$$

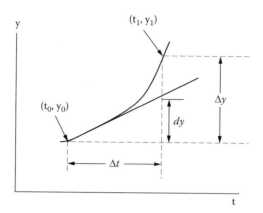

Figure A3.2.2

This process can then be repeated to obtain additional pairs of points. Usually, this process is carried out in a specified interval with prescribed spacing between the t coordinates. The spacing is usually taken to be uniform and equal to T. If it is desired to obtain n points in addition to the initial point, then T is given by the formula

$$T = \frac{b - t_0}{n}$$

where b is the value of the end of the desired range of t. When the sequence $\{y_i\}$ is plotted vs. $\{t_i\}$, hopefully it is very close to the solution curve.

The Euler method can be summarized as follows:

$$t_i = t_{i-1} + T \qquad i = 1, 2, 3, \cdots, n \qquad (2.3)$$

$$y_i = y_{i-1} + f(t_{i-1}, y_{i-1})T \qquad i = 1, 2, 3, \cdots, n \qquad (2.4)$$

Example A3.2.1: Use the Euler method to solve numerically the IVP problem

$$y' = x^2 + y^2, \qquad y(0) = 0, \qquad 0 \le t \le 1 \qquad (2.5)$$

Solution: If we desire $n = 5$, then the sampling time $T = (1 - 0)/5 = 0.2$. Hence, $t_2 = 0 + 0.2 = 0.2$. Thus, $y_1 = y_0 + (t_0^2 + y_0^2)0.2 = 0$, $y_2 = y_1 + (t_1^2 + y_1^2)0.2 = 0.008$, $y_3 = 0.040013$, and $y_4 = 0.112333$. ∎

Chapter 3:   Discrete systems

Book m-File for the Euler Method: exA_3_2_1

```
%exA_3_2_1 is the name of the m file to solve
%ExA3.2.1
T=(1-0)/5;
y(1)=0;
t=0:T:1;
for i=1:5
   y(i+1)=y(i)+(t(i)^2+y(i)^2)*T;
end;
```

chapter 4

Periodic continuous signals and their spectrums

The study of periodic functions has had a long history. Bernoulli, around the middle of the 18th century, suggested that the physical motion of a clamped string, such as those in stringed instruments, could be represented by linear combinations of normal modes (sinusoids). At about the same time, Lagrange strongly criticized the use of trigonometric series, arguing that it was not possible to represent functions that contain edges with such series. It was a half century later that Jean Batiste Joseph Fourier, in conducting his studies on heat diffusion and propagation, developed the series now carrying his name that mathematically addressed such problems. His revolutionary discoveries have had a major impact on the development of mathematics. Fourier series are used extensively today in the fields of science and engineering.

As we will see in this chapter, many types of nonsinusoidal periodic functions can be represented as the sum of periodic complex exponential functions, or sinusoids, a property noted in the next section. This important concept plays a significant role in our general studies. In particular, we will find that the output of a linear time-invariant (LTI) system to a periodic input signal composed of sinusoids is the sum of these same sinusoids, each of which is scaled and time shifted.

The representation of periodic functions in sampled form, a requirement when using a computer and digital signal processing, will also be examined. Further, the conditions necessary for an acceptable approximation that leads to the discrete Fourier transform (DFT) series will be established.

4.1 Complex functions

First, we shall consider the complex form of sinusoidal signals, which are useful as building blocks or basis functions from which we can construct other signals. We shall consider both continuous- and discrete-time functions.

Continuous-time signals

A general complex signal is of the form

$$y(t) = ae^{bt} \qquad (4.1)$$

where a and b are complex numbers in general. If a and b are real numbers, we have the well-known exponential functions. If a is a constant and b is a complex number, $b = c + jd$, we have the **complex exponential** of the form

$$y(t) = ae^{(c+jd)t} = ae^{ct}e^{jdt} \qquad (4.2)$$

A very important signal in our subsequent work is obtained by setting in (4.1) $a = 1$ and $b = j\omega_0$. This signal is

$$y(t) = e^{j\omega_0 t} = \cos\omega_0 t + j\sin\omega_0 t \qquad (4.3)$$

where use is made of the Euler relation (see Chapter 1). The signal $y(t)$ can also be written in the form

$$y(t) = \sqrt{\cos^2\omega_0 t + \sin^2\omega_0 t}\, e^{j\tan^{-1}(\sin\omega_0 t/\cos\omega_0 t)} \qquad (4.4)$$

The frequency is $\omega_0 = 2\pi f_0 = 2\pi/T$, where T is the **period**.

A feature of the exponential signal is made evident by writing $t = t + T$ in (4.3). Hence,

$$\begin{aligned}y(t+T) &= e^{j\omega_0(t+T)} = e^{j\omega_0 t}e^{j\omega_0 T} = e^{j\omega_0 t}e^{j2\pi}\\ &= e^{j\omega_0 t}(\cos 2\pi + j\sin 2\pi) = e^{j\omega_0 t} = y(t)\end{aligned} \qquad (4.5)$$

This shows that $y(t)$ is periodic with period T. Since T is the smallest time t that makes $\exp(j\omega_0 t)$ equal to 1, it is called the **fundamental period**. Further, if we plot $\exp(j\omega_0 t)$ on the complex plane as t varies, we observe a unit vector that rotates counterclockwise at the rate $\omega_0/2\pi$ rev/s. Figure 4.1.1 shows the function $\exp(j\omega_0 t)$ in the complex plane for the particular value $\omega_0 = 4\pi$ rad/s and for different values of t. Similarly, the function $\exp(-j\omega_0 t)$ is also periodic, but it rotates in the clockwise direction as the time t increases.

A signal that is closed related to the complex one is the sinusoid of the form

$$y(t) = A\cos(\omega_0 t + \phi) \qquad (4.6)$$

Chapter 4: Periodic continuous signals and their spectrums

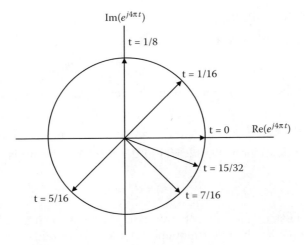

Figure 4.1.1 Complex function $e^{j4\pi t}$ at different values of t.

This can also be written in the form

$$y(t) = \frac{Ae^{j(\omega_0 t + \phi)} + Ae^{-j(\omega_0 t + \phi)}}{2} = A\,\text{Re}\left\{e^{j(\omega_0 t + \phi)}\right\} \quad (4.7)$$

where Re{} denotes the real part of the complex expression in the braces.

As already suggested, we can construct periodic nonsinusoidal signals using sinusoids at frequencies that are multiples of the **fundamental frequency** ω_0 (angular frequency). These sinusoids are of the form

$$y_n(t) = e^{jn\omega_0 t} \qquad n = 0, \pm 1, \pm 2, \cdots \quad (4.8)$$

We observe that for $n = 0$, $y(0) = 1$, a constant. Further, for any n, the signal exp $(jn\omega_0 t)$ has a frequency $|n|\omega_0/2\pi$ Hz and fundamental period $2\pi/(|n|\omega_0) = T/n$.

An important feature of the complex functions $\exp(jn\omega_0 t)$ for integral values of n ($-\infty < n < \infty$) is that these functions constitute an orthogonal set over a full period (see Chapter 1). This means that

$$\int_0^T e^{jn\omega_0 t}(e^{jm\omega_0 t})^* \, dt = \int_0^T e^{j(n-m)\omega_0 t} \, dt = \begin{cases} T & n = m \\ 0 & n \neq m \end{cases} \quad (4.9)$$

Also, $\sin n\omega_0 t$ and $\cos n\omega_0 t$ constitute orthogonal sets, with

$$\int_0^T \cos n\omega_0 t \cos m\omega_0 t \, dt = \begin{cases} \dfrac{T}{2} & n = m \\ 0 & n \neq m \end{cases} \quad (4.10)$$

$$\int_0^T \sin n\omega_0 t \sin m\omega_0 t \, dt = \begin{cases} \dfrac{T}{2} & n = m \\ 0 & n \neq m \end{cases} \qquad (4.11)$$

These properties are of importance in the development of the Fourier series.

Discrete-time signals

The equivalent exponential function in the discrete domain is

$$y(n) = e^{jn\tilde{\omega}} \qquad (4.12)$$

In this form, $\tilde{\omega}$ stands for discrete frequency and has the units of radians per unit. If we consider the new frequency $\tilde{\omega} + 2\pi$, we obtain

$$e^{j(\tilde{\omega}+2\pi)n} = e^{j\tilde{\omega}n} e^{jn2\pi} = e^{jn\tilde{\omega}} \qquad (4.13)$$

This result indicates that the exponential function has the same value at $\tilde{\omega}$ and at $\tilde{\omega} + 2\pi$. In fact, the functions are the same at $\tilde{\omega}$, $\tilde{\omega} + 2\pi$, $\tilde{\omega} + 4\pi$, and so forth. We thus conclude for discrete exponential signals that we need consider only the interval of length 2π, $0 \leq \tilde{\omega} \leq 2\pi$ or $-\pi \leq \tilde{\omega} < \pi$. This conclusion is different from that of the continuous case, in which we have a different function for any distinct frequency. Figure 4.1.2 plots the function $\cos(n\tilde{\omega})$ as $\tilde{\omega}$ increases. Observe that the function starts repeating itself as $\tilde{\omega}$ increases beyond the value 2π. Observe that the fourth and sixth plots are identical since $\pi/2$ corresponds to $5\pi/2 = 2\pi + (\pi/2) - 2\pi$.

Note: *As the frequency of an analog sinusoid increases to infinity, the undulations also increase to infinity. On the contrary, the digitized sinusoids repeat themselves every period. For $T = 1$, the period is 2π, and for any T, the period is $2\pi/T$.*

For the discrete function $\exp(j\tilde{\omega}n)$ with period $N > 0$, we must have

$$e^{j\tilde{\omega}(n+N)} = e^{j\tilde{\omega}n} e^{j\tilde{\omega}N} \qquad (4.14)$$

which, for periodicity, requires that

$$e^{j\tilde{\omega}N} = 1 \qquad (4.15)$$

This requires that

$$\tilde{\omega}N = n2\pi \quad \text{or} \quad \frac{\tilde{\omega}}{2\pi} = \frac{n}{N} \quad n = \text{integer} \qquad (4.16)$$

This expression indicates that the ratio $\tilde{\omega}/2\pi$ must be a rational number for the exponential function $\exp(j\tilde{\omega}n)$ to be periodic.

Chapter 4: Periodic continuous signals and their spectrums 161

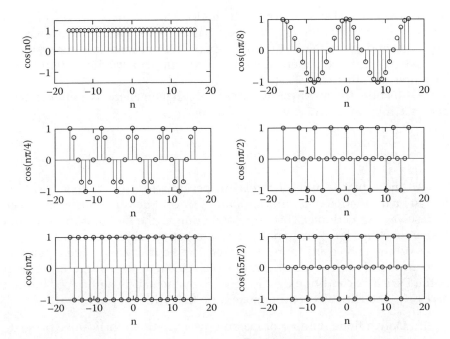

Figure 4.1.2 Discrete sinusoidal function for different frequencies.

When we wish to represent a continuous sinusoidal function by its discrete form representation, we must investigate whether the new function is periodic. For example, the continuous function exp($j\omega t$) can be digitized to the form

$$f(nT) = e^{j\omega nT} = e^{j(\omega T)n} \tag{4.17}$$

which selects values of $f(t)$ at time intervals $t = nT$.

Note: *The frequency ω has the units of rad/s. However, the the frequency T has the units rad/unit and is known as the **discrete frequency**. If the value of T is 1, then the frequency ω is automatically a discrete frequency.*

Comparing (4.17) and (4.15), we must conclude that the digitized function is periodic if

$$\frac{\omega T}{2\pi} = \text{rational number} \tag{4.18}$$

This expression shows that although exp($j\omega t$) is periodic, exp($j\omega nT$) may not be.

4.2 Fourier series of continuous functions

One of the most frequent features of natural phenomena is **periodicity**. A wide range of periodic phenomena exist, such as the audible note of a mosquito, the beautiful patterns of crystal structures, the periodic vibrations of musical instruments, and acoustic and electromagnetic waves of most types. In all these phenomena, there is a pattern of displacement that repeats itself over and over again in time or in space. In some instances, for example, in wave patterns, the pattern itself moves in time. A periodic pattern might be simple or rather complicated.

A function $f(t)$ (a physical pattern or phenomenon) is periodic with period T if its value satisfies the condition

$$f(t+T) = f(t) \quad \text{or} \quad f(t+nT) = f(t) \quad n = \pm 1, \pm 2, \pm 3, \cdots \quad (4.19)$$

for any t and n.

An important feature of a general periodic function is that it can be represented in terms of an infinite sum of sine and cosine functions, which themselves are periodic. This is possible due to the fact that the sine and cosine functions are orthogonal under integration within their period. This series of sine and cosine terms is known as a **Fourier series**. For a periodic function to be Fourier series transformable, it must possess properties known as the **Dirichlet conditions**, which ensure mathematical sufficiency but not necessity. The Dirichlet conditions require that within a period:

1. Only a finite number of maximums and minimums can be present.
2. The number of discontinuities must be finite.
3. The discontinuities must be bounded. That is, the function must be absolutely integrable, which requires that

$$\int_0^T |f(t)| \, dt < \infty.$$

Fourier series in complex exponential form

Any periodic signal $f(t)$ having period T and satisfying the Dirichlet conditions can be expressed in a series form given below:

$$
\begin{aligned}
f(t) &= \sum_{n=-\infty}^{\infty} a_n e^{jn\omega_0 t} = \sum_{n=-\infty}^{\infty} |a_n| e^{j(n\omega_0 t + \phi_n)} \quad &\text{a)} \\
a_n &= \frac{1}{T} \int_{t_0}^{t_0+T} f(t) e^{-jn\omega_0 t} dt = \text{complex constant} \\
&= |a_n| e^{j\phi_n} = |a_n| \cos\phi_n + j|a_n| \sin\phi_n \quad &\text{b)} \\
\omega_0 &= \frac{2\pi}{T} = \text{fundamental frequency} \\
t_0 &= \text{arbitrary value of } t \\
T &= \text{period}
\end{aligned} \quad (4.20)
$$

Observe that the amplitudes a_n of the component terms in the Fourier series expansion of $f(t)$ are determined from the given $f(t)$ using a range of t equal to the period T. Correspondingly, $f(t)$ is determined from the knowledge of the complex constants a_n. Often, these two related equations are referred to as a **Fourier series transform pair**; that is, knowing one equation, the second can be found, and vice versa. In addition, any periodic function that satisfies the Dirichlet conditions and is written in a Fourier series expansion converges at any point t to $f(t)$ provided that the function is continuous at t. If $f(t)$ is discontinuous at $t = t_0$, the function $f(t_0)$ will converge to $f(t_0) = [f(t_0+) + f(t_0-)]/2$, the mean value at the point of discontinuity (the arithmetic mean of the left-hand and right-hand limits). If $f(t)$ is real, then

$$a_{-n} = \frac{1}{T}\int_{t_0}^{t_0+T} f(t)e^{jn\omega_0 t}dt = \left[\frac{1}{T}\int_{t_0}^{t_0+T} f(t)e^{-jn\omega_0 t}dt\right]^* = a_n^* \qquad (4.21)$$

This result, when combined with (4.20a), yields

$$f(t) = a_0 + \sum_{n=1}^{\infty}[(a_n + a_n^*)\cos n\omega_0 t + j(a_n - a_n^*)\sin \omega_0 t] \qquad (4.22)$$

Example 4.2.1: Find the coefficients a_n for the periodic function shown in Figure 4.2.1a and plot $f(t)$ using different number of factors of the sum.

Solution: Apply (4.20b) to obtain

$$a_n = \frac{1}{3.5}\int_{-0.5}^{3.5} f(t)e^{-jn\omega_0 t}dt = \frac{1}{3.5}\left[\int_{-0.5}^{1} 1e^{-jn\omega_0 t}dt + \int_{1}^{3} 0 e^{-jn\omega_0 t}dt\right]$$

$$= \frac{1}{3.5(-jn\omega_0)}e^{-jn\omega_0 t}\bigg|_{-0.5}^{1} = \frac{1}{-j3.5n\omega_0}(e^{-jn\omega_0} - e^{j0.5n\omega_0})$$

Note: *The j in the integration is treated as another constant.*

For the values a_{-n}, the complex conjugate values of a_n, it is necessary to change the sign in front of each j in the equation for a_n. The result is

$$f(t) = \frac{3}{7} + \sum_{n=1}^{\infty}\left(\left[\frac{1}{-j3.5n\omega_0}(e^{-jn\omega_0} - e^{j0.5n\omega_0}) + \frac{1}{j3.5n\omega_0}(e^{jn\omega_0} - e^{-j0.5n\omega_0})\right]\cos n\omega_0 t\right.$$

$$\left.+ j\left[\frac{1}{-j3.5n\omega_0}(e^{-jn\omega_0} - e^{j0.5n\omega_0}) - \frac{1}{j3.5n\omega_0}(e^{jn\omega_0} - e^{-j0.5n\omega_0})\right]\sin n\omega_0 t\right)$$

$$= \frac{3}{7} + \sum_{n=1}^{\infty} \left[\frac{2}{3.5n\omega_0} (\sin n\omega_0 + \sin 0.5n\omega_0) \cos n\omega_0 t - \frac{2}{3.5n\omega_0} \right.$$

$$\left. (\cos n\omega_0 - \cos 0.5n\omega_0) \sin n\omega_0 t \right]$$

$$= \frac{3}{7} + \sum_{n=1}^{\infty} \left[\frac{4}{3.5n\omega_0} [(\sin 0.75n\omega_0 \cos 0.25n\omega_0) \cos n\omega_0 t \right.$$

$$\left. + (\sin 0.75n\omega_0 \sin 0.25n\omega_0) \sin n\omega_0 t] \right]$$

$$= \frac{3}{7} + \sum_{n=1}^{\infty} A_n \cos n\omega_0 t + B_n \sin n\omega_0 t \qquad (4.23)$$

The plot in Figure 4.2.1b shows the results of (4.23) for $n = 3$, $n = 15$, and $3 \leq n \leq 15$.

Book MATLAB m-file: four_ser

```
%four-ser=this is the name of an m-file to
%compute the fourier series;
%the reader must change a0,T,n,an,and bn for
%each case;
a0=3/7;
T=3.5;
om=2*pi/T;
n=5:15;
t=0:0.01:1.5*T;
an=(4./(3.5*n*om)).*(sin(0.75*n*om).*cos(0.25*n*om));
bn=(4./(3.5*n*om)).*(sin(0.75*n*om).*sin(0.25*n*om));
f=a0+cos(t'*n*om)*an'+sin(t'*n*om)*bn';
        %f is the desired function and to plot
        %it we just input plot(t,f);
        %note that cos (t'*n*om) is a matrix;
```

We observe from Figure 4.2.1b that an overshoot occurs at each point of the discontinuity. It was shown by Gibbs that the overshoot remains at about 10% at each edge, regardless of the number of terms included in the expansion, although the rest of the undulations decrease with increasing n. This property is known as the **Gibbs' phenomenon**. Observe that as we add more and more terms (higher frequencies), we recapture better the abrupt changes of the signal. This tells us that the high frequencies in the expansion contribute to building the discontinuities of the function. Thus, if a periodic

Chapter 4: Periodic continuous signals and their spectrums 165

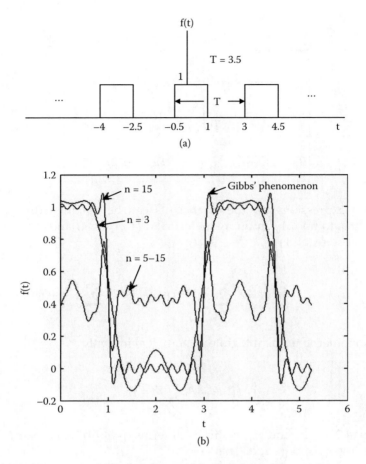

Figure 4.2.1 A square periodic function and its Fourier representation.

nonsinusoidal signal is applied to a system and if the higher frequencies in the expansion are attenuated more than the lower frequencies, the output signal will be smoother than the original. This is obvious from the plot for $n = 3$ and $n = 15$. This feature of frequency-selective attenuation is very important in our studies of signals and systems. This procedure is known as **filtering**. ∎

Fourier series in trigonometric form

The Fourier series expansion given by (4.22) can be written

$$f(t) = A_0 + \sum_{n=1}^{\infty} (A_n \cos n\omega_0 t + B_n \sin n\omega_0 t) \qquad (4.24)$$

where

$$A_0 = a_0 = \frac{1}{T}\int_{t_0}^{t_0+T} f(t)dt \qquad \text{a)}$$

$$A_n = (a_n + a_n^*) = \frac{2}{T}\int_{t_0}^{t_0+T} f(t)\cos n\omega_0 t\, dt \qquad \text{b)} \qquad (4.25)$$

$$B_n = j(a_n - a_n^*) = \frac{2}{T}\int_{t_0}^{t_0+T} f(t)\sin n\omega_0 t\, dt \qquad \text{c)}$$

These expressions for the coefficients can be obtained directly from (4.24). Thus, to find A_0, multiply all terms in (4.24) by dt and integrate over a period. The result is

$$\int_0^T f(t)dt = A_0 \int_0^T dt + A_n \sum_{n=1}^\infty \int_0^T \cos n\omega_0 t\, dt + B_n \sum_{n=1}^\infty \int_0^T \sin n\omega_0 t\, dt$$

The trigonometric terms integrate to zero, leaving only

$$\int_0^T f(t)dt = A_0 \int_0^T dt = A_0 T$$

which is (4.25a). To find A_n, multiply all terms in (4.24) by $\cos(m\omega_0 t)\, dt$ and integrate over a full period. This yields

$$\int_0^T f(t)\cos m\omega_0 t\, dt = A_0 \int_0^T \cos m\omega_0 t\, dt + \sum_{n=1}^\infty A_0 \int_0^T \cos n\omega_0 t \cos m\omega_0 t\, dt$$

$$+ \sum_{n=0}^\infty B_n \int_0^T \sin n\omega_0 t \cos m\omega_0 t\, dt$$

By the orthogonality relationships, every integral on the right-hand side becomes zero except the single expression for $n = m$ in the form

$$\int_0^T f(t)\cos n\omega_0 t\, dt = A_n \int_0^T \cos^2 n\omega_0 t\, dt = A_n \frac{T}{2}$$

Similarly, by multiplying all terms in (4.24) by $\sin(m\omega_0 t)dt$ and integrating over a full period, (4.25c) results.

Chapter 4: Periodic continuous signals and their spectrums

We can write (4.24) in a slightly different form:

$$f(t) = A_0 + \sum_{n=1}^{\infty} A_n \left(\cos n\omega_0 t + \frac{B_n}{A_n} \sin n\omega_0 t \right)$$

$$= A_0 + \sum_{n=1}^{\infty} A_n (\cos n\omega_0 t - \tan \phi_n \sin n\omega_0 t)$$

$$= A_0 + \sum_{n=1}^{\infty} \frac{A_n}{\cos \phi_n} \cos(n\omega_0 t + \phi_n) \quad (4.26)$$

$$= A_0 + \sum_{n=1}^{\infty} C_n \cos(n\omega_0 t + \phi_n)$$

where

$$\phi_n = -\tan^{-1}(B_n / A_n) \qquad C_n = (A_n^2 + B_n^2)^{1/2} \quad (4.27)$$

The foregoing results are summarized in Table 4.2.1.

Example 4.2.2: Develop the trigonometric forms of the Fourier series for the periodic function given in Example 4.2.1.

Solution: By (4.25), we find

$$A_n = \frac{2}{3.5} \int_{-0.5}^{3} f(t) \cos n\omega_0 t \, dt = \frac{2}{3.5} \int_{-0.5}^{1} \cos n\omega_0 t \, dt$$

$$= \frac{2}{3.5 n\omega_0} (\sin n\omega_0 + \sin 0.5 n\omega_0)$$

$$A_0 = \frac{2}{3.5} \int_{-0.5}^{1} 1 \, dt = \frac{3}{3.5}$$

$$B_n = \frac{2}{3.5} \int_{-0.5}^{1} \sin n\omega_0 t \, dt = \frac{2}{3.5 n\omega_0} (\cos 0.5 n\omega_0 - \cos n\omega_0)$$

Using the above values, we can easily verify (4.23). Next, using (4.27) we obtain

$$C_n = \left\{ \left[\frac{2}{3.5 n\omega_0} (\sin n\omega_0 + \sin 0.5 n\omega_0) \right]^2 + \left[\frac{2}{3.5} (\cos 0.5 n\omega_0 - \cos n\omega_0) \right]^2 \right\}^{1/2}$$

Table 4.2.1 Forms of Fourier Series for Real Functions

Complex Form	Trigonometric Forms
$f(t) = \sum_{n=1}^{\infty} a_n e^{jn\omega_0 t}$	$f(t) = A_0 + \sum_{n=1}^{\infty}(A_n \cos n\omega_0 t + B_n \sin n\omega_0 t)$
$= \sum_{n=-\infty}^{\infty} |a_n| e^{j(n\omega_0 t + \phi_n)}$	$f(t) = A_0 + \sum_{n=1}^{\infty} C_n \cos(n\omega_0 t + \phi_n)$

Formulas for the Coefficients	
$\omega_0 = \dfrac{2\pi}{T}$	$A_0 = a_0$
$a_n = |a_n| e^{j\phi_n}$	$A_n = a_n + a_n^* = 2\,\mathrm{Re}\{a_n\}$
$a_n = \dfrac{1}{T}\displaystyle\int_{t_0}^{t_0+T} f(t) e^{-jn\omega_0 t} dt$	$B_n = j(a_n - a_n^*) = -2\,\mathrm{Im}\{a_n\}$
$t_0 \le t \le t_0 + T$	$C_n = [A_n^2 + B_n^2]^{1/2}$
$A_n = \dfrac{2}{T}\displaystyle\int_{t_0}^{t_0+T} f(t)\cos m\omega_0 t\, dt$	$\phi_n = -\tan^{-1}[B_n/A_n]$
$n = 1, 2, 3, \cdots$	$A_n = C_n \cos\phi_n$
$B_n = \dfrac{2}{T}\displaystyle\int_{t_0}^{t_0+T} f(t)\sin n\omega_0 t\, dt$	$B_n = -C_n \sin\phi_n$
$n = 1, 2, 3, \cdots$	

$$\phi_n = -\tan^{-1}\left(\frac{\cos 0.5 n\omega_0 - \cos n\omega_0}{\sin n\omega_0 + \sin 0.5 n\omega_0}\right)$$

The above equations are ploted in the upper and lower parts of Figure 4.2.2, respectively. The coefficients C_n are known as the **amplitude spectrum**, and the phase ϕ_n is the **phase spectrum**. Specifying amplitude and phase spectrums is a very important concept.

Note: *The amplitude spectrum contains information about the energy content of a signal and the distribution of energy among the different frequencies. The phase spectrum indicates the phase shift for each frequency.*

The spectrum is important in many practical applications. For example, knowledge of the spectrum of an electrocardiogram of the heart or a signal from an engine may reveal abnormal operation and possible imminent failure.

■

Chapter 4: Periodic continuous signals and their spectrums

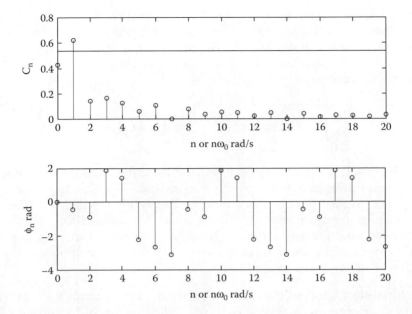

Figure 4.2.2 Amplitude and phase spectrums for the signals shown in Figure 4.2.1a.

Note: *The spectrum of periodic functions is made up of discrete harmonics; each one is a multiple of the fundamental harmonic, which is equal to $2\pi/T$. Therefore, the smallest frequency existing in a periodic function is equal to its fundamental harmonic.*

4.3 Features of periodic continuous functions

Parseval's formula

One important property of continuous periodic functions is the energy relation known as the **Parseval formula**. Consider a periodic function $f(t)$ that is represented in Fourier series expansion

$$f(t) = A_0 + \sum_{n=1}^{\infty} C_n \cos(n\omega_0 t + \phi_n) \tag{4.28}$$

The mean value of $f^2(t)$ (this represents instantaneous power if we assume that $f(t)$ is voltage and is multiplied by $f(t)/(1\text{-ohm resistor})$, which is current) over a period is given by

$$\frac{1}{T}\int_{t_0}^{t_0+T} f(t)^2 dt = \int_{t_0}^{t_0+T} \frac{1}{T}\left[A_0 + \sum_{n=1}^{\infty} C_n \cos(n\omega_0 t + \phi_n)\right] \\ \times \left[A_0 + \sum_{n=1}^{\infty} C_m \cos(m\omega_0 t + \phi_m)\right] dt \tag{4.29}$$

We can see that the terms in the integrand will be of the form $C_n C_m \cos(n\omega_0 t + \phi_n)\cos(m\omega_0 t + \phi_m)$. These terms are periodic with period T, and the value of the integral for $n \neq m$ will vanish. The remaining integrands for $m = n$ will be of the form $C_n^2 \cos^2(n\omega_0 t + \phi_n)$, and these integrals have the value $C_n^2/2$. Hence, (4.29) becomes

$$\boxed{\frac{1}{T}\int_{t_0}^{t_0+T} f(t)^2 dt = A_0^2 + \sum_{n=1}^{\infty} \frac{C_n^2}{2}} \qquad \text{Parseval formula} \qquad (4.30)$$

This result is the **energy relation**, and it can be considered the energy dissipated during one period by a 1-ohm resistor when a voltage source $f(t)$ is applied to it. If $f(t)$ denotes a current, this result specifies that the root mean square (rms) value of the current is the sum of the direct current (dc) component and the rms values of the alternating current (ac) components. In other words, the total power is the power of the dc component plus the power of all the ac components.

Example 4.3.1: Find the energy associated with the terms $n = 3$ to $n = \infty$ in the Fourier series expansion of the periodic function of Example 4.2.2.

Solution: Use (4.30) and the results of Example 4.2.2 to write

$$\frac{1}{3.5}\int_{-0.5}^{1} dt - \left(\frac{3}{7}\right)^2 - \frac{C_1^2}{2} - \frac{C_2^2}{2} = \sum_{n=3}^{\infty} \frac{C_n^2}{2} \qquad \text{or}$$

$$\sum_{n=3}^{\infty} \frac{C_n^2}{2} = \frac{1.5}{3.5} - \frac{9}{49} - \frac{1}{2}\left[\frac{2}{3.5(2\pi/3.5)}\left(\sin\frac{2\pi}{3.5} + \sin\frac{0.5 \times 2\pi}{3.5}\right)\right]^2$$

$$-\frac{1}{2}\left[\frac{2}{3.5(2\pi/3.5)}\left(\cos\frac{0.5 \times 2\pi}{3.5} - \cos\frac{2\pi}{3.5}\right)\right]^2$$

$$-\frac{1}{2}\left[\frac{2}{3.5 \times 2(2\pi/3.5)}\left(\sin\frac{4\pi}{3.5} + \sin\frac{0.5 \times 2 \times 2\pi}{3.5}\right)\right]^2$$

$$-\frac{1}{2}\left[\frac{2}{3.5 \times 2(2\pi/3.5)}\left(\cos\frac{0.5 \times 2 \times 2\pi}{3.5} - \cos\frac{4\pi}{3.5}\right)\right]^2 = 0.0445$$

This result shows that most of the energy is contained in the dc component and a few of the low-frequency harmonics. ∎

Symmetric functions

There are two very important symmetries of real periodic functions: zero-axis (even) and zero-point (odd) symmetry. An even function has symmetry

Chapter 4: Periodic continuous signals and their spectrums 171

about the axis $t = 0$, and an odd function has symmetry about the origin. These two features are illustrated in Figures 4.3.1a and b. Figure 4.3.1c shows a nonsymmetric periodic function having a mean value of zero.

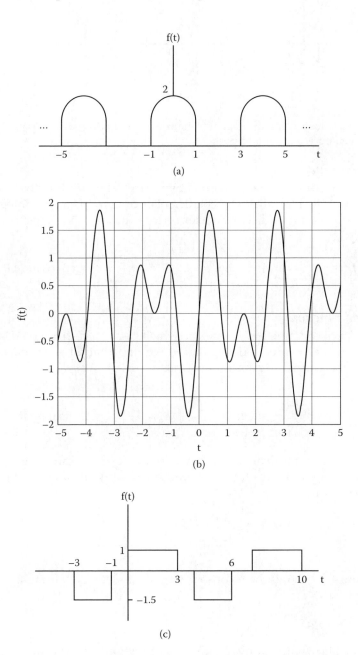

Figure 4.3.1 Types of symmetry: (a) even symmetry, (b) odd symmetry, and (c) zero-average value.

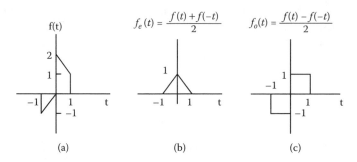

Figure 4.3.2 Construction of a signal from its even and odd parts: (a) given function, and (b) even part, and (c) odd part.

Even function

An even periodic function is defined by $f(-t) = f(t) = f_e(t)$. It follows from the definition of B_n that all these coefficients are zero since the integrand is (even × odd = odd) odd. The Fourier series expansion becomes

$$f_e(t) = A_0 + \sum_{n=1}^{\infty} A_n \cos n\omega_0 t \qquad A_n = \frac{2}{T}\int_{t_0}^{t_0+T} f_e(t)\cos n\omega_0 t\, dt \qquad (4.31)$$

Odd function

An odd periodic function is one for which $f(-t) = -f(-t) = f_o(t)$. Using the function in the Fourier series equation we find that all A_n coefficients are zero and the Fourier expansion becomes

$$f_o(t) = \sum_{n=1}^{\infty} B_n \sin n\omega_0 t \qquad B_n = \frac{2}{T}\int_{t_0}^{t_0+T} f_o(t)\sin n\omega_0 t\, dt \qquad (4.32)$$

It is important to realize that any periodic function can be resolved into even and odd parts. This follows because we can write any function $f(t)$ in equivalent form:

$$f(t) = \underbrace{\frac{f(t)+f(-t)}{2}}_{even} + \underbrace{\frac{f(t)-f(-t)}{2}}_{odd} \qquad (4.33)$$

Figure 4.3.2 shows the reconstruction of a function $f(t)$ from its even and odd parts.

*Finite signals

It can be shown that any continuous function $f(t)$ within an interval $t_0 \leq t \leq t_0 + T$ can be approximated by a trigonometric polynomial of the form given in (4.24) to any degree of accuracy specified in advance. To approximate

Chapter 4: Periodic continuous signals and their spectrums

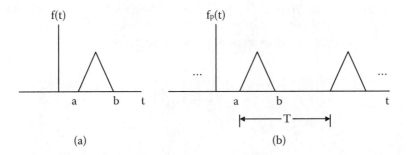

Figure 4.3.3 Nonperiodic function expanded in Fourier series.

the function $f(t)$ shown in Figure 4.3.3a, the periodic function $f_p(t)$ shown in Figure 4.3.3b is created. The expansion then proceeds as before for the periodic function, but the range of applicability is limited to the range of the original function. The solution is then specified as

$$f(t) = \begin{cases} \sum_{n=-\infty}^{\infty} a_n e^{jn\omega_0 t} & -a \leq t \leq b \\ 0 & t < -a \quad t > b \end{cases} \qquad a_n = \frac{1}{T}\int_{t_0}^{t_0+T} f_p(t) e^{-jn\omega_0 t} dt \qquad (4.34)$$

*Convolution

As a continuation of the discussion in Section 2.6, let us consider the special case when $f(t)$ and $h(t)$ are periodic functions with the **same** period. We will find that the convolution integral assumes special properties. **Periodic**, or **cyclic**, **convolution** is defined by the integral

$$\boxed{g(t) = \frac{1}{T}\int_0^T f(x)h(t-x)dx} \qquad (4.35)$$

where the integral is taken over a period. In this case, $g(t)$ is also periodic. To show that $g(t)$ is periodic, consider the integral

$$g_p(t) = \frac{1}{T}\int_c^{c+T} f(x)h(t-x)dx$$

and choose $c = nT + a$ with $0 \leq a \leq T$. The above equation then becomes

$$g_p(t) = \frac{1}{T}\int_{nT+a}^{nT+a+T} f(x)h(t-x)dx = \frac{1}{T}\int_0^T f(x'+nT+a)h[t-(x'+nT+a)]dx'$$

$$= \frac{1}{T}\int_0^T f(x')h(t-x')dx' = g(t)$$

where we set $x' = x - (nT + a)$, $dx = dx'$. Since $f(t)$ and $h(t)$ are periodic, it turns out that $g(t)$ is periodic as shown above. Now, since $g(t)$ is periodic, we can express it in a Fourier series expansion with coefficients

$$a_n = \frac{1}{T}\int_{-T/2}^{T/2} g(t)e^{-jn\omega_0 t}dt = \frac{1}{T^2}\int\int_{-T/2}^{T/2} f(x)h(t-x)e^{-jn\omega_0 t}dtdx$$

$$= \frac{1}{T}\int_{-T/2}^{T/2} f(x)e^{-jn\omega_0 x}dx \frac{1}{T}\int_{-T/2}^{T/2} h(t-x)e^{-jn\omega_0(t-x)}dt$$

Write $t - x = x'$ in the second integral, and so

$$a_n = \frac{1}{T}\int_{-T/2}^{T/2} f(x)e^{-jn\omega_0 x}dx \frac{1}{T}\int_{(-T/2)+x}^{(T/2)+x} h(x')e^{-jn\omega_0 x'}dx'$$

However, for periodic functions, the integration of the second integral is independent of the shift x. Then,

$$a_n = b_n c_n \qquad (4.36)$$

where the Fourier coefficients b_n and c_n belong to the functions $f(t)$ and $h(t)$, respectively. Therefore, we have that

$$\boxed{g(t) = \sum_{n=-\infty}^{\infty} a_n e^{jn\omega_0 t} = \sum_{n=-\infty}^{\infty} b_n c_n e^{jn\omega_0 t} = \frac{1}{T}\int_{-T/2}^{T/2} f(x)h(t-x)dx} \qquad (4.37)$$

Another important property of (4.35) is that cyclic convolution is commutative, so that

$$f(t) * h(t) = h(t) * f(t) \qquad (4.38)$$

To show this property, substitute $t - x = x'$ in (4.35). We write

$$g(t) = -\int_{t}^{t-T} f(t-x')h(x')dx' = -\int_{T}^{0} f(t-x')h(x')dx' = \int_{0}^{T} f(t-x')h(x')dx'$$

4.4 Linear systems with periodic inputs

There are many practical periodic signals that are required to pass through linear time-invariant (LTI) systems. The timing signal of a computer is a pulse periodic signal that passes through many computer circuits. The transducer

Chapter 4: Periodic continuous signals and their spectrums

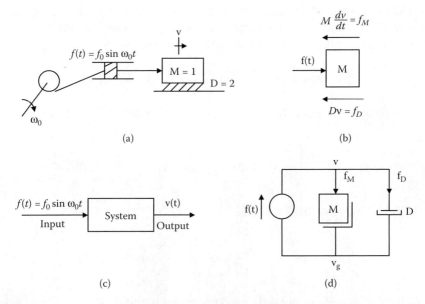

Figure 4.4.1 Linear time-invariant system excited by a sinusoidal time function.

that detects heartbeats is a linear time-invariant system whose input is the periodic heart signal and the output is a signal observed on an oscilloscope that is supposed to resemble the original signal (this may not be true). Therefore, it is important to find out what quantities we should know from both the periodic signals and the LTI systems, such that we can predict with extreme accuracy the output of LTI systems to periodic inputs.

Refer to the system shown in Figure 4.4.1a. It is assumed that the driving system produces a force $f(t) = f_0 \sin \omega_0 t$, where f_0 is the peak amplitude. To develop the equivalent circuit representation of the mechanical system, we draw two levels that correspond to velocity v of the mass and damper and to reference velocity v_g, or ground velocity. Now, we connect the mass element between these two levels. Finally, we connect the forces between v_g and v. The circuit representation is shown in Figure 4.4.1d. Figure 4.4.1b shows the force equilibrium diagram, and Figure 4.4.1c shows the input–output relationships of the system.

An application of D'Alembert's principle at node v yields

$$-f(t) + f_M + f_D = 0 \quad \text{or} \quad f_M + f_D = f(t) \qquad (4.39)$$

The resulting equation that describes the system behavior is ($M = D = 1$)

$$M\frac{dv(t)}{dt} + Dv(t) = f(t) \quad \text{or} \quad \frac{dv(t)}{dt} + 2v(t) = f_0 \sin \omega_0 t \qquad (4.40)$$

To find the particular solution (steady state) of this differential equation, we assume a solution of the form $v(t) = A\cos\omega_0 t + B\sin\omega_0 t$. Introduce this solution into (4.40), from which we obtain the identity

$$(2A + B\omega_0)\cos\omega_0 t + (2B - A\omega_0)\sin\omega_0 t = f_0 \sin\omega_0 t$$

Equating coefficients of like functions on both sides of this equation, we obtain

$$2A + B\omega_0 = 0 \qquad -\omega_0 A + 2B = f_0 \quad \text{or} \quad A = -\frac{f_0 \omega_0}{4 + \omega_0^2} \qquad B = \frac{2f_0}{4 + \omega_0^2}$$

Thus, the velocity of the system is given by

$$v(t) = -f_0 \frac{\omega_0}{4 + \omega_0^2}\left(\cos\omega_0 t - \frac{2}{\omega_0}\sin\omega_0 t\right) \tag{4.41}$$

If we set $\tan\phi_0 = 2/\omega_0$ in (4.41), we obtain the equation

$$v(t) = -\frac{\omega_0}{4 + \omega_0^2}\frac{f_0}{\cos\phi_0}(\cos\omega_0 t \cos\phi_0 - \sin\omega_0 t \sin\phi_0)$$

$$= \frac{\omega_0}{4 + \omega_0^2}\frac{f_0}{\cos\phi_0}\cos(\omega_0 t + \phi_0 + \pi) = \frac{\omega_0}{4 + \omega_0^2}\frac{f_0}{\cos\phi_0}\sin\left(\omega_0 t + \phi_0 + \frac{3\pi}{2}\right) \tag{4.42}$$

For the particular case for $\omega_0 = 1$ rad/s, the phase shift is 5.8195 rad and the amplitude factor is $0.4472 f_0$.

We observe from this LTI system that when the input is a sine function, the output is the same sine function, but with a phase shift of $\phi_0 + (3\pi/2)$ and with an amplitude change of $\omega_0/[(4 + \omega_0^2)\cos\phi_0]$. A similar result would be found if a cosine function was the input. This, of course, applies to all LTI systems.

Note: *When the input to an LTI system is a sinusoidal function, the output is also sinusoidal, but with different amplitude and phase. The phase and amplitude changes are functions of the input frequency. We can thus conclude that if the signal is made up of many sinusoidal functions of different frequencies, each component will experience different phase shifts and amplitude changes.*

Now refer to the system shown in Figure 4.4.2. The equation that describes this system is found by using the Kirchhoff node equation, and it is

$$C\frac{dv(t)}{dt} + \frac{v(t)}{R} = i(t)$$

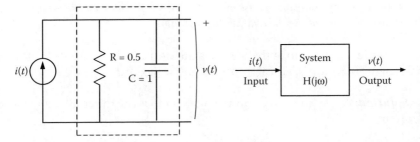

Figure 4.4.2 RC parallel circuit.

For the parameters given in the figure, the above equation becomes

$$\frac{dv(t)}{dt} + 2v(t) = i(t) \quad (4.43)$$

If we assume that the input is an exponential of the form $i(t) = i_0 e^{j\omega_0 t}$, the output will be of the same form (remember that the exponential function is the sum of two sinusoidal functions): $v(t) = v_0 e^{j\omega_0 t}$. Here, i_0 is assumed to be a real constant (without loss of generality), and so v_0 (a complex constant) specifies the output amplitude. In particular, if we include both forms in (4.43), we have

$$v_0 j\omega_0 e^{j\omega_0 t} + 2v_0 e^{j\omega_0 t} = i_0 e^{j\omega_0 t}$$

Note: *The system function H(jω) (also called transfer function) is defined as the ratio of the output to input of an LTI system when the excitation is sinusoidal.*

Therefore, the system function is

$$H(j\omega_0) \triangleq \frac{v_0 e^{j\omega_0 t}}{i_0 e^{j\omega_0 t}} = \frac{1}{2+j\omega_0} = \frac{1}{\sqrt{4+\omega_0^2}\, e^{j\tan^{-1}(\omega_0/2)}} = \frac{1}{\sqrt{4+\omega_0^2}} e^{-j\tan^{-1}(0.5\omega_0)} \quad (4.44)$$

$$= |H(j\omega_0)| e^{j\phi(\omega_0)}$$

and the output is given by

$$v(t) = \frac{i_0}{\sqrt{4+\omega_0^2}} e^{j(\omega_0 t - \tan^{-1}(0.5\omega_0))} \quad (4.45)$$

If we assume that the input is $i_0 \sin\omega_0 t = \text{Im}\{i_0 e^{j\omega_0 t}\}$, the output would be

$$v(t) = \text{Im}\{v(t)\} = \frac{i_0}{\sqrt{4+\omega_0^2}} \sin(\omega_0 t - \tan^{-1}(0.5\omega_0)) \quad (4.46)$$

Note: *The amplitude and phase of the output are dictated by the amplitude and phase characteristics of the system function H(jω).*

Example 4.4.1: Deduce the steady-state output (the current) for the system shown in Figure 4.4.3 for an input $v(t) = 3\cos\omega_0 t$.

Solution: Apply the Kirchhoff voltage law to write the controlling system equation

$$2\frac{di(t)}{dt} + i(t) + \int i(t)dt = v(t) \tag{4.47}$$

The input is written $v(t) = \text{Re}\{3e^{j\omega_0 t}\}$, and the output is assumed to be of the same form, $i(t) = \text{Re}\{i_0 e^{j\omega_0 t}\}$, where i_0 is an unknown complex number to be determined. By introducing the complex form of these quantities in (4.47) we find

$$2j\omega_0 i_0 e^{j\omega_0 t} + i_0 e^{j\omega_0 t} + \frac{1}{j\omega_0} i_0 e^{j\omega_0 t} = 3e^{j\omega_0 t}$$

from which the system function $H(j\omega_0)$ is (an admittance function)

$$H(j\omega_0) = \frac{i_0}{v_0} = \frac{1}{1+2j\omega_0 + \dfrac{1}{j\omega_0}} = \frac{j\omega_0}{1-2\omega_0^2 + j\omega_0}$$

Thus,

$$i_0 = H(j\omega_0)v_0 = \frac{3j\omega_0}{1-2\omega_0^2 + j\omega_0} = \frac{3\omega_0 e^{j\pi/2}}{\sqrt{(1-2\omega_0^2)^2 + \omega_0^2}\exp[j\tan^{-1}[\omega_0/(1-2\omega_0^2)]]}$$

$$= \frac{3\omega_0}{[(1-2\omega_0^2) + \omega_0^2]^{1/2}} e^{j[\pi/2 - \tan^{-1}[\omega_0/(1-2\omega_0^2)]]} \tag{4.48}$$

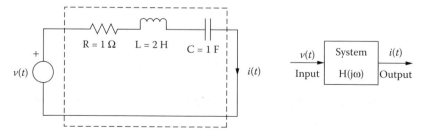

Figure 4.4.3 RLC series circuit.

Chapter 4: Periodic continuous signals and their spectrums

The output current is then

$$i(t) = \text{Re}\{i(t)\} = \text{Re}\{i_0 e^{j\omega_0 t}\}$$

$$= \frac{3\omega_0}{[(1-2\omega_0^2)^2 + \omega_0^2]^{1/2}} \cos\left(\omega_0 t + \frac{\pi}{2} - \tan^{-1}\frac{\omega_0}{1-2\omega_0^2}\right) \quad (4.49)$$

As in the previous discussion, we observe that the amplitude and phase of the output signal are dictated by the amplitude and phase of the transfer function (system function), a somewhat complicated function. ∎

We observe from our analysis that when the input is a sinusoidal signal $\exp(j\omega_0 t)$, the transfer function is $H(j\omega_0)$ and the output is $H(j\omega_0)\exp(j\omega_0 t)$. Of course, if the input frequency is $n\omega_0$ so that the complex input is $\exp(jn\omega_0)$, the transfer function is $H(jn\omega_0)$ and the output is $H(jn\omega_0)\exp(jn\omega_0 t)$. Clearly, when the input is the sum of sinusoids, the output will consist of each component sinusoid multiplied by the appropriately chosen transfer function. Thus, for an input

$$f_i(t) = \sum_{n=-\infty}^{\infty} a_n e^{jn\omega_0 t}$$

the output will be

$$\boxed{f_o(t) = \sum_{n=-\infty}^{\infty} a_n H(jn\omega_0) e^{jn\omega_0 t}} \quad (4.50)$$

Note: *If the input to an LTI system is a periodic nonsinusoidal function, the output is a periodic function with Fourier series coefficients defined by the expression $a_n H(jn\omega_0)$.*

In all these cases we imply that the input periodic function has been applied for a long time so that the transient effect has disappeared.

If we expand in sine and cosine form the input periodic function to an LTI system

$$f_i(t) = A_0 + \sum_{n=1}^{\infty}(A_n \cos n\omega_0 t + B_n \sin n\omega_0 t)$$

the output function will have the form

$$f_o(t) = A_0 H(j0) + \sum_{n=1}^{\infty} |H(jn\omega_0)| \qquad (4.51)$$
$$\times (A_n \cos[n\omega_0 t + \phi_n(n\omega_0)] + B_n \sin[n\omega_0 t + \phi_n(n\omega_0)])$$

Example 4.4.2: The periodic signal $f_i(t) = 2\sin(\pi t/100) + 0.1\sin(5\pi t/100)$ is applied to a system with transfer function $H(j\omega) = 1/(1 + j2\omega)$. Determine the output $f_o(t)$ of the system.

Solution: The amplitude and phase functions of the transfer function are

$$|H(j\omega)| = \frac{1}{\sqrt{1+4\omega^2}} \qquad \phi(\omega) = -\tan^{-1}(2\omega)$$

By (4.51) and with $A_0 = A_n = 0$, we obtain

$$f_o(t) = 2 \frac{1}{\sqrt{1+4\left(\frac{\pi}{100}\right)^2}} \sin\left(\frac{\pi}{100}t - \tan^{-1} 2\frac{\pi}{100}\right)$$

$$+ 0.1 \frac{1}{\sqrt{1+4\left(\frac{5\pi}{100}\right)^2}} \sin\left(\frac{5\pi}{100}t - \tan^{-1} 2\frac{5\pi}{100}\right)$$

■

Important definitions and concepts

1. Complex exponentials
2. Fundamental period of a periodic function
3. Fundamental frequency of a periodic function
4. Fundamental difference between continuous sinusoidal signals and discrete ones
5. Periodicity
6. Dirichlet conditions for periodic functions
7. Complex and sinusoidal forms of the Fourier series of periodic functions
8. Gibbs' phenomenon
9. Amplitude and phase spectrums of periodic signals
10. Parseval's formula
11. Energy relation of periodic signals
12. Fourier series of the convolution of periodic signals having the same period

13. Transfer (system) function and its influence on the output of an LTI system when the input is a periodic function
14. Half- and full-wave signals
15. Ripple factor of a half-wave (or full-wave) signal
16. Filtering

Chapter 4 Problems

Section 4.1

1. Draw the following functions given that c and a are real numbers: $y_1(t) = c\,e^{at}$ and $y_2(t) = c\,e^{-at}$.
2. Plot the real and imaginary parts of the signal $y(t) = c\,e^{-at}$ for $c = 2 + j$ and $a = 2 - 2j$.
3. Using Euler's relations a) and b), show the relations c) and d):

   a. $\cos\omega t = (e^{j\omega t} + e^{-j\omega t})/2$

   b. $\sin\omega t = (e^{j\omega t} - e^{-j\omega t})/2j$

   c. $\cos^2\omega t = \dfrac{1}{2}(1+\cos 2\omega t)$

   d. $\sin\omega_1 t\,\sin\omega_2 t = [\cos(\omega_1 - \omega_2)t - \cos(\omega_1 + \omega_2)t]/2$

4. Any complex number z can be represented by a point in the complex plane with $z = a + jb = \sqrt{a^2 + b^2}\exp(j\tan^{-1}(b/a)) \triangleq r\exp(j\theta) = r\cos\theta + jr\sin\theta$, as shown in Figure P4.1.4.

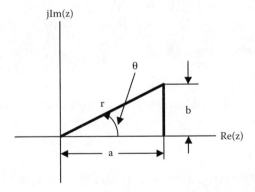

Figure P4.1.4

5. Corresponding to the complex number, $z = a + jb = r\exp(j\theta)$ is the **complex conjugate** $z^* = a - jb = r\exp(-j\theta)$. If $z_1 = a_1 + jb_1$, $z_2 = a_2 + jb_2$, and $z_3 = a_3 + jb_3$, determine the following relationships:

a. $z_1 z_1^*$
b. $z_2/(z_2 + z_2^*)$
c. $z_3 + z_3^*$
d. $(z_1 + z_2)^*$
e. $(z_1 + z_1^*)/(z_2 + z_2^*)$
f. $(3z_1 z_2)^*$

6. Express each relation by a complex number in its polar form ($re^{j\theta}$):

   a. $(2 + j4)/(1 + j)$
   b. $[2e^{j\pi/6}/(1 + j)]$
   c. $[2e^{j\pi/6}/(1 + j)]^*$
   d. $\sqrt{3}e^{j\pi/6}(2 + j6)^*$
   e. $j(3 + j4) * e^{j\pi/6}$
   f. $e^{-j\pi/2} - e^{j\pi}$
   g. $[(2 - j6)/(2 + j6)]^*$
   h. $(1 - j)^5$
   i. $(2 + j)*(2 - j)*e^{j\pi/4}$

7. The quantities z_1 and z_2 are complex numbers. Prove the following relationships:

   a. $|z_1| = r_1$

   b. $|z_1 z_2| = |z_1||z_2|$

   c. $(|z_1| - |z_2|)^2 \le |z_1 + z_2|^2 \le (|z_1|^2 + |z_2|^2)$

8. Prove the following orthogonality relations:

   a. $\int_0^T \cos n\omega_0 t \sin m\omega_0 t \, dt = 0 \qquad \omega_0 = 2\pi/T$

   b. $\int_0^T \cos n\omega_0 t \cos m\omega_0 t \, dt = \begin{cases} 0 & n \neq 0 \\ T/2 & n = m \end{cases}$

9. The continuous function $\cos \omega_0 t$ is sampled at $t = nT$. Specify the condition that enables $\cos n\omega_0 T$ to be periodic. Do the same for a sine signal.

Section 4.2

The reader should verify, wherever applicable, the analytical results using MATLAB.

1. Show the following relations:

   a. $A_0 = a_0$
   b. $A_n = 2 \operatorname{Re}\{a_n\}$
   c. $B_n = -2 \operatorname{Im}\{a_n\}$

2. Show that $C_n = \sqrt{A_n^2 + B_n^2}$.
3. Deduce the Fourier series in trigonometric form for the periodic function shown in Figure P4.2.3. Find and plot the amplitude and phase spectrums of the series. This function is known as the **half-wave rectified** signal. It is assumed that the figure extends from minus infinity to infinity.

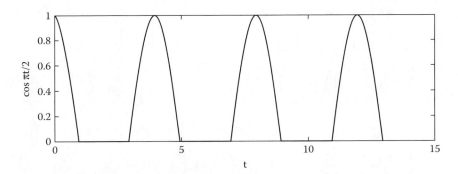

Figure P4.2.3

4. Find the Fourier series in trigonometric form of the full-wave rectified signal shown in Figure P4.2.4. Find and plot the amplitude and phase spectrums.

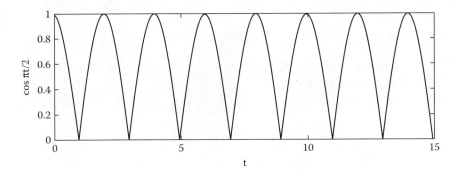

Figure P4.2.4

5. The **ripple factor** of a half-wave (or full-wave) rectified signal is defined by

$$r = \sqrt{(f_{rms}/f_{dc}) - 1},$$

where f_{rms} is the rms value of the ac components of the signal within a period and f_{dc} is the average value of the periodic signal. Show that the r values for the half- and full-wave signals are 1.21 and 0.48, respectively.

6. Find the Fourier series expansion in trigonometric form for the periodic signals shown in Figure P4.2.6. The signals extend from minus infinity to infinity.

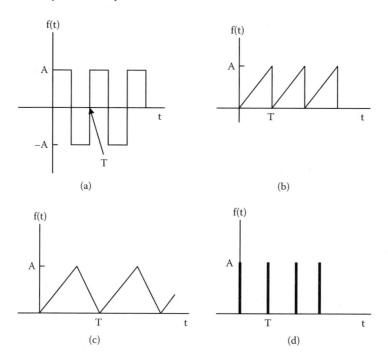

Figure P4.2.6

To plot the half- and full-wave rectifier, the following book MATLAB program was used:

```
for t=1:300
if cos(pi*0.05*t/2)>=0
s(t)=cos(pi*0.05*t/2);
else
s(t)=abs(cos(pi*0.05*t/2));%for half-wave s(t)=0
end;
end;
m=0:0.05:15-0.05;
plot(m,s)
```

7. Compute the Fourier series of the following functions:

   a. $f(t) = t^2 \quad -\pi \le t \le \pi, \; T = 2\pi$
   b. $f(t) = t \quad \;\; -1 \le t \le 1 \quad T = 2$

Chapter 4: Periodic continuous signals and their spectrums

8. Compute the Fourier series of the following functions:

   a. $f(t) = t \quad -t_0 \leq t \leq t_0 \quad f(t+2t_0) = f(t)$
   b. $f(t) = 1+e^t \quad -1 \leq t \leq 1 \quad T = 4$

Section 4.3

1. If the Fourier series expansion of a periodic function $f(t)$ is written in its exponential form

$$f(t) = \sum_{n=-\infty}^{\infty} a_n e^{jn\omega_0 t},$$

   show that

$$(1/T) \int_{t_0}^{t_0+T} f(t) f^*(t) dt = (1/T) \int_{t_0}^{t_0+T} |f(t)|^2 dt = \sum_{n=-\infty}^{\infty} |a_n|^2$$

   This is another form of Parseval energy relation.

2. Show that the Parseval energy relation can be written in the form

$$(1/T) \int_{t_0}^{t_0+T} |f(t)|^2 dt = A_0^2 + \sum_{n=1}^{\infty} \left(\frac{A_n^2}{2} + \frac{B_n^2}{2} \right)$$

3. Deduce the Fourier series expansions for the signals shown in Figure P4.3.3. Compare the amplitude and phase spectrums of the two functions in Figures P4.3.3b and c and state your observation.

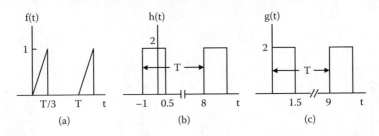

Figure P4.3.3

4. Find the Fourier series for the functions shown in Figure P4.3.4. Hint: Create a periodic function with any period.

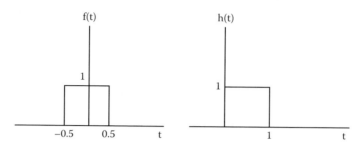

Figure P4.3.4

5. Determine the Fourier series of the periodic functions shown in Figure P4.2.6 if they are shifted a half period, $T/2$, to the right and compare their spectra.

Section 4.4

1. Determine the output for the systems shown in Figure P4.4.1 if the input is $\sin \omega_0 t$. State your observation about the output as ω_0 varies from zero to infinity.

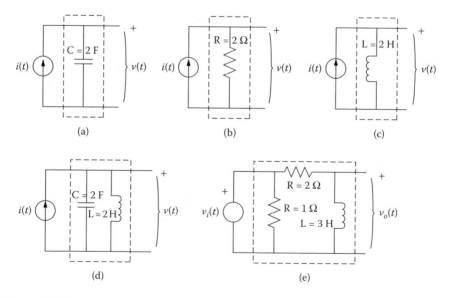

Figure P4.4.1

Chapter 4: Periodic continuous signals and their spectrums

2. Determine the output for the systems shown in Figure P4.4.2 if the input is $\sin \omega_1 t + \cos \omega_2 t$.

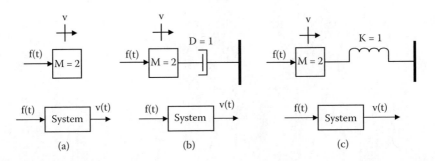

Figure P4.4.2

Hint: Remember that the systems are linear; thus, define the system function appropriately for each frequency.

chapter 5

Nonperiodic signals and their Fourier transform

Unlike the Fourier series, which is essentially oriented toward periodic functions, the Fourier integral permits a description of nonperiodic functions. The Fourier integral bears a close relationship to the Fourier series. The fact that a Fourier transform pair exists is of importance to our work. The transform pair permits a frequency domain function $F(\omega)$ to be deduced from a knowledge of the time domain function $f(t)$, and vice versa.

Note: *Every well-behaved function f(t) has a unique characteristic spectrum in the frequency domain.*

This chapter considers the properties of the Fourier transform, with examples of the calculations of some elementary but fundamental transform pairs. Among the important applications of the Fourier transform are analytically representing nonperiodic functions, solving differential equations, aiding in the analysis of linear time-invariant (LTI) systems, and analyzing and processing signals in engineering, medical, optical, metallographic, and seismic problems, just to mention a few.

For the case where signals are experimentally derived, analytic forms are usually not available for the integration involved. The approach now is to replace the continuous Fourier transform by an equivalent discrete Fourier transform (DFT), and then evaluate the DFT using discrete data. While the process is straightforward in principle, the number of calculations involved can become extremely large, even in relatively small problems. It is the use of the fast Fourier transform (FFT), a computational algorithm that greatly reduces the number of calculations in the use of the DFT, that makes the DFT a viable and indispensable procedure. The DFT will be discussed in detail in a later chapter.

5.1 Direct and inverse Fourier transform

The Fourier transform of a function $f(t)$, written $\mathcal{F}\{f(t)\} \triangleq F(\omega)$, and the inverse transform, written $\mathcal{F}^{-1}\{F(\omega)\} \triangleq f(t)$, are defined by the integral relations

$$\mathcal{F}\{f(t)\} \triangleq F(\omega) = \int_{-\infty}^{\infty} f(t) e^{-j\omega t} dt \qquad \text{a)}$$

direct, or forward, Fourier transform

$$\mathcal{F}^{-1}\{F(\omega)\} \triangleq f(t) = \frac{1}{2\pi} \int_{-\infty}^{\infty} F(\omega) e^{j\omega t} d\omega \qquad \text{b)}$$

inverse Fourier transform

(5.1)

$F(\omega)$ is known as the **spectrum** function. If $F(\omega)$ is a complex function of the form $F(\omega) = F_r(\omega) + jF_i(\omega)$ ($F_r(\omega)$, $F_i(\omega)$ are real functions of ω), we would ordinarily plot the absolute value

$$|F(\omega)| = \sqrt{F_r^2(\omega) + F_i^2(\omega)}$$

and argument $Arg\{F(\omega)\} = \tan^{-1}[F_i(\omega)/F_r(\omega)]$ vs. frequency ω, or the real and imaginary parts vs. frequency.

Not all functions are Fourier transformable. Sufficiency conditions for a function $f(t)$ to be Fourier transformable are called **Dirichlet conditions**, which are:

1. $\int_{-\infty}^{\infty} |f(t)| dt < \infty$.
2. $f(t)$ has finite maximums and minimums within any finite interval.
3. $f(t)$ has a finite number of discontinuities within any finite interval.

If these conditions are met, $f(t)$ can be transformed uniquely. Some functions exist that do not possess Fourier transforms in the strict sense since they violate one or another of the Dirichlet conditions. Yet, in many cases, it is still possible to deduce the Fourier transform if the function under consideration belongs to a set known as **generalized functions**. One such function is the **delta function** $\delta(t)$, which is considered below.

Example 5.1.1: Determine the Fourier transform of the function $f(t) = e^{-at}u(t)$ $a > 0$ and, from this result, plot the following vs. frequency: (1) the absolute value and phase spectrum and (2) the real and imaginary parts of its spectrum.

Chapter 5: Nonperiodic signals and their Fourier transform

Solution: By an application of (5.1a),

$$F(\omega) = \int_{-\infty}^{\infty} e^{-at} u(t) e^{-j\omega t} dt = \int_{0}^{\infty} e^{-(a+j\omega)t} dt = \frac{1}{a+j\omega} \qquad a>0$$

The real and imaginary parts of the spectrum are

$$F_r(\omega) = \text{Re}\left\{\frac{1}{a+j\omega}\right\} = \text{Re}\left\{\frac{a-j\omega}{a^2+\omega^2}\right\} = \frac{a}{a^2+\omega^2}$$

$$F_i(\omega) = \text{Im}\left\{\frac{1}{a+j\omega}\right\} = \text{Im}\left\{\frac{a-j\omega}{a^2+\omega^2}\right\} = -\frac{\omega}{a^2+\omega^2}$$

The function $F(\omega)$ can also be written in the form

$$F(\omega) = \frac{1}{\sqrt{a^2+\omega^2}} e^{j\tan^{-1}[F_i/F_r]} = \frac{1}{\sqrt{a^2+\omega^2}} e^{-j\tan^{-1}(\omega/a)}$$

Thus, the absolute value and phase spectrums are

$$|F(\omega)| = \frac{1}{\sqrt{a^2+\omega^2}} \qquad \text{Arg}F(\omega) = -\tan^{-1}(\omega/a)$$

Book m-File for example 5.1.1: ex5_1_1

```
%ex5_1_1, this is the m-file name for the Ex5.1.1;
t=0:.05:5;
f1=exp(-0.5*t);
f2=exp(-4*t);
w=-20:.05:20;
fr1=0.5./(0.25+w.^2);
fr2=4./(16+w.^2);
fi1=-w./(0.25+w.^2);
fi2=-w./(16+w.^2);
fabs1=1./sqrt(0.25+w.^2);
fabs2=1./sqrt(16+w.^2);
arg1=-atan(2*w);
arg2=-atan(w/4);
subplot(2,1,1);plot(t,f1,'k');hold on;plot(t,f2,'k');
```

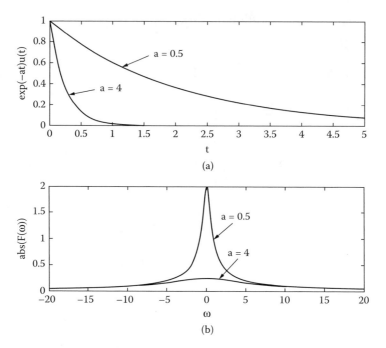

Figure 5.1.1 Time and frequency functions. (a) Time functions. (b) The absolute value of the spectrum.

```
xlabel('t');ylabel('exp(-at)');
subplot(2,1,2);plot(w,fabs1,'k');hold on;plot(w,fabs2,'k');
xlabel('\omega');ylabel('abs(F(\omega))');
%the last four lines can be changed appropriately
%to plot the rest of the figure;
```

Plots of several functions are shown in Figure 5.1.1. ∎

Example 5.1.2: Determine the spectrum of the following functions: (a) $f(t) = \delta(t)$ and (b) $g(t) = \delta(t - a)$ for $a > 0$.

Solution: Employ the properties of the delta function to obtain

$$F(\omega) = \int_{-\infty}^{\infty} \delta(t)e^{-j\omega t}dt = e^{-j\omega 0} = 1$$

$$G(\omega) = \int_{-\infty}^{\infty} \delta(t-a)e^{-j\omega t}dt = e^{-j\omega a} \qquad a > 0$$

The functions and their spectrums are shown in Figure 5.1.2. ∎

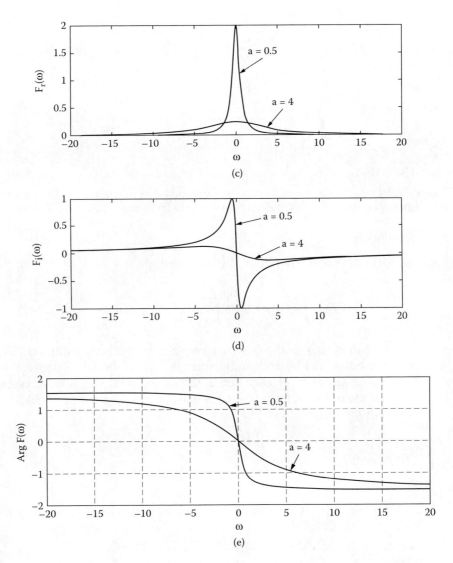

Figure 5.1.1 (continued). Time and frequency functions. (c) Corresponding real parts of their spectrums. (d) Corresponding imaginary parts of their spectrums. (e) Arguments of the spectrums.

Based on our observations of the Figure 5.1.2 spectrums we conclude:

Note: *A signal and its shifted version have the same absolute value of their spectrums but they differ on their phase spectrums.*

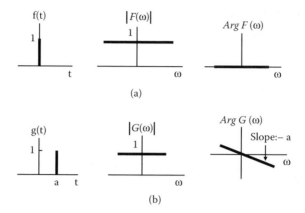

Figure 5.1.2 Delta function and its presentations. (a) Delta function at the origin and its Fourier spectrum. (b) Shifted delta function and its Fourier spectrum.

Real functions

Suppose that $f(t)$ is real. We obtain the relationship

$$F(-\omega) = \int_{-\infty}^{\infty} f(t)e^{j\omega t}dt = \left[\int_{-\infty}^{\infty} f(t)e^{-j\omega t}dt\right]^* = F^*(\omega) \qquad (5.2)$$

The above result is due to the fact that an integral can be substituted (for well-behaved functions) by an infinite sum. But since the conjugate of the sum is equal to the sum of each term, (5.2) is correct. The above result indicates that the reflected form of $F(\omega)$ is equal to the conjugate of $F(\omega)$. We can start from the right side of (5.2) to find

$$\left[\int_{-\infty}^{\infty} f(t)\cos\omega t\, dt - j\int_{-\infty}^{\infty} f(t)\sin\omega t\, dt\right]^*$$

$$= \int_{-\infty}^{\infty} f(t)\cos\omega t + j\int_{-\infty}^{\infty} f(t)\sin\omega t\, dt = \int_{-\infty}^{\infty} [f(t)e^{j\omega t}]dt = \int_{-\infty}^{\infty} f(t)e^{-j(-\omega)t}dt = F(-\omega)$$

This result, when applied to Example 5.1.1, shows that

$$F(-\omega) = \frac{1}{a-j\omega} = \left(\frac{1}{a+j\omega}\right)^* = F^*(\omega)$$

as it should be in accordance with the previous results.

Note: *A conjugation is performed on an expression by simply changing the signs of all j's that are present in the expression.*

Real and even functions

For $f(t)$ even and real, we obtain

$$F(\omega) = F_r(\omega) + jF_i(\omega) = \int_{-\infty}^{\infty} f(t)e^{-j\omega t}dt = \int_{-\infty}^{\infty} f(t)\cos \omega t\, dt$$
$$-j\int_{-\infty}^{\infty} f(t)\sin \omega t\, dt = \int_{-\infty}^{\infty} f(t)\cos \omega t\, dt \quad (5.3)$$

Here, the integrand $f(t)\sin \omega t$ is an odd function and the integral in the range from minus infinity to infinity is zero. By equating the real and imaginary parts at both sides of (5.3) we obtain

$$F_r(\omega) = \int_{-\infty}^{\infty} f(t)\cos \omega t\, dt = 2\int_{0}^{\infty} f(t)\cos \omega t\, dt; \quad F_i(\omega) = 0 \quad (5.4)$$

Example 5.1.3: Deduce and plot the Fourier spectrum of the even function $f(t) = \exp(-2|t|)$ illustrated in Figure 5.1.3a. Show that this spectrum verifies (5.4).

Solution: The Fourier transform of the given function is

$$F(\omega) = \int_{-\infty}^{\infty} e^{-2|t|}e^{-j\omega t}dt = \int_{-\infty}^{0} e^{2t}e^{-j\omega t}dt + \int_{0}^{\infty} e^{-2t}e^{-j\omega t}dt$$
$$= \int_{-\infty}^{0} e^{(2-j\omega)t}dt + \int_{0}^{\infty} e^{-(2+j\omega)t}dt = \frac{1}{2-j\omega} + \frac{1}{2+j\omega} = \frac{4}{4+\omega^2}$$

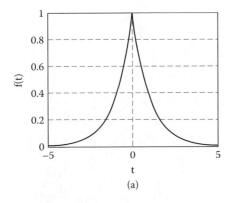
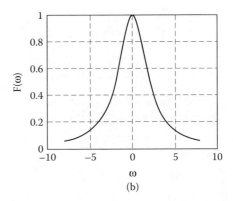

Figure 5.1.3 Illustration of Example 5.1.3.

This final expression shows that $F_i(\omega) = 0$, and the even function is shown in Figure 5.1.3b. Now, we examine

$$F_r(\omega) = 2\int_0^\infty e^{-2t}\cos\omega t\, dt = \int_0^\infty e^{-2t}(e^{j\omega t} + e^{-j\omega t})dt$$

$$= \frac{1}{2-j\omega} + \frac{1}{2+j\omega} = \frac{4}{4+j\omega}$$

which agrees with the above. ∎

Real and odd functions

For $f(t)$ odd and real, we obtain

$$F(\omega) = \int_{-\infty}^\infty f(t)e^{-j\omega t}dt = \int_{-\infty}^\infty f(t)\cos\omega t\, dt - j\int_{-\infty}^\infty f(t)\sin\omega t\, dt$$

$$= -j\int_{-\infty}^\infty f(t)\sin\omega t\, dt \tag{5.5}$$

Thus,

$$F_r(\omega) = 0; \quad F_i(\omega) = -j\int_{-\infty}^\infty f(t)\sin\omega t\, dt \tag{5.6}$$

5.2 Properties of Fourier transforms

Fourier transforms possess a number of very important properties. These properties are developed next.

Linearity

Consider the Fourier integral of the function $(af(t) + bh(t))$

$$\boxed{\mathcal{F}\{af(t)+bg(t)\} = a\mathcal{F}\{f(t)\}+b\mathcal{F}\{g(t)\} = aF(\omega)+bG(\omega)} \tag{5.7}$$

where a and b are constants. This property is a direct result of the linear operation of integration.

Symmetry

If $\mathcal{F}\{f(t)\} = F(\omega)$, then

$$\boxed{2\pi f(-\omega) = \int_{-\infty}^\infty F(t)e^{-j\omega t}dt} \tag{5.8}$$

Chapter 5: Nonperiodic signals and their Fourier transform

Proof: Begin with the inverse Fourier integral

$$2\pi f(t) = \int_{-\infty}^{\infty} F(\omega)e^{j\omega t}d\omega, \quad \text{or} \quad 2\pi f(\omega) = \int_{-\infty}^{\infty} F(t)e^{j\omega t}dt$$

where we interchanged the symbols ω and t in the first equation. By introducing the change from ω to $-\omega$, the result is (5.8). This symmetry property helps in extending the tables of Fourier transforms.

Example 5.2.1: Examine the symmetry property of the pulse function $p_a(t)$ and the impulse function (delta function) $\delta(t)$.

Solution: The Fourier transform of the pulse function is

$$\mathcal{F}\{p_a(t)\} = \int_{-\infty}^{\infty} p_a(t)e^{-j\omega t}dt = \int_{-a}^{a} e^{-j\omega t}dt = 2\frac{\sin \omega a}{\omega} = 2\sin c_a \omega \quad (5.9)$$

From the symmetry property, we write

$$2\pi p_a(-\omega) = \mathcal{F}\left\{2\frac{\sin at}{t}\right\} \quad (5.10)$$

The graphical representation of these formulas is given in Figure 5.2.1a. Remember also that the pulse function is symmetric and $p_a(-\omega) = p_a(\omega)$. To prove this result requires the use of integral tables, which give the following expression

$$2\int_0^{\infty} 2\frac{\sin at}{t}\cos \omega t\, dt = 2\pi \quad -a \leq \omega \leq a$$

For the delta function, we have

$$\mathcal{F}\{\delta(t)\} = \int_{-\infty}^{\infty} \delta(t)e^{-j\omega t}dt = e^{-j\omega 0} = 1 \triangleq \Delta(\omega) \quad (5.11)$$

By the symmetry property and because the delta function is even,

$$2\pi\delta(-\omega) = 2\pi\delta(\omega) = \mathcal{F}\{\Delta(t)\} = \mathcal{F}\{1\} \quad (5.12)$$

At this point we may ask: Can we physically produce an infinite number of sinusoids with each having an amplitude of 1, as (5.11) indicates? The answer is no. This is true since we will need an infinite amount of energy

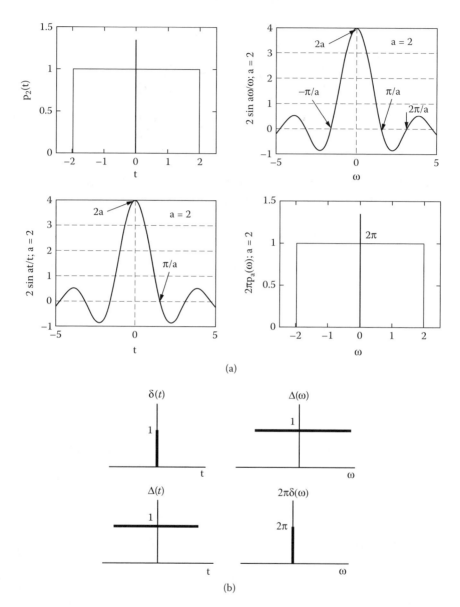

Figure 5.2.1 Symmetry properties. (a) Illustration of the symmetry property of the pulse function. (b) Illustration of the symmetry property of the delta function.

to produce the signals. Of course, no contradiction exists since the delta function is a mathematical expression and no physical mechanism exists that can produce it. The relationships of (5.11) and (5.12) are shown in Figure 5.2.1b. ∎

Chapter 5: Nonperiodic signals and their Fourier transform 199

Time shifting

For any real-time t_0

$$\mathcal{F}\{f(t \pm t_0)\} = e^{\pm j\omega t_0}\mathcal{F}\{f(t)\} = e^{\pm j\omega t_0} F(\omega) \qquad (5.13)$$

Example 5.2.2: Find the Fourier transform of the function $f(t - t_0) = e^{-(t-t_0)} u(t - t_0)$.

Solution: Use (5.13) to write (see also Example 5.1.1)

$$\mathcal{F}\{f(t-t_0)\} = e^{-j\omega t_0}\mathcal{F}\{f(t)\} = e^{-j\omega t_0}\mathcal{F}\{e^{-t}u(t)\} = e^{-j\omega t_0}\frac{1}{1+j\omega}$$

The effect of shifting is shown in Figure 5.2.2 for $t_0 = 0$ and $t_0 = 4$. Observe that only the phase spectrum is modified. ∎

Note: *When a signal is shifted, only its phase spectrum changes.*

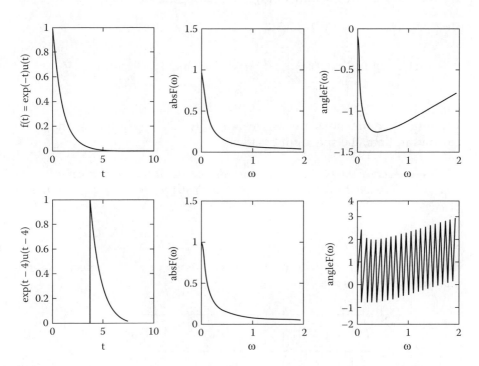

Figure 5.2.2 The effect of shifting in the time domain the function $\exp(-t)u(t)$.

Scaling

If $\mathcal{F}\{f(t)\} = F(\omega)$, then

$$\mathcal{F}\{f(at)\} = \frac{1}{|a|} F\left(\frac{\omega}{a}\right) \tag{5.14}$$

and from this, by setting $a = -1$, we see that

$$\mathcal{F}\{f(-t)\} = F(-\omega) \tag{5.15}$$

This equation indicates that if a time function is reflected, its Fourier spectrum is also reflected.

Example 5.2.3: Discuss the Fourier transform of the pulse function $f_b(t) = p_a(bt)$, where $b > a$.

Solution: By Example 5.2.1, the Fourier transform of $f_a(t) = p_a(t)$ is $2(\sin a\omega / \omega) \triangleq 2\sin c_a(\omega)$. By (5.14), the Fourier transform of $f_b(t) = p_a(bt)$ is

$$\mathcal{F}\{f_b(t)\} = \mathcal{F}\{p_a(bt)\} = \frac{1}{|b|}\mathcal{F}\{f_a(t)\} = \frac{1}{|b|}F_a(\omega)\Big|_{\omega=\omega/b} = \frac{1}{|b|}\frac{2\sin\left(a\frac{\omega}{b}\right)}{\frac{\omega}{b}}$$

The respective functions are shown in Figure 5.2.3. Note that as the pulse narrows in time, the spectrum broadens. ∎

Note: *From Example 5.2.3 we conclude that signals with low variations in time are rich in low frequencies and signals with sharp variations are rich in high frequencies.*

Central ordinate

By setting $\omega = 0$ and $t = 0$ in the two equations of (5.1), respectively, the relating expressions are

$$F(0) = \int_{-\infty}^{\infty} f(t)dt \qquad f(0) = \frac{1}{2\pi}\int_{-\infty}^{\infty} F(\omega)d\omega \tag{5.16}$$

The first of these equations shows that the area under the $f(t)$ curve is equal to the central ordinate of the Fourier transform. The second of these equations

Chapter 5: Nonperiodic signals and their Fourier transform

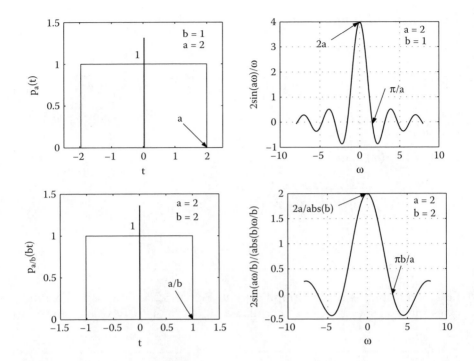

Figure 5.2.3 The effect on the spectrum due to time scaling.

shows that the area under the $F(\omega)$ curve is 2π times the value of the function at $t = 0$.

Example 5.2.4: Find $F(0)$ and $f(0)$ for the function $f(t) = 2p_a(t)$ using (5.16).

Solution: From Example 5.2.1, the Fourier integral of $f(t)$ is $F(\omega) = 4\sin(a\omega)/\omega$. For $\omega = 0$, by application of L'Hospital's rule, $F(0) = 4a$. From the above equation we have

$$F(0) = \int_{-\infty}^{\infty} 2p_a(t)dt = 2\int_{-a}^{a} dt = 4a$$

For $f(0)$, we use the inverse Fourier integral

$$f(0) = \frac{1}{2\pi}\int_{-\infty}^{\infty} 4\frac{\sin n\omega}{\omega}d\omega = \frac{2}{\pi}\int_{-\infty}^{\infty} \frac{\sin n\omega}{a\omega}d(a\omega) = \frac{2}{\pi}\int_{-\infty}^{\infty} \frac{\sin x}{x}dx = \frac{2}{\pi}\times\pi = 2$$

where use has been made of mathematical tables for the value of the last integral. ∎

Frequency shifting

If $\mathcal{F}\{f(t)\} = F(\omega)$, then

$$\mathcal{F}\{e^{\pm j\omega_0 t} f(t)\} = F(\omega \mp \omega_0) \tag{5.17}$$

Modulation

If $\mathcal{F}\{f(t)\} = F(\omega)$, then

$$\begin{aligned} \mathcal{F}\{f(t)\cos\omega_0 t\} &= \frac{1}{2}[F(\omega+\omega_0) + F(\omega-\omega_0)] &\text{a)} \\ \mathcal{F}\{f(t)\sin\omega_0 t\} &= \frac{1}{2j}[F(\omega-\omega_0) - F(\omega+\omega_0)] &\text{b)} \end{aligned} \tag{5.18}$$

Equation (5.18a) is shown graphically in Figure 5.2.4. Equation (5.18) indicates that the spectrum of a modulated signal $f(t)$ is halved and shifted to

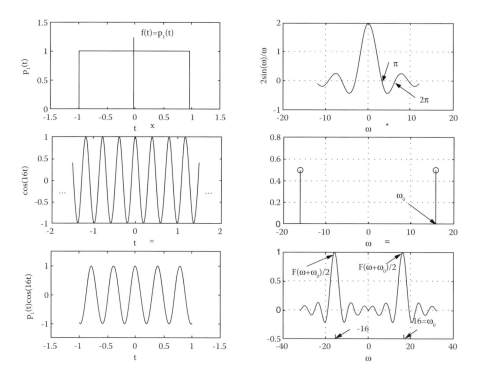

Figure 5.2.4 Modulation. Illustration of the spectrum shift of a modulated signal $f(t) = p_1(t)$.

Chapter 5: Nonperiodic signals and their Fourier transform

the carrier frequency ω_0. These formulas are easily derived by expanding the cosine and sine terms into their equivalent exponential forms (see Chapter 1) and then using (5.17).

The results of (5.18) constitute the fundamental properties of modulation and are basic to the field of communications. In Figure 5.2.5 we have also indicated that the resulting spectrum is equal to the convolution of the spectrums of the product functions, as we will see in the property below on *frequency convolution*.

Example 5.2.5: Modulation. Find the spectrum of the modulated signal $f_m(t) = f(t)\cos\omega_c t$ shown in Figure 5.2.5.

Solution: Initially, we find the Fourier spectrums of $\cos\omega_c t$ and $f(t)$. These are

$$\mathcal{F}\{\cos\omega_c t\} = \mathcal{F}\left\{\frac{e^{j\omega_c t} + e^{-j\omega_c t}}{2}\right\} = \frac{1}{2}\int_{-\infty}^{\infty} e^{-j(\omega-\omega_c)t}dt + \frac{1}{2}\int_{-\infty}^{\infty} e^{-j(\omega+\omega_c)t}dt$$

$$= \pi\delta(\omega-\omega_c) + \pi\delta(\omega+\omega_c)$$

$$\mathcal{F}\{e^{-|t|}\} = \int_{-\infty}^{\infty} e^{-|t|}e^{-j\omega t}dt = \int_{-\infty}^{0} e^{t}e^{-j\omega t}dt + \int_{0}^{\infty} e^{-t}e^{-j\omega t}dt = \frac{2}{1+\omega^2}$$

The spectrums are shown in Figure 5.2.5. The spectrum of the modulated function $f_m(t)$ is deduced as follows:

$$F_m(\omega) \triangleq \mathcal{F}\{e^{-|t|}\cos\omega_c t\} = \mathcal{F}\left\{\frac{e^{-|t|}e^{j\omega_c t}}{2} + \frac{e^{-|t|}e^{-j\omega_c t}}{2}\right\}$$

$$= \frac{1}{2}\int_{-\infty}^{0} e^{-(j\omega-j\omega_c-1)t}dt + \frac{1}{2}\int_{0}^{\infty} e^{-(j\omega+j\omega_c+1)t}dt$$

$$+ \frac{1}{2}\int_{-\infty}^{0} e^{-(j\omega+j\omega_c-1)t}dt + \frac{1}{2}\int_{0}^{\infty} e^{-(j\omega-j\omega_c+1)t}dt$$

$$= -\frac{1}{2}\frac{1}{j(\omega-\omega_c)-1} + \frac{1}{2}\frac{1}{j(\omega+\omega_c)+1}$$

$$-\frac{1}{2}\frac{1}{j(\omega+\omega_c)-1} + \frac{1}{2}\frac{1}{j(\omega-\omega_c)+1}$$

$$= \frac{1}{1+(\omega-\omega_c)^2} + \frac{1}{1+(\omega+\omega_c)^2}$$

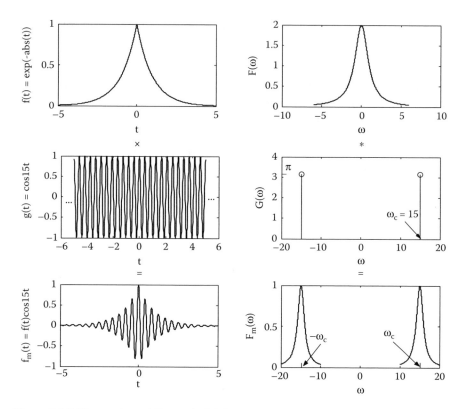

Figure 5.2.5 Illustration of Example 5.2.5.

The spectrum of $F_m(\omega)$ is also shown in Figure 5.2.5. Observe from Figure 5.2.4 and Figure 5.2.5 that when a signal is modulated by a sinusoidal function of frequency ω_c, the **carrier frequency**, the spectrum of the signal is shifted in the frequency domain by ω_c. This process, known as amplitude modulation (AM), is used in communications. Many of us listen to AM radio stations, which use this type of modulation for their broadcasting. ∎

The need for modulation systems arises from practical considerations. For transmission through space, the question of the antenna length is an important consideration. Without some form of modulation, the antenna problem is impracticable. A second consideration arises if we want to use a given **channel** to transmit a number of messages simultaneously without interference. We are familiar with the many broadcasting stations within the allowable broadcast band that use free space as the channel for transmission. In fact, this same free-space channel is used for frequency modulation (FM), TV, and many other forms of communication.

In the amplitude-modulated (AM) system, the carrier signal is easily produced by a stable oscillator with power amplifiers to establish the desired power level. The carrier signal is of the form $v_c(t) = V_c \cos \omega_c t$, where ω_c is

Chapter 5: Nonperiodic signals and their Fourier transform

the carrier frequency. The modulating signal in its simplest form (tone modulation) is $v_m(t) = V_m \cos \omega_m t$ and may denote one component of a complex signal pattern, where ω_m is the modulating frequency and $\omega_m \ll \omega_c$. The general form of the modulation process is specified by the equation

$$v(t) = V_c[1 + mv_m(t)]\cos \omega_c t \qquad (5.19)$$

where m is the **modulation index** and is defined as $m = V_m/V_c$. The maximum value of $mv_m(t)$ must be less than unity if distortion of the envelope is to be avoided. For the case of simple sinusoidal modulation, the equation can be expanded to

$$v(t) = V_c \cos \omega_c T + V_c m \cos \omega_c t \cos \omega_m t$$

or

$$v(t) = V_c \left\{ \cos \omega_c t + \frac{m}{2}[\cos(\omega_c + \omega_m)t + \cos(\omega_c - \omega_m)t] \right\} \qquad (5.20)$$

The amplitude spectrums of the three waveforms of (5.20) are given in Figure 5.2.6a. Figure 5.2.6b shows the system of an AM generator in schematic form. Observe from Figure 5.2.6 that the amplitude-modulated signal has twice the bandwidth of the original signal and, in addition, that the carrier is also present. From the symmetry property we have shown that the Fourier transform of 1 is $2\pi\delta(\omega)$. Therefore, when we want to find the Fourier transform of $\cos\omega_c t = (e^{j\omega_c t} + e^{-j\omega_c t})/2$, we end up finding the Fourier transforms of 1, but with frequencies $(\omega + \omega_c)$ and $(\omega - \omega_c)$, respectively. Hence, the result will be that shown in Figure 5.2.6a. Observe that in the figure we presented only the positive side of the spectrum.

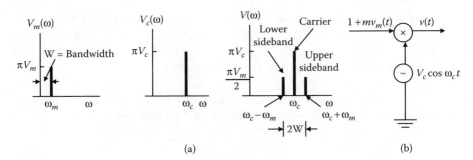

Figure 5.2.6 Amplitude modulation. (a) Frequency spectrum of an AM signal modulated by $V_m \cos \omega_m t$. (b) A schematic representation of an ideal generator of an AM signal.

Suppose that we had used a modulated signal of the form

$$v(t) = mv_m(t)\cos\omega_c t = m\cos\omega_m t \cos\omega_c t = \frac{m}{2}[\cos(\omega_c + \omega_m)t + \cos(\omega_c - \omega_m)t]$$

In such a system, the carrier frequency is not present. This type of modulation is known as **double-sideband suppressed carrier** (DSBSC) modulation. While such modulation does save energy of the carrier in the transmitted signal and the carrier component carries no information, subsequent extraction of the information component is rather difficult. Moreover, the information is contained redundantly in each sideband; hence, it is desirable to remove one or the other sideband to yield **single-sideband** (SSB) modulation.

The ability to shift the frequency spectrum of the information-carrying signal by means of amplitude modulation allows us to transmit many different signals simultaneously through a given channel, such as a transmission line, telephone line, microwave link, radio broadcast, and the like. This scheme is known as **frequency division multiplexing** (FDM). Of course, the ability to extract a given signal from the channel is equally essential, and a variety of **demodulation** or **detection schemes** exist. The sum of the frequency bands of the transmitted signals plus frequency guard bands separating the signals must be less than or equal to the bandwidth of the channel for undistorted transmission.

For purposes of classification, the frequency spectrum is roughly divided as follows: 3 to 300 kHz for telephony, navigation, industrial communication, and long-range navigation; 0.3 to 30 MHz for AM broadcasting, military communication, and amateur and citizen-band radio; 30 to 300 MHz for FM broadcasting, TV broadcasting, and land transportation; and 0.3 to 3 GHz for UHF TV, radar, and military applications. The frequencies above 30 GHz are used mostly for research purposes and radio astronomy.

The process of separating a modulating signal from a modulated signal is called **demodulation** or **detection**.

Example 5.2.6: Demodulation. Show that the spectrum of the modulated signal can be conveniently retranslated back to the original position by multiplying the modulated signal by $\cos\omega_c t$ at the receiving end. We assume that the modulating signal is band limited.

Solution: Figure 5.2.7a shows the schematic representation of the demodulation process. Let the modulated signal be expressed as

$$f(t) = m(t)\cos\omega_c t \qquad (5.21)$$

Then, as shown in Figure 5.2.7a, at the receiver we multiply the received signal $f(t)$, assuming that no noise was added, with the carrier $\cos\omega_c t$ to obtain, by the use of trigonometric identity,

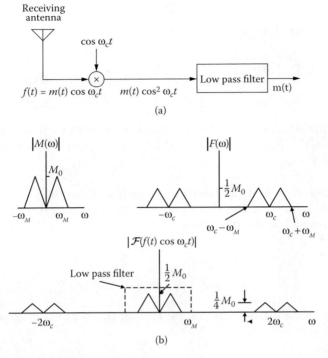

Figure 5.2.7 The demodulation process

$$f(t)\cos\omega_c t = m(t)\cos^2\omega_c t = m(t)\frac{1}{2}(1+\cos 2\omega_c t) \tag{5.22}$$

$$= \frac{1}{2}m(t) + \frac{1}{2}m(t)\cos 2\omega_c t$$

Now if

$$\mathcal{F}\{m(t)\} = M(\omega) \quad \text{and} \quad M(\omega) = 0 \quad \text{for} \quad |\omega| > \omega_M$$

where ω_M is the highest frequency of the modulating signal $m(t)$, then we obtain

$$\mathcal{F}\{f(t)\cos\omega_c t\} = \mathcal{F}\{m(t)\cos^2\omega_c t\} = \mathcal{F}\left\{\frac{1}{2}m(t)\right\} + \mathcal{F}\left\{\frac{1}{2}m(t)\cos 2\omega_c t\right\} \tag{5.23}$$

$$= \frac{1}{2}M(\omega) + \frac{1}{4}M(\omega - 2\omega_c) + \frac{1}{4}M(\omega + 2\omega_c)$$

The bottom part of Figure 5.2.7b indicates that we can recover the original signal $m(t)$ using a low-pass filter that passes the spectrum up to the highest frequency of $m(t)$. The low-pass filter is shown by the dotted lines. ∎

Derivatives

If $\mathcal{F}\{f(t)\} = F(\omega)$, then

$$\mathcal{F}\left\{\frac{df(t)}{dt}\right\} = j\omega F(\omega); \quad \mathcal{F}\left\{\frac{d^n f(t)}{dt^n}\right\} = (j\omega)^n F(\omega) \quad (5.24)$$

Example 5.2.7: Find the transformed input–output relationship of the system shown in Figure 5.2.8a, with the block diagram representation shown in Figure 5.2.8b.

Solution: It is first necessary to write the differential equation of the system in the time domain. This equation is the result of applying Kirchhoff's node equation. The resulting equation is

$$\frac{v(t)}{R} + C\frac{dv(t)}{dt} = i(t)$$

Because of the linearity property, the Fourier transform of both sides of this equation becomes

$$\left(\frac{1}{R} + j\omega C\right)V(\omega) = I(\omega)$$

from which,

$$\underbrace{V(\omega)}_{output} = \underbrace{\frac{1}{(1/R) + j\omega C}}_{system\ function}\underbrace{I(\omega)}_{input} = \frac{1}{Y(\omega)}I(\omega) = Z(\omega)I(\omega) = \underbrace{H(\omega)}_{system\ function}\underbrace{I(\omega)}_{input}$$

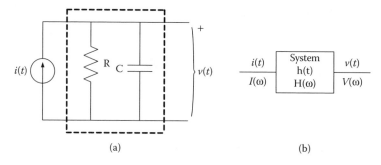

Figure 5.2.8 An electric system and its schematic representation.

Chapter 5: Nonperiodic signals and their Fourier transform

This expression shows that the input–output relationship for any LTI system is given by

$$\underbrace{G(\omega)}_{output} = \underbrace{H(\omega)}_{\substack{system\\function}} \underbrace{F(\omega)}_{input} \quad or \quad H(\omega) = system\ function = \frac{output}{input} = \frac{G(\omega)}{F(\omega)} \quad (5.25)$$

Note: When we find the system function (or equivalently, transfer function), all initial conditions are set to zero.

If the input to the system is an impulse function (delta function), the impulse response $h(t)$ of the system is deduced from the differential equation

$$\frac{h(t)}{R} + C\frac{dh(t)}{dt} = \delta(t) \quad (5.26)$$

Note: Observe that the units of $h(t)$ are volts, as they are in the node equation above.

Our first step is to find $h(t)$ at $t = 0$, following the procedure discussed in Section 2.7. Thus, we integrate from $t = 0-$ to $t = 0+$, which yields

$$\frac{1}{R}\int_{0-}^{0+} h(t)dt + \int_{0-}^{0+} \frac{dh(t)}{dt}dt = \int_{0-}^{0+} \delta(t)dt \quad or \quad \frac{1}{R}\times 0 + C[h(0+) - h(0-)] = 1$$

since $h(t)$ is assumed to be continuous at $t = 0$. This condition is true because $h(t)$ is the response of a physical system and $h(t)$ is not expected to be an impulse. Further, for a causal system, $h(0-) = 0$; hence, $h(0+) = 1/C$. Next, the solution to the homogeneous equation

$$\frac{dh(t)}{dt} + \frac{1}{RC}h(t) = 0$$

is easily found to be (see Chapter 2)

$$h(t) = Ae^{-t/RC} \quad t > 0$$

where A is an unknown constant. However, we have already found that $h(0+) = 1/C$; therefore, from the last equation, $A = 1/C$. Hence,

$$h(t) = \frac{1}{C}e^{-t/RC} \quad t > 0$$

The Fourier transform of this function is given by

$$H(\omega) = \frac{1}{C}\int_0^\infty e^{-t/RC} e^{-j\omega t} dt = -\frac{1}{C}\frac{1}{j\omega + \frac{1}{RC}} e^{-(j\omega+1/RC)t}\Big|_0^\infty = \frac{1}{\frac{1}{R}+j\omega C}$$

which is identical with the transfer function above, but which was found by a different approach.

We conclude from this example that the impulse response of a system can be deduced by the following steps:

1. Fourier-transform the differential equation that describes the system assuming a delta function input and zero initial conditions:

$$\mathcal{F}\left\{\frac{h(t)}{R} + C\frac{dh(t)}{dt}\right\} = \mathcal{F}\{\delta(t)\} \quad \text{or} \quad \left(\frac{1}{R}+j\omega C\right)H(\omega) = 1$$

2. Solve for the output function:

$$H(\omega) = \frac{R}{1+j\omega RC} = \frac{1}{C}\frac{1}{\frac{1}{RC}+j\omega}$$

3. Take the inverse Fourier transform:

$$\mathcal{F}^{-1}\{H(\omega)\} \triangleq h(t) = \frac{1}{2\pi C}\int_{-\infty}^\infty \frac{1}{\frac{1}{RC}+j\omega} e^{j\omega t} d\omega = \frac{1}{C}e^{-t/RC}$$

This result is obtained from the table in Appendix 5.1. ∎

Example 5.2.8: Deduce the impulse response of the system shown in Figure 5.2.9a from a knowledge of the system response to the voltage $v(t) = \exp(j\omega t)$. Indicate the response to a general voltage $v(t)$.

Solution: By an application of Kirchhoff's voltage law, we write

$$\frac{di(t)}{dt} + \frac{R}{L}i(t) = \frac{1}{L}v(t)$$

If the input is $v(t) = e^{j\omega t}$, the current (output) is the same sinusoidal (exponential form), but modified by change of its amplitude and phase; thus, we have

Chapter 5: Nonperiodic signals and their Fourier transform

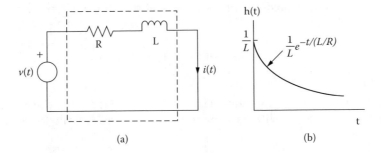

Figure 5.2.9 Simple linear time-invariant system.

$$i(t) = H(\omega)e^{j\omega t} = |H(\omega)|e^{j[\omega t + \tan^{-1}(H_i(\omega)/H_r(\omega))]}$$

where H_r and H_i are real functions of omega. Observe that $i(t)$ is a complex function. Introducing this value of $i(t)$ in the first equation we obtain

$$j\omega H(\omega)e^{j\omega t} + \frac{R}{L}H(\omega)e^{j\omega t} = \frac{1}{L}e^{j\omega t} \quad \text{or} \quad H(\omega) = \frac{1}{L}\frac{1}{\frac{R}{L}+j\omega}$$

This value of the transfer function is the value that would be obtained by using the Fourier transform on the controlling differential equation. The impulse response is obtained by taking the inverse Fourier transform of the transfer function:

$$h(t) \triangleq \mathcal{F}^{-1}\{H(\omega)\} = \frac{1}{2\pi}\frac{1}{L}\int_{-\infty}^{\infty}\frac{1}{(R/L)+j\omega}e^{j\omega t}d\omega = \frac{1}{L}e^{-t/(L/R)} \quad t > 0$$

The function is shown in Figure 5.2.9b. Thus, the solution for any input voltage $v(t)$ is given by

$$i(t) = h(t) * v(t) = \frac{1}{L}e^{-t/(L/R)}\int_{-\infty}^{t} v(x)e^{x/(L/R)}dx$$

∎

The result of these examples indicates that the transform of a differential equation that describes the physical system will be of the form

$$\sum_{n=0}^{N} a_n \frac{d^n g(t)}{dt^n} = \sum_{m=0}^{M} b_m \frac{d^m f(t)}{dt^m} \tag{5.27}$$

Taking the Fourier transform of both sides of the above equation, we find

$$\sum_{n=0}^{N} a_n (j\omega)^n G(\omega) = \sum_{m=0}^{M} b_m (j\omega)^m F(\omega) \qquad (5.28)$$

which is an algebraic equation. We will use similar procedures in later chapters when dealing with Laplace and z-transforms. Solving for the output we find

$$\boxed{\underbrace{G(\omega)}_{ouput} = \underbrace{\dfrac{\sum_{m=0}^{M} b_m (j\omega)^m}{\sum_{n=0}^{N} a_n (j\omega)^n}}_{system\ function} \underbrace{F(\omega)}_{input} = H(\omega)F(\omega)} \qquad (5.29)$$

The output in the time domain then follows from the inverse transform, although many times it is not an easy task to find a closed solution. Hence,

$$g(t) = \dfrac{1}{2\pi} \int_{-\infty}^{\infty} H(\omega)F(\omega)e^{j\omega t}\,d\omega \qquad (5.30)$$

Example 5.2.9: Deduce the transfer (system) function for the system shown in Figure 5.2.10.

Solution: By an application of Kirchhoff's equation to the two loops,

$$L\dfrac{di_1(t)}{dt} + R_1 i_1(t) - L\dfrac{di_2(t)}{dt} = v(t); \quad L\dfrac{di_1(t)}{dt} - (R_2 + R_3)i_2(t) - L\dfrac{di_2(t)}{dt} = 0$$

The Fourier transform of these equations yields

$$(j\omega L + R_1)I_1(\omega) - j\omega L I_2(\omega) = V(\omega); \quad j\omega L I_1(\omega) - (R_2 + R_3 + j\omega L)I_2(\omega) = 0$$

Solve this system of equations to find $I_2(\omega)$. The result is

$$I_2(\omega) = \dfrac{j\omega L}{(R_1 + j\omega L)(R_2 + R_3 + j\omega L) + \omega^2 L^2} V(\omega)$$

Since the output $V_o(\omega) = R_3 I_2(\omega)$, then

$$\dfrac{V_o(\omega)}{V(\omega)} \triangleq H(\omega) = \dfrac{j\omega L R_3}{R_1(R_2 + R_3) + j\omega L(R_1 + R_2 + R_3)}$$

Chapter 5: Nonperiodic signals and their Fourier transform

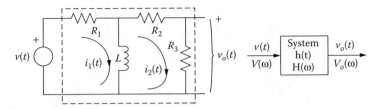

Figure 5.2.10 Illustration of Example 5.2.9.

As an interesting exercise, we now solve this problem when it is cast in the form given in (5.27). We subtract the first two equations deduced by an application of Kirchhoff's laws. This yields

$$i_1(t) = -\frac{R_2 + R_3}{R_1} i_2(t) + \frac{v(t)}{R_1}$$

Differentiate this equation and then substitute the results into the second of Kirchhoff's equations. Then, we have

$$L(R_2 + R_3 + R_1)\frac{di_2(t)}{dt} + R_1(R_2 + R_3)i_2(t) = L\frac{dv(t)}{dt}$$

which has the form (5.27). The Fourier transform of this equation is

$$[j\omega L(R_1 + R_2 + R_3) + R_1(R_2 + R_3)]I_2(\omega) = j\omega L V(\omega)$$

Multiplying the above equation by R_3 and setting $I_2 R_3 = V_o$, we obtain a transfer function identical to the one already found. ∎

Parseval's theorem

Parseval's relation is given by

$$\boxed{E = \int_{-\infty}^{\infty} |f(t)|^2 \, dt = \frac{1}{2\pi}\int_{-\infty}^{\infty} |F(\omega)|^2 \, dt} \tag{5.31}$$

Proof: Proceed as follows:

$$\int_{-\infty}^{\infty} |f(t)|^2 \, dt = \int_{-\infty}^{\infty} f(t)f^*(t)\,dt = \int_{-\infty}^{\infty}\left[\frac{1}{2\pi}\int_{-\infty}^{\infty} F(\omega)e^{j\omega t}\,d\omega\right]f^*(t)\,dt$$

$$= \frac{1}{2\pi} \int_{-\infty}^{\infty} F(\omega) \left[\int_{-\infty}^{\infty} f(t)e^{-j\omega t} dt \right]^* d\omega$$

$$= \frac{1}{2\pi} \int_{-\infty}^{\infty} F(\omega)F^*(\omega) d\omega = \frac{1}{2\pi} \int_{-\infty}^{\infty} |F(\omega)|^2 d\omega$$

This shows that the energy of the time domain signal is equal to the energy of the frequency domain. This is a statement of the conservation of energy.

If the **power density** spectrum of a signal is defined by

$$\boxed{W(\omega) = \frac{1}{2\pi} |F(\omega)|^2 = \text{spectrum power density}} \qquad (5.32)$$

then the energy in an infinitesimal band of frequencies $d\omega$ is $W(\omega)d\omega$, and the energy contained within a band of frequencies is

$$\Delta E = \int_{\omega_1}^{\omega_2} \frac{1}{2\pi} |F(\omega)|^2 d\omega \qquad \omega_1 \leq \omega \leq \omega_2 \qquad (5.33)$$

The fraction of the total energy that is contained with a band is

$$\frac{\Delta E}{E} = \frac{\text{energy in band}}{\text{total energy}} = \frac{\int_{\omega_1}^{\omega_2} \frac{1}{2\pi} |F(\omega)|^2 d\omega}{\int_{-\infty}^{\infty} \frac{1}{2\pi} |F(\omega)|^2 d\omega} = \frac{\int_{\omega_1}^{\omega_2} |F(\omega)|^2 d\omega}{\int_{-\infty}^{\infty} |F(\omega)|^2 d\omega} \qquad (5.34)$$

The interpretation of energy and power of signals in this manner is possible because $f(t)$ may be thought of as a voltage, so that $f(t)/(1 - \text{ohm resistor})$ is current, and thus $f^2(t)$ is proportional to power.

Example 5.2.10: Determine the total energy associated with the function $f(t) = e^{-t}u(t)$.

Solution: From (5.31), the total energy is

$$E = \int_{-\infty}^{\infty} (e^{-t})^2 u(t) dt = \int_0^{\infty} e^{-2t} dt = \frac{1}{2}$$

By using the results of Example 5.1.1, we can also proceed from a frequency point of view:

$$E = \frac{1}{2\pi} \int_{-\infty}^{\infty} \left(\frac{1}{1+j\omega} \frac{1}{1-j\omega} \right) d\omega = \frac{1}{2\pi} \int_{-\infty}^{\infty} \frac{1}{1+\omega^2} d\omega = \frac{1}{\pi} \int_0^{\infty} \frac{1}{1+\omega^2} d\omega = \frac{1}{\pi} \frac{\pi}{2} = \frac{1}{2}$$

Chapter 5: Nonperiodic signals and their Fourier transform

where we used the symmetry property of the integrand. Using mathematical tables, the last integral is equal to $\tan^{-1}(\omega)$ evaluated from zero to infinity. ∎

Another important form follows using the spectrum between input and output specified by the definition of the transfer function. Hence,

$$G(\omega) = F(\omega)H(\omega)$$

so that

$$|G(\omega)|^2 = |F(\omega)H(\omega)|^2 = F(\omega)H(\omega)F^*(\omega)H^*(\omega) = |F(\omega)|^2 |H(\omega)|^2 \quad (5.35)$$

In general, the transfer function is a complex function and can be written in the polar form as follows:

$$H(\omega) = \sqrt{H_r^2(\omega) + H_i^2(\omega)} e^{j\tan^{-1}[H_i(\omega)/H_r(\omega)]} = H_0(\omega)e^{j\theta(\omega)} \quad (5.36)$$

where H_r, H_i, and H_0 are real functions of frequency. Therefore,

$$\boxed{|G(\omega)|^2 = H_0^2(\omega)|F(\omega)|^2} \quad (5.37)$$

Note: *The power density spectrum of the response of an LTI system is the product of the power density spectrum of the input function and the square amplitude function of the system. The phase characteristics of the system do not affect the energy density of the output.*

Example 5.2.11: Find the frequency response of an LTI system given that the input is $f(t) = e^{-t}u(t)$ and the output is $g(t) = 2e^{-2t}u(t)$. Determine the impulse response of the system.

Solution: From (5.25), we write the relation

$$H(\omega) = \frac{\text{output spectrum}}{\text{input spectrum}} = \frac{2/(2+j\omega)}{1/(1+j\omega)} = 2\frac{1+j\omega}{2+j\omega} = 2 - \frac{2}{2+j\omega}$$

The impulse response is the inverse Fourier transform of the above equation, and hence

$$h(t) = 2\delta(t) - 2e^{-2t} \qquad t \geq 0$$

∎

Time convolution

If $F(\omega)$ and $H(\omega)$ denote the Fourier transforms of $f(t)$ and $h(t)$, respectively, then

$$\boxed{\mathcal{F}\{f(t)*h(t)\} = \mathcal{F}\left\{\int_{-\infty}^{\infty} f(x)h(t-x)dx\right\} = F(\omega)H(\omega)} \quad (5.38)$$

Proof: The Fourier transform

$$\mathcal{F}\{f(t)*h(t)\} = \int_{-\infty}^{\infty}[f(t)*h(t)]e^{-j\omega t}dt = \int_{-\infty}^{\infty}f(x)dx\int_{-\infty}^{\infty}h(t-x)e^{-j\omega t}dt$$

$$= \int_{-\infty}^{\infty}f(x)e^{-j\omega x}dx\int_{-\infty}^{\infty}h(s)e^{-j\omega s}ds = F(\omega)H(\omega)$$

where we have written $t - x = s$. This result agrees with (5.25). Furthermore, if we assume that $f(t)$ is the input to an LTI system, $h(t)$ is its impulse response, and $g(t)$ its output, then (5.38) is written in the form

$$g(t) = \int_{-\infty}^{\infty} f(x)h(t-x)dx = \mathcal{F}^{-1}\{F(\omega)H(\omega)\} = \frac{1}{2\pi}\int_{-\infty}^{\infty}F(\omega)H(\omega)e^{j\omega t}d\omega \quad (5.39)$$

This expression shows that the output of an LTI system in the time domain is equal to the inverse Fourier transform of the product of the transfer function and the Fourier transform of its input.

Example 5.2.12: Determine the Fourier transform of the function $f(t) = te^{-t}u(t)$.

Solution: We observe that $f(t)$ is the convolution of $f_1(t) = e^{-t}u(t)$ and $f_2(t) = e^{-t}u(t)$. However, the Fourier transform of $e^{-t}u(t)$ is $1/(1 + j\omega)$, so that $F(\omega) = 1/(1 + j\omega)^2$. ∎

Example 5.2.13: Determine the frequency spectrum of the output for the system shown in Figure 5.2.11.

Solution: By an application of Kirchhoff's current law at a node, the controlling differential equation is

$$\frac{dv_o(t)}{dt} + 5v_o(t) = 5p(t)$$

Chapter 5: Nonperiodic signals and their Fourier transform

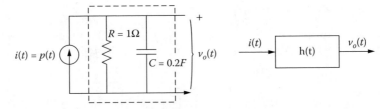

Figure 5.2.11 Illustration of Example 5.2.13.

The impulse response of the system is obtained from the equation

$$\frac{dh(t)}{dt} + 5h(t) = \delta(t)$$

The Fourier transform of this equation is

$$j\omega H(\omega) + 5H(\omega) = 1 \quad \text{or} \quad H(\omega) = 1/(5+j\omega)$$

The inverse Fourier transform of the above equation is

$$h(t) = e^{-5t}u(t)$$

However, the output is the convolution of the system input and its impulse response, and the frequency spectrum of its output is given by the product

$$V_o(\omega) = \mathcal{F}\{5p(t)\}\mathcal{F}\{e^{-5t}u(t)\} = 10\frac{\sin\omega}{\omega}\frac{1}{5+j\omega}$$

We would also proceed by Fourier-transforming the first equation. This yields

$$j\omega V_o(\omega) + 5V_o(\omega) = 10\frac{\sin\omega}{\omega} \quad \text{or} \quad V_o(\omega) = 10\frac{\sin\omega}{\omega}\frac{1}{5+j\omega}$$

which is identical to our previous result. ∎

Frequency convolution

If $f(t)$ and $h(t)$ are transformable functions, then

$$\boxed{\mathcal{F}\{f(t)h(t)\} = \frac{1}{2\pi}\int_{-\infty}^{\infty} F(x)H(\omega - x)dx = \frac{1}{2\pi}F(\omega) * H(\omega)} \quad (5.40)$$

Proof: We proceed by writing

$$\mathcal{F}^{-1}\{\mathcal{F}\{f(t)h(t)\}\} \triangleq f(t)h(t) = \mathcal{F}^{-1}\left\{\frac{1}{2\pi}F(\omega)*H(\omega)\right\}$$

$$= \frac{1}{2\pi}\frac{1}{2\pi}\int\int_{-\infty}^{\infty}F(x)H(\omega-x)e^{j\omega t}dxd\omega$$

$$= \frac{1}{2\pi}\int_{-\infty}^{\infty}F(x)e^{jtx}dx\frac{1}{2\pi}\int_{-\infty}^{\infty}H(s)e^{jts}ds$$

where we have written $\omega - x = s$, that is $\omega = x + s$ and $d\omega = ds$.

Example 5.2.14: Verify the results of Example 5.2.5 by using the convolution property of the Fourier transform.

Solution: The Fourier transform of the function $\cos\omega_c t$ is given by

$$\mathcal{F}\{\cos\omega_c t\} = \mathcal{F}\left\{\frac{e^{j\omega_c t}+e^{-j\omega_c t}}{2}\right\} = \frac{1}{2}\int_{-\infty}^{\infty}e^{-j(\omega-\omega_c)t}dt + \frac{1}{2}\int_{-\infty}^{\infty}e^{-j(\omega+\omega_c)t}dt$$

$$= \pi\delta(\omega-\omega_c) + \pi\delta(\omega+\omega_c)$$

Further, we know that $\mathcal{F}\{e^{-|t|}\} = 2/(1+\omega^2)$. Next, use (5.40) and apply the delta function properties (the value of the integral is found by substituting the value of the position in which the delta function is located to the rest of the functions in the integrand). Thus,

$$\mathcal{F}\{\cos\omega_c t\, e^{-|t|}\} = \frac{1}{2\pi}\int_{-\infty}^{\infty}[\pi\delta(x-\omega_c)+\pi\delta(x+\omega_c)]\frac{2}{1+(\omega-x)^2}dx$$

$$= \int_{-\infty}^{\infty}\delta(x-\omega_c)\frac{1}{1+(\omega-x)^2}dx + \int_{-\infty}^{\infty}\delta(x+\omega_c)\frac{1}{1+(\omega-x)^2}dx$$

$$= \frac{1}{1+(\omega-\omega_c)^2} + \frac{1}{1+(\omega+\omega_c)^2}$$

∎

Summary of continuous-time Fourier properties

Linearity $\qquad\qquad af(t)+bh(t) \xleftrightarrow{\mathcal{F}} aF(\omega)+bH(\omega)$

Time shifting $\qquad\quad f(t\pm t_0) \xleftrightarrow{\mathcal{F}} e^{\pm j\omega t_0}F(\omega)$

Chapter 5: Nonperiodic signals and their Fourier transform

Symmetry
$$\begin{cases} F(t) \xleftrightarrow{\mathcal{F}} 2\pi f(-\omega) \\ 1 \xleftrightarrow{\mathcal{F}} 2\pi \delta(-\omega) = 2\pi \delta(\omega) \end{cases}$$

Time scaling
$$f(at) \xleftrightarrow{\mathcal{F}} \frac{1}{|a|} F\left(\frac{\omega}{a}\right)$$

Time reversal
$$f(-t) \xleftrightarrow{\mathcal{F}} F(-\omega) \text{ (real time functions)}$$

Frequency shifting
$$e^{\pm j\omega_0 t} f(t) \xleftrightarrow{\mathcal{F}} F(\omega \mp \omega_0)$$

Modulation
$$\begin{cases} f(t)\cos\omega_0 t \xleftrightarrow{\mathcal{F}} \frac{1}{2}[F(\omega+\omega_0) + F(\omega-\omega_0)] \\ f(t)\sin\omega_0 t \xleftrightarrow{\mathcal{F}} \frac{1}{2j}[F(\omega-\omega_0) - F(\omega+\omega_0)] \end{cases}$$

Time differentiation
$$\frac{d^n f(t)}{dt^n} \xleftrightarrow{\mathcal{F}} (j\omega)^n F(\omega)$$

Frequency differentiation
$$\begin{cases} (-jt)f(t) \xleftrightarrow{\mathcal{F}} \frac{dF(\omega)}{d\omega} \\ (-jt)^n f(t) \xleftrightarrow{\mathcal{F}} \frac{d^n F(\omega)}{d\omega^n} \end{cases}$$

Time convolution
$$f(t)*h(t) = \int_{-\infty}^{\infty} f(x)h(t-x)dx \xleftrightarrow{\mathcal{F}} F(\omega)H(\omega)$$

Frequency convolution
$$f(t)h(t) \xleftrightarrow{\mathcal{F}} \frac{1}{2\pi} F(\omega)*H(\omega) = \frac{1}{2\pi} \int_{-\infty}^{\infty} F(x)H(\omega-x)dx$$

Autocorrelation
$$f(t) \odot f(t) = \int_{-\infty}^{\infty} f(x)f^*(x-t)dx \xleftrightarrow{\mathcal{F}} F(\omega)F^*($$
$$= |F(\omega)|^2$$

Central ordinate
$$f(0) = \frac{1}{2\pi} \int_{-\infty}^{\infty} F(\omega)d\omega \qquad F(0) = \int_{-\infty}^{\infty} f(t)dt$$

Parseval's theorem
$$E = \int_{-\infty}^{\infty} |f(t)|^2 dt \qquad E = \frac{1}{2\pi} \int_{-\infty}^{\infty} |F(\omega)|^2 d\omega$$

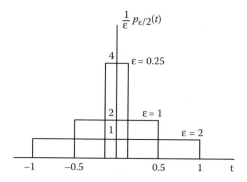

Figure 5.3.1 A delta function sequence.

*5.3 Some special Fourier transform pairs

To find the Fourier transform of the delta function, although it does not satisfy the Dirichlet conditions, we use a limiting process. We consider some special functions below.

Example 5.3.1: Find the Fourier transform of the delta function illustrated in Figure 5.3.1 and defined here by the limiting formula for different positive real values of ε.

$$\delta(t) = \lim_{\varepsilon \to 0} \frac{1}{\varepsilon} p_{\varepsilon/2}(t)$$

Solution: The Fourier transform of this equation is

$$\mathcal{F}\{\delta(t)\} = \mathcal{F}\left\{\lim_{\varepsilon \to 0} \frac{1}{\varepsilon} p_{\varepsilon/2}(t)\right\} = \lim_{\varepsilon \to 0} \mathcal{F}\left\{\frac{1}{\varepsilon} p_{\varepsilon/2}(t)\right\} = \lim_{\varepsilon \to 0} \frac{\sin(\varepsilon \omega/2)}{(\varepsilon \omega/2)} = 1 \quad (5.41)$$

where the limit is found using L'Hospital's rule on limits. ∎

Example 5.3.2: Find the Fourier transform of the function sgn(t) shown in Figure 5.3.2a.

Solution: We proceed to write the signum function as $\lim_{\varepsilon \to 0}$ exp $[-\varepsilon|t|]\mathrm{sgn}(t)$, as shown in Figure 5.3.2b. The procedure is now direct and yields

$$\mathcal{F}\{\mathrm{sgn}(t)\} = \mathcal{F}\left\{\lim_{\varepsilon \to 0} e^{-\varepsilon|t|} \mathrm{sgn}(t)\right\} = \lim_{\varepsilon \to 0} \int_{-\infty}^{\infty} e^{-\varepsilon|t|} \mathrm{sgn}(t) e^{-j\omega t} dt$$

$$= \lim_{\varepsilon \to 0} \left[\int_{-\infty}^{0} -e^{(\varepsilon-j\omega)t} dt + \int_{0}^{\infty} e^{-(\varepsilon+j\omega)t} dt \right] \quad (5.42)$$

Chapter 5: Nonperiodic signals and their Fourier transform

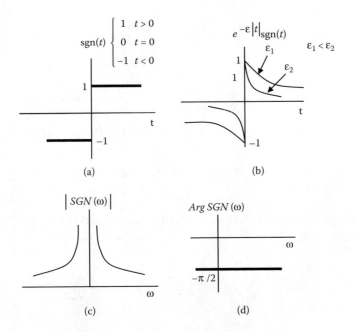

Figure 5.3.2 The sgn function and its Fourier representation.

$$= \lim_{\varepsilon \to 0}\left(-\frac{1}{\varepsilon - j\omega} + \frac{1}{\varepsilon + j\omega}\right) = \frac{2}{j\omega} = \frac{2}{\omega}e^{-j\pi/2} \triangleq SGN(\omega)$$

The Fourier amplitude and phase spectrums of sgn(t) are shown in Figures 5.3.2c and d. In this proof, we changed the order of limit and integration. Such an interchange is not always appropriate and requires mathematical justification, which is beyond the scope of this text. ∎

Example 5.3.3: Find the Fourier transform of the unit step function $u(t)$.

Solution: Begin by writing $u(t)$ in its equivalent representation involving the sgn(t) function:

$$u(t) = \frac{1}{2} + \frac{1}{2}\text{sgn}(t) \tag{5.43}$$

Therefore, the Fourier transform is

$$U(\omega) \triangleq \mathcal{F}\{u(t)\} = \frac{2\pi}{2}\delta(\omega) + \frac{2}{2j\omega} = \pi\delta(\omega) + \frac{1}{j\omega} \tag{5.44}$$

The real and imaginary parts of its spectrum are shown in Figure 5.3.3. ∎

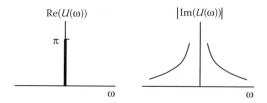

Figure 5.3.3 Fourier transform representation of the unit step function u(t).

Example 5.3.4: Find the Fourier transform of the $comb_T(t)$ function shown in Figure 5.3.4a.

Solution: Carrying out this problem requires that the $comb_T(t)$ function first be represented in its Fourier series expression, which is

$$comb_T(t) = \sum_{n=-\infty}^{\infty} \delta(t-nT) \triangleq \sum_{n=-\infty}^{\infty} a_n e^{jn\omega_0 t} = \frac{1}{T} \sum_{n=-\infty}^{\infty} e^{jn\omega_0 t} \qquad (5.45)$$

where

$$a_n = \frac{1}{T} \int_{-T/2}^{T/2} \delta(t) e^{-jn\omega_0 t} dt = \frac{1}{T} \qquad \omega_0 = \frac{2\pi}{T}$$

The Fourier transform is then

$$\mathcal{F}\{comb_T(t)\} = \frac{1}{T} \sum_{n=-\infty}^{\infty} \mathcal{F}\{e^{jn\omega_0 t}\} = \frac{1}{T} \sum_{n=-\infty}^{\infty} 2\pi \delta(\omega - n\omega_0)$$

$$= \frac{2\pi}{T} \sum_{n=-\infty}^{\infty} \delta\left(\omega - n\frac{2\pi}{T}\right) \triangleq \frac{2\pi}{T} COMB_{\omega_0}(\omega) \qquad (5.46)$$

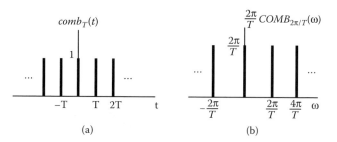

Figure 5.3.4 Fourier transform of the $comb_T(t)$ function.

Chapter 5: Nonperiodic signals and their Fourier transform

The foregoing shows that any periodic function of the form

$$f(t) = \sum_{n=-\infty}^{\infty} a_n e^{jn\omega_0 t} \qquad (5.47)$$

has a Fourier transform of the form

$$\mathcal{F}\{f(t)\} \triangleq F(\omega) = 2\pi \sum_{n=-\infty}^{\infty} a_n \delta(\omega - n\omega_0) \qquad (5.48)$$

∎

Example 5.3.5: Pulse amplitude modulation (PAM). Find the spectrum of the pulse amplitude modulation (PAM) case, if $m(t)$ (see Figures 5.3.5a and b) is a band-limited modulating signal and $g(t)$ is a train of periodic rectangular pulses of width $2a$ seconds and repeating every T seconds (see Figures 5.3.5c and d).

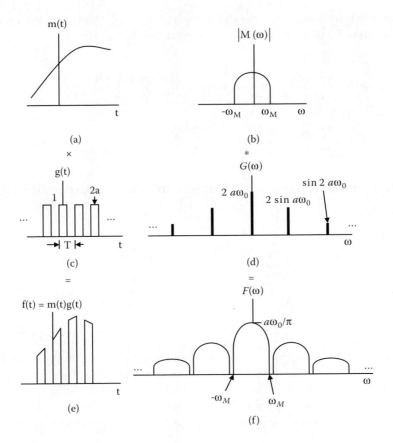

Figure 5.3.5 Pulse amplitude modulation.

Solution: Let

$$\mathcal{F}\{m(t)\} = M(\omega) \quad M(\omega) = 0 \text{ for } |\omega| > \omega_M \tag{5.49}$$

According to the frequency convolution theorem (5.40), the Fourier transform of $f(t) = m(t)g(t)$ is

$$F(\omega) \triangleq \mathcal{F}\{f(t)\} = \mathcal{F}\{m(t)g(t)\} = \frac{1}{2\pi} M(\omega) * G(\omega), \quad G(\omega) = \mathcal{F}\{g(t)\} \tag{5.50}$$

The signal $g(t)$ can be constructed by the convolution of the $comb_T(t)$ and the one-period signal (centered at the origin) extracted from $g(t)$, which we can call $g_1(t)$. Hence,

$$G(\omega) = \mathcal{F}\{comb_T(T) * g_1(t)\} = \mathcal{F}\{comb_T(t)\}\mathcal{F}\{g_1(t)\} \tag{5.51}$$

The Fourier transform of the comb function is given by (5.46), and the Fourier of the centered pulse of width $2a$ is given by (5.9). Hence,

$$G(\omega) = \frac{2\sin a\omega}{\omega} \frac{2\pi}{T} \sum_{n=-\infty}^{\infty} \delta(\omega - n\omega_0) = \frac{4\pi}{T} \sum_{n=-\infty}^{\infty} \frac{\sin an\omega_0}{n\omega_0} \delta(\omega - n\omega_0)$$

$$= 2\sum_{n=-\infty}^{\infty} \frac{\sin an\omega_0}{n} \delta(\omega - n\omega_0), \quad \omega_0 = \frac{2\pi}{T} \tag{5.52}$$

and it is plotted in Figure 5.3.5d. Therefore, introducing (5.52) into (5.50), we obtain

$$F(\omega) = \frac{1}{2\pi} 2\sum_{n=-\infty}^{\infty} \frac{\sin an\omega_0}{n} \int_{-\infty}^{\infty} \delta(x - n\omega_0) M(\omega - x) dx$$

$$= \frac{1}{\pi} \sum_{n=-\infty}^{\infty} \frac{\sin an\omega_0}{n} M(\omega - n\omega_0) \tag{5.53}$$

We observe from the figure and the analysis above that the output spectrum is the convolution of the spectrum of the periodic function and the spectrum of the modulating function. From Figure 5.3.5f we observe that we can recover exactly the signal $m(t)$ if we carefully select the sampling time T and let $f(t)$ pass through an ideal low-pass filter with bandwidth from $-\omega_M$ to ω_M. ∎

*5.4 Effects of truncation and Gibbs' phenomenon

From the symmetry property we have found out that the spectrum of a pulse is a sinc function. Then, if we create a time function from this sinc function, its Fourier transform is a pulse. If a pulse is symmetric about the origin and has a width of 2 and height of 1, then its Fourier transform is $2\sin\omega/\omega$. We then change this function to a time function and divide it by 2π to obtain

$$f(t) = \frac{\sin t}{\pi t} \qquad -\infty < t < \infty$$

The inverse Fourier transform is the pulse. Next, instead of taking the whole length for its inversion, we first take a portion from −5 to 5. What this, in theory, does is multiply the sinc function with a pulse of width 10 and make it symmetric about the origin. Hence, when we take the Fourier transform of the truncated sinc function, it is equivalent to taking the transform of the product functions, which are the sinc function and the pulse. But from the product property of the Fourier transform we know that the spectrum is the convolution of these two spectrums. Hence, we do not expect an exact reproduction of the pulse, but an approximation to it. Figure 5.4.1 shows the results when the truncation is from −5 to 5 and −30 to 30. Observe that as we increase the time bandwidth, the resemblance to the pulse form improves. However, even if we increase further the time bandwidth, the undulations in the middle will be diminished, but at the edge the undulation will remain about 9%. This is known as **Gibbs' phenomenon**.

The same happens if we truncate the spectrum and try to find its time function by using the inverse Fourier transform. We have found that the Fourier transform of $f(t) = \exp(-|t|)$ is $2/(1+\omega^2)$ and the range is from minus infinity to infinity. Next we use the following two frequency ranges, $-0.5 \le \omega \le 0.5$ and $-6 \le \omega \le 6$, respectively, and proceed to find the inverse transforms.

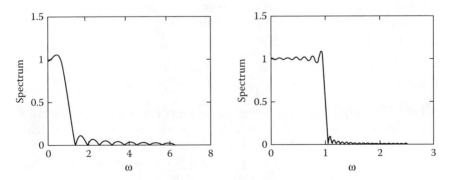

Figure 5.4.1 The effect of truncation in the time domain.

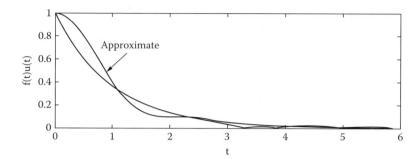

Figure 5.4.2 Truncation in the frequency domain.

The results are shown in Figure 5.4.2. Observe that if the range is not infinity, an approximate time function is found (we plotted here the positive time only). Since we truncated the spectrum, in effect, we multiplied the exact spectrum with a rectangular window spectrum. From the time convolution property of the Fourier transform, we know that the approximate time function is equal to the convolution of the original time function with the inverse Fourier transform of the pulse in the frequency domain. The inverse transform of the pulse is a sinc function in the time domain. Hence, the result is an approximation to the exact function since any convolution creates a signal that is not exactly either of the two that were involved in the convolution process.

*5.5 Linear time-invariant filters

The Fourier transform discussed in the previous sections has important uses in the study of the input–output relationships of linear time-invariant (LTI) systems. It is also important in determining the frequency spectra of LTI systems. We have already found that the output of an LTI system is related to the input through a convolution integral. Moreover, we have shown that this relationship becomes a simple algebraic one in the frequency domain: the frequency spectrum of the output is equal to the product of the spectrum of the input and the spectrum of the impulse response (inverse Fourier transform of the transfer function) of the system.

Note: *The output spectrum in the most general terms is a filtered version of the input.*

Filters, which for our purposes are systems that influence a signal in a prescribed way, have many useful applications and are used in diverse fields — for example, filtering of mechanical vibrations, fluctuations of economic data, electrical signals, biological signals, and two-dimensional signals (images), just to name a few. Their construction is such that when filters are included in a given system, the detected output will have some desired characteristics. For example, when we purchase an audio amplifier (a hi-fi

Chapter 5: Nonperiodic signals and their Fourier transform

system), we expect that is equipped with both bass and treble controls so that we can modify to our liking the low- and high-frequency responses of the program to which we are listening. What is actually provided in the amplifier is essentially two variable filters. In the case of AM communications, we may wish to transmit the upper sideband only, since it contains all the information of the transmitted signal. This might require inserting a high-pass filter in the system so that only those frequencies above the carrier frequency are passed. Filtering is used extensively to eliminate undesired low- and high-frequency noise. In later chapters we will discuss how to build analog and digital filters to accomplish the appropriate filtering of signals at hand.

As already discussed, the output spectrum of $g(t)$ is equal to the product of the spectrums of the input $f(t)$ and the impulse response $h(t)$ of the system (filter) and is given by the relation

$$\boxed{G(\omega) = F(\omega)H(\omega)} \tag{5.54}$$

In general, the system function is a complex function and can be set in the form

$$H(\omega) \triangleq |H(\omega)| e^{j\theta_h(\omega)} = \sqrt{H_r^2 + H_i^2}\, e^{j\tan^{-1}(H_i/H_r)} \tag{5.55}$$

where $|H(\omega)|$ is the **amplitude transfer function** and $\theta_h(\omega)$ is the **phase transfer function**. Similarly, we can write the input signal spectrum function in the form

$$F(\omega) \triangleq |F(\omega)| e^{j\theta_f(\omega)} = \sqrt{F_r^2 + F_i^2}\, e^{j\tan^{-1}(F_i/F_r)} \tag{5.56}$$

where $|F(\omega)|$ is the amplitude spectrum of the input signal and $\theta_f(\omega)$ is its phase spectrum. H_r, H_i, F_r, and F_i are all real functions of the frequency. Combining the last three equations, we obtain the relation

$$\boxed{\begin{aligned} G(\omega) &\triangleq |G(\omega)| e^{j\theta_g(\omega)} = |H(\omega)||F\omega| e^{j[\theta_h(\omega)+\theta_f(\omega)]} \\ |G(\omega)| &= |H(\omega)||F\omega| \\ \theta_g(\omega) &= \theta_h(\omega) + \theta_f(\omega) \end{aligned}} \tag{5.57}$$

Note: *The amplitude of the output spectrum is the product of the amplitude spectrum of the input and the amplitude spectrum of the filter (system function or transfer function). The phase spectrum of the output is equal to the sum of the phase spectrum of the input and the phase spectrum of the filter.*

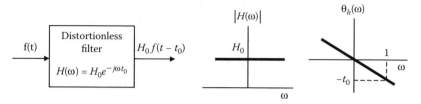

Figure 5.5.1 The transfer function of a nondistortion filter.

Distortionless filter

A filter is distortionless when the form of the output is identical to that of the input, except perhaps for a change in the amplitude and a possible time lag. Under these conditions, the output is specified as

$$g(t) = H_o f(t - t_0) \tag{5.58}$$

where $f(t)$ is the input signal and H_o is the constant amplitude response of the filter, which may be less or equal to unity. Taking the Fourier transform of this equation, we obtain

$$G(\omega) = H_o e^{-j\omega t_0} F(\omega) \tag{5.59}$$

Hence, the system function of a distortionless filter (see 5.54) is

$$\boxed{H(\omega) = H_o e^{-j\omega t_0} \triangleq \text{Transfer function of distortionless filter}} \tag{5.60}$$

A graphical representation of this function is shown in Figure 5.5.1.

Example 5.5.1: Find the conditions under which the filter shown in Figure 5.5.2 is of the distortionless type.

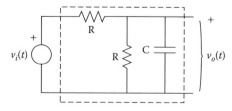

Figure 5.5.2 Illustration of Example 5.5.1.

Chapter 5: Nonperiodic signals and their Fourier transform

Solution: From the voltage division rule, or applying the node Kirchhoff's equation and then taking the Fourier transform, we obtain

$$H(\omega) = \frac{V_o(\omega)}{V_i(\omega)} = \frac{R/(1+jRC\omega)}{R + R/(1+jRC\omega)} = \frac{1}{2+jRC\omega} = \frac{1}{\sqrt{4+(RC\omega)^2}} e^{-j\tan^{-1}(RC\omega/2)}$$

Since

$$\tan^{-1}(RC\omega/2) = (RC\omega/2) - \frac{(RC\omega/2)^3}{3!} + \frac{(RC\omega/2)^5}{5!} - \cdots$$

for $(RC\omega/2) \ll 1$, the arctangent is approximated by $RC\omega/2$, and hence the transfer function becomes

$$H(\omega) \cong \frac{1}{2} e^{-j(RC/2)\omega} \qquad \omega \ll \frac{2}{RC}$$

■

Ideal low-pass filter

An ideal low-pass filter is one that has a transfer function of the form

$$\boxed{H(\omega) = H_0 p_{\omega_0}(\omega) e^{-j\omega t_0} \triangleq \text{Ideal low-pass filter}} \qquad (5.61)$$

The characteristics of this filter are shown in Figure 5.5.3a. Clearly, the ideal low-pass filter is one that permits ideal transmission over a specified band of frequencies and completely excludes all other frequencies. Such ideal filters are not physically realizable, as discussed below, but the concept does help in an understanding of different types of physical filters whose response may be approximated by those of the ideal filters.

The impulse response of the ideal low-pass filter specified by (5.61) is found by taking the inverse Fourier transform of $H(\omega)$. This is

$$h(t) = \frac{1}{2\pi} \int_{-\infty}^{\infty} H_0 p_{\omega_0}(\omega) e^{-j\omega t_0} e^{j\omega t} d\omega = \frac{1}{2\pi} H_0 \int_{-\omega_0}^{\omega_0} e^{j\omega(t-t_0)} d\omega$$

or

$$\boxed{h(t) = \frac{H_0}{\pi} \frac{\sin[\omega_0(t-t_0)]}{t-t_0} = \frac{H_0}{\pi} \sin c_{\omega_0}(t-t_0) \\ \triangleq \text{Impulse response of ideal low-pass filter}} \qquad (5.62)$$

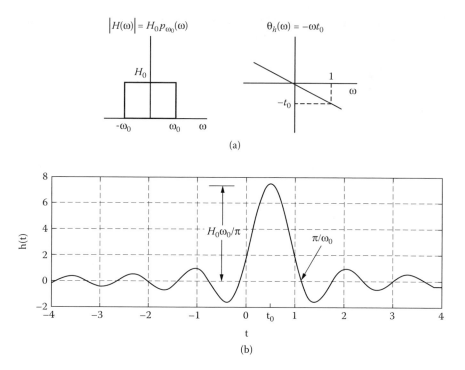

Figure 5.5.3 The impulse response characteristics of an ideal low-pass filter.

The form dictated by this expression is shown in Figure 5.5.3b. This figure shows that the impulse response is not identically zero for time equal to or less than zero. This would suggest that when a delta source is applied at $t = 0$, the system anticipates the input and a signal occurs at the output of the system even before the source has been applied. Of course, this is not true for physical systems. This is proof that ideal filters are not physically realizable; that is, it is not possible to build an ideal filter with any combination of resistors, capacitors, and inductors.

Ideal high-pass filter

The frequency characteristic of an ideal high-pass filter is given by

$$H(\omega) = [H_0 - H_0 p_{\omega_0}(\omega)] e^{-j\omega t_0} \tag{5.63}$$

The corresponding impulse response function, which is obtained by taking the inverse Fourier transform of this equation, is

$$\boxed{h(t) = H_0 \delta(t - t_0) - \frac{H_0}{\pi} \frac{\sin[\omega_0 (t - t_0)]}{t - t_0}} \tag{5.64}$$

Chapter 5: Nonperiodic signals and their Fourier transform 231

Important definitions and concepts

1. Uniqueness of spectrum for each signal
2. The Fourier transform and its inverse constitute a pair
3. The shifting of a function affects only the phase spectrum of the signal
4. The Fourier transform of delta function $\delta(t)$ is 1, and the Fourier transform of a constant A is $2\pi A \delta(\omega)$
5. The faster a signal varies, per unit time, the richer it is in high frequencies
6. Modulation succeeds to shift the spectrum of the modulating function to the carrier frequency
7. A channel is any medium through which an information signal is transmitted.
8. The opposite of a modulation process is called demodulation or detection
9. The ratio of the output spectrum of an LTI system to the input, assuming sinusoidal excitation, is known as the transfer or system function
10. Parseval's identity indicates that the *total* energy of the signal can be found from its time domain or frequency domain

    Note: *Parseval's identity tells us about the total energy of the signal, but it does not tell us how the energy of a signal contained at a specified range in the time domain is distributed in the frequency domain. The same is true if we consider the transformation from the frequency domain to the time domain. This is one major drawback of the Fourier transform operation.*

11. Power spectrums densities
12. Pulse amplitude modulation
13. Effects of truncation in time and frequency domains and Gibbs' phenomenon
14. Ideal filters

Chapter 5 Problems

Section 5.1

1. Deduce the spectrum functions of the time functions shown in Figure P5.1.1a. Deduce the spectrum function of Figure P5.1.1b by (a) using the Fourier integral and (b) combining the results found in Figure P5.1.1a. (Note that the pulse function must be multiplied by A.)
2. Refer to Figure P5.1.2. (a) Deduce the even and odd functions $F_r(\omega)$ and $F_i(\omega)$ for this function. (b) Decompose the given function in the time domain into two functions with even and odd parts and again find $F_r(\omega)$ and $F_i(\omega)$. Compare the results with the results in part (a).

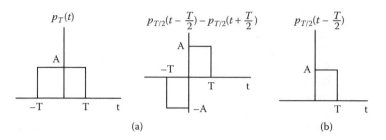

Figure P5.1.1

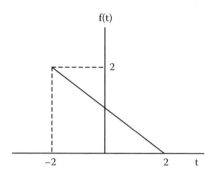

Figure P5.1.2

3. Deduce the Fourier transform of the odd function

$$f(t) = \begin{cases} e^{-t}u(t) \\ -e^{t}u(-t) \end{cases}$$

and verify the odd and real function properties of the Fourier transform.

4. Determine the Fourier transform of the following functions and give the value of the $|F(\omega)|$ as $\omega \to \infty$.

   a. $f(t) = p_{0.5}(t) + \delta(t+1)$

   b. $f(t) = e^{-t}u(t) + \delta(t-2)$

Section 5.2

1. Prove (5.13), (5.14), (5.17), and (5.18).
2. Find the Fourier transform of the following functions and sketch their amplitude and phase spectrums:

a. $f(t) = p_3(t-2)$

b. $f(t) = p_2(t-2) + p_2(t+2)$

c. $f(t) = e^{-|t|} \cos \omega_0 t$

d. $f(t) = e^{j3t} p_1(t)$

e. $f(t) = e^{j2t} p_2(3t)$

3. Find the Fourier transform of the following functions:

   a. $f(t) = 1/(a+jt)$

   b. $f(t) = 1/(1+t^2)$

   c. $f(t) = \sin(at)/t$

   d. $f(t) = t^2$

4. A signal $f(t) = e^{-t} u(t)$ is applied to a system that has an impulse response $h(t) = e^{-3t} u(t)$. Specify the system output and find its spectrum.

5. Systems are given in block diagram form in Figure P5.2.5. Find the transfer function of these systems and the Fourier outputs.

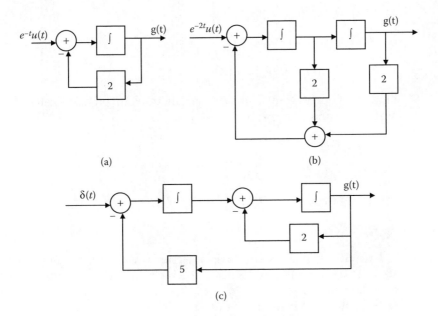

Figure P5.2.5

6. Deduce the transfer function for the systems shown in Figure P5.2.6 by (a) a direct application of Fourier transform and (b) first finding $h(t)$ and then using the Fourier transform to find the transfer function. Compare the results. The systems are relaxed at $t = 0$.

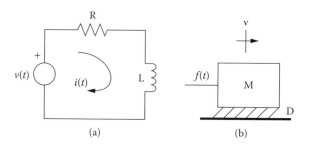

Figure P5.2.6

7. The impulse response of a network is given as $h(t) = p_2(t) - 0.5 p_4(t)$. Deduce the transfer function and, from the spectrum, indicate the effect on the spectrums of the input signals.
8. Prove the following identities, where a, b, and ω_0 are constants:

   a. $\mathcal{F}\{f(t) * h(t) * g(t)\} = F(\omega)H(\omega)G(\omega)$

   b. $\mathcal{F}^{-1}\{F(\omega) * H(\omega)\} = 2\pi f(t)h(t)$

   c. $\mathcal{F}\{f(t)\delta(t-a)\} = f(a)e^{-j a \omega}$

   d. $\mathcal{F}\{f(t) * \delta[(t/a) - b]\} = |a| F(\omega) e^{-jab\omega}$

9. Find the Fourier transforms of the functions given and sketch their amplitude and phase spectrums.

   a. $f(t) = p_1(t) * p_3(t-3)$

   b. $f(t) = p_2(t+2) * p_1(t-2)$

   c. $f(t) = \operatorname{sinc}(t)\sin c_3(t)$ $(\sin c_3(t) = \sin 3t/t)$;

   d. $f(t) = e^{-|t|} \cos \omega t$

10. A 1 μs (10^{-6} s) pulse modulates a carrier wave with frequency $\omega_c = 10^9$ rad/s. Locate the frequencies at which the resulting output spectrum becomes zero.
11. Deduce the spectrum of the output for the systems shown in Figure P5.2.11 for a delta function input.

Chapter 5: Nonperiodic signals and their Fourier transform

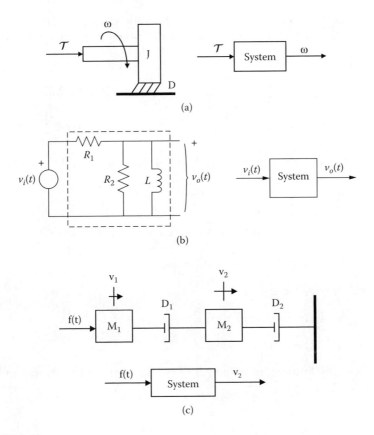

Figure P5.2.11

12. Compare the amplitude and phase spectrums of the following sets of functions:

    a. $p(t), p(t-2), p(t+2)$
    b. $p(t), p(2t-2)$

13. The impulse response of a system is $h(t) = e^{-(2-j2)t}u(t)$. Find the spectrum of the output if the input is $p(t)$.

14. **Square-law modulator.** Assume that a nonlinear resistive device has the voltage–current characteristics given by $v = a_1 i + a_2 i^2$. Suppose that the input is $i = I_m \cos\omega_0 t + Bm(t)$, where $m(t)$ has a band-limited spectrum. Find the spectrum of $v(t)$.

15. Using the convolution property, find the frequency spectrum of the function $p_a(t) * p_a(t) = 2a\Lambda_{2a}(t)$. Verify your result by evaluating

$$2a\mathcal{F}\{\Lambda_{2a}(t)\}; \quad \Lambda_{2a}(t) = \begin{cases} 1 - \dfrac{|t|}{2a} & -2a \le t \le 2a \\ 0 & \text{otherwise} \end{cases}.$$

16. Determine the Fourier transform of the function $f(t) = p(t)\cos 2\pi t$.

Section 5.3

1. Find the Fourier transform of the following functions:

   a. $f(t) = \sin\omega_0(t - t_0) + \cos\omega_0(t - 3t_0)$

   b. $f(t) = e^{-t}\sin\omega_0 t\, u(t)$

   c. $f(t) = \cos\omega_1 t \cos\omega_2 t$

   d. $f(t) = (1 + \cos\omega_m t)\cos\omega_c t$

2. Find the Fourier transform of the periodic functions shown in Figure P5.3.2.

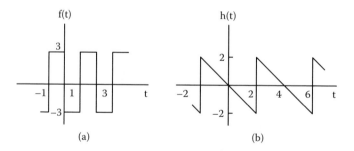

Figure P5.3.2

3. We may often want to find the spectrum of a function taken from an oscilloscope trace. Also, there are certain closed-form functions for which the Fourier transform cannot be found. One approach is to approximate the function by piecewise polynomial approximations or straight-line segments. Compare the Fourier spectrums of the function $p(t)$ with those of $(1 - t^2)$ and $(1 - t^4)$ for $-1 \le t \le 1$. Discuss your results.

Chapter 5: Nonperiodic signals and their Fourier transform

4. Prove the frequency differentiation property of the Fourier transform

$$\mathcal{F}\{(-jt)f(t)\} = \frac{dF(\omega)}{d\omega} \qquad \mathcal{F}\{(-jt)^n f(t)\} = \frac{d^n F(\omega)}{d\omega^n}$$

5. Determine the Fourier transform of the signals shown in Figure P5.3.5 by employing Fourier transform properties in conjunction with $f_0(t) = \exp(-t)u(t)$. All the curves shown in the figure decay as $\exp(-t)$.

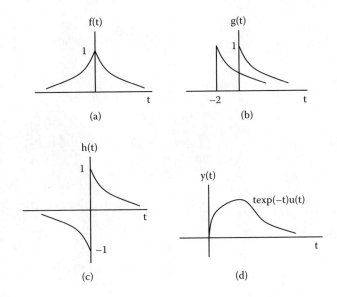

Figure P5.3.5

Appendix 5.1

Table A.5.1 Table of Fourier Transform Pairs

$$f(t) = \frac{1}{2\pi}\int_{-\infty}^{\infty} F(\omega)e^{j\omega t}d\omega \qquad F(\omega) = \int_{-\infty}^{\infty} f(t)e^{-j\omega t}dt$$

1.	$f(t) = \begin{cases} 1 & \|t\| \leq a \\ 0 & \text{otherwise} \end{cases}$		$F(\omega) = 2\dfrac{\sin a\omega}{\omega}$
2.	$f(t) = Ae^{-at}u(t)$		$F(\omega) = \dfrac{A}{a+j\omega}$
3.	$f(t) = Ae^{-a\|t\|}$		$F(\omega) = \dfrac{2aA}{a^2+\omega^2}$

Table A.5.1 (continued) Table of Fourier Transform Pairs

4. $f(t) = \begin{cases} A\left(1 - \dfrac{|t|}{a}\right) & |t| \le a \\ 0 & \text{otherwise} \end{cases}$ $F(\omega) = Aa\left[\dfrac{\sin(a\omega/2)}{(a\omega/2)}\right]^2$

5. $f(t) = Ae^{-a^2 t^2}$ $F(\omega) = A\dfrac{\sqrt{\pi}}{a} e^{-(\omega/2a)^2}$

6. $f(t) = \begin{cases} A\cos\omega_0 t\, e^{-at} & t \ge 0 \\ 0 & t < 0 \end{cases}$ $F(\omega) = A\dfrac{a + j\omega}{(a + j\omega)^2 + \omega_0^2}$

7. $f(t) = \begin{cases} A\sin\omega_0 t\, e^{-at} & t \ge 0 \\ 0 & t < 0 \end{cases}$ $F(\omega) = \dfrac{A\omega_0}{(a + j\omega)^2 + \omega_0^2}$

8. $f(t) = \begin{cases} A\cos\omega_0 t & |t| \le a \\ 0 & \text{otherwise} \end{cases}$ $F(\omega) = A\dfrac{\sin a(\omega - \omega_0)}{\omega - \omega_0}$

   $\qquad + A\dfrac{\sin a(\omega + \omega_0)}{\omega + \omega_0}$

9. $f(t) = A\delta(t)$ $F(\omega) = A$

10. $f(t) = \begin{cases} A & t > 0 \\ 0 & \text{otherwise} \end{cases}$ $F(\omega) = A\left[\pi\delta(\omega) - j\dfrac{1}{\omega}\right]$

11. $f(t) = \begin{cases} A & t > 0 \\ 0 & t = 0 \\ -A & t < 0 \end{cases}$ $F(\omega) = -j2A\dfrac{1}{\omega}$

12. $f(t) = A$ $F(\omega) = 2\pi A\delta(\omega)$

13. $f(t) = A\cos\omega_0 t$ $F(\omega) = \pi A[\delta(\omega - \omega_0) + \delta(\omega + \omega_0)]$

14. $f(t) = A\sum_{n=-\infty}^{\infty}\delta(t - nT)$ $F(\omega) = \dfrac{2\pi A}{T}\sum_{n=-\infty}^{\infty}\delta\left(\omega - n\dfrac{2\pi}{T}\right)$

15. $f(t) = \dfrac{\sin at}{\pi t}$ $F(\omega) = p_a(\omega)$

16. $f(t) = \dfrac{2\sin^2(at/2)}{\pi t^2}$ $F(\omega) = \begin{cases} 1 - \dfrac{|\omega|}{a} & |\omega| \le a \\ 0 & \text{otherwise} \end{cases}$

chapter 6

Sampling of continuous signals

The sampling of continuous signals at periodic intervals has become a very important and practical mathematical operation. One of the main concerns when signals are sampled is the accuracy with which the sampled signal is represented by its sampled values. Also, what type must the signal be and what must the sampling interval be in order that an optimum recovery of the original signal can be accomplished from the sampled values?

A very important fact in the sampling operation is that all physical engineering systems have frequency response limitations; that is, they respond only to some upper frequency limit, which we call the **Nyquist frequency**. As a result, the output signal of these systems is band limited. We have already seen that the ordinary house telephone signal has an upper frequency limit of about 4 kHz and the television signals have an upper frequency limit of 4 MHz. A very important consequence of a finite bandwidth signal is that it can be accurately represented by a narrow time duration sampling sequence, with samples taken at discrete and periodic instants. As already noted, the time space between the samples of one signal can be used to accommodate (**time multiplex**) without interference the samples of a different signal when transmitted through some transmitting channel. Often the samples are digitized, and this digitization is readily accomplished with an analog-to-digital (A/D) converter, the output being amplitude information in digital form. It is in this form that sampled and digitized signals enter a computer for further processing.

6.1 Fundamentals of sampling

The values of the function at the sampling points are called **sampled values**. These values are the exact values (infinite precision) of the signal at those corresponding sampling times. The time that separates the sampling points is called the **sampling interval** (T_s), and the reciprocal of the sampling interval is the **sampling frequency** ($F_s = 1/T_s$) or **sampling rate**. The value of any continuous function $f(t)$ at the point nT_s is specified by

$$f(t)\delta(t - nT_s) = f(nT_s)\delta(t - nT_s) \tag{6.1}$$

This relationship is easily proved by integrating both sides of the equation. The sampling interval is chosen to be a real positive number and constant, and $n = 0, \pm 1, \pm 2, \pm 3, \ldots$. The choice of T_s is critical in recapturing the original signal, a matter to be discussed in detail below. The sampled signal (see also Figure 1.2.10) is

$$f_s(t) \triangleq f(t) \sum_{n=-\infty}^{\infty} \delta(t - nT_s) = f(t) \mathrm{comb}_{T_s}(t) = \sum_{n=-\infty}^{\infty} f(nT_s)\delta(t - nT_s) \tag{6.2}$$

where the equality (6.1) was used. The Fourier transform of the above equation is

$$\boxed{F_s(\omega) \triangleq \mathcal{F}\{f_s(t)\} = \sum_{n=-\infty}^{\infty} f(nT_s) \mathcal{F}\{\delta(t - nT_s)\} = \sum_{n=-\infty}^{\infty} f(nT_s) e^{-j\omega nT_s}} \tag{6.3}$$

where the delta function properties for the Fourier integral were used. Note that the Fourier transform operates on functions having continuous time t as their independent variable.

Now, consider the Fourier transform of the term $f_s(t) = f(t) \mathrm{comb}_{T_s}(t)$, which appears in (6.2) and which is the equivalent of the quantity involving the shifted delta function. By (5.40) (frequency convolution Fourier property) and (5.46) (the Fourier transform of the comb function), we obtain

$$F_s(\omega) = \mathcal{F}\{f(t)\mathrm{comb}_{T_s}(t)\} = \frac{1}{2\pi} F(\omega) * \mathcal{F}\{\mathrm{comb}_{T_s}(t)\}$$

$$= \frac{1}{2\pi} F(\omega) * \left[\frac{2\pi}{T_s} \sum_{n=-\infty}^{\infty} \delta(\omega - n\omega_s) \right]$$

$$= \frac{1}{T_s} \sum_{n=-\infty}^{\infty} \int_{-\infty}^{\infty} F(x) \delta(\omega - x - n\omega_s) dx \tag{6.4}$$

$$= \frac{1}{T_s} \sum_{n=-\infty}^{\infty} F(\omega - n\omega_s)$$

Chapter 6: Sampling of continuous signals

By (6.3) and (6.4),

$$\boxed{F_s(\omega) = \sum_{n=-\infty}^{\infty} f(nT_s)e^{-j\omega nT_s} = \frac{1}{T_s}\sum_{n=-\infty}^{\infty} F(\omega - n\omega_s) \qquad \omega_s = \frac{2\pi}{T_s} = 2\pi F_s} \qquad (6.5)$$

where ω_s is the sampling frequency in rad/s and F_s is the sampling frequency in s^{-1}.

This discussion shows that if we know the Fourier transform of $f(t)$, the Fourier transform of its sampled version is uniquely determined. Furthermore, if we set $\omega = \omega - m\omega_s$ in (6.5), we obtain

$$F_s(\omega - m\omega_s) = \frac{1}{T_s}\sum_{n=-\infty}^{\infty} F[\omega - m\omega_s - n\omega_s] = \frac{1}{T_s}\sum_{n=-\infty}^{\infty} F[\omega - (m+n)\omega_s]$$

$$= \frac{1}{T_s}\sum_{k=-\infty}^{\infty} F[\omega - k\omega_s] = F_s(\omega) \qquad (6.6)$$

where we set $n + m = k$. Observe that for any m, as n increases to plus minus infinity, the values of k also increase to plus minus infinity.

Note: *The spectrum of the sampled signal $f_s(t)$ is an infinite sum of shifted spectrums of the original signal $f(t)$. The spectrum of the sampled signal is periodic with period ω_s (see Figure 6.1.1).*

When the function $f(t)$ is causal (positive time), $f(t) = 0$ for $t < 0$, then

$$f_s(t) = \sum_{n=0}^{\infty} f(nT_s)\delta(t - nT_s) \qquad (6.7)$$

which can be shown to yield

$$F_s(\omega) = \sum_{n=0}^{\infty} f(nT_s)e^{-j\omega nT_s} = \frac{f(0+)}{2} + \sum_{n=-\infty}^{\infty} F(\omega - n\omega_s) \qquad (6.8)$$

Example 6.1.1: Find the Fourier transform of the sampled functions

a. $f_s(t) = e^{-|t|} comb_{T_s}(t)$

b. $f_s(t) = e^{-t} u(t) comb_{T_s}(t)$

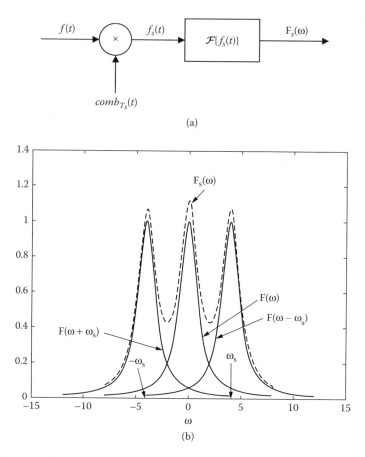

Figure 6.1.1 Fourier spectrum of a sampled signal with only three elements. The $F(\omega)$ is present for $n\omega_s$ shifts with $-\infty < n < \infty$.

Solution: By (6.5) and (6.8) and Appendix 5.1, we obtain, respectively,

$$\mathcal{F}\{e^{-|t|}comb_{T_s}(t)\} \triangleq F_s(\omega) = \frac{1}{T_s}\sum_{n=-\infty}^{\infty}\frac{2}{1+(\omega+n\omega_s)^2} \quad \omega_s = \frac{2\pi}{T_s} \quad (6.9)$$

$$\mathcal{F}\{e^{-t}u(t)comb_{T_s}(t)\} \triangleq F_s(\omega) = \frac{1}{2} + \frac{1}{T_s}\sum_{n=-\infty}^{\infty}\frac{1}{1+j(\omega+n\omega_s)} \quad (6.10)$$

∎

Example 6.1.2: Consider three functions, $f(t)$, $h(t)$, and $g(t)$, with respective frequency characteristics $F(\omega)$, $H(\omega)$, and $G(\omega)$, as shown in Figure 6.1.2b. Find the maximum sampling interval T_s in order that the function $y(t) = f(t) +$

Chapter 6: Sampling of continuous signals

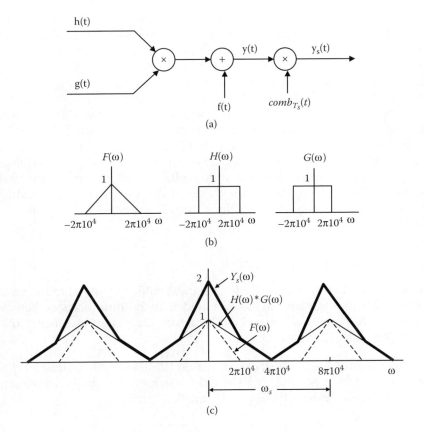

Figure 6.1.2 Illustration of Example 6.1.2.

$h(t)g(t)$ shown in Figure 6.1.2a is recovered from its sampled version $y_s(t)$ using a low-pass filter.

Solution: The Fourier transform of the sampled function is given by

$$Y_s(\omega) = \mathcal{F}\{f(t)comb_{T_s}(t) + h(t)g(t)comb_{T_s}(t)\}$$

From the frequency convolution property of the Fourier transform, we find

$$Y_s(\omega) = \frac{1}{2\pi}F(\omega) * \frac{2\pi}{T_s}COMB_{\omega_s}(\omega) + \frac{1}{2\pi}\mathcal{F}\{h(t)g(t)\} * \frac{2\pi}{T_s}COMB_{\omega_s}(\omega)$$

$$= \frac{1}{T_s}F(\omega) * COMB_{\omega_s}(\omega) + \left(\frac{1}{2\pi}\right)^2 H(\omega) * G(\omega) * \frac{2\pi}{T_s}COMB_{\omega_s}(\omega)$$

The convolution of $F(\omega)$ and $COMB_{\omega_s}(\omega)$ gives us a periodic repetition of the spectrum $F(\omega)$. The convolution $H(\omega)*G(\omega)$ with $COMB_{\omega_s}(\omega)$ gives us a periodic repetition of the spectrum $H(\omega)*G(\omega)$. However, the spectrum width of $H(\omega)*G(\omega)$ is equal to the sum of the spectral widths of $H(\omega)$ and $G(\omega)$; hence, in the present case, $\omega_N = \omega_{NH} + \omega_{NG} = 2\pi 10^2 + 2\pi 10^2 = 4\pi 10^2$. The spectrum $Y_s(\omega)$ is shown in Figure 6.1.2c. We observe that the minimum sampling frequency ω_s, in order that the spectrum of $H(\omega)*G(\omega)$ does not overlap, is $8\pi 10^4$ or, equivalently, the maximum $T_s = 2\pi/8\pi 10^4 = 0.25 \times 10^{-4}$. Because the spectral width of $H(\omega)*G(\omega)$ is greater than the spectral width of $F(\omega)$, the value of the sampling time is determined by the spectral width of $H(\omega)*G(\omega)$. However, if the spectral width of $F(\omega)$ were greater than the spectral width of $H(\omega)*G(\omega)$, the value of the sampling time would be determined from the spectral width of $F(\omega)$. The spectrums in Figure 6.1.2c have been normalized to unity. ∎

6.2 The sampling theorem

We next show that it is possible for a band-limited signal $f(t)$ to be exactly specified by its sampled values provided that the time distance between sampled values does not exceed a critical sampling interval. The sampling theorem is stated as follows:

Theorem 6.2.1: A finite energy function $f(t)$ having a band-limited Fourier spectrum, that is, $F(\omega) = 0$ for $|\omega| \geq \omega_N$, can be completely reconstructed from its sampled values $f(nT_s)$ (see Figure 6.2.1) with

$$\boxed{f(t) = \sum_{n=-\infty}^{\infty} T_s f(nT_s) \frac{\sin[\omega_s(t - nT_s)/2]}{\pi(t - nT_s)} \qquad \omega_s = \frac{2\pi}{T_s}} \qquad (6.11)$$

provided that the sampling time is selected to satisfy

$$\frac{2\pi}{\omega_s} \triangleq T_s = \frac{\pi}{\omega_N} = \frac{\pi}{2\pi f_N} = \frac{1}{2f_N} \triangleq \frac{T_N}{2} \quad \text{also} \quad \omega_s = 2\omega_N \qquad (6.12)$$

where ω_N, known as the **Nyquist frequency**, is the highest frequency of the signal.

This theorem states that no loss of information is incurred through the sampling process if the signal is sampled at a sampling frequency that is at least twice as fast as the highest frequency contained in the signal. Equivalently, the sampling time must be less than or equal to one half of the Nyquist time T_N. For band-limiting signals, the sampling process introduces no error, since, in theory, we can recover the original continuous time signal from its sampled version. The sinc function in (6.11) is known as the **interpolation function** to indicate that it allows an interpolation between the sampled values to find $f(t)$ for all t.

Chapter 6: Sampling of continuous signals

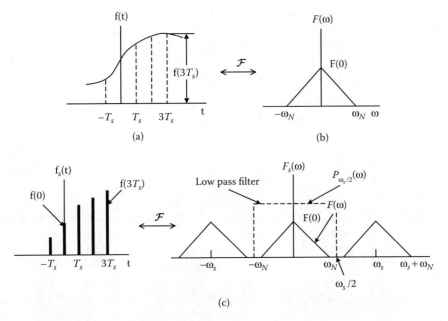

Figure 6.2.1 Sampling theorem.

Proof: Employ (6.5) and Figure 6.2.1b to write ($n = 0$)

$$F(\omega) = P_{\omega_s/2}(\omega)T_s F_s(\omega) \tag{6.13}$$

By (6.3), this equation becomes

$$F(\omega) = P_{\omega_s/2}(\omega)T_s \sum_{n=-\infty}^{\infty} f(nT_s)e^{-j\omega nT_s} \tag{6.14}$$

from which we have that

$$f(t) = \mathcal{F}^{-1}\{F(\omega)\} = \mathcal{F}^{-1}\left\{ P_{\omega_s/2}(\omega)T_s \sum_{n=-\infty}^{\infty} f(nT_s)e^{-j\omega nT_s} \right\}$$

$$= T_s \sum_{n=-\infty}^{\infty} f(nT_s)\mathcal{F}^{-1}\{P_{\omega_s/2}(\omega)e^{-j\omega nT_s}\} \tag{6.15}$$

Note that the inverse Fourier transform operates only on the function with the independent variable ω. By the application of the frequency-shift property of Fourier transform, it is seen that this equation proves the theorem. ∎

For the case when $\omega_s = 2\omega_N$, (6.11) becomes

$$f(t) = \sum_{n=-\infty}^{\infty} f(nT_s) \frac{\sin[\omega_N(t - nT_s)]}{\omega_N(t - nT_s)} \quad (6.16)$$

and the spectrum of the sampled function is a periodic one with the successive replicas of the spectrum of $f(t)$ just touching.

The sampling time

$$\boxed{T_s = \frac{T_N}{2} = \frac{1}{2f_N}} \quad (6.17)$$

is called the **Nyquist interval**. It is the longest time interval that can be used for sampling a band-limited signal while still permitting us to recover the signal without distortion. Figure 6.2.2 shows how a band-limited signal can be reconstructed from its samples using (6.11). Observe that the sinc functions tend to cancel between the sampling times and reinforce at the sampling points. Although we used a Gaussian function that has an infinite spectrum, a unit sampling time, and only three sinc functions, we observe that the approximation is very good. Of course, for a band-limited function and an infinite number of shifted sinc functions the recovery is complete and not approximate.

If we select the sampling time to be larger than half the Nyquist time or, equivalently, the sampling frequency to be less than twice the largest

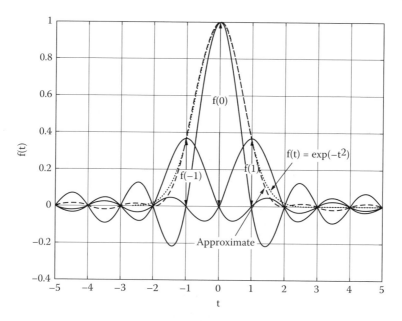

Figure 6.2.2 Reconstruction of a signal from its samples.

Chapter 6: Sampling of continuous signals

frequency (Nyquist frequency), an overlapping of the spectrums occurs. This overlapping is known as **aliasing**.

The Fourier transform of a sampled function is

$$F_s(\omega) = \mathcal{F}\{f(t)comb_{T_s}(t)\} = \frac{1}{2\pi}F(\omega) * \mathcal{F}\{comb_{T_s}(t)\} \equiv \text{frequency convolution}$$

$$= \frac{1}{2\pi}F(\omega) * \left[\frac{2\pi}{T_s}\sum_{n=-\infty}^{\infty}\delta(\omega - n\omega_s)\right] = \frac{1}{T_s}F(\omega) * \left[\sum_{n=-\infty}^{\infty}\delta(\omega - n\omega_s)\right]$$

$$= \frac{1}{T_s}F(\omega) * COMB_{\omega_s}(\omega) \tag{6.18}$$

If the sampling frequency is not appropriate (is less than twice the Nyquist frequency), then aliasing will take place and the spectrum will look like the one shown in Figure 6.2.3a. If the sampling time diminishes at least to the value $T_N/2$, then the spectrum of the sampled function will look like the one shown in Figure 6.2.3b. It is clear from Figure 6.2.3a that there is no filter available that is capable of extracting the frequency spectrum of the signal without including additional frequencies produced by the shifted spectrums.

The effect of aliasing can be used to our advantage in some cases. Suppose that we want to stop optically a repetitive action, for example, the wing undulation of a bee or the turning of a wheel. To accomplish this optical effect, we flash the object with a strobe light. If we adjust the repetition of the strobe flash to equal the wing repetition rate or the wheel turning rate, these events will appear to be stationary. If the strobe frequency is much higher than twice that of the periodic phenomenon, the speed of the phenomenon does not appear to change. However, if the strobe flashes are less than twice the frequency of the phenomenon under observation, the repetition slows down; thus, we observe a slow-flying bee or a slow-rotating wheel. This phenomenon is commonly observed in movies when we observe a moving stagecoach, with wheels appearing stationary or turning slowly

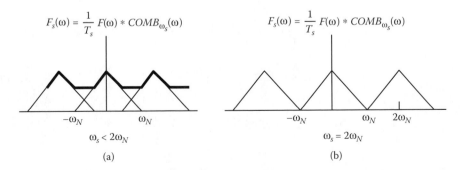

Figure 6.2.3 Aliasing and nonaliasing situation.

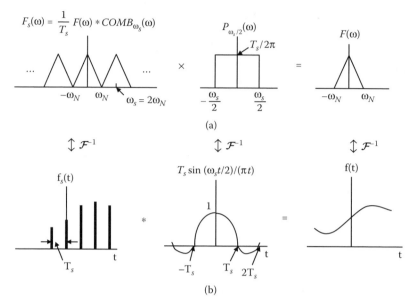

Figure 6.2.4 Delta sampling, representation, and recovery of signals.

(sometimes backwards). In the movies, the sampling rate is 1/20 s because the frame rate of the film is about 20 frames per second.

Sampling the function at twice the largest frequency available in the signal (Nyquist frequency), the signal spectrums just touch each other and extend from minus infinity to infinity. This situation is shown to the left of Figure 6.2.4a. To isolate the spectrum of the signal, we must multiply the sampled spectrum with a ideal low-pass filter, which is shown in the center of Figure 6.2.4a. The resulting spectrum of this multiplication is the spectrum of the signal $F(\omega)$. Hence, to recapture the signal intact, we write

$$\mathcal{F}^{-1}\{F(\omega)\} \triangleq f(t) = \mathcal{F}^{-1}\left\{\left[\frac{1}{T_s}F(\omega)*COMB_{\omega_s}(\omega)\right]\frac{T_s}{2\pi}P_{\omega_s/2}(\omega)\right\}$$

$$= \mathcal{F}^{-1}\left\{\frac{1}{T_s}F(\omega)*COMB_{\omega_s}(\omega)\right\} * \frac{T_s}{2\pi}\mathcal{F}^{-1}\{P_{\omega_s/2}(\omega)\}$$

$$= f(t)comb_{T_s}(t) * \frac{T_s}{2\pi}\frac{2\sin(\omega_s t/2)}{t}$$

$$= \frac{T_s}{\pi}\left[\sum_{n=-\infty}^{\infty}f(nT_s)\delta(t-nT_s)\right] * \frac{\sin(\omega_s t/2)}{t}$$

Chapter 6: Sampling of continuous signals

$$= \frac{T_s}{\pi} \sum_{n=-\infty}^{\infty} f(nT_s) \int_{-\infty}^{\infty} \delta(t-x-nT_s) \frac{\sin(\omega_s x/2)}{x} dx$$

$$= T_s \sum_{n=-\infty}^{\infty} f(nT_s) \frac{\sin(\omega_s(t-nT_s)/2)}{\pi(t-nT_s)}$$

Example 6.2.1: Show the aliasing phenomenon by decreasing ω_s or, equivalently, by increasing the sampling time T_s, associated with a pure cosine function $f(t) = \cos\omega_0 t$. Use a low-pass filter of bandwidth ω_s in the output.

Solution: The Fourier transform of $f(t)$ is $F(\omega) = \pi\delta(\omega-\omega_0) + \pi\delta(\omega+\omega_0)$. Thus, the Fourier transform of the sampled function $f_s(t)$ is (see also (6.18))

$$F_s(\omega) = \frac{1}{T_s} F(\omega) * COMB_{\omega_s}(\omega)$$

$$= \frac{\pi}{T_s}[\delta(\omega-\omega_0) * COMB_{\omega_s}(\omega) + \delta(\omega+\omega_0) * COMB_{\omega_s}(\omega)]$$

Figure 6.2.5a shows the spectrum of the sampled function when $\omega_s \gg \omega_0$ or, equivalently, when $T_s \ll T_0 = 2\pi/\omega_0$. Figure 6.2.5b is the result of the convolution of $\delta(\omega-\omega_0)$ and $COMB_{\omega_s}(\omega)$. Figure 6.2.5d is the convolution of $\delta(\omega+\omega_0)$ and $COMB_{\omega_s}(\omega)$. By adding the spectrums of Figure 6.2.5b and d, we obtain the total spectrum of $F_s(\omega)$ as specified by the above equation. The spectrum of the sampled function is shown in Figure 6.2.5e. If we incorporate a filter having a frequency bandwidth of $\omega_s/2$, we regain our signal since

$$\frac{1}{2\pi}\int_{-\infty}^{\infty} T_s P_{\omega_s}(\omega)e^{j\omega t}d\omega = \frac{1}{2\pi}\int_{-\infty}^{\infty} T_s\left[\frac{\pi}{T_s}\delta(\omega+\omega_0) + \frac{\pi}{T_s}\delta(\omega-\omega_0)\right]e^{j\omega t}d\omega$$

$$= \frac{1}{2}[e^{j\omega_0 t} + e^{-j\omega_0 t}] = \cos\omega_0 t$$

We follow the same procedure as shown in Figure 6.2.5, with the difference that in this case the sampling frequency was taken to be less than twice the Nyquist frequency, which in this case is ω_0. We observe, in Figure 6.2.6, that the low-pass filter will produce a cosine function, but with a different frequency, equal to $\omega_s - \omega_0$, which is the alias of ω_0. ∎

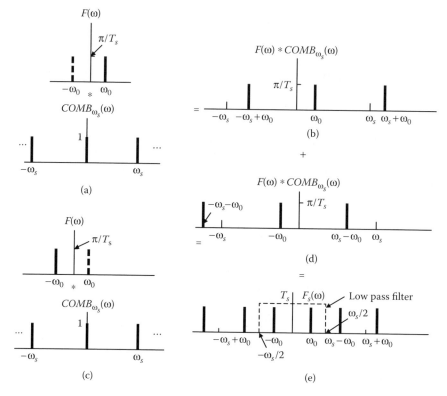

Figure 6.2.5 Nonaliasing problem using a pure sinusoid function.

Example 6.2.2: Consider the analog signal $f(t) = e^{-0.5t}u(t)$.

a. Find and plot the magnitude of its spectrum.
b. Using MATLAB, plot the magnitude spectrum of the sampled version of $f(t)$ for $T_s = 0.6$ and $T_s = 0.4$.

Solution: The Fourier transform of $f(t)$ is easily found using the Fourier integral (see also Appendix 5.1) and it is equal to $F(\omega) = 1/(0.5 + j\omega)$. From (6.3) we find that the Fourier transform of its sampled version is

$$F_s(\omega) = \sum_{n=0}^{\infty} e^{-0.5nT_s} e^{-j\omega nT_s} = \sum_{n=0}^{\infty} e^{-(0.5T_s + j\omega T_s)n} = 1 + e^{-(0.5T_s + j\omega T_s)} + e^{-(0.5T_s + j\omega T_s)2} + \cdots$$

$$= \frac{1}{1 - e^{-(0.5T_s + j\omega T_s)}} = \frac{1}{1 - e^{-0.5T_s} e^{-j\omega T_s}}$$

where the formula for the infinite geometric series was used: $1/(1-x) = 1 + x + x^2 + x^3 + \cdots$. The value of x must be less than 1 for the series to converge.

Chapter 6: Sampling of continuous signals 251

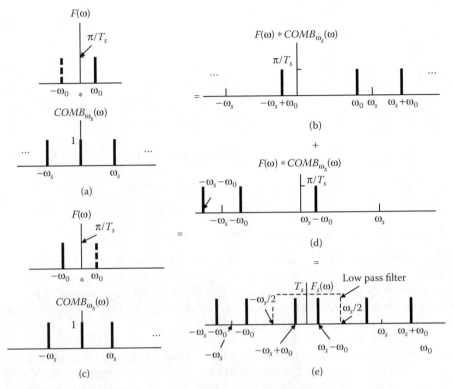

Figure 6.2.6 Aliasing problem using pure sinusoidal function.

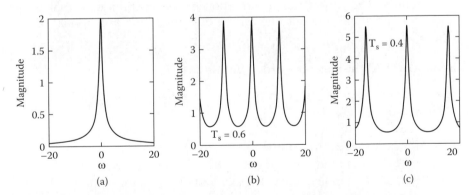

Figure 6.2.7 Illustration of Example 6.2.2.

Figure 6.2.7a shows the magnitude of the spectrum $F(\omega)$ and Figures 6.2.7b and c shows the spectrum of the sampled function for $T_s = 0.6$ and 0.4, respectively.

Book MATLAB m-file: ex6_2_2

```
%ex6_2_2 is the m-file for illustrating Ex 6.2.2
w=-10:.01:10;
fw=abs(1./(0.5+j*w));
fs1w=abs(1./(1-exp(-0.5*0.6)*exp(-j*w*0.6)));
fs2w=abs(1./(1-exp(-0.5*0.1)*exp(-j*w*0.1)));
subplot(2,3,1);plot(w,fw,'k');xlabel('\omega');ylabel...
    ('magnitude');
subplot(2,3,2);plot(w,fs1w,'k');xlabel('\omega');ylabel...
    ('magnitude');
subplot(2,3,3);plot(w,fs2w,'k');xlabel('\omega');ylabel...
    ('magnitude');
```
■

Construction of analog signal from its sampled values

If a continuous signal is band limited, we can reconstruct it by using the sampling theorem. This requires an infinite number of samples — not a practical way to accomplish its reconstruction. Furthermore, this approach does not lead to real-time processing, which is needed in communications and many other areas of signal processing.

The simplest way to construct $f(t)$, for $nT_s < t \leq (n + 1)T_s$, from $f(nT_s)$ is to hold the value of $f(nT_s)$ until the arrival of the next sample value $f((n + 1)T_s)$. This process is known as the **zero-order hold**. If we had connected the values of the samples by a straight line, then we would talk about a **first-order hold**. The MATLAB function *plot* performs this process. Figure 6.2.8a shows the sample values of continuous signal. Figure 6.2.8b shows the zero-order hold, and Figure 6.2.8c shows the construction of the continuous signal using linear interpolation (straight line can be described by a polynomial of t of degree 1).

Book MATLAB m-file: zero_order_hold

```
%zero_order_hold is a m-file that creates zero-hold and
%linear interpolation;
t=0:1:10;
f=exp(-0.2*t)-exp(-0.8*t);
subplot(1,3,1);stem(t,f,'k');xlabel('nT_s');ylabel('f(nT_s)');
subplot(1,3,2);stairs(t,f,'k');xlabel('t');ylabel...
    ('zero-order hold');
subplot(1,3,3);plot(t,f,'k');xlabel('t');ylabel('linear...
    interpolation')
```

It is apparent that when we use the zero-order hold to reconstruct $f(t)$, we create errors. These errors will become smaller the shorter the sampling time used. In addition, cost and performance specifications should also be considered.

Chapter 6: Sampling of continuous signals

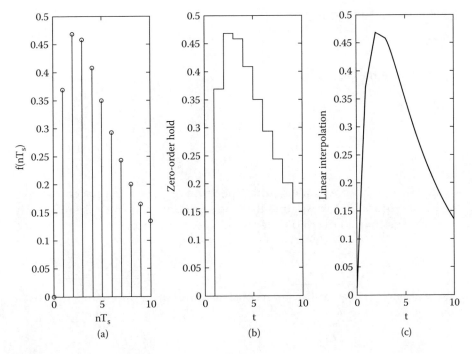

Figure 6.2.8 First-order hold and linear interpolation.

Important definitions and concepts

1. Importance of sampling
2. Sampled values
3. Sampling interval and sampling frequency
4. Sampling rate
5. Ideal sampling with the comb function
6. The spectrum of the sampled signal
7. The sampling theorem
8. The Nyquist frequency of a signal
9. The interpolation function
10. Aliasing
11. Sampling of pure sinusoids
12. Construction of continuous signals from their sampled values
13. Zero-order hold and linear interpolation

Chapter 6 Problems
Section 6.1

1. Prove the identity:

$$\sum_{n=-\infty}^{\infty} \delta(\omega - n\omega_s) = (1/\omega_s) \sum_{n=-\infty}^{\infty} e^{j\omega nT_s} \qquad \omega_s = 2\pi/T_s$$

   Hint: Expand in Fourier series.

2. Sketch the amplitude Fourier spectrum of $f_s(t)$ for the systems shown in Figure P6.1.2. Assume that the sampling frequency is sufficiently large that no overlap of repeated spectrums occurs.

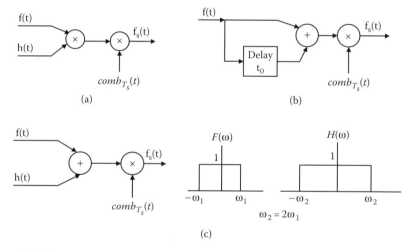

Figure P6.1.2

3. Verify (6.1) and (6.3).
4. Determine the Fourier transform of the product $f_1(t)f_2(t)f_3(t)$. Based on this answer, deduce the Fourier transform of the product $f_1(t)f_2(t)\cdots f_n(t)$.
5. Find the Fourier transform of the function

$$f(t) = \begin{cases} e^{-0.5t} & 0 \le t \le 3 \\ 0 & \text{otherwise} \end{cases}$$

   for $T_s = 0.5$ and $T_s = 0.05$. Use (6.3) and the following Book MATLAB function:

Chapter 6: Sampling of continuous signals

```
%p6_1_5 m-file solves the request of Prob6.1.5
Ts=0.25;
n=0:3/Ts;
w=0:.05:200;
f=exp(-0.5*Ts*n)*exp(-j*0.5*Ts*n'*w);
%f is an nxw matrix, first we add the rows
%using the MATLAB command sum(f,1) and then plot the results;
sf=sum(f,1);
plot(abs(sf)*Ts)
```

Section 6.2

1. Suppose that we choose a sampling plan that will contain 99% of the energy of $F(\omega)$ of the time function $f(t) = \exp(-t)u(t)$. To find the Nyquist frequency, and hence to find the sampling frequency, use the relationship

$$\int_0^{\omega_N} |F(\omega)|^2 \, d\omega = 0.99 \int_0^\infty |F(\omega)|^2 \, d\omega \quad \text{or} \quad \int_0^{\omega_N} \frac{1}{1+\omega^2} \, d\omega = 0.99 \int_0^\infty \frac{1}{1+\omega^2} \, d\omega$$

2. Verify the sampling theorem using (6.15).
3. Expand the band-limited $F(\omega)$, with $|\omega| \leq \omega_N$, in a Fourier series and prove the sampling theorem.
4. A signal $f(t)$ is sampled with pulses. The sampled signal is represented by $f_s(t) = f(t)s(t)$, where

$$s(t) = \sum_{n=-\infty}^{\infty} p_{a/2}(t - nT_s).$$

Use the Fourier transformation of $s(t)$ to find the Fourier spectrum of $f_s(t)$. Sketch the Fourier spectrum of $f_s(t)$ if $f(t)$ is a band-limited signal.

5. Find the Nyquist frequency for each of the following signals:
   a. $f(t) = 10 \cos(100\pi t)$
   b. $f(t) = p_2(t)$
   c. $f(t) = 2\sin(2t)/t$
   d. $f(t) = [\sin(100)/t]\cos(30000\pi t)$

6. Find the value of the RC combination such that the low-pass filter shown in Figure P6.2.6 is used as an anti-aliasing filter. The magnitude square of the transfer function must be equal to $(0.01)^2$ at $20{,}000\pi$ rad/s. Plot also the transfer function in a linear-log scale to observe the roll-off with frequency. In practice, sharper filters are used.

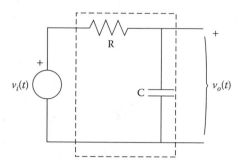

Figure P6.2.6

chapter 7

Discrete-time transforms

In the previous chapters we have included discussion of the use of Fourier transform techniques in continuous-time signal processing and signal sampling studies. However, in many problems, the signal may be experimentally derived and analytic functions are not available for the integrations involved. There are two general approaches that might be adopted in such cases. One method calls for approximating the functions and carrying out the integrations by numerical means. A second method, and one that is used extensively, calls for replacing the continuous Fourier transform by an equivalent **discrete Fourier transform** (DFT) and then evaluating the DFT using the discrete data. Now, instead of integrations, the direct solution of DFT requires, for each sample, N complex multiplications, N complex additions, and the access of N complex exponentials that appear in the DFT. Hence, with 10^4 samples (a small number in many cases), more than 10^8 mathematical operations are required in the solution. It was the development of the **fast Fourier transform** (FFT), a computational technique that reduces the number of mathematical operations of the DFT to $N \log_2 N$, that made DFT an extremely useful transform in many fields of science and engineering.

7.1 Discrete-time Fourier transform (DTFT)

Approximating the Fourier transform

The Fourier transform of a continuous-time function is given by

$$F(\Omega) = \int_{-\infty}^{\infty} f(t) e^{-j\Omega t} dt \qquad (7.1)$$

where we have used the capital omega to represent the continuous frequency independent variable in rad/s. We can approximate the above integral by introducing a short time interval T_s such that essentially all the radian frequency content of $F(\Omega)$ lies in the interval $-\pi/T_s \leq \Omega \leq \pi/T_s$. With this approximation, we express the Fourier transform integral (7.1) in the form

$$F(\Omega) = \sum_{n=-\infty}^{\infty} \int_{nT_s}^{nT_s+T_s} f(t)e^{-j\Omega t}dt \cong \sum_{n=-\infty}^{\infty} T_s f(nT_s)e^{-j\Omega nT_s} \qquad T_s \to 0 \qquad (7.2)$$

If we set

$$f(n) \equiv T_s f(nT_s)$$
$$\omega \equiv \Omega T_s \qquad \text{rad/unit} \qquad (7.3)$$

in the above equation, we obtain

$$\boxed{\mathcal{F}_{DT}\{f(n)\} \triangleq F(\omega) \triangleq F(e^{j\omega}) = \sum_{n=-\infty}^{\infty} f(n)e^{-j\omega n}} \qquad (7.4)$$

which is known as the DTFT.

Note: *The discrete frequency ω is equal to ΩT_s and has the units of radians per unit length, where Ω has units of radians per second. If T_s is unity time, then the discrete frequency and the continuous frequency have the same values but different units. Both frequencies Ω and ω are continuous independent variables.*

Therefore, the following steps can be taken to approximate the continuous-time Fourier transform:

1. Select the time sampling T_s such that $F(\Omega) \cong 0$ for all $|\Omega| > \pi/T_s$.
2. Sample $f(t)$ at times nT_s to obtain $f(nT_s)$.
3. Compute the DTFT using the sequence $\{T_s f(nT_s)\}$.
4. The resulting approximation is then $F(\Omega) \cong F(\omega)$ for $-\pi/T_s \leq \Omega \leq \pi/T_s$.

If we introduce the value of $\omega + 2\pi$ in (7.4), we obtain

$$F(\omega + 2\pi) \triangleq F(e^{j(\omega+2\pi)}) = F(e^{j\omega}) = F(\omega) \qquad (7.5)$$

since $\exp(j2\pi) = 1$. Therefore, the function $F(\omega)$ is periodic with a period of 2π. This property contrasts with the Fourier transforms of continuous functions whose range is infinite. Thus, due to periodicity, the frequency range of any discrete-time signal is limited to the range $(-\pi, \pi]$ or $(0, 2\pi]$ for $T_s = 1$; any frequency outside these intervals is equivalent to a frequency within these intervals. If we had included the sampling time, (7.5) would have taken the form

$$F\left(e^{j(\Omega+\frac{2\pi}{T_s})T_s}\right) = F\left(e^{j(\Omega T_s)}\right) \qquad (7.6)$$

Chapter 7: Discrete-time transforms

which indicates that the spectrum is periodic every $2\pi/T_s$.

Because $F(\omega)$ is periodic, it can be expanded in a Fourier series. To accomplish this, multiply both sides of (7.4) by $e^{j\omega m} d\omega$ and integrate the expression within the period. This yields

$$\int_{-\pi}^{\pi} F(e^{j\omega}) e^{j\omega m} d\omega = \int_{-\pi}^{\pi} \left[\sum_{n=-\infty}^{\infty} f(n) e^{-jn\omega} \right] e^{jm\omega} d\omega \qquad (7.7)$$

But taking into consideration that the integral of any sinusoidal signal is zero when it is integrated within a period (or multiple period), we obtain

$$\sum_{n=-\infty}^{\infty} f(n) \int_{-\pi}^{\pi} e^{j\omega(m-n)} d\omega = \begin{cases} 2\pi f(n) & m = n \\ 0 & m \neq n \end{cases} \qquad (7.8)$$

Then (7.7) becomes (only one element from the sum remains)

$$f(n) = \frac{1}{2\pi} \int_{-\pi}^{\pi} F(e^{j\omega}) e^{j\omega n} d\omega \qquad (7.9)$$

Thus, the *DT* Fourier transform pair of discrete-time signals is

$$\boxed{\begin{aligned} F(e^{j\omega}) &= \left| F(e^{j\omega}) \right| e^{j\phi(\omega)} = \sum_{n=-\infty}^{\infty} f(n) e^{-j\omega n} \\ f(n) &= \frac{1}{2\pi} \int_{-\pi}^{\pi} F(e^{j\omega}) e^{j\omega n} d\omega \end{aligned}} \qquad (7.10)$$

Example 7.1.1: Determine the DTFT of the signal $f(t) = \exp(-t)u(t)$, for $T_s = 1$ and $T_s = 0.1$.

Solution:

a. From (7.10), we write

$$F(e^{j\omega}) = \sum_{n=0}^{\infty} e^{-n} e^{-j\omega n} = \sum_{n=0}^{\infty} e^{-(1+j\omega)n} = \frac{1}{1 - e^{-(1+j\omega)}}$$

since the summation is an infinite geometric series.

b. Taking into consideration the sampling time in (7.10), we obtain

$$F(e^{j\Omega T_s}) = T_s \sum_{n=0}^{\infty} f(nT_s)e^{-j\Omega T_s n} = 0.1\sum_{n=0}^{\infty} e^{-nT_s}e^{-j\Omega T_s n}$$

$$= 0.1\sum_{n=0}^{\infty} e^{-(0.1+j\Omega 0.1)n} = \frac{0.1}{1-e^{-0.1}e^{-j0.1\Omega}}$$

Since $\{f(n)\}$ and $f(nT_s)$ are absolutely summable,

$$\sum_{n=0}^{\infty}\left|e^{-(1+j\omega)n}\right| = \sum_{n=0}^{\infty}\left|e^{-n}\right| = \sum_{n=0}^{\infty} e^{-n} = \frac{1}{1-e^{-1}} < \infty$$

$$0.1\sum_{n=0}^{\infty}\left|e^{-(0.1+j0.1\Omega)n}\right| = 0.1\sum_{n=0}^{\infty}\left|e^{-0.1n}\right| = \frac{0.1}{1-e^{-0.1}} < \infty$$

the series converge (we used the complex identity $|z_1 z_2| = |z_1||z_2|$).

The magnitude and phase of the second case, for example, are given by

$$F(e^{j\Omega 0.1}) = \frac{0.1}{(1-e^{-0.1}\cos 0.1\Omega) + j(e^{-0.1}\sin 0.1\Omega)}$$

$$= \frac{0.1}{\sqrt{(1-e^{-0.1}\cos 0.1\Omega)^2 + (e^{-0.1}\sin 0.1\Omega)^2}} e^{-j\tan^{-1}\frac{e^{-0.1}\sin 0.1\Omega}{1-e^{-0.1}\cos 0.1\Omega}}$$

If we set $\Omega = -\Omega$ in the equation above, we observe that the magnitude of the spectrum is an even function and the phase is an odd function.

Note: For real functions, the amplitude spectrum is an **even** function and the phase spectrum is an **odd** function. Thus, the representation of the frequency spectrum within the range $0 \leq \Omega \leq \pi$ will suffice.

The reader should plot the magnitude of the spectrums given above to observe the differences. Furthermore, if we set $T_s = 0.01$, the spectrum will closely approximate the exact one to 100π, and its magnitude at zero frequency is $0.01/(1-\exp(-0.01)) = 1.0050$, which is very close to 1, as it should be. ∎

7.2 Summary of DTFT properties

Linearity $\quad\quad\quad af(n) + bh(n) \xleftrightarrow{\mathcal{F}_{DT}} aF(e^{j\omega}) + bH(e^{j\omega})$

Time shifting $\quad\quad f(n-m) \xleftrightarrow{\mathcal{F}_{DT}} e^{-j\omega m}F(e^{j\omega})$

Chapter 7: Discrete-time transforms 261

Time reversal $\quad f(-n) \xleftrightarrow{\mathcal{F}_{DT}} F(e^{-j\omega})$

Convolution $\quad f(n)*h(n) \xleftrightarrow{\mathcal{F}_{DT}} F(e^{j\omega})H(e^{j\omega})$

Frequency shifting $\quad e^{j\omega_0 n}f(n) \xleftrightarrow{\mathcal{F}_{DT}} F(e^{j(\omega-\omega_0)})$

Modulation $\quad f(n)\cos\omega_0 n \xleftrightarrow{\mathcal{F}_{DT}} \dfrac{1}{2}F(e^{j(\omega+\omega_0)})+\dfrac{1}{2}F(e^{j(\omega-\omega_0)})$

Correlation $\quad f(n)\odot h(n) \xleftrightarrow{\mathcal{F}_{DT}} F(e^{j\omega})H(e^{-j\omega})$

Parseval's formula $\quad \displaystyle\sum_{n=-\infty}^{\infty}|f(n)|^2 = \dfrac{1}{2\pi}\int_{-\pi}^{\pi}|F(e^{j\omega})|^2\,d\omega$

The proofs of these properties are given in Appendix 7.1.

Example 7.2.1: Find the DTFT of the function $g(n) = 0.9^n u(n)\cos 0.1\pi n$.

Solution: We observe that the function $0.9^n u(n)$ is a modulating function and thus

$$\mathcal{F}\{0.9^n u(n)\cos 0.1\pi n\} = \mathcal{F}\left\{\dfrac{0.9^n u(n)e^{j0.1\pi n}}{2} + \dfrac{0.9^n u(n)e^{-j0.1\pi n}}{2}\right\}$$

$$= \dfrac{1}{2}\sum_{n=0}^{\infty}(0.9e^{-j(\omega-0.1\pi)})^n + \dfrac{1}{2}\sum_{n=0}^{\infty}(0.9e^{-j(\omega+0.1\pi)})^n = \dfrac{1}{2}\dfrac{1}{1-0.9e^{-j(\omega-0.1\pi)}}$$

$$+\dfrac{1}{2}\dfrac{1}{1-0.9e^{-j(\omega+0.1\pi)}} = \dfrac{1}{2}G(e^{j(\omega-0.1\pi)})+\dfrac{1}{2}G(e^{j(\omega+0.1\pi)})$$

We must always have in mind that the spectrum of discrete signals is infinitely periodic. Therefore, the magnitude of $G(e^{j\omega})$ is a double periodic structure, one shifted to the left by 0.1π and the other shifted to the right by 0.1π. Since in this case $T_s = 1$, the spectrum can be plotted from $-\pi/1$ to $\pi/1$.

Book MATLAB m-file: ex7_2_1

```
%ex7_2_1 is an m file for illustrating Ex 7.2.1
w=-pi:0.01:pi;
fw1=abs(0.5*(1./(1-0.9*exp(-j*(w-0.1*pi)))));
fw2=abs(0.5*(1./(1-0.9*exp(-j*(w+0.1*pi)))));
plot(w,fw1,w,fw2);
```
∎

Note: *In contrast with the Fourier transform, where we had continuous functions in both the time and frequency domains, in the case of DTFT we have discrete functions in time domain and continuous functions in the frequency domain.*

Example 7.2.2: Verify Parseval's formula using the function

$$F(e^{j\omega}) = \begin{cases} 1 & -\pi < \omega < \pi \\ 0 & \text{otherwise} \end{cases}$$

Solution: The time sequence corresponding to the given $F(e^{j\omega})$ is

$$f(n) = \frac{1}{2\pi} \int_{-\pi}^{\pi} e^{j\omega n} d\omega = \frac{\sin n\pi}{n\pi}$$

But the relations

$$\sum_{n=0}^{\infty} \left(\frac{\sin n\pi}{n\pi} \right)^2 = \frac{1}{\pi^2} \sum_{n=0}^{\infty} \left(\frac{\sin n\pi}{n} \right)^2 = \frac{1}{\pi^2}(\pi^2 + 0 + 0 + \cdots) = 1$$

and

$$\frac{1}{2\pi} \int_{-\pi}^{\pi} d\omega = 1$$

prove the theorem. ∎

7.3 DTFT of finite time sequences

Practical considerations usually dictate that we deal with truncated series. Therefore, the spectrum of a truncated series requires special attention. When we observe a finite number of data points, $f(0), f(1), f(2), \cdots, f(N-1)$, how do we account for the unobserved time series elements that lie outside the measured interval $0 \leq n \leq N - 1$? We must consider the effect of the missing data since a one-sided Fourier transform, for example, requires the entire set of time series elements $f(n)$ for the interval $0 \leq n < \infty$.

The one-sided truncated DTFT is defined by

$$\boxed{F_N(e^{j\omega}) = \sum_{n=0}^{N-1} f(n) e^{-j\omega n}} \tag{7.11}$$

Chapter 7: Discrete-time transforms 263

We introduce the DTFT of $f(n)$ in the above expression so that

$$F_N(e^{j\omega}) = \sum_{n=0}^{N-1}\left[\frac{1}{2\pi}\int_{-\pi}^{\pi}F(e^{j\omega'})e^{j\omega'n}d\omega'\right]e^{-j\omega n}$$

$$= \frac{1}{2\pi}\int_{-\pi}^{\pi}F(e^{j\omega'})\sum_{n=0}^{N-1}e^{j(\omega-\omega')n}d\omega' = \frac{1}{2\pi}\int_{-\pi}^{\pi}F(e^{j\omega'})W(e^{j(\omega-\omega')})d\omega' \quad (7.12)$$

$$= \frac{1}{2\pi}F(e^{j\omega})*W(e^{j\omega})$$

where

$$W(e^{j\omega}) = \sum_{n=0}^{N-1}e^{-j\omega n} = \frac{1-(e^{-j\omega})^N}{1-e^{-j\omega}} = e^{-j\omega(N-1)/2}\frac{\sin(\omega N/2)}{\sin(\omega/2)} \quad (7.13)$$

and the finite geometric series formula was used. The transform function $W(e^{j\omega})$ is the DTFT of the rectangular window, since it is the transform of the time function $w(n) = u(n) - u(n-N) = p_{N/2}(n-N/2)$ for even N. We observe that with a finite sequence a convolution appears in the frequency domain. From (7.12) we observe that to find the exact spectrum $F(e^{j\omega})$ we require a Fourier-transformed window $W(e^{j\omega})$ equal to a delta function $\delta(\omega)$ in the interval $-\pi \leq \omega \leq \pi$. However, the magnitude of $|W(e^{j\omega})| = \sin(\omega N/2)/\sin(\omega/2)$ has the properties of a delta function and approaches it as $N \to \infty$. Therefore, the longer the time–data sequence we observe, the less distortion that will occur in the spectrum of the signal, $F(e^{j\omega})$.

We observe that

$$\cos\omega_0 n = \mathcal{F}_{DT}^{-1}\{\pi\delta(\omega-\omega_0)+\pi\delta(\omega+\omega_0)\} \quad (7.14)$$

and hence the DTFT of

$$f(n) = \cos(\omega_0 n) + \cos[(\omega_0 + \Delta\omega_0)n] \quad (7.15)$$

is

$$F(e^{j\omega}) = \pi[\delta(\omega-\omega_0)+\delta(\omega-\omega_0)+\delta(\omega-\omega_0-\Delta\omega_0)+\delta(\omega+\omega_0+\Delta\omega_0)] \quad (7.16)$$

The convolution of (7.16) with (7.13) gives the following expression:

$$F_N(e^{j\omega}) = \frac{1}{2} e^{-j(\omega-\omega_0)(N-1)/2} \frac{\sin\frac{\omega-\omega_0}{2}N}{\sin\frac{\omega-\omega_0}{2}} + \frac{1}{2} e^{-j(\omega+\omega_0)(N-1)/2} \frac{\sin\frac{\omega+\omega_0}{2}N}{\sin\frac{\omega+\omega_0}{2}}$$

$$+ \frac{1}{2} e^{-j(\omega-\omega_0-\Delta\omega_0)(N-1)/2} \frac{\sin\frac{\omega-\omega_0-\Delta\omega_0}{2}N}{\sin\frac{\omega-\omega_0-\Delta\omega_0}{2}} \quad (7.17)$$

$$+ \frac{1}{2} e^{-j(\omega+\omega_0+\Delta\omega_0)(N-1)/2} \frac{\sin\frac{\omega+\omega_0+\Delta\omega_0}{2}N}{\sin\frac{\omega+\omega_0+\Delta\omega_0}{2}}$$

We observe that the DTFT of two finite sinusoids is made up of four sinc functions, two on the right side and two on the left side. We will plot the magnitude of (7.17) for different values of $\Delta\omega_0$. This will show us how the value of N affects the resolution capabilities of the DTFT for two sinusoids whose frequencies are very close to each other. We shall plot only half of the sprectum for convenience. Figure 7.3.1a shows the magnitude of the spectrum

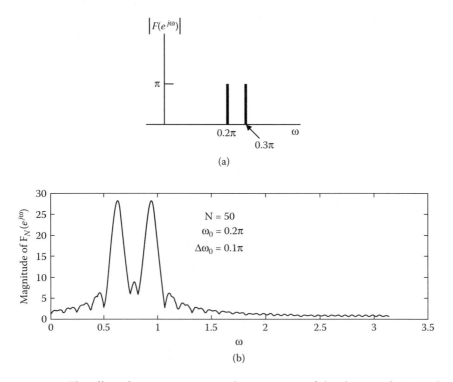

Figure 7.3.1 The effect of time truncation on the separation of closely spaced sinusoids.

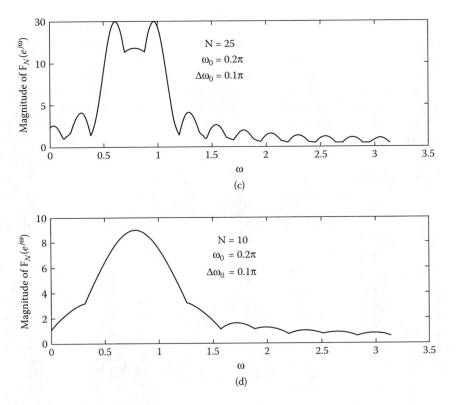

Figure 7.3.1 (continued)

for two sinusoids with $\omega_0 = 0.2\pi$, $\Delta\omega_0 = 0.1\pi$ and an infinite number of terms. Figure 7.3.1b shows the magnitude spectrum for the case

$$\Delta\omega_0 = 0.1\pi \gg \frac{2\pi}{N} = \frac{2\pi}{50}.$$

Figure 7.3.1c shows the magnitude spectrum for the case

$$\Delta\omega_0 = 0.1\pi > \frac{2\pi}{N} = \frac{2\pi}{25},$$

and finally, Figure 7.3.1d shows the magnitude spectrum for the case

$$\Delta\omega_0 = 0.1\pi < \frac{2\pi}{N} = \frac{2\pi}{10}.$$

Since the limit value is $2\pi/N$, then as $N \to \infty$, the value of $2\pi/N$ approaches zero. This indicates that we can separate two sinusoids with infinitesimal difference if we take enough signal values.

Windowing

It turns out that the Fourier transform spectrum will depend on the type of window used. It is appropriate that the window functions used are such that the smoothed version of the spectrum resembles the exact spectrum as closely as possible. Typically it is found that for a given value N, the smoothing effect is directly proportional to the width of the main lobe of the window, and the effects of ripple decrease as the relative amplitude of the main lobe and the largest side lobe diverge. A few important discrete windows are:

1. von Hann or Hanning:

$$w(n) = \frac{1}{2}\left[1 - \cos\left(\frac{2\pi n}{N-1}\right)\right] \qquad 0 \leq n < N \qquad (7.18)$$

2. Bartlett:

$$w(n) = \begin{cases} \dfrac{2n}{N-1} & 0 \leq n \leq \dfrac{N-1}{2} \\ 2 - \dfrac{2n}{N-1} & \dfrac{N-1}{2} \leq n < N \end{cases} \qquad (7.19)$$

3. Hamming:

$$w(n) = 0.54 - 0.46 \cos\left(\frac{2\pi n}{N-1}\right) \qquad 0 \leq n < N \qquad (7.20)$$

4. Blackman:

$$w(n) = 0.4 - 0.5 \cos\left(\frac{2\pi n}{N-1}\right) + 0.08 \cos\left(\frac{4\pi n}{N-1}\right) \qquad 0 \leq n < N \qquad (7.21)$$

Example 7.3.1: Find the DTFT of the discrete function $f(n) = 0.95^n u(n)$ using $N = 21$ and compare that spectrum with the spectrum using the Hamming window instead of the rectangular one.

Solution: Figure 7.3.2a shows the spectrum for the rectangular window, and Figure 7.3.2b shows the spectrum using the Hamming window. For both cases we used $N = 21$. Figure 7.3.2c is the exact spectrum.

Chapter 7: Discrete-time transforms 267

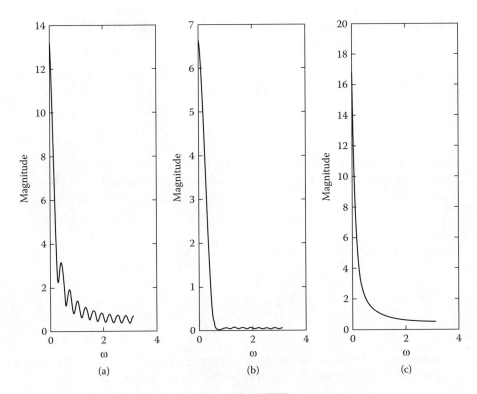

Figure 7.3.2 The effect of windowing on the DTFT spectrum.

Book MATLAB m-file: ex7_3_1

```
%ex7_3_1 is an m file that illustrates Ex 7.3.1
N=21;
n=0:N-1;
w=0:.01:pi;
wh=0.54-0.46*cos(2*pi*n'/(N-1));%Nx1 vector;
f1=abs((0.95.^n')'*exp(-j*n'*w));
f1s=sum(f1,1);%sums the columns of the Nxlength(w) matrix f1;
f2h=abs((0.95.^n'.*wh)'*exp(-j*n'*w));
f2hs=sum(f2h,1);
subplot(1,3,1);plot(w,f1s,'k');xlabel('\omega');ylabel...
     ('magnitude')
subplot(1,3,2);plot(w,f2hs,'k');xlabel('\omega');ylabel...
     ('magnitude')
fwe=abs(1./(1-0.95*exp(-j*w)));
subplot(1,3,3);plot(w,fwe,'k');xlabel('\omega');ylabel...
     ('magnitude')
```

■

7.4 Frequency response of linear time-invariant (LTI) discrete systems

In Chapter 3 we studied the discrete systems, which are described by difference equations. The general difference equation for the first-order system is given by

$$y(n) = b_0 x(n) + b_1 x(n-1) - a_1 y(n-1) \tag{7.22}$$

We assume that the DTFT of $y(n)$, $x(n)$ and the system impulse response $h(n)$ all exist. Taking the DTFT of both sides of the above equation, we obtain

$$Y(e^{j\omega}) = b_0 X(e^{j\omega}) + b_1 e^{-j\omega} X(e^{j\omega}) - a_1 e^{-j\omega} Y(e^{j\omega}) \tag{7.23}$$

from which we write the system function

$$H(e^{j\omega}) = \frac{Y(e^{j\omega})}{X(e^{j\omega})} = \frac{b_0 + b_1 e^{-j\omega}}{1 + a_1 e^{-j\omega}} = b_0 \frac{e^{j\omega} + \frac{b_1}{b_0}}{e^{j\omega} + a_1} \triangleq b_0 \frac{z - z_1}{z - p_1}\bigg|_{z=e^{j\omega}} \tag{7.24}$$

If we set $\omega + 2\pi$ in the above equation, we find that $H(e^{j(\omega+2\pi)}) = H(e^{j\omega})$, which indicates that the transfer function is periodic with period 2π ($T_s = 1$).

Example 7.4.1: Determine the system function of the system specified by

$$y(n) - 0.5 y(n-1) = x(n) \tag{7.25}$$

Solution: Comparing (7.25) to (7.22), we observe that the constants are $a_1 = -0.5$, $b_0 = 1$, and $b_1 = 0$. Hence, (7.24) gives the following transfer function:

$$H(e^{j\omega}) = \frac{1}{1 - 0.5 e^{-j\omega}} = \frac{e^{j\omega}}{e^{j\omega} - 0.5} \tag{7.26}$$

We know that the DTFT of $0.5^n u(n)$ is

$$\sum_{n=0}^{\infty} 0.5^n e^{-j\omega n} = \frac{1}{1 - 0.5 e^{-j\omega}} = H(e^{j\omega})$$

Therefore, the impulse response of the system is

$$h(n) = 0.5^n u(n) \tag{7.27}$$

■

7.5 The discrete Fourier transform (DFT)

We have shown in Chapter 6 that if a time function is sampled uniformly in time, its Fourier spectrum is a periodic function. Therefore, corresponding to any sampled function in the frequency domain, a periodic function exists in the time domain. As a result, the sampled signal values can be related in both domains.

As a practical matter, we are only able to manipulate a certain length of signal. That is, suppose that the data sequence is available only within a finite time window from $n = 0$ to $n = N - 1$. The transform is discretized for N values by taking samples at the frequencies $2\pi/NT_s$, where T_s is the time interval between sample points in the time domain. Hence, we define the **discrete Fourier transform** (DFT) of a sequence of N samples $\{f(nT_s)\}$ by the relation

$$F(k\Omega_b) \triangleq F(e^{jk\Omega_b}) = \mathcal{F}_D\{f(nT_s)\} = T_s \sum_{n=0}^{N-1} f(nT_s) e^{-j\Omega_b T_s kn}$$

$$\Omega_b = \frac{2\pi}{NT_s}; \ 0 \leq n \leq N-1$$

(7.28)

where

N = number of samples (even number)

T_s = sampling time interval

$(N-1)T_s$ = signal length in time domain

$\Omega_b = \dfrac{\omega_s}{N} = \dfrac{2\pi}{NT_s}$ = the frequency sampling interval (frequency bin)

$e^{-j\Omega_b T_s}$ = Nth principal root of unity

The inverse DFT (IDFT) is related to DFT in much the same way that the Fourier transform is related to its inverse Fourier transform. The IDFT is given by

$$f(nT_s) \triangleq \mathcal{F}_D^{-1}\{F(k\Omega_b)\} = \frac{1}{NT_s} \sum_{k=0}^{N-1} F(k\Omega_b) e^{j\Omega_b T_s nk}$$

$$\Omega_b = \frac{2\pi}{NT_s}; \ 0 \leq k \leq N-1$$

(7.29)

Proof: We write

$$\frac{1}{NT_s}\sum_{k=0}^{N-1}F(k\Omega_b)e^{j\Omega_b T_s nk} = \frac{1}{NT_s}\sum_{k=0}^{N-1}\left[T_s\sum_{m=0}^{N-1}f(mT_s)e^{-j\Omega_b T_s mk}\right]e^{j\Omega_b T_s nk}$$

$$= \frac{1}{N}\sum_{m=0}^{N-1}f(mT_s)\sum_{k=0}^{N-1}e^{-j\Omega_b T_s (m-n)k}$$

However, using the finite number geometric series formula, we obtain (see Problem 7.5.1)

$$\sum_{k=0}^{N-1}e^{-j\Omega_b T_s (m-n)k} = \begin{cases} N & m = n \\ 0 & m \neq n \end{cases}$$

and hence the right side of the above equation becomes equal to $f(nT_s)$, as it should. Therefore, (7.29) and (7.28) constitute a pair of the DFT.

Note: *The sequences in time domain and in frequency domain are periodic with period N (see Problem 7.5.2). This indicates that when we take N terms from a time sequence and use the DFT, we automatically create a periotic time function and the corresponding periodic frequency function.*

Note: *In DFT both the time domain and the frequency domain are discrete sequences.*

In general, $F(e^{jk\Omega_b})$ is complex function and can be written in the form

$$F(e^{jk\Omega_b}) = |F(e^{jk\Omega_b})|e^{j\phi(k\Omega_b)} \qquad (7.30)$$

where $|F(e^{jk\Omega_b})|$ and $\phi(k\Omega_b)$ are discrete frequency functions. The plots of these functions vs. $k\Omega_b$ are referred to as the **amplitude** and **phase** spectrums, respectively, of the signal $f(nT_s)$.

Example 7.5.1: Find the DFT of the function $f(n) = 0.9^n u(n)$ for $N = 4$.

Solution: Since the sampling time is unity, (7.28) becomes

$$F\left[0\left(\frac{2\pi}{4}\right)\right] = \sum_{n=0}^{4-1}0.9^n e^{-j\frac{2\pi}{4}0n} = 1 + 0.9 + 0.9^2 + 0.9^3 = 3.4390$$

Chapter 7: Discrete-time transforms

$$F\left[1\left(\frac{2\pi}{4}\right)\right] = \sum_{n=0}^{4-1} 0.9^n e^{-j\frac{2\pi}{4}1n} = 1 + 0.9e^{-j\frac{2\pi}{4}1\times 1} + 0.9^2 e^{-j\frac{2\pi}{4}1\times 2} + 0.9^3 e^{-j\frac{2\pi}{4}1\times 3}$$

$$= 0.1900 - 0.1710j$$

$$F\left[2\left(\frac{2\pi}{4}\right)\right] = \sum_{n=0}^{4-1} 0.9^n e^{-j\frac{2\pi}{4}2n} = 1 + 0.9e^{-j\frac{2\pi}{4}2\times 1} + 0.9^2 e^{-j\frac{2\pi}{4}2\times 2} + 0.9^3 e^{-j\frac{2\pi}{4}2\times 3}$$

$$= 0.1810 - 0.0000j$$

$$F\left[3\left(\frac{2\pi}{4}\right)\right] = \sum_{n=0}^{4-1} 0.9^n e^{-j\frac{2\pi}{4}3n} = 1 + 0.9e^{-j\frac{2\pi}{4}3\times 1} + 0.9^2 e^{-j\frac{2\pi}{4}3\times 2} + 0.9^3 e^{-j\frac{2\pi}{4}3\times 3}$$

$$= 0.1900 + 0.1710j$$

Note: *The magnitude of the spectrum for k = 3 and k = 1 are the same. The frequency $\pi/T_s = \pi/1$ is known as the **fold-over** frequency.*

The amplitude spectrum is 3.4390, 0.2556, 0.1810, 0.2556 for k = 0, 1, 2, and 3, respectively. The phase spectrum is 0.0000, −0.7328, 0, and 0.7328 for k = 0, 1, 2, and 3, respectively. The phase spectrum is given in radians. Since the frequency bin is $\Omega_b = 2\pi/NT_s = 2\pi/4 \times 1 = \pi/2$, the amplitude and frequency spectrums are located at 0, $\pi/2$, π (fold-over frequency), and $3\pi/2$ rad/unit length. ■

7.6 Summary of the DFT properties

Below we give a summary of the DFT properties. Their proof is given in Appendix 7.2. It is convenient to abbreviate the notation of (7.28) and (7.29) by writing

$$F(k\Omega_b) = F(k); \quad f(nT) = f(n); \quad e^{-j2\pi/N} = W \qquad (7.31)$$

Using this notation, the DFT pair takes the form

$$\boxed{\begin{aligned} F(k) &= \sum_{n=0}^{N-1} f(n) W^{nk} = \sum_{n=0}^{N-1} f(n) e^{-j2\pi nk/N} \quad k = 0, 1, 2, \cdots, N-1 \\ f(n) &= \frac{1}{N} \sum_{k=0}^{N-1} F(k) W^{-nk} = \frac{1}{N} \sum_{k=0}^{N-1} F(k) e^{j2\pi nk/N} \quad n = 0, 1, 2, \cdots, N-1 \end{aligned}} \qquad (7.32)$$

Linearity	$af(n)+bh(n) \xleftrightarrow{\mathcal{F}_{DF}} aF(k)+bH(k)$
Symmetry	$\frac{1}{N}F(n) \xleftrightarrow{\mathcal{F}_{DF}} f(-k)$
Time shifting	$f(n-i) \xleftrightarrow{\mathcal{F}_{DF}} F(k)e^{-j2\pi ki/N} = F(k)W^{ki}$
Frequency shifting	$f(n)e^{jni} \xleftrightarrow{\mathcal{F}_{DF}} F(k-i)$
Time convolution	$y(n) \triangleq f(n)*h(n) \xleftrightarrow{\mathcal{F}_{DF}} F(k)H(k)$
Frequency convolution	$f(n)h(n) \xleftrightarrow{\mathcal{F}_{DF}} \frac{1}{N}\sum_{x=0}^{N-1}F(x)H(n-x)$
Parseval's theorem	$\sum_{n=0}^{N-1}|f(n)|^2 = \frac{1}{N}\sum_{k=0}^{N-1}|F(k)|^2$
Time reversal	$f(-n) \xleftrightarrow{\mathcal{F}_{DF}} F(-k)$
Delta function	$\delta(n) \xleftrightarrow{\mathcal{F}_{DF}} 1$
Central ordinate	$f(0) = \frac{1}{N}\sum_{k=0}^{N-1}F(k); \quad F(0) = \sum_{n=0}^{N-1}f(n)$

Example 7.6.1: Shifting property. Deduce the DFT of the two sequences shown in Figure 7.6.1. Observe that $h(n)$ is a shifted function of $f(n)$.

Solution: From (7.28) we obtain, respectively,

$$F\left(k\frac{2\pi}{4}\right) = \sum_{n=0}^{4-1}f(n)e^{-j2\pi kn/4}; \quad H\left(k\frac{2\pi}{4}\right) = \sum_{n=3}^{6}f(n)e^{-j2\pi kn/4}$$

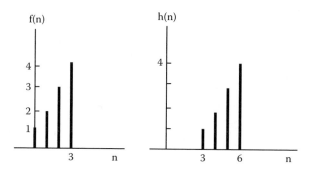

Figure 7.6.1 Illustration of Example 7.6.1.

Chapter 7: Discrete-time transforms

The specific expansions are

$$F\left(0\frac{\pi}{2}\right) = 1\cos\left(\frac{\pi}{2}0\times 0\right) + 2\cos\left(\frac{\pi}{2}0\times 1\right) + 3\cos\left(\frac{\pi}{2}0\times 2\right) + 4\cos\left(\frac{\pi}{2}0\times 3\right)$$

$$-j\left[1\sin\left(\frac{\pi}{2}0\times 0\right) + 2\sin\left(\frac{\pi}{2}0\times 1\right) + 3\sin\left(\frac{\pi}{2}0\times 2\right) + 4\sin\left(\frac{\pi}{2}0\times 3\right)\right]$$

$$= 10 - j0$$

$$F\left(1\frac{\pi}{2}\right) = 1\cos\left(\frac{\pi}{2}1\times 0\right) + 2\cos\left(\frac{\pi}{2}1\times 1\right) + 3\cos\left(\frac{\pi}{2}1\times 2\right) + 4\cos\left(\frac{\pi}{2}1\times 3\right)$$

$$-j\left[1\sin\left(\frac{\pi}{2}1\times 0\right) + 2\sin\left(\frac{\pi}{2}1\times 1\right) + 3\sin\left(\frac{\pi}{2}1\times 2\right) + 4\sin\left(\frac{\pi}{2}1\times 3\right)\right]$$

$$= -2 + j2$$

We can proceed the same way to find the rest of the values. However, we can use the following Book MATLAB m-file to find the values easily.

Book MATLAB m-files: ex7_6_1, ex7_6_1b

```
%ex7_6_1 is an m-file to find the values of the summations
N=4;
n=0:N-1;
fn=[1 2 3 4];
k=0:N-1;
F=exp(-j*k'*n*2*pi/N)*fn';
subplot(2,1,1);stem(k,abs(F));subplot(2,1,2);stem(k,angle(F));
%after you call the m-file, type F and return and you will
%be presented with the values of the DFT
```

The values of F are:

10.000
−2.0000 + 2.0000i
−2.0000 − 0.0000i
−2.0000 − 2.0000i

If we call the m-file ex7_6_1b, we obtain the DFT of the shifted function $h(n)$. The values are:

10.0000
−2.0000 − 2.0000i
2.0000 + 0.0000i
−2.0000 + 2.0000i

Observe that the magnitude of the spectrum stays the same, but the phases have changed. This is the practical verification of the shifting property.

∎

Example 7.6.2: Time convolution. Consider the two periodic sequences $f(n) = \{1, -1, 4\}$ and $h(n) = \{0, 1, 3\}$. Verify the time convolution property by showing that for $T_s = 1$,

$$y(2) = \sum_{x=0}^{N-1} f(x)h(2-x) = \mathcal{F}_{DF}^{-1}\{F(k)H(k)\} = \frac{1}{N}\sum_{k=0}^{N-1} F(k)H(k)e^{j2\pi k2/N}$$

Solution: We first find the summation

$$\sum_{x=0}^{2} f(x)h(2-x) = f(0)h(2) + f(1)h(1) + f(2)h(0) = 1 \times 3 + (-1)1 + 4 \times 0 = 2$$

Next, we obtain the DFT of the sequences $f(n)$ and $h(n)$, which are

$$F(0) = \sum_{n=0}^{2} f(n)e^{-j2\pi 0 n/3} = f(0) + f(1) + f(2) = 1 - 1 + 4 = 4$$

$$F(1) = \sum_{n=0}^{2} f(n)e^{-j2\pi 1 n/3} = 1 - e^{-j2\pi 1 \times 1/3} + 4e^{-j2\pi 1 \times 2/3}$$

$$F(2) = \sum_{n=0}^{2} f(n)e^{-j2\pi 2 n/3} = 1 - e^{-j2\pi 2 \times 1/3} + 4e^{-j2\pi 2 \times 2/3}$$

Similarly, we obtain

$$H(0) = \sum_{n=0}^{2} h(n)e^{-j2\pi 0 n/3} = h(0) + h(1) + h(2) = 0 + 1 + 3 = 4$$

$$H(1) = \sum_{n=0}^{2} h(n)e^{-j2\pi 1 n/3} = 0 + e^{-j2\pi 1 \times 1/3} + 3e^{-j2\pi 1 \times 2/3}$$

$$H(2) = \sum_{n=0}^{2} h(n)e^{-j2\pi 2 n/3} = 0 + e^{-j2\pi 2 \times 1/3} + 3e^{-j2\pi 2 \times 2/3}$$

Chapter 7: Discrete-time transforms

The second summation given above becomes

$$\frac{1}{3}\sum_{k=0}^{2} F(k)H(k)e^{jk2\pi 2/3} = \frac{1}{3}[F(0)H(0) + F(1)H(1)e^{j4\pi/3} + F(2)H(2)e^{j8\pi/3}]$$

$$= 2.0000 - 0.0000j$$

For convenience, the reader can use MATLAB directly. ■

Note: *To obtain the circular convolution of two sequences f(n) and h(n) we follow the following steps:*

1. *Calculate the DFT of these sequences to obtain F(k) and H(k).*
2. *Multiply element by element these DFTs to obtain G(k) = F(k)H(k).*
3. *Calculate the IDFT of G(k).*

The result is the circular convolution, and for the two given in the example above, g(n) = {1 13 2}.

Note: *To obtain the linear convolution of f(n) with N elements and h(n) with M elements, we add zeros at the end of each sequence such that its total length is Q = N + M − 1.*

Example 7.6.3: Frequency convolution. Use the values of $F(k)$ and $H(k)$ of Example 7.6.1 to verify the frequency convolution property.

Solution: Figure 7.6.2 shows the circular convolution of $F(k)$ and $H(k)$, with the values obtained from Example 7.6.2. From the frequency convolution property we obtain the periodic sequence

$$y(n) = \mathcal{F}_{DF}^{-1}\{Y(k)\} = \mathcal{F}_{DF}^{-1}\left[\frac{1}{N}\sum_{i=0}^{N-1} F(i)H(k-i)\right]$$

or

$$\{y(n)\} = \{f(n)h(n)\} = \{0, -1, 12\}$$

Thus, from the results of Figure 7.6.2, we obtain

$$y(0) = \mathcal{F}_{DF}^{-1}\{Y(k)\} = \frac{1}{3}\sum_{k=0}^{N-1} Y(k)e^{jk(2\pi/3)0}$$

$$= \frac{1}{3}\left[11 - \frac{11}{2} - \frac{11}{2} + j13\frac{\sqrt{3}}{2} - j13\frac{\sqrt{3}}{2}\right] = 0$$

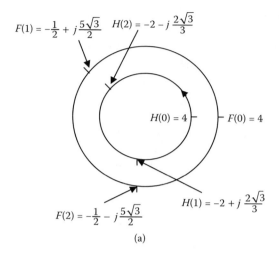

Figure 7.6.2 Circular convolution in the frequency domain.

$$y(1) = \mathcal{F}_{DF}^{-1}\{Y(k)\} = \frac{1}{3}\left[11 + \left(-\frac{11}{2} + j13\frac{\sqrt{3}}{2}\right)\left(\cos\frac{2\pi}{3} + j\sin\frac{2\pi}{3}\right)\right.$$

$$\left. + \left(-\frac{11}{2} - j13\frac{\sqrt{3}}{2}\right)\left(\cos\frac{4\pi}{3} + j\sin\frac{4\pi}{3}\right)\right] = -1$$

$$y(2) = \mathcal{F}_{DF}^{-1}\{Y(k)\} = \frac{1}{3}\left[11 + \left(-\frac{11}{2} + j13\frac{\sqrt{3}}{2}\right)\left(\cos\frac{4\pi}{3} + j\sin\frac{4\pi}{3}\right)\right.$$

$$\left. + \left(-\frac{11}{2} - j13\frac{\sqrt{3}}{2}\right)\left(\cos\frac{8\pi}{3} + j\sin\frac{8\pi}{3}\right)\right] = 12$$

From Figure 7.6.2 we obtain

$$Y(0) = \frac{[F(0)H(0) + F(1)H(2) + F(2)H(1)]}{3} = 11,$$

$$Y(1) = \frac{[F(0)H(1) + F(1)H(0) + F(2)H(2)]}{3} = -\frac{11}{2} + j13\frac{\sqrt{3}}{2}$$

$$Y(2) = \frac{[F(0)H(2) + F(1)H(1) + F(2)H(0)]}{3} = -\frac{11}{2} - j13\frac{\sqrt{3}}{2}$$

The results obtained verify the frequency convolution property. ∎

Chapter 7: Discrete-time transforms

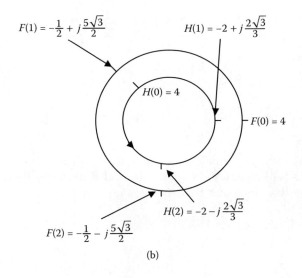

(b)

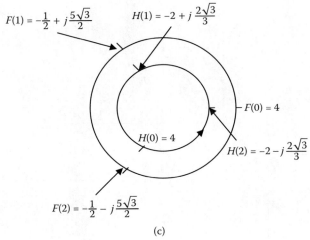

(c)

Figure 7.6.2 (continued).

Example 7.6.4: Parseval's theorem. Verify Parseval's theorem using the sequence $f(n) = \{1, -1, 4\}$.

Solution: We have directly that

$$\sum_{n=0}^{3-1} f^2(n) = 1 + 1 + 16 = 18.$$

The values of F(k) for this sequence are given in Example 7.6.2, so that

$$\frac{1}{3}\sum_{k=0}^{2}|F(k)|^2 = \frac{1}{3}[16 + |-0.5 + j5 \times 0.8660|^2 + |-0.5 - j5 \times 0.8660|^2$$

$$= \frac{1}{3}[16 + 19 + 19] = 18$$

∎

The above examples elucidate some of the DFT properties. The following two examples will clarify the similarities and differences between the DFT of continuous functions and their Fourier transform.

Example 7.6.5: Deduce the DFT of the function shown in Figure 7.6.3a discretized with $T_s = 1$ and $T_s = 0.5$.

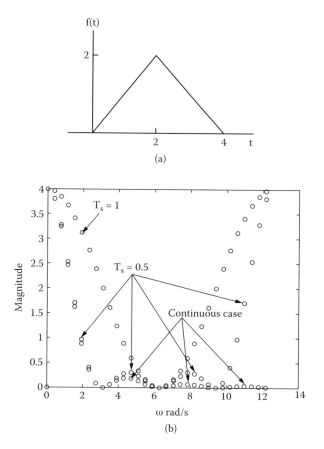

Figure 7.6.3 Illustration of Example 7.6.5.

Chapter 7: Discrete-time transforms

Solution: This signal is a shifted triangle by 2 s. The centered signal is equal to the convolution of two pulses of total width 2 s each: $f(t) = p(t) * p(t)$. From the time convolution property, the FT of $f(t)$ is equal to $F(\omega) = P(\omega)P(\omega)$. But the FT of a pulse is equal to

$$P(\omega) = \frac{2\sin\omega}{\omega}.$$

Hence, the FT of the symmetric signal is

$$F(\omega) = \frac{4\sin^2\omega}{\omega^2}.$$

Taking into consideration the time-shifting property, the FT of the shifted signal is given by

$$F(\omega) = e^{-j2\omega}\frac{4\sin^2\omega}{\omega^2}.$$

For the case of $T_s = 1$ we have the following discrete signal: $f(nT_s) = \{0\ 1\ 2\ 1\ 0\}$ for times $t = \{0\ 1\ 2\ 3\ 4\}$ s, respectively. Let us further define $NT_s = 16$. Since we set $T_s = 1$ and $NT_s = 16$, we find that $N = 16$. This value specifies the bins in the frequency domain that are located at every $\Omega_b = 2\pi/NT_s = 2\pi/16 = \pi/8$ rad/s. For the case of $T_s = 0.5$ s and $NT_s = 16$ we obtain $N = 32$. However, because $NT_s = 16$, the bins at the frequency domain are located every $\pi/8$. But since in this case $N = 32$, the fold-over frequency (the useful frequency extent) is at $(\pi/8)32 = 4\pi$ rad/s. This can be compared with the first case, whose fold-over frequency is $(\pi/8)16 = 2\pi$. Comparing the two results, we observe that by decreasing the sampling time by 2 we expand the fold-over frequency (the useful frequency) by 2. The spectrums can be found following Example 7.5.1. Here we use MATLAB for convenience.

Book MATLAB m-file: ex7_6_5

```
%ex7_6_5 is an m-file for the Ex 7.6.5
N1=16;
T1=1;
T2=0.5;
f1=[0 1 2 1 0 zeros(1,27)];
f2=[0 0.5 1 1.5 2 1.5 1 0.5 0 zeros(1,23)];
fd1=fft(f1);
wb1=0:2*pi/(N1*T1):4*pi-2*pi/(N1*T1);
fd2=T2*fft(f2);%the fft must be multiplied by sampling time;
```

```
ftf=4*(sin(wb1).^2./(wb1+eps).^2);
plot(wb1,abs(fd1),'ko');hold on; plot(wb1,abs(fd2),'ko');...
   hold on;plot(wb1,abs(ftf),'ko');
```

The reader can change the plotting function by incorporating 'kx', for example, instead of 'ko'.

Note: *We observe that by **decreasing** the time sampling we succeed in extending the useful frequency range (the fold-over frequency), and within that range the approximation becomes better.* ∎

Example 7.6.6: Deduce the DFT transform of the function

$$f(t) = \begin{cases} 2 & 0 \leq t \leq 6 \\ 0 & \text{otherwise} \end{cases}$$

for the following three cases

a. $T_s = 1.0$, $NT_s = 16$
b. $T_s = 1.0$, $NT_s = 32$
c. $T_s = 0.2$, $NT_s = 16$

Solution: We readily find the magnitude of the FT of the continuous-time function to be

$$|F(\omega)| = \left|\frac{4 \sin 3\omega}{\omega}\right|$$

Cases a and b: Setting $T_s = 1$ and $NT_s = 16$ we find that $N = 16$. The frequency bins are apart by

$$\Omega_b = \frac{2\pi}{NT_s} = \frac{2\pi}{16} = \frac{\pi}{8} \quad \text{rad/s.}$$

The fold-over frequency is at

$$\frac{\pi}{8} \times \frac{N}{2} = \frac{\pi}{8} 8 = \pi \quad \text{rad/s.}$$

Figure 7.6.4a shows the absolute value of the difference (error) of their magnitude between the continuous and DFT case at the same points of the frequency axis. For this case, the distance is $\pi/8$. The discrete-time function

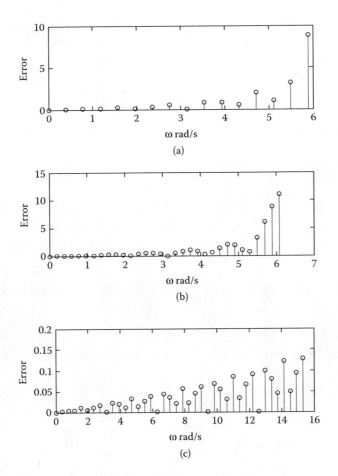

Figure 7.6.4 Illustration of Example 7.6.6.

is $f(n) = [2\ 2\ 2\ 2\ 2\ \text{zeros}(1, 10)]$. Below are nine magnitude values of the DFT spectrum for *case a*:

12.0000 9.4713 3.6955 1.3776 2.8284 0.9205 1.5307 1.8840 0

In *case b* we set $T_s = 1$ and $NT_s = 32$ to obtain $N = 32$ and the frequency bin

$$\Omega_b = \frac{2\pi}{NT_s} = \frac{2\pi}{32} = \frac{\pi}{16} \text{ rad/s.}$$

The fold-over frequency is

$$\frac{\pi}{16} \times \frac{N}{2} = \frac{\pi}{16} 16 = \pi,$$

which is identical to that of *case a* above. The discrete-time function for this case is $f(n) = [2\ 2\ 2\ 2\ 2\ 2\ \text{zeros}(1,\ 26)]$. Note that in *case a* we added 10 zeros and for *case b* we added 26 zeros so that the numbers N in the two cases have the correct value. Although the fold-over frequency is the same, the number of frequency bins is twice as many as in *case a*. Comparing the nine above frequency values with the nine numbers below for *case b*

12.0000 11.3362 9.4713 6.7574 3.6955 0.8277 1.3776 2.6213 2.8284

we observe that new values appeared in the second case between every $\pi/8$ distance. We note that at $\pi/8$ frequency bins the values of the spectrum for both cases are the same.

Note: *By adding zeros to a discrete sequence, the accuracy of the spectrum does not increase. However, within any range, the number of points increases and gives us a better overall view of the spectrum.*

Figure 7.6.4b depicts the error between the discrete *case b* and the exact spectrum.

Case c: For this case we set $T_s = 0.2$ and $NT_s = 16$. Hence, the number of elements of the sequence is $N = 16/0.2 = 80$. The frequency bin is

$$\Omega_b = \frac{2\pi}{16} = \frac{\pi}{8}.$$

The fold-over frequency is

$$\Omega_b \frac{N}{2} = \frac{\pi}{8} 40 = 5\pi \ \text{rad/s},$$

which is five times the fold-over frequency of *cases a* and *b*.

Note: *By decreasing the sampling frequency, we extend the useful frequency (fold-over) to a larger range.*

The discrete-time function created from the continuous signal is $f(n0.2) =$ [2 2 2 2 2 2 2 2 2 2 2 2 2 2 2 2 2 2 2 2 2 2 2 2 2 2 2 2 2 2 zeros (1,50)]. The first nine elements of DFT magnitude are

12.0000 9.4130 3.6050 1.3023 2.5570 0.7846 1.2116 1.3614 0.0000

The exact first nine frequency values for the continuous case are

12.0000 9.4106 3.6013 1.2993 2.5465 0.7796 1.2004 1.3444 0.0000

Chapter 7: Discrete-time transforms

If we compare *case c* with *case a*, we observe that *case c* results are more accurate (they are closer to the continuous case). If we compare *case c* with *case a* for the frequency $3(\pi/8)$ rad/s, we obtain 0.0030 for *case c* and 0.0783 for *case a*. Figure 7.6.4c shows the error between the DFT for *case c* and the exact values.

Note: *By decreasing the sampling time, the discrete signal spectrum approximates better the exact spectrum of the continuous signal up to the folding frequency. The approximation increases as the time sampling decreases further.* ∎

Because the DFT uses a finite number of samples, we must be concerned about the effect that truncation has on the Fourier spectrum. Specifically, if the signal $f(t)$ extends beyond the total sampling $(N-1)T_s$ time, the resulting frequency spectrum is an approximation of the exact one. If, for example, we take the DFT of a truncated sinusoidal signal, we find that an error is present in the Fourier spectrum. If N is small and the sampling covers neither a large number of periods nor an integral number of cycles of the signal, a large error may occur, known as the **leakage** problem. Since the truncated signal is equal to $f(t)p_a(t)$, its Fourier spectrum is the convolution of the exact spectrum with the spectrum of the window. Since the truncation preassumes a rectangular window, its spectrum consists of a main lobe and side lobes. The side lobes affect the spectra and sometimes completely hide weak spectra lines. Since the magnitudes of the side lobes do not decrease with N, it is recommended to use other type of windows that may remedy the problem. However, the main lobe for other windows is wider, and we must justify any effect that it may introduce.

Example 7.6.7: Find the DFT of the exponential function $f(t) = e^{-t}u(t)$ that is truncated at times $t = 0.8$ s and $t = 1.6$ s. Assume the sampling time $T_s = 0.02$ for both cases.

Solution: The results are shown in Figure 7.6.5. The discrete spectrum was plotted in the continuous format to produce a better visual picture of the differences. The following Book m-file produces the results shown in the figure.

Book MATLAB m-file: ex7_6_7

```
%ex7_6_7 is an m file for the Ex7.6.7
t=0:0.02:0.8-.02;
f=[exp(-t) zeros(1,40)];
t1=0:0.02:1.6-0.02;
f1=exp(-t1);
df=0.02*fft(f);
df1=0.02*fft(f1);
```

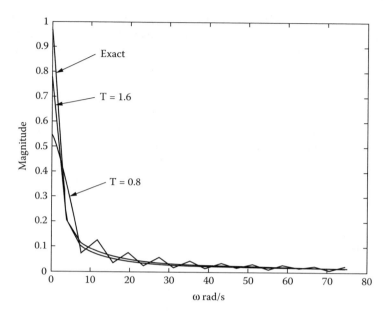

Figure 7.6.5 The time truncation effect.

```
w=0:2*pi/(80*0.02):100*pi-(2*pi/(80*0.02));
fcw=1./(1+w);
plot(w(1,1:20),fcw(1,1:20),'k');
hold on;plot(w(1,1:20),abs(df1(1,1:20)),'k');
hold on;plot(w(1,1:20),abs(df(1,1:20)),'k');
```
■

*7.7 Multirate digital signal processing

There are many applications where the signal of a given sampling rate must be converted into an equivalent signal with a different sampling rate. In other words, we sample digital signals. The sampling can be achieved by using the **down sampler (decimation)**, a device (or process) that creates another discrete signal from the original by skipping a specific number of elements, and by using the **up sampler (interpolation)**, a device (or process) that pads with a specific number of zeros in all the spaces between the original discrete-time signal.

Down sampling (or decimation)

The down-sampling operation by an **integer** $D > 1$ on a discrete-time signal $x(n)$ consists of keeping every D^{th} sample of $x(n)$ and removing $D - 1$ in-between samples. Hence, the output of such an operation is (see also Figure 7.7.1)

Chapter 7: Discrete-time transforms

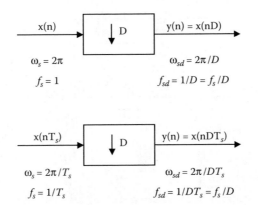

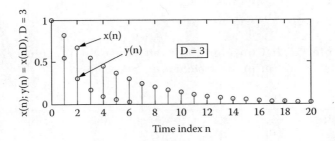

Figure 7.7.1 Block diagram representation of down sampling.

Figure 7.7.2 Illustration of down-sampling process.

$$y(n) = x(nD) \qquad (7.33)$$

Since the initial sampling time of $x(t)$ was assumed to be $T_s = 1$ producing $x(n)$, the sampling time of $x(nD)$ is $T_s D = D$. This indicates that the initial sampling rate $\omega_s = 2\pi/T_s = 2\pi$ for $x(n)$ changes to $\omega_{sd} = 2\pi/DT_s = 2\pi/D$, which is $(1/D)^{th}$ of that of $x(n)$. By decreasing the sampling frequency, we produce a periodic frequency spectrum whose period is $\omega_{sd} < \omega_s$. Down sampling produces aliasing. Figure 7.7.2 illustrates the effect of down sampling in the time domain, and Figure 7.7.1 represents the down-sampling operation in block diagram form.

Book MATLAB function for down sampling

```
function[y,n1]=ssdownsampling(x,D)
    %x=input sequence to be downsampled assuming
    %that Ts=1;
    %D=down-sampling factor;
```

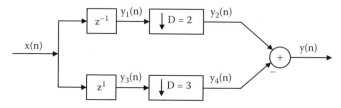

Figure 7.7.3 A multirate system.

```
n1=0:floor(length(x)/D)-1;
    %floor(x) rounds the elements of x to the
    %nearest integers towards minus infinity;
    %if desired, n1 can be used as the x-axis
    %to plot y from the origin;
y=x(1,1:D:D*floor(length(x)/D));
```

Example 7.7.1: Consider the system shown in Figure 7.7.3. The sequences that appear in the system are given below.

n:	0	1	2	3	4	5	6
$x(n)$	$x(0)$	$x(1)$	$x(2)$	$x(3)$	$x(4)$	$x(5)$	$x(6)$
$y_1(n)$	$x(-1)$	$x(0)$	$x(1)$	$x(2)$	$x(3)$	$x(4)$	$x(5)$
$y_3(n)$	$x(1)$	$x(2)$	$x(3)$	$x(4)$	$x(5)$	$x(6)$	$x(7)$
$y_2(n)$	$x(-1)$	$x(1)$	$x(3)$	$x(5)$	$x(7)$	$x(9)$	$x(11)$
$y_4(n)$	$x(1)$	$x(4)$	$x(7)$	$x(10)$	$x(13)$	$x(16)$	$x(19)$
$y(n)$	$x(-1)-x(1)$	$x(1)-x(4)$	$x(3)-x(7)$	$x(5)-x(10)$	$x(7)-x(13)$	$x(9)-x(16)$	$x(11)-x(19)$

■

The sampling rate conversion can also be understood from the point of view of digital resampling of the same analog signal. If the analog signal $x(t)$ is sampled at the rate $1/T_s$ to generate $x(n)$, the analog signal must be sampled at the rate $1/(DT_s)$ to generate $y(n)$.

Frequency domain of down-sampled signals

It is instructive to develop the spectrums of the input and output sequences of a down sampler. The spectrum of the output signal can be found after being transformed in the DTFT domain. Hence, we write

$$y(e^{j\omega}) = \sum_{n=-\infty}^{\infty} y(n)e^{-j\omega n} = \sum_{n=-\infty}^{\infty} x(nD)e^{-j\omega n} \qquad -\infty < n < \infty \qquad (7.34)$$

Chapter 7: Discrete-time transforms

If we set $nD = k$, then $n = k/D$ and (7.34) becomes (n is an integer)

$$Y(e^{j\omega}) = \sum_{k=-\infty}^{\infty} x(k)e^{-j\omega k/D} \qquad k = 0, \pm D, \pm 2D, \cdots \qquad (7.35)$$

The above equation suggests that we must take the DTFT of the values of $x(n)$ at every D numbers apart so that k/D is an integer. Since only every D values of $x(n)$ are used (7.35) is not in the form of the definition of DTFT. Therefore, we must write (7.35) in the form

$$Y(e^{j\omega}) = \sum_{n=-\infty}^{\infty} c(n)x(n)e^{-j\omega n/D} \qquad n = 0, \pm 1, \pm 2, \cdots \qquad (7.36)$$

$$c(n) = \begin{cases} 1 & n = 0, \pm D, \pm 2D, \cdots \\ 0 & \text{otherwise} \end{cases} \qquad (7.37)$$

And thus the DTFT applies. The function $x(n)$ is a discrete periodic function, a comb function with impulses every D apart and zero otherwise; it can be represented by the relation (see Problem 7.7.1)

$$c(n) = \frac{1}{D}\sum_{k=0}^{D-1} e^{j2\pi kn/D} = \frac{1}{D}\sum_{k=0}^{D-1} W^{-kn} \qquad W = e^{-j2\pi/D} \qquad (7.38)$$

which is known as the discrete sampling function. Note that (7.38) is the IDF transform. Introducing (7.38) in (7.36) we obtain

$$Y(e^{j\omega}) = \frac{1}{D}\sum_{k=0}^{D-1}\sum_{n=-\infty}^{\infty} W^{-kn} x(n) e^{-j\omega n/D} = \frac{1}{D}\sum_{k=0}^{D-1}\sum_{n=-\infty}^{\infty} \left(W^k e^{j\omega/D}\right)^{-n} x(n)$$

$$= \frac{1}{D}\sum_{k=0}^{D-1} X\left(W^k e^{j\omega/D}\right) = \frac{1}{D}\sum_{k=0}^{D-1} X\left(e^{j\frac{\omega-2\pi k}{D}}\right) \qquad (7.39)$$

which indicates that the spectrum of the down-sampled signal is the sum of the D uniformly shifted and stretched version of the spectrum $X(e^{j\omega})$ and scaled down by D. Graphically (see Figure 7.7.4) the above formula is interpreted as follows: (a) stretch $X(e^{j\omega})$ by D to obtain $X(e^{j\omega/D})$, (b) create $D-1$ copies of this stretched version by shifting it uniformly in successive amounts of 2π, and (c) add all these shifted versions to $X(e^{j\omega/D})$ and divide by D.

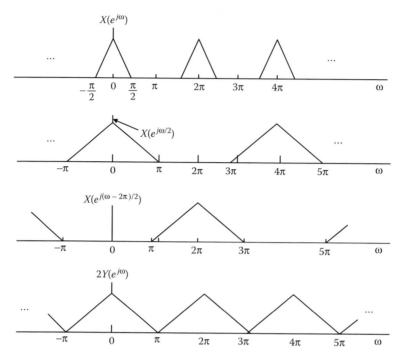

Figure 7.7.4 Frequency spectra of a down-sampled signal without aliasing.

Let the spectrum of the sequence $\{x(n)\}$ be that shown at the top of Figure 7.7.4. Let us further set $D = 2$. Then (7.39) becomes

$$Y(e^{j\omega}) = \frac{1}{2}[X(e^{j\omega/2}) + X(e^{-j\pi}e^{j\omega/2})] = \frac{1}{2}[X(e^{j\omega/2}) + X(e^{j(\omega-2\pi)/2})] \quad (7.40)$$

Because there is a multiplication by $1/D = 0.5 < 1$ of ω, the implication is that the function $X(e^{j\omega})$ is stretched by a factor of 2. Figure 7.7.4 shows the spectrum of a digital signal and the nonaliased spectrum at the output of a decimator with $D = 2$. By judiciously assuming the Nyquist frequency of $\{x(n)\}$ to be $\omega_N = \pi/2$ and $D = 2$, we avoided the aliasing phenomenon. If, however, the Nyquist frequency was in the range $\pi/2 < \omega < \pi$, a decimator with $D = 2$ would produce aliasing at its output, as shown in Figure 7.7.5. In general, aliasing is avoided if the Nyquist frequency of $\{x(n)\}$ is less than π/D.

Example 7.7.2: Figure 7.7.6a shows the frequency response of a discrete signal. Figure 7.7.6b shows the frequency response when the corresponding time signal is down sampled by a factor of 2. Figure 7.7.6c shows the frequency response when the time signal is down sampled by a factor of 4, and finally, Figure 7.7.6d shows the spectrum when the time signal is down

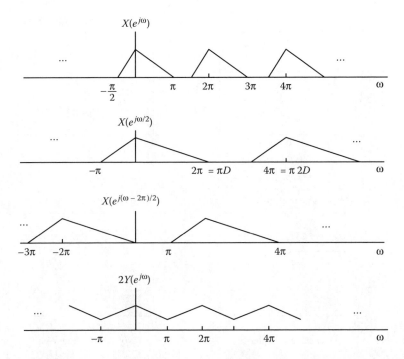

Figure 7.7.5 Frequency spectra of a down-sampled signal with aliasing, D = 2.

sampled by a factor of 6. Observe the distortion of the spectrum in d due to aliasing. Observe, also, that the magnitudes in b and c are 1/2 and 1/4 of the original spectrum, which verifies (7.39). The following program produces Figure 7.7.6.

Book MATLAB m-file: ex7_7_2

```
%ex7_7_2 is an m file for the EX 7.7.2
f=[0 1/8 1/4 1/2 3/4 1];%normalized frequency bins;
m=[1 1 0 0 0 0];%magnitude spectrum corresponding
         %to the above frequency bins;
h=fir2(200,f,m);%based on the frequency magnitude m it returns
         %the time function (filter coefficients);
[Hz,w]=freqz(h,1,512);%evaluation of the spectrum;
subplot(2,2,1);plot(w/pi,abs(Hz));grid on;
xlabel('Normalized frequency');ylabel('Magnitude');
title('Input spectrum');
D=2;y=h(1,1:D:length(h));
[Yz,w]=freqz(y,1,512);subplot(2,2,2);plot(w/pi,abs(Yz));grid on;
```

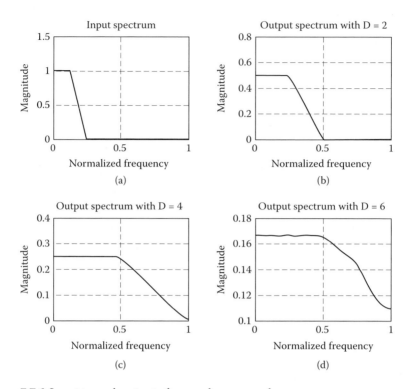

Figure 7.7.6 Input to and outputs from a down sampler.

```
xlabel('Normalized frequency');ylabel('Magnitude');
title('Output spectrum with D=2');
D1=4;y1=h(1,1:D1:length(h));
[Y1z,w]=freqz(y1,1,512);subplot(2,2,3);plot(w/pi,abs(Y1z));
    grid on;
xlabel('Normalized frequency');ylabel('Magnitude');
title('Output spectrum with D=4');
D2=6;y2=h(1,1:D2:length(h));
[Y2z,w]=freqz(y2,1,512);subplot(2,2,4);plot(w/pi,abs(Y2z));
    grid on;
xlabel('Normalized frequency');ylabel('Magnitude');
title('Output spectrum with D=6');                           ∎
```

Note: *To avoid aliasing, we must first reduce the bandwidth of the signal by π/D and then down sample it by a factor D.*

Figure 7.7.7 shows graphically the relationship among the sampled form of $x(t)$ ($T_s = 1$), the function $c(n)$, and the output function $y(n)$ of the decimation

Chapter 7: Discrete-time transforms   291

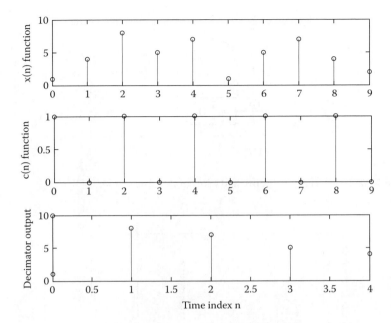

Figure 7.7.7 The decimation operation with a factor $D = 2$.

process with factor $D = 2$. Note how the different functions shown are registered in the time domain.

Let us assume that we have a signal $x(t)$ with $\pm 200\pi$ rad/s bandwidth. If we sample the signal at $T_s = 0.001$ seconds, the sampled function will occupy the frequency range $\pm 200\pi \times 0.001 = \pm 0.2\pi$ rad. This indicates that we can down sample up to $D = 5$ without aliasing. We can also see that half of the sampling frequency is 1000π rad/s, and hence the signal bandwidth can be widened up to *five* times.

Interpolation (up sampling) by a factor U

Interpolation or up sampling is the process of increasing the sampling rate of a signal by an **integer** factor $U > 1$, which results in adding $U - 1$ zero samples between two consecutive samples of the input sequence $\{x(n)\}$. Figure 7.7.8 shows the block diagram representation of an up-sampling operation (interpolation). Figure 7.7.9 shows the form of a discrete signal before and after the up sampler (interpolator) with $U = 3$. The following Book MATLAB will produce the output of an up sampler:

Book MATLAB function

```
function[y]=ssupsampling(x,U)
    %y=output of the up-sampler;U=up-sampling factor
    %x=input to the up-sampler;
```

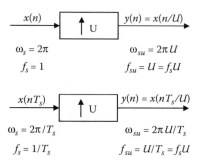

Figure 7.7.8 Block diagram representation of an up sampler.

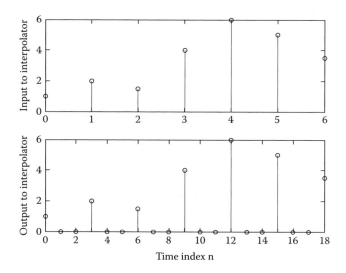

Figure 7.7.9 The up-sampling (interpolation) operation.

```
y1=zeros(1,U*length(x));
y1(1,1:U:length(y1))=x;
y=y1(1,1:length(y1)-U+1);
```

Frequency domain characterization of up-sampled signals

The input and output relationship of the signals of an interpolator is

$$y(n) = \begin{cases} x(n/U) & n = 0, \pm U, \pm 2U, \cdots \\ 0 & n \neq \text{multiple of } U \end{cases} \quad (7.41)$$

Chapter 7: Discrete-time transforms

Note: *The sampling rate of y(n) is U times larger than that of the original sequence {x(n)}.*

The DTFT of (7.41) is given by

$$Y(e^{j\omega}) = \sum_{n=-\infty}^{\infty} y(n)e^{-j\omega n} = \sum_{n=-\infty}^{\infty} x(n/U)e^{-j\omega n} = \sum_{m=-\infty}^{\infty} x(m)e^{-j\omega m U}$$

$$= \sum_{n=-\infty}^{\infty} x(m)(e^{j\omega U})^{-m} = X(e^{j\omega U})$$

(7.42)

The above equation states that we take the DTFT of the input sequence to the interpolator and then we substitute every $e^{j\omega}$ with $e^{j\omega U}$.

Let the spectrum of $\{x(n)\}$, given by $X(e^{j\omega})$, be the one shown in Figure 7.7.10a. The spectrum of the output of the interpolator is given by $X(e^{j\omega U})$. If we set $\omega = \omega + (2\pi k/U)$ in $Y(e^{j\omega})$, we find that the output is periodic for $k = \pm U, \pm 2U, \cdots$. Furthermore, $U > 1$ implies that the ω-axis is **compressed** by a factor of U. By adding $U - 1$ zeros between the values of the input sequence $\{x(n)\}$, we increase the sampling frequency, which results in a signal whose spectrum $Y(e^{j\omega})$ is a U-fold periodic repetition of the input signal spectrum $X(e^{j\omega})$. This effect on the input spectrum is shown in Figure 7.7.10b.

Since only frequency components of this $\{x(n)\}$ in the range $0 \le \omega \le \pi/U$ are unique for the reproduction of the sequence $\{x(n)\}$, the images above $\omega = \pi/U$ should be eliminated using a low-pass filter with ideal characteristics

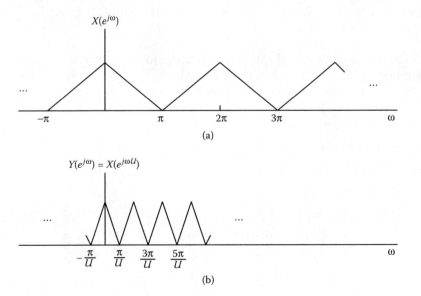

Figure 7.7.10 Spectra of $x(n)$ and $y(n)$ of an up sampler.

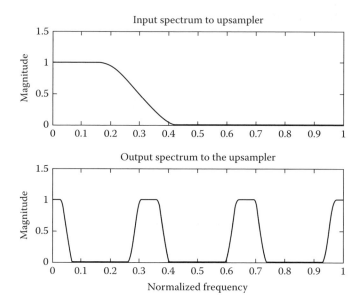

Figure 7.7.11 Input and output spectra of an interpolatior (up sampler) with factor $U = 6$.

$$H(e^{j\omega}) = \begin{cases} A & 0 \leq \omega \leq \pi/U \\ 0 & \text{otherwise} \end{cases} \qquad (7.43)$$

where A is a scale factor to be used for the normalization of the output sequence $\{y(n)\}$, which is U.

Example 7.7.3: Let the input spectrum to an interpolator with up-sampling factor $U = 4$ be that shown in Figure 7.7.11a. The Book MATLAB m-file given below produces the spectrum shown in Figure 7.7.11b.

Book MATLAB m-file: ex7_7_3

```
%ex7_7_3 an m file for the EX7.7.3
freq=[0 0.2 0.3 0.4 1];%frequency range must run from 0 to 1;
mag=[1 1 0.5 0 0];
x=fir2(59,freq,mag);%given the magnitude vector of a spectrum
          %and its corresponding frequency vector,
              %fir2(N,freq,mag) produces N+1 filter
              %coefficients (time function);
[xw,w]=freqz(x,1,512);%input frequency spectrum;
[y]=ssupsampling(x,6);
```

Chapter 7: Discrete-time transforms 295

```
[yw,w]=freqz(y,1,512);
subplot(2,1,1);plot(w/pi,abs(xw),'k');ylabel('Magnitude');
title('Input spectrum to upsampler');
subplot(2,1,2);plot(w/pi,abs(yw),'k');xlabel('Normalized...
    frequency');
ylabel('Magnitude');title('Output spectrum of the upsampler');
```
∎

Important definitions and concepts

1. Discrete-time Fourier transform (DTFT)
2. Discrete frequency
3. Sampling time
4. Sampling frequency
5. Amplitude and phase spectra of DTFT
6. Properties of DTFT
7. The DTFT is a continuous periodic function
8. Effect on spectrum by truncating a sequence
9. Windowing of sequences
10. Frequency response of LTI discrete systems
11. The discrete Fourier transform (DFT)
12. The DFT presupposes a periodic time sequence and its corresponding discrete frequency sequence
13. The π/T_s ratio is known as the fold-over frequency, and it is the largest frequency that is useful to compare the spectrum of the analog signal with the spectrum of its sampled form
14. DFT properties
15. Down sampling of discrete sequences
16. Up sampling of discrete sequences
17. Spectrums of the down-sampled and up-sampled discrete sequences

Chapter 7 Problems

Section 7.1

1. Given the two sequences

$$f(n) = \begin{cases} \left(\dfrac{1}{3}\right)^n & n \geq 0 \\ 0 & n < 0 \end{cases} \qquad h(n) = \begin{cases} \left(-\dfrac{1}{3}\right)^n & n \geq 0 \\ 0 & n < 0 \end{cases}$$

sketch the corresponding $|F(\omega)|$ for these sequences.

2. Deduce the DTFT of the following functions for $n \geq 0$:

   a. $e^{-n} \sin n$
   b. $e^{-2n} \cos 3n$

3. The frequency response of a system is shown in Figure P7.1.3. Determine the impulse response of the system.

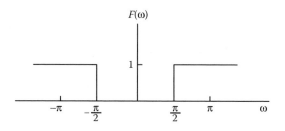

Figure P7.1.3

4. Deduce the transfer function and plot its magnitude if the following difference equation characterizes the system:

$$y(n) + y(n-1) - 0.5y(n-2) = x(n) + 0.8x(n-1) - 0.1x(n-2)$$

5. Prove the time multiplication property of DTFT, which is

$$\mathcal{F}_{DT}\{nf(n)\} = -e^{j\omega} \frac{dF(e^{j\omega})}{d(e^{j\omega})}$$

Section 7.2

1. Write several terms to verify the relationship

$$\sum_{n=0}^{N-1} e^{-j[2\pi(m-k)n]/N} = \begin{cases} N & m = k \\ 0 & m \neq k \end{cases}$$

2. The function $f(t) = \cos t \, u(t)$ is sampled at $T_s = 1/4$ s intervals. Write the DFT of $N = 16$ and $N = 32$. Compare and discuss the results.
3. Show that $F(k\Omega_b)$ and $f(nT_s)$ are periodic and determine their periods.
4. Deduce the DFT of the function $f(nT_s) = \cos(n\omega_0 T_s)$. Compare the results with the Fourier transform of $f(t) = \cos(\omega_0 t)$ when $\omega_0 = m2\pi/NT_s$.

5. Deduce the DFT of the signals shown in Figure P7.2.5.

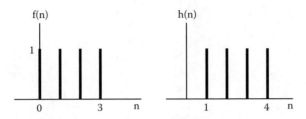

Figure P7.2.5

6. Deduce the DFT of the signals shown in Figure P7.2.6.

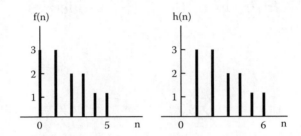

Figure P7.2.6

7. Using FFT, plot the DFT of $f(n) = \sin(0.1\pi n)$ for $0 \le n \le N-1$ for $N = 8, 16, 32$, and 64, and explain your results.
8. If $f(n) = 0.9^n$ $0 \le n \le 3$, find the DFT of $f(n-3)$ and compare your result with the DFT shifting property.
9. Find the IDFT of the spectrum $F(k-2)$, which is found in Problem 7.2.8. Compare it with the DFT definition of signals.
10. Given two sequences, $f(n) = \{1\ 0\ 1\ 1\}$ and $h(n) = \{1\ 1\ 0\ 1\}$, verify the time convolution property by considering each side of the equation

$$y(n) = \sum_{m=0}^{3} f(m)h(n-m) = \frac{1}{N}\sum_{k=0}^{3} F(k)H(k)e^{j(2\pi nk)/N}$$

11. Show that Parseval's theorem holds for the discretized function shown in Figure P7.2.11. Show also that the central ordinate property is true.

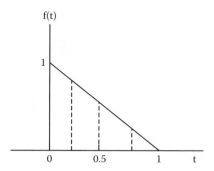

Figure P7.2.11

Appendix 7.1: Proofs of the DTFT properties

Linearity
Proof:

$$\sum_{n=-\infty}^{\infty}[af(n)+bh(n)]e^{-j\omega n} = a\sum_{n=-\infty}^{\infty}f(n)e^{-j\omega n} + b\sum_{n=-\infty}^{\infty}h(n)e^{-j\omega n} = aF(e^{j\omega})+bH(e^{j\omega})$$

Time shifting
Proof:

$$F_{sh}(e^{j\omega}) = \sum_{n=-\infty}^{\infty}f(n-m)e^{-j\omega n} = e^{-j\omega m}\sum_{k=-\infty}^{\infty}f(k)e^{-j\omega k} = e^{-j\omega m}F(e^{j\omega})$$

where we set $n - m = k$. Observe that k is a dummy variable.

Time reversal
Proof:

$$\sum_{n=-\infty}^{\infty}f(-n)e^{-j\omega n} = \sum_{m=\infty}^{-\infty}f(m)e^{-j(-\omega)m} = F(e^{-j\omega})$$

Time convolution
Proof:

$$\sum_{n=-\infty}^{\infty}\left(\sum_{m=-\infty}^{\infty}f(m)h(n-m)\right)e^{-j\omega n} = \sum_{m=-\infty}^{\infty}f(m)\sum_{n=-\infty}^{\infty}h(n-m)e^{-j\omega n}$$

Chapter 7: Discrete-time transforms

$$= \sum_{m=-\infty}^{\infty} f(m)e^{-j\omega m} \sum_{k=-\infty}^{\infty} h(k)e^{-j\omega k} = F(e^{j\omega})H(e^{j\omega})$$

where we set $n - m = k$. Observe that both m and k are dummy variables and can take any other name, e.g., n.

Frequency shifting

Proof:

$$\sum_{n=-\infty}^{\infty} e^{\pm j\omega_0 n} f(n)e^{-j\omega n} = \sum_{n=-\infty}^{\infty} f(n)e^{-j(\omega \mp \omega_0)n} = F(e^{j(\omega \mp \omega_0)})$$

Time multiplication

$$\sum_{n=-\infty}^{\infty} nf(n)e^{-j\omega n} = -\sum_{n=-\infty}^{\infty} e^{j\omega} f(n) \frac{d(e^{j\omega})^{-n}}{d(e^{j\omega})} = -e^{j\omega} \frac{d}{d(e^{j\omega})} \sum_{n=-\infty}^{\infty} f(n)e^{-j\omega n}$$

$$= -e^{j\omega} \frac{d}{d(e^{j\omega})} F(e^{j\omega})$$

Example 1: Find the DTFT of $y(n) = na^n$, $a < 1$ and $0 \le n < \infty$.

Solution: The DTFT of $f(n) = a^n$ for $n = 0, 1, 2, \ldots$ is equal to

$$F(e^{j\omega}) = \sum_{n=0}^{\infty} a^n e^{-j\omega n} = \sum_{n=0}^{\infty} (a\, e^{-j\omega})^n = \frac{1}{1 - ae^{-j\omega}} = \frac{e^{j\omega}}{e^{j\omega} - a}$$

and hence

$$-e^{j\omega} \frac{d}{d(e^{j\omega})} \left(\frac{e^{j\omega}}{e^{j\omega} - a} \right) = a \frac{e^{j\omega}}{(e^{j\omega} - a)^2}$$

∎

Modulation

Proof:

$$\sum_{n=-\infty}^{\infty} f(n)\cos\omega_0 n\, e^{-j\omega n} = \sum_{n=-\infty}^{\infty} \left(\frac{f(n)e^{j\omega_0 n}}{2} + \frac{f(n)e^{-j\omega_0 n}}{2} \right) e^{-j\omega n}$$

$$= \frac{1}{2} \sum_{n=-\infty}^{\infty} f(n)e^{-j(\omega-\omega_0)n} + \frac{1}{2} \sum_{n=-\infty}^{\infty} f(n)e^{-j(\omega+\omega_0)n} = \frac{1}{2} F(e^{j(\omega-\omega_0)}) + \frac{1}{2} F(e^{j(\omega+\omega_0)})$$

Correlation

Proof:

$$\sum_{n=-\infty}^{\infty}\left(\sum_{m=-\infty}^{\infty} f(m)h(m-n)\right)e^{-j\omega n} = \sum_{n=-\infty}^{\infty}\left(\sum_{m=-\infty}^{\infty} f(m)h[-(n-m)]\right)e^{-j\omega n}$$

$$= \sum_{m=-\infty}^{\infty} f(m) \sum_{n=-\infty}^{\infty} h[-(n-m)]e^{-j\omega n} = \sum_{m=-\infty}^{\infty} f(m)e^{j\omega m}H(e^{-j\omega}) = F(e^{j\omega})H(e^{-j\omega})$$

where in the last step we used the time reversal and shifting properties.

Parseval's formula

Proof: We start from the right-hand side of the identity. Hence,

$$\frac{1}{2\pi}\int_{-\pi}^{\pi}\left[\sum_{n=-\infty}^{\infty} f(n)e^{-j\omega n}\right]F^*(e^{j\omega})d\omega = \sum_{n=-\infty}^{\infty} f(n)\frac{1}{2\pi}\int_{-\pi}^{\pi} F^*(e^{j\omega})e^{-j\omega n}d\omega$$

$$= \sum_{n=-\infty}^{\infty} f(n)\left[\frac{1}{2\pi}\int_{-\pi}^{\pi} F(e^{j\omega})e^{j\omega n}d\omega\right]^* = \sum_{n=-\infty}^{\infty} f(n)f^*(n) = \sum_{n=-\infty}^{\infty} |f(n)|^2$$

Appendix 7.2: Proofs of DFT properties

Note: For simplicity, we have deleted the constant $2\pi/N$ in the exponent.

Linearity

Proof: See the linearity proof in Appendix 7.1.

Symmetry

Proof: Set $n = -n$ in the IDFT. Hence,

$$f(-n) = \frac{1}{N}\sum_{n=0}^{N-1} F(k)e^{jk(-n)}$$

Next, interchange the parameters n and k to find

$$f(-k) = \frac{1}{N}\sum_{n=0}^{N-1} F(n)e^{-jkn} = DFT\left\{\frac{1}{N}F(n)\right\}$$

Time shifting

Proof: Substitute $m = n - i$ into IDFT so that the equation

$$f(m) = \frac{1}{N}\sum_{k=0}^{N-1} F(k)e^{jkm}$$

becomes

$$f(n-i) = \frac{1}{N}\sum_{k=0}^{N-1} F(k)e^{j(n-i)k} = \frac{1}{N}\sum_{k=0}^{N-1}[F(k)e^{-jik}]e^{jnk} = IDFT\{F(k)e^{-jik}\}$$

Frequency shifting

Proof: We write the DFT in the form

$$F(m) = \sum_{n=0}^{N-1} f(n)e^{-jmn}$$

Next, we set $m = k - i$ in the expression to find

$$F(k-i) = \sum_{k=0}^{N-1} f(n)e^{-j(k-i)n} = \sum_{k=0}^{N-1}[f(n)e^{jin}]e^{-jnk} = DFT\{f(n)e^{jin}\}$$

Time convolution

Proof: Start with the function

$$\sum_{i=0}^{N-1} f(i)h(n-i) = \sum_{i=0}^{N-1} \frac{1}{N}\sum_{k=0}^{N-1} F(k)e^{jik} \times \frac{1}{N}\sum_{m=0}^{N-1} H(m)e^{j(n-i)m}$$

This is rearranged to

$$= \frac{1}{N}\sum_{k=0}^{N-1}\sum_{m=0}^{N-1} F(k)H(m)e^{jmn}\left[\frac{1}{N}\sum_{i=0}^{N-1} e^{jik}e^{-jim}\right]$$

But, using the finite geometric series formula, we can prove that

$$\frac{1}{N}\sum_{i=0}^{N-1} e^{jik}e^{-jim} = \begin{cases} n & n = m \\ 0 & n \neq m \end{cases}$$

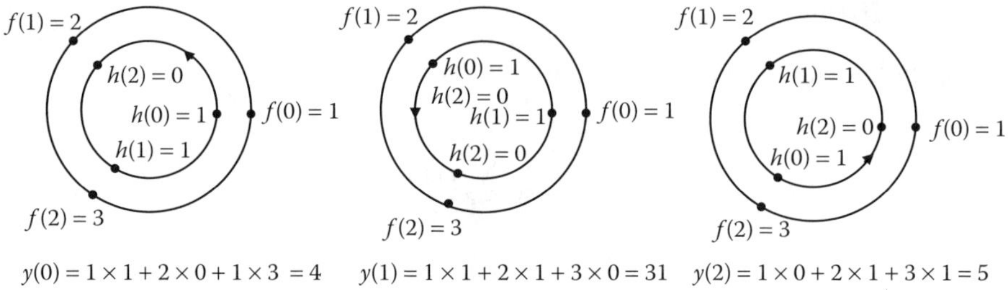

$f(1) = 2$ ⟨$h(2) = 0$, $h(0) = 1$, $h(1) = 1$⟩ $f(0) = 1$ $f(1) = 2$ ⟨$h(0) = 1$, $h(2) = 0$, $h(1) = 1$, $h(2) = 0$⟩ $f(0) = 1$ $f(1) = 2$ ⟨$h(1) = 1$, $h(2) = 0$, $h(0) = 1$⟩ $f(0) = 1$

$f(2) = 3$ $f(2) = 3$ $f(2) = 3$

$y(0) = 1 \times 1 + 2 \times 0 + 1 \times 3 = 4$ $y(1) = 1 \times 1 + 2 \times 1 + 3 \times 0 = 31$ $y(2) = 1 \times 0 + 2 \times 1 + 3 \times 1 = 5$

Figure A7.2.1

Hence, for $n = m$ in the second sum, we finally find

$$y(n) = \sum_{i=0}^{N-1} f(i)h(n-i) = \frac{1}{N} \sum_{k=0}^{N-1} F(k)H(k)e^{jnk} \triangleq IDFT\{F(k)H(k)\}$$

Since we have shown that the DFT automatically supposes periodic sequences in the time and frequency domain, the periodic convolution for the two specific signals, $f(n) = \{1\ 2\ 3\}$ and $h(n) = \{1\ 1\ 0\}$, are shown in Figure A7.2.1.

Another approach to the evaluation of cyclic convolution is to cast the convolution form in matrix form. For the case of the periodic sequences $f(n)$ and $h(n)$, each of three elements, as in Figure A7.2.1, we write

$$y(0) = f(0)h(0) + f(1)h(2) + f(2)h(1)$$
$$y(1) = f(0)h(1) + f(1)h(0) + f(2)h(2)$$
$$y(2) = f(0)h(2) + f(1)h(1) + f(2)h(0)$$

This set can be written in matrix form:

$$[y]^T = [f(0)\ f(1)\ f(2)] \begin{bmatrix} h(0) & h(1) & h(2) \\ h(2) & h(0) & h(1) \\ h(1) & h(2) & h(0) \end{bmatrix}$$

For this specific example,

$$[y(0)\ y(1)\ y(2)] = [1\ 2\ 3] \begin{bmatrix} 1 & 1 & 0 \\ 0 & 1 & 1 \\ 1 & 0 & 1 \end{bmatrix} = [4\ 3\ 5]$$

To produce a linear convolution for this case, each sequence must have $N = 3 + 3 - 1 = 5$ elements. But since each sequence has only three elements, we pad them with zeros. Hence, for $f(n) = \{1\ 2\ 3\ 0\ 0\}$ and $h(n) = \{1\ 1\ 1\ 0\ 0\}$, the result is $f*h = \{0\ 1\ 2\ 1\ 12\ 0\ 0\ 0\ 0\}$.

Frequency convolution

Proof: Substitute known forms into

$$\sum_{i=0}^{N-1} F(i)H(k-i) = \sum_{i=0}^{N-1}\left[\sum_{m=0}^{N-1} f(m)e^{-jmi}\right]\left[\sum_{n=0}^{N-1} h(n)e^{-jn(k-i)}\right]$$

$$= \sum_{m=0}^{N-1}\sum_{n=0}^{N-1} f(m)h(n)e^{-jkn}\left[\sum_{i=0}^{N-1} e^{-jmi}e^{jni}\right]$$

The bracketed term is the orthogonality relationship and is equal to N if $m = k$ and zero if m is different than k. Therefore,

$$\sum_{i=0}^{N-1} F(i)H(k-i) = N\sum_{n=0}^{N-1} f(n)h(n)e^{-jkn} \triangleq N \times DFT\{f(n)h(n)\}$$

from which the property is verified if we take the inverse DFT of both sides of the above equation. Because $F(k)$ and $H(k)$ are periodic, their convolution is a circular convolution in the frequency domain.

Parseval's theorem

Using the equation that characterizes the frequency convolution property, we obtain

$$\sum_{n=0}^{N-1} f(n)f(n)e^{-jkn} = \frac{1}{N}\sum_{i=0}^{N-1} F(i)F(k-i)$$

If we set $k = 0$ in this expression, we find

$$\sum_{n=0}^{N-1} f^2(n) = \frac{1}{N}\sum_{i=0}^{N-1} F(i)F(-i) = \frac{1}{N}\sum_{i=0}^{N-1} |F(i)|^2$$

From this result we may **define a discrete energy spectral density** or **periodogram spectral estimate**:

$$S(k) = |F(k)|^2 \qquad 0 \leq k \leq N-1$$

Time reversal

Setting $n = -n$ in the IDFT, we find that

$$f(-n) = \frac{1}{N}\sum_{k=0}^{N-1} F(k)e^{-jnk}$$

Next, set $k = -m$ on the right side. Then

$$f(-n) = \frac{1}{N}\sum_{m=0}^{-(N-1)} F(-m)e^{jnm}$$

Because of the periodic nature of $F(-m)$ and $\exp(jmn)$, the sum over $-(N-1)$ to 0 and $(N-1)$ to 0 is the same. Thus,

$$f(-n) = \frac{1}{N}\sum_{m=0}^{N-1} F(-m)e^{jnm} \triangleq IDFT\{F(-m)\}$$

which verifies the time reversal property.

Appendix 7.3: Fast Fourier transform (FFT)

Decimation in time procedure

The DFT of a sequence $\{x(n)\}$ is given by

$$F_N(k) = \sum_{n=0}^{N-1} f(n)W_N^{kn}, \quad W_N = e^{-j2\pi/N}, \quad k = 0, 1, \cdots, N-1 \quad (3.1)$$

Let us assume that $N = 2^r$ for r integer. This indicates that N is always an even number. Because of N being even, we can decompose the summation of the above equation into two terms, one with even indices and one with odd indices. Hence,

$$F_N(k) = \sum_{even} f(n)W_N^{kn} + \sum_{odd} f(n)W_N^{kn} \quad (3.2)$$

We can write the even indices as $n = 2m$ and the odd indices as $n = 2m + 1$, with $m = 0, 1, \ldots, (N/2) - 1$. Therefore, (3.2) becomes

Chapter 7: Discrete-time transforms

$$F_N(k) = \sum_{m=0}^{(N/2)-1} f(2m)W_N^{2mk} + W_N^k \sum_{m=0}^{(N/2)-1} f(2m)W_N^{2mk} \qquad (3.3)$$

In the second summation we used the identity

$$W_N^{(2m+1)k} = W_N^k W_N^{2mk}$$

Observe that

1. $W_N^2 = e^{-j2\pi/(N/2)} = W_{N/2}$
2. $W_N^N = e^{-j2\pi} = 1$
3. $W_N^{N/2} = e^{-j\pi} = -1$

Furthermore, the two summations in (3.3) are DFTs themselves: the DFT of the even samples $f(0), f(2), \ldots, f(N-2)$ and the DFT of the odd samples $f(1)$, $f(3), \ldots, f(N-1)$. Therefore, we write

$$F_N(k) = F_{N/2}^{(e)}(k) + W_N F_{N/2}^{(o)}(k) \qquad k = 0, 1, \cdots, N-1 \qquad (3.4)$$

where

$$F_{N/2}^{(e)}(k) = DFT\{f(0), f(2), \cdots, f(N-2)\} = \sum_{m=0}^{(N/2)-1} f(2m)W_{N/2}^{km} \qquad (3.5)$$

$$F_{N/2}^{(o)}(k) = DFT\{f(1), f(3), \cdots, f(N-1)\} = \sum_{m=0}^{(N/2)-1} f(2m+1)W_{N/2}^{km} \qquad (3.6)$$

We observe that the DFT of the N-point sequence has been decomposed into two DFTs of $N/2$ samples each.

For example, when $N = 4 = 2^2$ we need to compute $F_4(0), F_4(1), F_4(2), F_4(3)$, on the basis of $F_2^{(e)}(0), F_2^{(e)}(1)$ and $F_2^{(o)}(0), F_2^{(o)}(1)$. Hence, for $k = 0,1$ we write

$$F_4(k) = F_2^{(e)}(k) + W_4^k F_2^{(o)}(k) \qquad k = 0, 1 \qquad (3.7)$$

For the two indices, $k = 2, 3$, we use the fact that the DFT of a 2-point sequence is periodic with period 2. Hence, for $k = 0, 1$, we find

$$F_4(k+2) = F_2^{(e)}(k) + W_4^{2+k} F_2^{(o)}(k) \qquad k = 0, 1 \qquad (3.8)$$

because $F_2(k) = F_2(k+2)$. Because $W_4^2 = (e^{-j2\pi/4})^2 = e^{-j\pi} = -1$, (3.7) and (3.8) become

$$\begin{aligned} F_4(k) &= F_2^{(e)}(k) + W_N^k F_2^{(o)}(k) & k = 0, 1 \\ F_4(k+2) &= F_2^{(e)}(k) - W_N^k F_2^{(o)}(k) & k = 0, 1 \end{aligned} \qquad (3.9)$$

The above equation can be generalized for any sequence of even-number N samples in the form

$$\begin{aligned} F_N(k) &= F_{N/2}^{(e)}(k) + W_N^k F_{N/2}^{(o)}(k) & k = 0, 1, \cdots, (N/2)-1 \\ F_N(k+2) &= F_{N/2}^{(e)}(k) - W_N^k F_{N/2}^{(o)}(k) & k = 0, 1, \cdots, (N/2)-1 \end{aligned} \qquad (3.10)$$

The above equation is known as a **butterfly**. Figure A7.3.1 depicts (3.10). Figure A7.3.2 shows the FFT for $N = 8$.

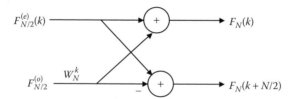

Figure A7.3.1

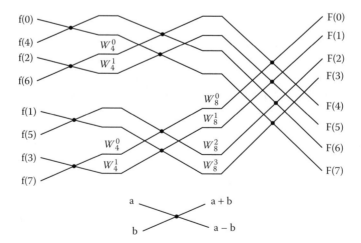

Figure A7.3.2

Chapter 7: Discrete-time transforms

Next, the question is how efficient the FFT is. Let us look at the case when $N = 8$. In Figure A7.3.1 we see that the first stage is two 4-order DFTs. Each one of these DFTs can be split in two 2-order FFTs. As we can easily see, a 2-order DFT is just a butterfly, since

$$W_2 = e^{-j2\pi/2} = -1 \quad \text{and} \quad W_2^k = (-1)^k = \pm 1$$

when $k = 0, 1$. For $N = 8$ we find that we obtain $3 = \log_2(8)$ layers of butterflies with $8/2$ butterflies to be computed in each layer. Since each butterfly involves one multiplication and two additions, we require about $(8/2)\log_2(8)$ complex multiplications and $8\log_2(8)$ complex additions. Hence, the complexity of DFT is $O(8\log_2(8))$. If we substitute N for 8, we find that the FFT complexity is of the order $N\log_2(N)$. This must be compared with N^2 operations that must be performed by the DFT. For $N = 1024 = 2^{10}$, the DFT needs $(1024)^2 = 1{,}048{,}576$ operations, whereas the FFT needs only $1024 \times 10 = 10{,}240$ operations.

Observe that to obtain the unscrambled output from the FFT operation, we must scramble the input as shown in Figure A7.3.2. To accomplish this, we first write the numbers, in this case 0 to 7, in their 3-bit representation. Next, we reverse the bits and find the scrambled input. This process is shown below:

$000 = 0$, $001 = 1$, $010 = 2$, $011 = 3$, $100 = 4$, $101 = 5$, $110 = 6$, $111 = 7$

$000 = 0$, $100 = 4$, $010 = 2$, $110 = 6$, $001 = 1$, $101 = 5$, $011 = 3$, $111 = 7$

chapter 8

Laplace transform

The use of Fourier transform (FT) in systems analysis involves the decomposition of the excitation function $f(t)$ into a function $F(\omega)$ over an infinite band of frequencies. The excitation function $F(\omega)$, together with the appropriate system function $H(\omega)$, leads, through the inverse Fourier transform, to the response of the system to the prescribed excitation or excitations. Despite its general importance in systems functions, the Fourier integral is not generally useful in determining the transient response of networks (systems).

The discussion of the FT showed that we could represent a function $f(t)$ by a continuous sum of weighted exponential functions of the form $f(t)e^{j\omega t}$. The values of the exponential are restricted along the imaginary axis of the complex plane as omega varies from minus infinity to infinity. This restriction proves to be undesirable in many cases. We can remove this restriction by representing $f(t)$ by a continuous sum of weighted damped exponential functions of the form $f(t)e^{-st}$, where $s = \sigma + j\omega$ with some real constant σ. The choice of s moves the values of the exponential function from the $j\omega$-axis to a parallel line off the $j\omega$-axis in the complex plane. The Laplace transform is well adapted to linear time domain systems analysis. Another feature of the Laplace transform is that it automatically provides for initial conditions in the systems.

8.1 One-sided Laplace transform

In systems problems, it is usually possible to restrict considerations to positive time functions. The reason is that the response of physical systems can be determined for all $t \geq 0$ from a knowledge of the input for $t \geq 0$ and the initial conditions. The one-sided Laplace transform is defined by

$$F(s) = \int_0^\infty f(t)e^{-st}dt \triangleq \mathcal{L}\{f(t)\} \qquad (8.1)$$

provided that the function is defined in the range of integration and that the integral exists for all values of s greater than a specific value s_0.

In our studies we will consider piecewise continuous functions (a function is piecewise continuous on an interval if the interval can be subdivided into a finite number of subintervals, in each of which the function is continuous and has finite left- and right-hand limits) and those functions for which

$$\lim_{t \to \infty} f(t)e^{-ct} = 0 \qquad c = \text{real constant} \qquad (8.2)$$

Functions of this type are known as functions of **exponential order** c. Also, from the expression

$$\int_0^\infty |f(t)e^{-st}| dt = \int_0^\infty |f(t)e^{-\sigma t}e^{-j\omega t}| dt = \int_0^\infty |f(t)||e^{-\sigma t}||e^{-j\omega t}| dt = \int_0^\infty |f(t)|e^{-\sigma t} dt$$

$$= \int_0^\infty |f(t)|e^{-ct}e^{-(\sigma-c)t} dt$$

we observe that if $f(t)$ is of exponential order, the integral converges for $\text{Re}\{s\} = \sigma > c$ (c is the abscissa of convergence) since the integrand goes to zero as $t \to \infty$. The abscissa of convergence may be positive, negative, or zero depending on the function. The importance of this result is that a finite number of infinite discontinuities are permissible so long as they have finite areas under them. In addition, the convergence is also uniform, which permits us to alter the order of integration in multiple integrals without affecting the results. The restriction in this equation, namely, $\text{Re}\{s\} = \sigma > c$, indicates that when we wish to find the inverse Laplace transform, we must choose an appropriate path of integration in the complex plane. By doing so, it is guaranteed that the time function so obtained is unique. The left-hand infinite space with its boundary the abscissa of convergence is known as the **region of convergence** (ROC).

Example 8.1.1: Find the Laplace transform of the unit step function $f(t) = u(t)$ and establish the region of convergence.

Solution: By (8.1), we find

$$U(s) = \int_0^\infty u(t)e^{-st} dt = \int_0^\infty e^{-st} dt = -\frac{e^{-st}}{s}\bigg|_0^\infty = \frac{1}{s} \qquad (8.3)$$

The region of convergence is found from the expression $|e^{-\sigma t} e^{-j\omega t}| = |e^{-\sigma t}| < \infty$, which is true only if $\sigma > 0$. ∎

Example 8.1.2: Find the Laplace transform of $f(t) = e^{-2t}u(t)$ and establish the ROC.

Chapter 8: Laplace transform

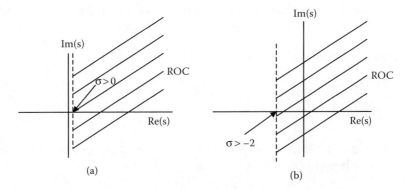

Figure 8.1.1 (a) ROC for the signal $f(t) = u(t)$. (b) ROC for the signal $f(t) = e^{-2t} u(t)$.

Solution: The Laplace transform is

$$F(s) = \int_0^\infty e^{-2t} e^{-st} dt = \int_0^\infty e^{-(s+2)t} dt = \frac{1}{-(s+2)} e^{-(s+2)t} \Big|_0^\infty = \frac{1}{s+2} \qquad (8.4)$$

and the ROC is found from

$$\left| e^{-(\sigma+j\omega+2)t} \right| = \left| e^{-(\sigma+2)t} \right| \left| e^{-j\omega t} \right| = e^{-(\sigma+2)t}$$

which results in $\sigma > -2$. Figure 8.1.1a shows the ROC for the function $f(t) = u(t)$, and Figure 8.1.1b for the function $f(t) = e^{-2t} u(t)$. These are found in Examples 8.1.1 and 8.1.2, respectively. ∎

Example 8.1.3: Find the Laplace transform of the delta function and the ROC.

Solution: The Laplace transform is

$$\Delta(s) = \int_0^\infty \delta(t) e^{-st} dt = e^{-s0} \int_{0-}^{0+} \delta(t) dt = 1 \qquad (8.5)$$

and the ROC is found from

$$\int_{0-}^\infty \left| \delta(t) e^{-st} \right| dt = \int_{0-}^\infty \left| \delta(t) e^{-s0} \right| dt = \int_{0-}^\infty \delta(t) dt = 1 < \infty$$

which indicates that the result is independent of σ, and hence the ROC is the entire s-plane. ∎

Example 8.1.4: Find the Laplace transform of the function $f(t) = 2u(t) + e^{-t}u(t)$.

Solution: From (8.1) we find

$$F(s) = \int_0^\infty [u(t) + e^{-t}]e^{-st}dt = \int_0^\infty e^{-st}dt + \int_0^\infty e^{-(s+1)t}dt = \frac{1}{s} + \frac{1}{s+1}$$

The ROC is $\sigma > 0$. Observe that the ROC, which is accepted, is the region of the s-plane where the two ROC overlap. ∎

8.2 Summary of the Laplace transform properties

1. Linearity $\qquad af(t) + bh(t) \xleftrightarrow{\mathcal{L}} aF(s) + bF(s)$

2. Time derivative $\dfrac{df(t)}{dt} \xleftrightarrow{\mathcal{L}} sF(s) - f(0)$

$\dfrac{d^n f(t)}{dt^n} \xleftrightarrow{\mathcal{L}} s^n F(s) - s^{n-1}f(0) - s^{n-2}f^{(1)}(0)$

$\qquad - \cdots - f^{(n-1)}(0), \quad f^{(m)}(0) \triangleq \dfrac{d^m f(t)}{dt^m}\bigg|_{t=0}$

3. Integral $\int_0^t f(x)dx \xleftrightarrow{\mathcal{L}} \dfrac{F(s)}{s}$ zero initial conditions

$\int_{-\infty}^t f(x)dx \xleftrightarrow{\mathcal{L}} \dfrac{F(s)}{s} + \dfrac{\int_{-\infty}^0 f(t)dt}{s}$ nonzero initial conditions

4. Multiplication by exponential $\quad e^{-at}f(t) \xleftrightarrow{\mathcal{L}} F(s+a)$
$\qquad a =$ positive constant

5. Multiplication by t $\qquad tf(t) \xleftrightarrow{\mathcal{L}} -\dfrac{dF(s)}{ds}$

6. Time shifting $\qquad f(t-a)u(t-a) \xleftrightarrow{\mathcal{L}} e^{-as}F(s) \qquad a > 0$

7. Complex frequency shift $\quad e^{s_0 t}f(t) \xleftrightarrow{\mathcal{L}} F(s - s_0)$

8. Scaling $\qquad f(at) \xleftrightarrow{\mathcal{L}} \dfrac{1}{a}F\left(\dfrac{s}{a}\right) \qquad a > 0$

9. Time convolution $\qquad f(t) * h(t) \xleftrightarrow{\mathcal{L}} F(s)H(s)$

Chapter 8: Laplace transform

10. **Initial value** $\lim_{t \to 0} f(t) = \lim_{s \to \infty} sF(s)$ provided that no delta function exists at $t = 0$

11. **Final value** $\lim_{t \to \infty} f(t) = \lim_{s \to 0} sF(s)$ provided that $sF(s)$ is analytic on the $j\omega$-axis and in the right half of the s-plane

Example 8.2.1: Time derivative. Find the Laplace transform of the differential equation characterizing an RL series circuit with a voltage $v(t)$ as input.

Solution: From Kirchhoff voltage law (KVL) we obtain the differential equation

$$L\frac{di(t)}{dt} + Ri(t) = v(t).$$

Using the linearity property we obtain

$$\mathcal{L}\left\{L\frac{di(t)}{dt} + Ri(t)\right\} = \mathcal{L}\{v(t)\} \text{ or } L\mathcal{L}\left\{\frac{di(t)}{dt}\right\} + R\mathcal{L}\{i(t)\} = \mathcal{L}\{v(t)\}$$

or

$$LsI(s) - Li(0) + RI(s) = V(s)$$

∎

Example 8.2.2: Integral. Find the Laplace transform of the integrodifferential equation of a series RC circuit with a voltage input. There is an initial voltage $v_C(0)$ across the capacitor.

Solution: From the KVL we obtain the equation

$$Ri(t) + \frac{1}{C}\int_{-\infty}^{t} i(x)dx = v(t).$$

Applying the linearity property, we can write the Laplace transform of the above equation as

$$R\mathcal{L}\{i(t)\} + \mathcal{L}\left\{\frac{1}{C}\int_{-\infty}^{t} i(x)dx\right\} = \mathcal{L}\{v(t)\} \text{ or } RI(s) + \frac{1}{C}\frac{I(s)}{s} + \frac{1}{C}\frac{\int_{-\infty}^{0} i(x)dx}{s} = V(s)$$

From the integration

$$\int_{-\infty}^{0} i(x)dx = q(0)$$

we obtain the charge accumulated at the initial time $t = 0$. But $q(0)/C$ is the initial voltage across the capacitor, and hence the transformed equation becomes

$$\left(R + \frac{1}{Cs}\right) I(s) = -\frac{v_C(0)}{s} + V(s)$$

∎

Example 8.2.3: Multiplication by an exponential. Find the Laplace transform (LT) of the function $g(t) = e^{-at} \cos \omega_0 t \, u(t)$.

Solution: To find the transformation, we need to find the LT of the cosine function. Therefore, we write

$$\mathcal{L}\{\cos \omega_0 t \, u(t)\} = \int_0^\infty \left[\frac{e^{-j\omega_0 t} + e^{j\omega_0 t}}{2}\right] e^{-st} dt = \frac{1}{2}\int_0^\infty e^{-(j\omega_0+s)t} dt + \frac{1}{2}\int_0^\infty e^{-(-j\omega_0+s)t} dt$$

$$= \frac{1}{2}\frac{1}{-(j\omega_0+s)} e^{-(j\omega_0+s)t}\bigg|_0^\infty + \frac{1}{2}\frac{1}{-(-j\omega_0+s)} e^{-(-j\omega_0+s)t}\bigg|_0^\infty$$

$$= \frac{1}{2}\frac{1}{(j\omega_0+s)} + \frac{1}{2}\frac{1}{(-j\omega_0+s)} = \frac{s}{s^2+\omega_0^2}$$

Thus, the answer is

$$G(s) = \frac{s+a}{(s+a)^2 + \omega_0^2}$$

∎

Example 8.2.4: Multiplication by t. Find the LT of the function $g(t) = t \sin \omega_0 t \, u(t)$. From the property, we need only to find the LT of the sine signal. Hence, following the procedure of the previous example we find

$$F(s) = \int_0^\infty \frac{e^{j\omega_0 t} - e^{-j\omega_0 t}}{2j} e^{-st} dt = \frac{\omega_0}{s^2 + \omega_0^2}$$

Thus,

$$G(s) = -\frac{d}{ds}\left(\frac{\omega_0}{s^2+\omega_0^2}\right) = \frac{2s\omega_0}{(s^2+\omega_0^2)^2}$$

∎

Example 8.2.5: Time shifting. Find the LT of the function $f_s(t) = (t-a)u(t-a) \quad a > 0$.

Chapter 8: Laplace transform

Solution: We first must find the LT of the unshifted function, which is

$$F(s) = \int_0^\infty te^{-st}dt = \frac{1}{-s}\int_0^\infty t\frac{de^{-st}}{dt}dt = \frac{1}{-s}\int_0^\infty td(e^{-st}) = \frac{1}{-s}\left[te^{-st}\Big|_0^\infty - \int_0^\infty e^{-st}dt\right] = \frac{1}{s^2}$$

Hence, applying the time-shifting property we obtain

$$F_s(s) = e^{-sa}F(s) = e^{-sa}\frac{1}{s^2}$$

∎

Example 8.2.6: Frequency shift. Find the LT of the function $g(t) = e^{-as}f(t) = e^{-as}\cos\omega_0 t\, u(t)$.

Solution: Above, we have already found the LT of the cosine function. Hence, the solution is

$$G(s) = \frac{s+a}{(s+a)^2 + \omega_0^2}$$

∎

Example 8.2.7: Scaling. Verify the LT of the function $f(\omega_0 t) = \cos\omega_0 t\, u(t)$ by using the scaling properties and the LT of the function $f(t) = \cos t$.

Solution: The LT of the signal $f(t)$ is equal to

$$\frac{s}{s^2+1}$$

Applying the scaling property we find

$$\frac{1}{\omega_0}F\left(\frac{s}{\omega_0}\right) = \frac{1}{\omega_0}\frac{s/\omega_0}{(s/\omega_0)^2+1} = \frac{s}{s^2+\omega_0}.$$

∎

Example 8.2.8: Time convolution. Find the LT of the convolution of the following two functions:

$$f(t) = tu(t),\ h(t) = \cos t\, u(t)$$

Solution:

$$\mathcal{L}\{f(t)*h(t)\} = F(s)H(s) = \frac{1}{s^2}\frac{s}{s^2+1} = \frac{1}{s}\frac{1}{s^2+1}$$

∎

Example 8.2.9: Initial value. Apply the initial value property to the following functions:

$$F(s) = \frac{s}{s^2+3}, \quad H(s) = \frac{s^2+s+3}{s^2+3}$$

Solution:

a. $\lim_{s \to \infty} sF(s) = \lim_{s \to \infty} \frac{s^2}{s^2+3} = \lim_{s \to \infty} \frac{1}{1+(3/s^2)} = 1$

b. If we expand the division of $H(s)$, we obtain

$$H(s) = 1 + \frac{s}{s^2+3}$$

This shows the presence of an impulse at $t = 0$ and, therefore, we cannot find the initial value of the time function. ∎

Example 8.2.10: Final value. Apply the final value property to the following two functions:

a. $F(s) = \dfrac{s+a}{(s+a)^2+b^2}$

b. $H(s) = \dfrac{s}{s^2+b}$

Solution:

a. $\lim_{s \to 0} sF(s) = \lim_{s \to 0} \dfrac{s(s+a)}{(s+a)^2+b^2} = 0$

b. In this case, the property is not applicable because the function has singularities on the imaginary axis at $s = \pm jb$. ∎

8.3 Systems analysis: transfer functions of LTI systems

The transfer function or system function, $H(s)$, of a linear time-invariant (LTI) system, an almost essential entity in system analysis, is defined as the ratio of the LT of the output to the LT of the input. Thus, if the input is $f(t)$ and the output is $y(t)$, then

$$H(s) = \frac{Y(s)}{F(s)} = \text{transfer function or system function} \quad (8.6)$$

Chapter 8: Laplace transform

The output may be a voltage or a current anywhere in the system, and $H(s)$ is then appropriate to the selected output for a specified input. That is, $H(s)$ may be an impedance, an admittance, or a transfer entity in the given problem. The transfer function $H(s)$ describes the properties of the system alone. That is, the system is assumed to be in its quiescent state (zero state); hence, the initial conditions are assumed zero. The output time function is given by

$$y(t) = \mathcal{L}^{-1}\{H(s)F(s)\} \tag{8.7}$$

If the input is a delta function, $f(t) = \delta(t)$, the output in the s-domain is $H(s)$ since the LT of the delta function is 1. The impulse response of the system is found by taking the Inverse Laplace transform (ILT) of $H(s)$. Hence, we find

$$h(t) = \mathcal{L}^{-1}\{H(s)\} \tag{8.8}$$

The above relation shows that the impulse response of a system is the ILT of the transfer function $H(s)$, and the transfer function is equal to the LT of the system's impulse response.

An important feature of the LT method is that it is not necessary to isolate and identify the system transfer function since $H(s)$ appears automatically through the transform of the differential equation and is included in the mathematical operations. The situation changes considerably in those cases when the system consists of a number of subsystems that are interconnected to form the completed system. Next, one must take due account of whether the subsystems are interconnected in cascade, parallel, or feedback configuration.

The following examples illustrate how we can find the transfer function of a system from its time domain representation. Often, it is convenient to draw a block diagram representation of the system and then use the properties of block diagram reduction, as is given in Section 2.5.

Example 8.3.1: Determine the transfer function of the system shown in Figure 8.3.1a.

Solution: The differential equation describing the system is

$$L\frac{di(t)}{dt} + Ri(t) = v_i(t)$$

which we write

$$\frac{L}{R}\frac{dv_o(t)}{dt} + v_o(t) = v_i(t); \quad v_o(t) = Ri(t)$$

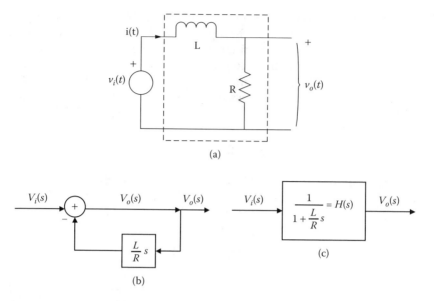

Figure 8.3.1 Electrical system and its block diagram representation: (a) circuit diagram, (b) block diagram, and (c) transfer function in block diagram representation.

The LT of both sides of the differential equation, invoking the linearity and differentiation properties of LT plus zero initial conditions (remember, the transfer function is defined only with zero initial conditions), is

$$\frac{L}{R}sV_o(s) + V_o(s) = V_i(s); \qquad H(s) = \frac{V_o(s)}{V_i(s)} = \frac{R/L}{s+(R/L)}$$

To obtain the block diagram representation of the system, the above equation is written in the following forms:

a. $V_o(s) = V_i(s) - \dfrac{L}{R}sV_o(s);$

b. $V_o(s) = \dfrac{1}{1+\dfrac{L}{R}s}V_i(s)$

The first equation is represented in block diagram form in Figure 8.3.1b, and the second equation is represented in block diagram form in Figure 8.3.1c. ∎

Example 8.3.2: Find the transfer function for the system shown in Figure 8.3.2a.

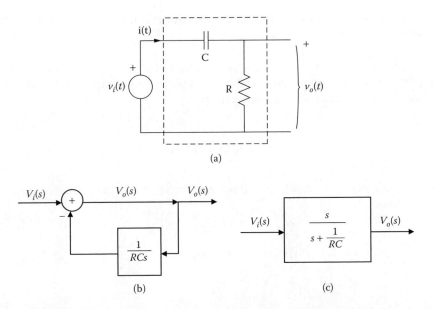

Figure 8.3.2 Electrical system and its block diagram representation: (a) circuit diagram, (b) block diagram, and (c) input–output block diagram.

Solution: The differential equation describing the system is

$$v_i(t) = Ri(t) + \frac{1}{C}\int_0^t i(x)dx \quad \text{(zero initial conditions)}$$

The above equation, by multiplying by R and dividing by R, becomes

$$V_i(s) = V_o(s) + \frac{V_o(s)}{RCs} \quad \text{or} \quad H(s) = \frac{V_o(s)}{V_i(s)} = \frac{s}{s + \frac{1}{RC}}$$

The transfer function is shown in block form in Figure 8.3.2c. Further, to obtain the block diagram format, we write the transformed equation in the form

$$V_o(s) = V_i(s) - \frac{1}{RCs}V_o(s)$$

The block diagram representation of the above equation is shown in Figure 8.3.2b. ∎

Example 8.3.3: Determine the transfer function, $H(s) = V(s)/F(s)$, of the system shown in Figure 8.3.3a.

Solution: To find the differential equation describing the system, we add algebraically the inertia force, the two frictional forces, and the input force. Since the forces are collinear, we obtain

$$M\frac{dv(t)}{dt} + D_1 v(t) + D_2 v(t) = f(t)$$

The LT of the above equation and its transfer function, assuming zero initial conditions, are

$$MsV(s) + (D_1 + D_2)V(s) = F(s), \quad H(s) = \frac{1}{M}\frac{1}{s + \frac{D}{M}} \quad D = D_1 + D_2$$

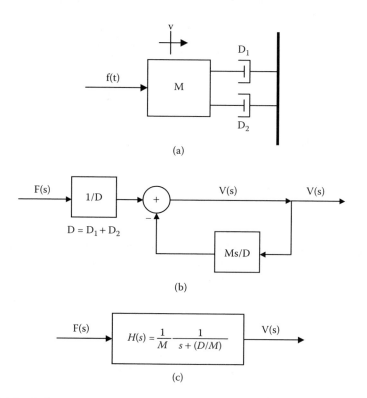

Figure 8.3.3 Mechanical system and its block diagram representation: (a) mechanical system, (b) block diagram, and (c) input–output block diagram.

Chapter 8: Laplace transform

To obtain the block diagram representation, we write the transformed equation in the form

$$V(s) = \frac{1}{D}F(s) - \frac{Ms}{D}V(s)$$

The block diagram for the above equation is shown in Figure 8.3.3b, and the transfer function is shown in Figure 8.3.3c ∎

*Example 8.3.4: Mechanical system. Find the transfer function $H(s) = V_1(s)/F(s)$ of the system shown in Figure 8.3.4.

Solution: We first develop the network equivalent diagram. It is shown in Figure 8.3.4b, where the velocities v_1 and v_2 are specified relative to ground as a fixed frame of reference. Observe that the force moves with velocity v_1; hence, this source is connected between ground and level v_1 as shown in Figure 8.3.4c. Observe also that the mass M_1 moves with velocity v_1 and is therefore connected between v_1 and v_g. The damper D_1 moves with relative velocity specified by v_1 and v_2; hence, it is connected between these two velocities. Lastly, since both M_2 and D_2 move with velocity v_2, they are connected in parallel between v_2 and v_g. A rearrangement of the resulting geometry yields Figure 8.3.4d, a familiar circuit configuration for the mechanical system.

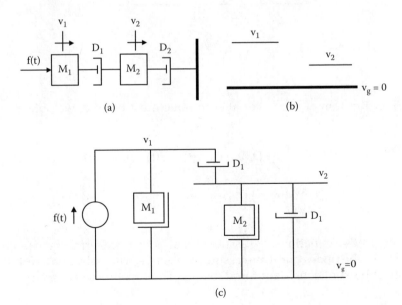

Figure 8.3.4 Mechanical system: (a) physical model, (b to d) steps in developing the network diagram, (e) block diagram, and (f) block diagram showing input–output relation.

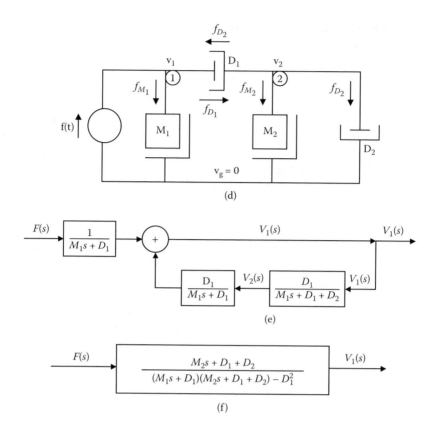

Figure 8.3.4 (continued).

Since the sum of the forces at each node must be zero, we obtain

$$M_1 \frac{dv_1(t)}{dt} + D_1[v_1(t) - v_2(t)] - f(t) = 0 \quad \text{a)}$$
$$M_2 \frac{dv_2(t)}{dt} + D_1[v_2(t) - v_1(t)] + D_2 v_2(t) = 0 \quad \text{b)}$$

(8.9)

In writing these equations, we assumed that nonsource terms were pointing away from the nodes and were assumed positive. This selection is arbitrary and must be kept consistent for all nodes. The LT of these equations is

$$(M_1 s + D_1)V_1(s) - D_1 V_2(s) = F(s) \quad \text{a)}$$
$$-D_1 V_1(s) + (M_2 s + D_1 + D_2)V_2(s) = 0 \quad \text{b)}$$

(8.10)

Chapter 8: Laplace transform

To draw the block diagram of the system, we rearrange these equations to

$$V_1(s) = \frac{D_1}{M_1s+D_1}V_2(s) + \frac{1}{M_1s+D_1}F(s), \qquad V_2(s) = \frac{D_1}{M_2s+D_1+D_2}V_1(s)$$

These equations are readily seen in Figure 8.3.4e. Substituting $V_2(s)$ from the second equation in the first we obtain

$$V_1(s) = \frac{D_1}{M_1s+D_1}\frac{D_1}{M_2s+D_1+D_2}V_1(s) + \frac{1}{M_1s+D_1}F(s)$$

or

$$\frac{V_1(s)}{F(s)} = \frac{\dfrac{1}{M_1s+D_1}}{1-\dfrac{D_1^2}{M_1s+D_1}\dfrac{1}{M_2s+D_1+D_2}} = \frac{1}{(M_1s+D_1)-\dfrac{D_1^2}{M_2s+D_1+D_2}}$$

$$= \frac{M_2s+D_1+D_2}{(M_1s+D_1)(M_2s+D_1+D_2)}$$

The final configuration is shown in Figure 8.3.4f. ∎

*Example 8.3.5: Electromechanical system. Find the transfer function $H(s) = \Omega(s)/E(s)$ of the rotational electromechanical transducer shown in Figure 8.3.5a. Mechanical and air friction damping are taken into consideration by the damping constant D. The movement of the cylinder–pointer combination is restrained by a spring with spring constant K_s. The moment of inertia of the coil assembly is J. There are N turns in the coil (in this example, e's define voltages and v's define velocities).

Solution: Because there are N turns on the coil, there are $2N$ conductors of length l perpendicular to the magnetic field at a distance a from the center of rotation. Therefore, the electrical torque is

$$\mathcal{T}_e = f_t a = (2NBli)a = (2NBla)i = K_e i \qquad (8.11)$$

where K_e is a constant depending on the physical and geometrical properties of the apparatus. The spring develops an equal and opposite torque, which is written as follows

$$\mathcal{T}_s = K_s \theta \qquad K_s = \text{(newton-meter)/degree} \qquad (8.12)$$

Owing to the movement of the coil in the magnetic field, a voltage is generated in the coil. The voltage is given by

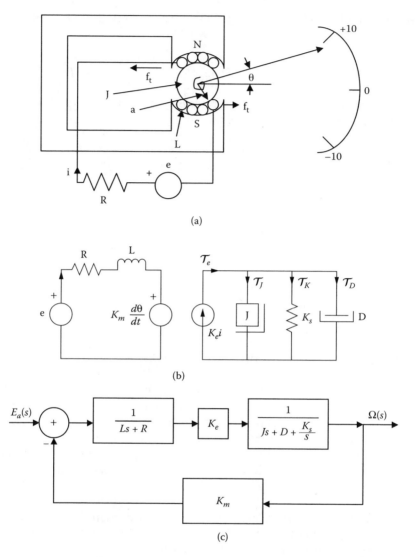

Figure 8.3.5 Rotational electromechanical transducer: (a) physical model, (b) circuit model, and (c) block diagram representation.

$$e_m = 2NBlv = (2NBla)\frac{d\theta}{dt} = K_m \frac{d\theta}{dt} = K_m \omega \quad \omega = \text{angular velocity} \quad (8.13)$$

From (8.11) and (8.13) we observe that K_e and K_m are equal. From the equivalent circuit representation of the system shown in Figure 8.3.5b, we obtain the equations

Chapter 8: Laplace transform

$$L\frac{di}{dt} + Ri + K_m\omega = e \qquad \text{Kirchhoff voltage law} \qquad \text{a)}$$
$$J\frac{d\omega}{dt} + D\omega + K_s\int \omega\, dt = K_e i \qquad \text{D'Alembert's principle} \qquad \text{b)}$$
(8.14)

The LT of these equations yields

$$(Ls+R)I(s) + K_m\Omega(s) = E(s) \qquad \text{a)}$$
$$(Js+D+\frac{K_s}{s})\Omega(s) - K_e I(s) = 0 \qquad \text{b)}$$
(8.15)

Substituting the value of $I(s)$ from the second of these equations into the first and solving for the ratio $\Omega(s)/E(s)$ we obtain

$$H(s) \triangleq \frac{\Omega(s)}{E(s)} = \frac{K_e}{(Ls+R)\left(Js+D+\frac{K_s}{s}\right) + K_e K_m} \qquad (8.16)$$

A block diagram representation of this system is shown in Figure 8.3.5c. To obtain the block diagram, we first observe that (8.15a) gives the error signal (the signal just after the summer) equal to

$$E(s) - K_m\Omega(s) = (Ls+R)I(s)$$

From (8.15b) we find the relationship

$$\Omega(s) = K_e \frac{1}{Ls+R} \frac{1}{Js+D+\frac{K_s}{s}} [(Ls+R)I(s)]$$

which indicates that the error signal must be multiplied by the three front factors shown in the above equation to obtain the output. These are shown as a combination of three systems in series. ∎

***Example 8.3.6: Bioengineering.** Determine the transfer function $H(s) \triangleq \Theta(s)/T(s)$ for the mechanical system (pendulum) shown in Figure 8.3.6a. Draw a block diagram of the system and use block diagram reductions (see Chapter 2) to deduce the transfer function.

Solution: By an application of D'Alembert's principle, which requires that the algebraic sum of torques be zero at a node, we write

$$T = T_g + T_D + T_J \qquad (8.17)$$

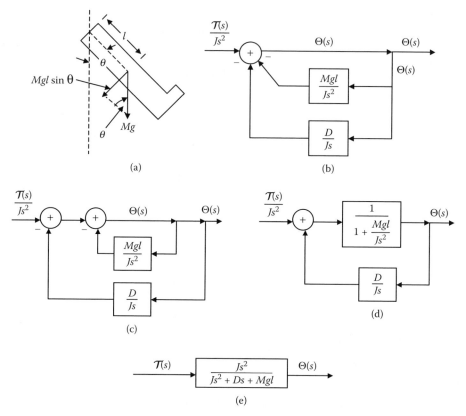

Figure 8.3.6 Mechanical system with steps in reducing the transfer function $H(s) = H(s) = \Theta(s)/T(s)$.

where

$$T = \text{input torque}$$

$$T_g = \text{gravity torque} = Mgl\sin\theta$$

$$T_D = \text{frictional torque} = D\omega = D\frac{d\theta}{dt}$$

$$T_J = \text{inertial torque} = D\frac{d\omega}{dt} = J\frac{d^2\theta}{dt^2}$$

We have characterized the problem as a bioengineering one because we can think of the physical presentation shown in Figure 8.3.6a as an idealized stiff human limb, with the idea to assess as an initial step the passive control of the locomotive action. Therefore, the equation that describes the system is (see Figure 8.3.6a)

Chapter 8: Laplace transform

$$J\frac{d^2\theta(t)}{dt^2} + D\frac{d\theta(t)}{dt} + Mgl\sin\theta(t) = \mathcal{T}(t) \tag{8.18}$$

This equation is nonlinear owing to the presence of the $\sin\theta(t)$ in the differential equation. However, if we assume small deflections, less than 20° to 30°, we can approximate the sine function with $\theta(t)$. Under these conditions the above equation becomes

$$J\frac{d^2\theta(t)}{dt^2} + D\frac{d\theta(t)}{dt} + Mgl\,\theta(t) = \mathcal{T}(t) \tag{8.19}$$

The LT of this equation is

$$Js^2\Theta(s) + Ds\Theta(s) + Mgl\Theta(s) = \mathcal{T}(s) \tag{8.20}$$

To obtain the block diagram representation we write the above equation in the form

$$\Theta(s) = -\frac{D}{Js}\Theta(s) - \frac{Mgl}{Js^2}\Theta(s) + \frac{1}{Js^2}\mathcal{T}(s)$$

This result is shown in Figure 8.3.6b. Rearrange the block diagram as in Figure 8.3.6c and then reduce the innermost feedback loop. This inner loop has the value $1/(1 + Mgl/Js^2)$. The block diagram simplifies to that shown in Figure 8.3.6d. Now, reduce this feedback loop to finally obtain

$$\frac{\Theta(s)}{\mathcal{T}(s)} = \frac{1/(1+Mgl/Js^2)}{\dfrac{Js^2}{1+\dfrac{D}{Js}\dfrac{1}{(1+Mgl/Js^2)}}} \quad \text{or} \quad \frac{\Theta(s)}{\mathcal{T}(s)} \triangleq H(s) = \frac{1}{Js^2 + Ds + Mgl}$$

The transfer function is shown in Figure 8.3.6e. ∎

*Example 8.3.7: Determine the transfer function $H(s) \triangleq V_o(s)/V_i(s)$ of the system shown in Figure 8.3.7.

Solution: We have seen in the LT operation that the operator d/dt in a differential equation is replaced by s and the operator $\int dt$ is replaced by $1/s$ in problems with zero initial conditions (see previous examples). Therefore, we can write Kirchhoff' voltage law in Laplace form by direct reference to Figure 8.3.7b. The equations are

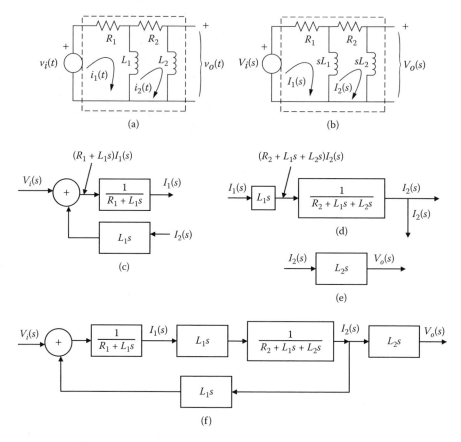

Figure 8.3.7 Illustration of Example 8.3.7.

$$(R_1 + L_1 s)I_1(s) - L_1 s I_2(s) = V_i(s) \qquad \text{a)}$$
$$-L_1 s I_1(s) + (R_2 + L_1 s + L_2 s)I_2(s) = 0 \qquad \text{b)} \qquad (8.21)$$
$$L_2 s I_2(s) = V_o(s) \qquad \text{c)}$$

These equations are shown in Figures 8.3.7c, d, and e, respectively. When the parts are combined, the resulting block diagram is that shown in Figure 8.3.7f. Using the rules of block reduction (see Chapter 2), the transfer function is easily determined. ∎

8.4 Inverse Laplace transform (ILT)

As already discussed, the LT is the integral that converts $F(s)$ into the equivalent $f(t)$. To perform the inverse transformation requires that the integration be performed in the complex plane along a path that ensures that $f(t)$ is

Chapter 8: Laplace transform

unique. Since this type of integration is beyond the level of this text, we will concentrate on the one-to-one correspondence between the direct and inverse transforms, as expressed by the pair

$$F(s) = \mathcal{L}\{f(t)\} \qquad f(t) = \mathcal{L}^{-1}\{F(s)\} \qquad (8.22)$$

Consequently, for this level of mathematical knowledge we must refer to Table 8.3.1 to write $f(t)$ appropriate to a given $\mathcal{L}^{-1}\{F(s)\}$. The following examples illustrate some of the usual methods used in finding inverse Laplace transforms.

Example 8.4.1: Separate roots. Find the inverse LT of the function

$$F(s) = \frac{s-3}{s^2 + 5s + 6} \qquad (8.23)$$

Solution: Observe that the denominator can be factored in the form $(s + 2)(s + 3)$. Thus, $F(s)$ is written in partial fraction form:

$$F(s) = \frac{s-3}{(s+2)(s+3)} = \frac{A}{s+2} + \frac{B}{s+3} \qquad (8.24)$$

where A and B are constants to be determined. To evaluate A, multiply both sides of (8.24) by $(s + 2)$ and set $s = -2$. The result is

$$\left.\frac{s-3}{(s+2)(s+3)}(s+2)\right|_{s=-2} = A + \left.\frac{(s+2)B}{s+3}\right|_{s=-2} \quad \text{or} \quad A = -5$$

We proceed in the same manner to reduce the constant B. By multiplying both sides by $(s + 3)$ and setting $s = -3$, we obtain $B = 6$. Hence, the inverse LT is given by

$$\mathcal{L}^{-1}\{F(s)\} = -5\mathcal{L}^{-1}\left\{\frac{1}{s+2}\right\} + 6\mathcal{L}^{-1}\left\{\frac{1}{s+3}\right\} = -5e^{-2t} + 6e^{-3t} \quad t \geq 0$$

where Table 8.3.1, entry 4, was used.

MATLAB *function residue*

We can find the partial fraction expansion of the rational functions by invoking the residue command of MATLAB: **[r,p,k]=residue(A,B)**, where **B** and **A** are row vectors specifying the coefficients of the numerator and denominator polynomials in descending powers of s. The residues are returned in the column vector **r**, the poles in column vector p, and the direct

Table 8.3.1 Table of Elementary Laplace Transform Pairs

Entry No.	$f(t)$	$F(s) = \int_0^\infty f(t)e^{-st}dt$
1	$\delta(t)$	1
2	$u(t)$	$\dfrac{1}{s}$
3	$t^n \quad n = 1, 2, 3, \ldots$	$\dfrac{n!}{s^{n+1}}$
4	e^{-at}	$\dfrac{1}{s+a}$
5	$t^n e^{-at}$	$\dfrac{n!}{(s+a)^{n+1}}$
6	$\sin \omega t$	$\dfrac{\omega}{s^2+\omega^2}$
7	$\cos \omega t$	$\dfrac{s}{s^2+\omega^2}$
8	$e^{-at} \sin \omega t$	$\dfrac{\omega}{(s+a)^2+\omega^2}$
9	$e^{-at} \cos \omega t$	$\dfrac{s+a}{(s+a)^2+\omega^2}$
10	$\sinh \omega t$	$\dfrac{\omega}{s^2-\omega^2}$
11	$\cosh \omega t$	$\dfrac{s}{s^2-\omega^2}$
12	$t \sin \omega t$	$\dfrac{2\omega s}{(s^2+\omega^2)^2}$
13	$t \cos \omega t$	$\dfrac{s^2-\omega^2}{(s^2+\omega^2)^2}$
14	$\dfrac{at+e^{-at}-1}{a^2}$	$\dfrac{1}{s^2(a+s)}$
15	$\dfrac{bat+(b-a)e^{-at}+(a-b)}{a^2}$	$\dfrac{b+s}{s(a+s)^2}$

Chapter 8: Laplace transform

Table 8.3.1 (continued) Table of Elementary Laplace Transform Pairs

Entry No.	$f(t)$	$F(s) = \int_0^\infty f(t)e^{-st}dt$
16	$\dfrac{(b^2+\omega^2)^{1/2}\sin(\omega t+\phi)}{\omega}$ $\phi = \tan^{-1}\dfrac{\omega}{b}$	$\dfrac{s+b}{s^2+\omega^2}$
17	$\dfrac{b-(b^2+\omega^2)^{1/2}\cos(\omega t+\phi)}{\omega^2}$ $\phi = \tan^{-1}\dfrac{\omega}{b}$	$\dfrac{s+b}{s(s^2+\omega^2)}$
18	$\dfrac{(a^2+b^2+c^2-2ac)^{1/2}e^{-at}\sin(bt+\phi)}{b}$ $\phi = \tan^{-1}\dfrac{b}{c-a}$	$\dfrac{s+c}{(s+a)^2+b^2}$
19	$\dfrac{b+(a^2+b^2)^{1/2}e^{-at}\sin(bt-\phi)}{b(a^2+b^2)}$ $\phi = -\tan^{-1}\dfrac{b}{a}$	$\dfrac{1}{s[(s+a)^2+b^2)]}$

terms in row vector k. If, for example, k = [1 –5], we add in the expression the function s – 5. For the case above, we write at the command window

    [r,p,k]=residue([0 1 -3],[1 5 6]);

where r = residues in column form, p = poles in column form, and k = some function of s or constant.
The results of our example above are:

        r   p   k
        6   -3   []
       -5   -2

With these results, we write

$$F(s) = \frac{6}{s-(-3)} + \frac{-5}{s-(-2)} + 0.$$

This result is identical to the one found above. ∎

Example 8.4.2: Find the inverse LT of the function

$$F(s) = \frac{s+1}{[(s+2)^2+1](s+3)}$$

Solution: This function is written in the form

$$F(s) = \frac{A}{s+3} + \frac{Bs+C}{[(s+2)^2+1]} = \frac{s+1}{[(s+2)^2+1]}$$

The value of A is evaluated by multiplying both sides of the equation by $s+3$ and then setting $s = -3$. The result is

$$A = (s+3)F(s) = \frac{-3+1}{(-3+2)^2+1} = -1$$

To evaluate B and C, combine the two equations

$$\frac{-1[(s+2)^2+1]+(s+3)(Bs+C)}{[(s+2)^2+1](s+3)} = \frac{s+1}{[(s+2)^2+1](s+3)}$$

from which it follows that

$$-(s^2+4s+5)+Bs^2+(C+3B)s+3C = s+1$$

Combine like-powered terms of s to write

$$(-1+B)s^2+(-4+C+3B)s+(-5+3C) = s+1$$

Equating the coefficients of equal power of s, we then have $-1+B = 0$; $-4+C+3B = 1$; and $-5+3C = 1$. From these equations, we obtain $B = 1$ and $C = 2$. Hence, the function is written in the equivalent form

$$F(s) = \frac{-1}{s+3} + \frac{s+2}{(s+2)^2+1}$$

Now, using Table 8.3.1, the result is

$$f(t) = -e^{3t} + e^{-2t}\cos t \qquad t \geq 0$$

Using the residue MATLAB function, `[r,p,k]=residue(([0 0 1 1],[1 7 17 15]);` we obtain

```
    r                    p                    k
-1.0000              -3.0000
 0.5000-0.0000i      -2.0000 + 1.0000i
 0.5000+0.0000i      -2.0000 - 1.0000i       []
```

Therefore, the function becomes

$$F(s) = \frac{-1}{s-(-3)} + \frac{0.5}{s-(-2+j)} + \frac{0.5}{s-(-2-j)} + 0$$

or

$$f(t) = -e^{-3t} + 0.5e^{-2t+jt} + 0.5e^{-2t-jt} = -e^{-3t} + 0.5e^{-2t}(e^{jt} + e^{-jt}) = -e^{-3t} + e^{-2t}\cos t$$

which is identical to the value found above. ∎

In many cases, $F(s)$ is the quotient of two polynomials with real coefficients. If the numerator polynomial is of the same or higher degree than the denominator polynomial, we must first divide the numerator polynomial by the denominator polynomial, the division carried forward until the numerator polynomial is of a degree one less than the denominator polynomial. This procedure results in a polynomial of s plus a **proper function**. The proper function can be expanded into a partial fraction expansion. The result of such an expansion is an expression in the form

$$F'(s) = B_0 + B_1 s + B_2 s^2 + \cdots + \frac{A_1}{s-s_1} + \frac{A_2}{s-s_2} + \frac{A_3}{s-s_3} + \cdots + \frac{A_{p1}}{s-s_p}$$

$$+ \frac{A_{p2}}{(s-s_p)^2} + \frac{A_{p3}}{(s-s_p)^3} + \cdots + \frac{A_{pr}}{(s-s_p)^r} \quad (8.25)$$

This expression has been written in a form to show three types of terms:

1. Polynomial
2. Simple partial fraction, including all terms with distinct roots
3. Partial fraction appropriate to multiple roots

To find the constants $A_1, A_2, A_3, \ldots$, the polynomial terms are removed, leaving the proper fraction

$$F(s) = F'(s) - (B_0 + B_1 s + B_3 s^2 + \cdots) \quad (8.26)$$

where $F(s)$ is the partial fraction expansion containing all the A's. To find the constants A_k, which in complex variable terminology are the residues of the function $F(s)$ at the simple poles s_k, it is only necessary to note that as $s \to s_k$, the term $A_k/(s-s_k)$ will become large compared with all other terms. In the limit,

$$A_k = \lim_{s \to s_k}[(s-s_k)F(s)] \qquad (8.27)$$

Therefore, for each simple pole, upon taking the inverse transform, the result will be a simple exponential of the form

$$\mathcal{L}^{-1}\left\{\frac{A_k}{s-s_k}\right\} = A_k e^{s_k t} \qquad (8.28)$$

Note also that since $F(s)$ contains only real coefficients, if s_k is a complex pole with A_k, there exists also a conjugate pole s_k^* with residue A_k^*. For such complex poles,

$$\mathcal{L}^{-1}\left\{\frac{A_k}{s-s_k} + \frac{A_k^*}{s-s_k^*}\right\} = A_k e^{s_k t} + A_k^* e^{s_k^* t}$$

These terms can be combined in the following way:

$$\text{response} = (a_k + jb_k)e^{(\sigma_k + j\omega_k)t} + (a_k - jb_k)e^{(\sigma_k - j\omega_k)t}$$
$$= e^{\sigma_k t}[(a_k + jb_k)(\cos\omega_k t + j\sin\omega_k t) + (a_k - jb_k)(\cos\omega_k t - j\sin\omega_k t)]$$
$$= 2e^{\sigma_k t}(a_k \cos\omega_k t - b_k \sin\omega_k t) = 2|A_k|e^{\sigma_k t}\cos(\omega_k t + \theta_k) \qquad (8.29)$$

$$|A_k| = \sqrt{a_k^2 + b_k^2}, \quad \theta = \tan^{-1}\left(\frac{b_k}{a_k}\right)$$

When the proper fraction contains a multiple pole of order r, the coefficients in the partial fraction expansion, which are involved in the terms

$$\frac{A_{p1}}{(s-s_p)} + \frac{A_{p2}}{(s-s_p)^2} + \frac{A_{p3}}{(s-s_p)^3} + \cdots + \frac{A_{pr}}{(s-s_p)^r}$$

must be evaluated. A simple application of (8.27) is not adequate. Now the procedure is to multiply both sides of (8.26) by $(s-s_p)^r$, which gives

Chapter 8: Laplace transform

$$(s-s_p)^r F(s) = (s-s_p)^r \left(\frac{A_1}{s-s_1} + \frac{A_2}{s-s_2} + \cdots + \frac{A_k}{s-s_k} \right) + A_{p1}(s-s_p)^{r-1} \qquad (8.30)$$

$$+ A_{p2}(s-s_p)^{r-2} + \cdots + A_{p(r-1)}(s-s_p) + A_{pr}$$

In the limit $s = s_p$, all terms on the right vanish with the exception of A_{pr}. Suppose now that this equation is differentiated once with respect to s. The constant A_{pr} will vanish in the differentiation, but $A_{p(r-1)}$ will be determined by setting $s = s_p$. This procedure is continued to find each of the coefficients A_{pk}. Specifically, this procedure is quantified by

$$\boxed{A_{pk} = \frac{1}{(r-k)!} \frac{d^{r-k}}{ds^{r-k}} [F(s)(s-s_p)^r]\bigg|_{s=s_p} \qquad k=1, 2, \cdots, r \qquad 0!=1} \qquad (8.31)$$

Example 8.4.3: Find the inverse transform of the following function:

$$F'(s) = \frac{s^3 + 2s^2 + 3s + 1}{s^2(s+1)}$$

Solution: This fraction is not a proper one. The numerator polynomial is divided by the denominator polynomial by simple long division. The result is

$$F'(s) = 1 + \frac{s^2 + 3s + 1}{s^2(s+1)}$$

The proper fraction is expanded into partial fraction form:

$$F(s) = \frac{s^2 + 3s + 1}{s^2(s+1)} = \frac{A_{11}}{s} + \frac{A_{12}}{s^2} + \frac{A_2}{s+1}$$

The value of A_2 is deduced by using (8.27):

$$A_2 = [(s+1)F(s)]_{s=-1} = \frac{s^2 + 3s + 1}{s^2(s+1)}\bigg|_{s=-1} = -1$$

To find A_{11} and A_{12}, we proceed as specified by (8.31):

$$A_{12} = s^2 F(s)\big|_{s=0} = \frac{s^2 + 3s + 1}{s+1}\bigg|_{s=0} = 1$$

$$A_{11} = \frac{1}{(2-1)!} \left[\frac{d^{2-1}}{ds^{2-1}} s^2 F(s) \right]_{s=0} = \frac{d}{ds}\left(\frac{s^2 + 3s + 1}{s+1} \right)\bigg|_{s=0} = \frac{s^2 + 2s + 2}{(s+1)^2}\bigg|_{s=0} = 2$$

Therefore,

$$F'(s) = 1 - \frac{1}{s+1} + 2\frac{1}{s} + \frac{1}{s^2} \quad \text{or} \quad f'(t) = \delta(t) - e^{-t} + 2 + t \quad \text{for} \quad t \geq 0$$

where Table 8.3.1 was used. ∎

We can use MATLAB function `[r,p,k]=residue(b,a)` for multiple roots. For example, to invert the function

$$F(s) = \frac{1}{s(s+1)^4} = \frac{1}{s^5 + 4s^4 + 6s^3 + 4s^2 + s}$$

we proceed as follows: At the command sign >> we write `[r,p,k]=residue([0 0 0 0 0 1],[1 4 6 4 1 0]);`. At the enter command we find the following values:

```
  r          p          k
-1.0000    -1.0000     []
-1.0000    -1.0000
-1.0000    -1.0000
-1.0000    -1.0000
 1.0000     0
```

Based on these values we write

$$F(s) = \frac{-1}{s-(-1)} + \frac{-1}{[s-(-1)]^2} + \frac{-1}{[s-(-1)]^3} + \frac{-1}{[s-(-1)]^4} + \frac{1}{s-(0)}$$

Its inverse LT, consulting Table 8.3.1, is

$$f(t) = -e^{-t} - \frac{te^{-t}}{1!} - \frac{t^2 e^{-t}}{2!} - \frac{t^3 e^{-t}}{3!} + u(t)$$

8.5 Problem solving with Laplace transform

It is instructive to present several examples and their solutions using the LT method.

Example 8.5.1: Study the changes in time of the current in an *RL* series circuit with initial condition $i(0)$ = constant and input voltage source $v(t)$.

Chapter 8: Laplace transform

Solution: The differential equation describing the system (KVL) is

$$L\frac{di(t)}{dt} + Ri(t) = v(t) \tag{8.32}$$

With the help of the LT properties (see Section 8.2), we find the LT of the above equation and the unknown $I(s)$ to be

$$LsI(s) - Li(0) + RI(s) = V(s) \tag{8.33}$$

or

$$I(s) = \frac{L}{Ls+R}i(0) + \frac{1}{Ls+R}V(s) \triangleq I_{zi}(s) + I_{zs}(s)$$

where

$$I_{zi}(s) = \frac{L}{Ls+R}i(0) = \textbf{zero-input solution}\ (\text{dependent on initial conditions and independent of inputs}) \tag{8.34}$$

and

$$I_{zs}(s) = \frac{1}{Ls+R}V(s) = \textbf{zero-state solution}\ (\text{dependent on inputs and independent of initial conditions}) \tag{8.35}$$

Therefore, the total solution in the transform and time domains is

$$I(s) = I_{zi}(s) + I_{zs}(s) \quad \text{or} \quad i(t) = i_{zi}(t) + i_{zs}(t) \tag{8.36}$$

Impulse response

To find the impulse response of the system, we set the initial condition equal to zero and a delta function as the input voltage. This implies that the zero-input solution is zero. Therefore, the **system function** for this problem, which is the zero-state solution in the transform domain, and the **impulse response** are

$$H(s) = \frac{1}{L}\frac{1}{s+(R/L)}; \quad h(t) = \mathcal{L}^{-1}\left\{\frac{1}{L}\frac{1}{s+(R/L)}\right\} = \frac{1}{L}e^{-(R/L)t}$$

Superposition

Let us assume that $v(t) = v_1(t) + v_2(t)$. Then the zero-state solution will be

$$I_{zs}(s) = \frac{1}{Ls+R}[V_1(s)+V_2(s)] = \frac{1}{Ls+R}V_1(s) + \frac{1}{Ls+R}V_2(s) \quad (8.37)$$
$$= I_{zs1}(s) + I_{zs2}(s)$$

Note: *The input voltages affect only the zero-state part of the solution.*

Time invariance

Suppose that the source $v(t)$ is shifted by t_0. Then the zero-state solution is

$$I_{0zs}(s) = \frac{1}{Ls+R}\mathcal{L}\{v(t-t_0)u(t-t_0)\} = \frac{1}{Ls+R}V(s)e^{-st_0} = I_{zs}(s)e^{-st_0}$$

This indicates that the current due to a shifted source is of the same form but shifted by t_0 due to $\exp(-st_0)$. Hence,

$$i_{0zs}(t) = i_{zs}(t-t_0)u(t-t_0) \quad (8.38)$$

Note: *A delay of the input results in an equal delay of the zero-state solution.*

Step response

If $v(t) = u(t)$ is a unit step signal applied at $t = 0$ to a system with zero initial conditions, then the resulting zero-state response is called **step response**. This step response will be denoted by $u_{sr}(t)$. Since the LT of $u(t)$ is $1/s$, then (8.35) yields

$$I_{zs}(s) = \frac{1}{Ls+R}\frac{1}{s} = \frac{1}{R}\frac{1}{s} - \frac{1}{R}\frac{1}{s+(R/L)}$$

Hence,

$$i_{zs}(t) \triangleq u_{sr}(t) = \frac{1}{R}\left(1 - e^{-(R/L)t}\right)u(t) \quad (8.39)$$

Let, for example, the values of the circuit elements be $R = 2$ and $L = 1$. We assume a voltage input of the form shown in Figure 8.5.1a and zero initial condition. The input voltage is

$$v(t) = 2u(t) + u(t-1) - 2u(t-2)$$

Chapter 8: Laplace transform

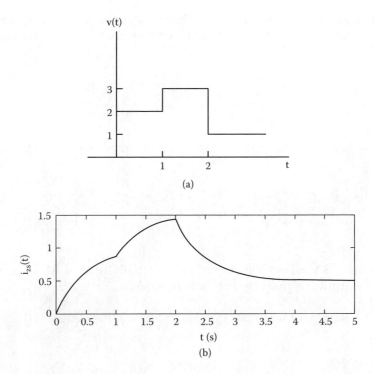

Figure 8.5.1 Zero-state response of an RL series circuit to multistep input.

Therefore (superposition),

$$i_{zs}(t) = 2u_{sr}(t) + u_{sr}(t-1) - 2u_{sr}(t-2)$$

But

$$u_{sr}(t) = \frac{1}{R}\left(1 - e^{-(R/L)t}\right)u(t) = \frac{1}{2}(1 - e^{-2t})u(t)$$

and hence the total current is

$$i_{zs}(t) = \begin{cases} (1 - e^{-2t}) & t \geq 0 \\ \frac{1}{2}(1 - e^{-2(t-1)}) & 1 \leq t < \infty \\ -(1 - e^{-2(t-2)}) & 2 \leq t < \infty \end{cases}$$

This current is plotted in Figure 8.5.1b. However, if we had also assumed initial condition $i(0)$, then the general LT format would have been as follows:

$$I(s) = \underbrace{\frac{1}{R}\left(\frac{1}{s} - \frac{1}{s+(R/L)}\right)}_{\text{zero-state response}} + \underbrace{\frac{1}{s+(R/L)}i(0)}_{\text{zero-input response}}$$

Using Table 8.3.1, the inverse transform of the above equation is

$$i(t) = \underbrace{\frac{1}{R}(1-e^{-(R/L)t})}_{\text{zero-state response}} + \underbrace{i(0)e^{-(R/L)t}}_{\text{zero-input response}} = \underbrace{\left[i(0)-\frac{1}{R}\right]e^{-(R/L)t}}_{\text{transient response}} + \underbrace{\frac{1}{R}}_{\substack{\text{steady-state}\\\text{response}}} \quad t \geq 0$$

■

Example 8.5.2: Find the current in an initially relaxed RL series circuit when the input is a pulse shown in Figure 8.5.2a.

Solution: The differential equation describing the system is

$$L\frac{di(t)}{dt} + Ri(t) = v(t)$$

The input signal is decomposed into two unit step functions, as shown in Figure 8.5.2:

$$v(t) = u(t) - u(t-1)$$

Because the system is LTI, we can use the superposition property and the time-invariant properties discussed above. Hence, we write

$$i_{zs}(t) = u_{sr}(t) - u_{sr}(t-1)$$

But $u_{sr}(t)$ is given by (8.39), and thus the current is given by

$$i_{zs}(t) = \frac{1}{R}\left(1-e^{-(R/L)t}\right)u(t) - \frac{1}{R}\left(1-e^{-(R/L)(t-1)}\right)u(t-1)$$

For $L = 1$ and $R = 2$ the current is shown in Figure 8.5.2c. ■

Example 8.5.3: A force is applied to a relaxed mechanical system of negligible mass, as shown in Figure 8.5.3a. Find the velocity of the system.

Chapter 8: Laplace transform

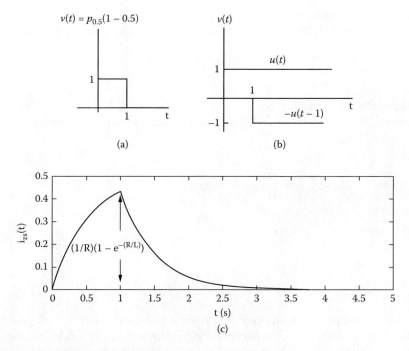

Figure 8.5.2 Illustration of Example 8.5.2, the principle of superposition and time invariance.

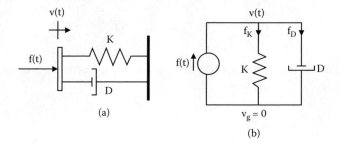

Figure 8.5.3 Simple mechanical system: (a) physical model and (b) circuit equivalent form.

Solution: From Figure 8.5.3b, we write

$$Dv + K\int_0^t v(x)dx = au(t)$$

Now, define a new variable

$$y(t) = \int_0^t v(x)dx;$$

this equation takes the form

$$D\frac{dy(t)}{dt} + Ky = au(t)$$

Observe that the solution is equal to $au_{sr}(t)$, where $u_{sr}(t)$ is the step response solution. Therefore, the LT of the above equation and its inverse are (see Example 8.5.1)

$$Y(s) = \frac{a}{D}\frac{1}{[s+(K/D)]s}; \quad y(t) = \frac{a}{K}(1-e^{-(K/D)t}) \quad t \geq 0$$

Hence, the velocity is given by

$$v(t) \triangleq \frac{dy(t)}{dt} = \frac{a}{D}e^{-(K/D)t} \quad t \geq 0$$

Next, we approach this problem from a different point of view, namely, the use of convolution. This approach requires that we find the impulse response of the system in the time and LT domains, which are ($h(t)$ has the same units as the velocity)

$$Dh(t) + K\int_0^t h(x)dx = \delta(t); \quad DH(s) + \frac{K}{s}H(s) = 1$$

from which the transfer function and its impulse response are

$$H(s) = \frac{1}{D} - \frac{K}{D^2}\frac{1}{s+(K/D)}; \quad h(t) = \frac{1}{D}\delta(t) - \frac{K}{D^2}e^{-(K/D)t} \quad t \geq 0$$

For a step function input $au(t)$, the output is

$$v(t) = au(t) * h(t) = \frac{a}{D}\int_{-\infty}^{\infty}\delta(x)u(t-x)dx - \frac{Ka}{D^2}\int_{-\infty}^{\infty}e^{-(K/D)x}u(x)u(t-x)dx$$

$$= \frac{a}{D}u(t) - \frac{Ka}{D^2}\int_0^t e^{-(K/D)x}dx = \frac{a}{D}u(t) + \frac{a}{D}(e^{-(K/D)t}-1)u(t) = \frac{a}{D}e^{-(K/D)t}u(t)$$

This result is precisely what we found above using a different approach. This makes sense if we remember that the derivative of a step function is a delta function.

We could have also taken the LT of the defining equation, which yields

$$DV(s) + K\frac{V(s)}{s} = a\frac{1}{s}$$

Solving for the unknown $V(s)$ and taking the inverse LT, we obtain, as before, the desired solution

$$V(s) = \frac{a}{D}\frac{1}{s+(K/D)}; \quad v(t) = \frac{a}{D}e^{-(K/D)T} \quad t \geq 0$$

■

Example 8.5.4: Refer to Figure 8.5.4a, which shows the switching of an inductor into a circuit with an initial current. Prior to switching, the circuit inductance is L_1; after switching, the total circuit inductance is $L_1 + L_2$. The switching occurs at $t = 0$. Determine the current in the circuit for $t \geq 0$.

Solution: The current just before switching is

$$i(0-) = \frac{V}{R}$$

To find the current after switching at $t = 0+$, we employ the law of conservation of flux linkages (see (2.54)). We can write over the switching period,

$$L_1 i(0-) = (L_1 + L_2)i(0+) \quad \text{or} \quad i(0+) = \frac{L_1}{(L_1+L_2)}i(0-) = \frac{L_1}{(L_1+L_2)}\frac{V}{R}$$

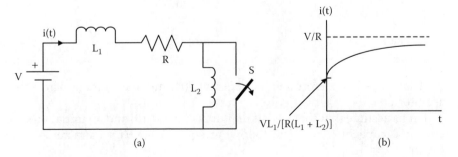

Figure 8.5.4 Switching L in a circuit with initial current: (a) circuit configuration and (b) response of the RL circuit after L is switched.

The differential equation that governs the circuit response, after the switch S is closed, is

$$(L_1 + L_2)\frac{di(t)}{dt} + Ri(t) = Vu(t)$$

The LT of this equation yields

$$I(s) = \frac{V}{R}\left[\frac{1}{s} - \frac{1}{s + R/(L_1 + L_2)}\right] + \frac{1}{s + R/(L_1 + L_2)}i(0+)$$

Include the value of $i(0+)$ in this expression and then take the inverse LT. The result is

$$i(t) = \frac{V}{R}\left(1 - \frac{L_2}{L_1 + L_2}e^{-Rt/(L_1+L_2)}\right) \qquad t \geq 0$$

The form of the current variation is shown in Figure 8.5.4b. ∎

*Example 8.5.5: Second-order systems: The series RLC circuit.** Let us analyze a series *RLC* circuit driven by a voltage source $v_i(t)$, as shown in Figure 8.5.5a. Specifically, we will be interested in the step response of the system. The output is the voltage across the capacitor. In our study the following important parameters are needed

Critical resistance: $R_c = 2\sqrt{\frac{L}{C}}$ Damping: $\alpha = \frac{R}{2L}$

Resonant frequency: $\omega_r = \frac{1}{\sqrt{LC}}$ Natural frequency: $\beta = \sqrt{\omega_r^2 - \alpha^2}$

Q-factor: $Q = \frac{\omega_r}{2\alpha}$ Damping ratio: $\varsigma = \frac{\alpha}{\omega_r} = \frac{R}{R_c}$

which are used to determine the current $i(t)$ and the voltage across the capacitor $v(t)$.

Let us assume that there exists an initial current $i(0)$ and an initial voltage across the capacitor $v(0)$. The following equation (KVL) characterizes the system

$$Ri(t) + L\frac{di(t)}{dt} + \frac{1}{C}\int_0^t i(x)dx + v(0) = v_i(t) \qquad (8.40)$$

Chapter 8: Laplace transform

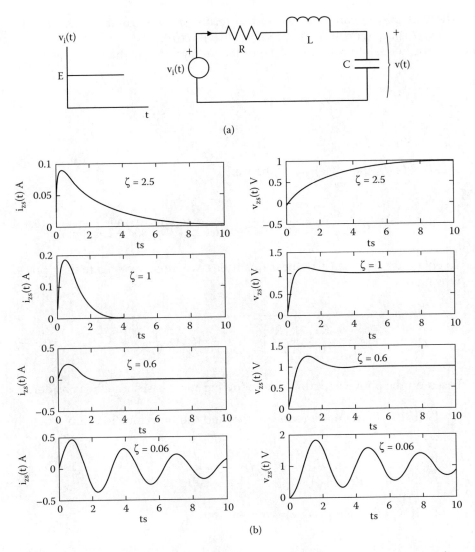

Figure 8.5.5 The zero-state current and zero-state voltage response across a capacitor of an *RLC* series circuit to a unit step function input.

Taking the transform of both sides, we obtain

$$RI(s) + LsI(s) - Li(0) + \frac{I(s)}{Cs} + \frac{v(0)}{s} = V_i(s)$$

Therefore,

$$I(s) = \underbrace{\frac{V_i(s)}{R + Ls + (1/Cs)}}_{\text{zero-state response}} + \underbrace{\frac{Li(0) - (v(0)/s)}{R + Ls + (1/Cs)}}_{\text{zero-input response}} \triangleq I_{zs}(s) + I_{zi}(s) \quad (8.41)$$

The voltage $v(t)$ can be expressed in terms of the current $i(t)$. Since $i(t) = C(dv(t)/dt)$, we conclude that $I(s) = C[sV(s) - v(0)]$.

The zero-state response $V_{zs}(s)$ of the voltage $V(s)$ is therefore given by

$$V_{zs}(s) = I_{zs}(s)\frac{1}{Cs} = \frac{V(s)}{LCs^2 + RCs + 1} \tag{8.42}$$

Step response

If $v_i(t) = Eu(t)$, then $V_i(s) = E/s$; hence,

$$I_{zs}(s) = \frac{E/s}{R + Ls + (1/Cs)} = \frac{E/L}{s^2 + 2\alpha s + \omega_r^2}; \quad V_{zs}(s) = \frac{E\omega_r^2}{s(s^2 + 2\alpha s + \omega_r^2)} \tag{8.43}$$

To obtain the inverse LT of these quantities, we must investigate their poles, which are

$$p_{1,2} = -\frac{R}{2L} \pm \sqrt{\left(\frac{R}{2L}\right)^2 - \frac{1}{LC}} = -\alpha \pm \omega_r\sqrt{\left(\frac{R}{R_c}\right)^2 - 1} \tag{8.44}$$

Based on the above equation, the following three cases may be considered:

1. If $R > R_c$, then the roots are real and the time functions are

$$i_{zs}(t) = \frac{E}{L(p_1 - p_2)}(e^{p_1 t} - e^{p_2 t}); \quad v_{zs}(t) = E + \frac{E\omega_r^2}{(p_1 - p_2)}\left(\frac{1}{p_1}e^{p_1 t} - \frac{1}{p_2}e^{p_2 t}\right) \tag{8.45}$$

2. If $R = R_c$, then $p_1 = p_2 = -\alpha = -\omega_r$ and

$$i_{zs}(t) = \frac{E}{L}te^{-\alpha t}; \quad v_{zs}(t) = E - E(1 + \omega_r t)e^{-\alpha t} \tag{8.46}$$

3. If $R < R_c$, then the roots are complex, $p_{1,2} = -\alpha \pm j\beta$, and the time functions are

$$i_{zs}(t) = \frac{E}{\beta L}e^{-\alpha t}\sin\beta t; \quad v_{zs}(t) = E - Ee^{-\alpha t}\left(\cos\beta t + \frac{\alpha}{\beta}\sin\beta t\right) \tag{8.47}$$

We note that if $R \ll R_c$, then $\alpha \ll \beta \cong \omega_r$, and (8.47) yields

Chapter 8: Laplace transform

$$i_{zs}(t) \cong E\sqrt{\frac{C}{L}} e^{-\alpha t} \sin \omega_r t; \qquad v_{zs}(t) \cong E - Ee^{-\alpha t} \cos \omega_r t \qquad (8.48)$$

It is instructive to find out how the current and voltage behave when the damping ratio ς varies. Let us assume the following circuit element values: $v_i(t) = u(t)$, $L = 1$, $C = 0.25$. The four cases, which we investigate, are shown in Figure 8.5.5b from top to bottom. The parameters for these four cases are:

a. $R = 10$, $\varsigma = 2.5$, $\omega_r = 2$, $\alpha = 5$, roots: -9.5826, -0.4174

b. $R = 4$, $\varsigma = 1.0$, $\omega_r = 2$, $\alpha = 2$, roots: -2.0000, -2.0000

c. $R = 2.4$, $\varsigma = 0.6$, $\omega_r = 2$, $\alpha = 1.2$, roots: $-1.2000 + j1.6000$, $-1.2000 - j1.6000$

d. $R = 0.24$, $\varsigma = 0.06$, $\omega_r = 2$, $\alpha = 0.12$, roots: $-0.1200 + j1.9964$, $-0.1200 - j1.9964$

There exists the MATLAB function **step(num,den)** where **num** is a vector containing the coefficients of the numerator and **den** is a vector containing the coefficients of the denominator. The variable, which we want to find, must be given in a Laplace-transformed rational fraction form. To plot the output we write: `>>sr=step(num,den); plot(sr);`. ∎

Example 8.5.6: Find the velocity of the system shown in Figure 8.5.6a when the applied force is $f(t) = \exp(-t)u(t)$. Use the LT method and assume zero initial conditions. Solve the same problem by means of the convolution technique. The input is the force and the output is the velocity.

Solution: From Figure 8.5.6b, we write the controlling equation

$$M\frac{dv(t)}{dt} + Dv(t) + K\int_0^t v(x)dx = f(t) \quad \text{or} \quad \frac{dv(t)}{dt} + 5v(t) + 4\int_0^t v(x)dx = e^{-t}u(t)$$

Laplace-transform these equations and then solve for $V(s)$. This yields

$$H(s) = \frac{s}{Ms^2 + Ds + K}; \quad V(s) = \frac{s}{(s+1)(s^2 + 5s + 4)} = \frac{s}{(s+1)^2(s+4)}$$

Write the expression in the form

$$V(s) = \frac{A}{s+4} + \frac{B}{s+1} + \frac{C}{(s+1)^2}$$

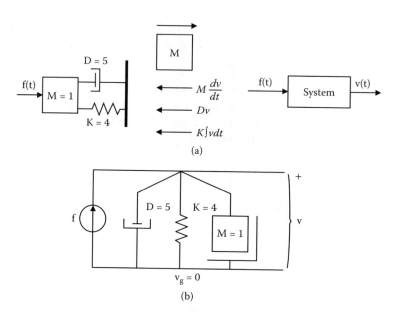

Figure 8.5.6 Illustration of Example 8.5.6: (a) mechanical system and (b) network representation.

where

$$A = \left.\frac{s}{(s+1)^2}\right|_{s=-4} = -\frac{4}{9};\ B = \left.\frac{1}{1!}\frac{d}{ds}\left(\frac{s}{s+4}\right)\right|_{s=-1} = \frac{4}{9};\ C = \left.\frac{s}{s+4}\right|_{s=-1} = -\frac{1}{3}$$

The inverse transform of $V(s)$ is given by

$$v(t) = -\frac{4}{9}e^{-4t} + \frac{4}{9}e^{-t} - \frac{1}{3}te^{-t} \qquad t \geq 0$$

To find the zero-state solution, $v(t)$, by the use of the convolution integral, we must first find the impulse response of the system, $h(t)$. The quantity is specified by

$$\frac{dh(t)}{dt} + 5h(t) + 4\int_0^t h(x)dx = \delta(t)$$

Because we want to find the impulse response, the system is assumed to have zero initial conditions. The LT of the equation yields and its inverse are

$$H(s) = \frac{s}{s^2+5s+4} = \frac{4}{3}\frac{1}{s+4} - \frac{1}{3}\frac{1}{s+1};\ h(t) = \frac{4}{3}e^{-4t} - \frac{1}{3}e^{-t} \qquad t \geq 0$$

Therefore, the output of the system to the input $\exp(-t)u(t)$ is written

$$v(t) = \int_{-\infty}^{\infty} h(x)f(t-x)dx = \int_{-\infty}^{\infty} e^{-(t-x)}u(t-x)\left(\frac{4}{3}e^{-4x} - \frac{1}{3}e^{-x}\right)u(x)dx$$

$$= e^{-t}\left(\frac{4}{3}\int_0^t e^{-3x}dx - \frac{1}{3}\int_0^t dx\right) = e^{-t}\left(\frac{4}{3}\left(\frac{1}{-3}\right)e^{-3x}\Big|_0^t - \frac{1}{3}t\right)$$

$$= -\frac{4}{9}e^{-4t} + \frac{4}{9}e^{-t} - \frac{1}{3}te^{-t} \quad t \geq 0$$

This result is identical with that found using the LT technique, as it should be. ∎

*Example 8.5.7: Environmental engineering. Let us assume that at tome $t = 0$ there are a certain number of some species, say $n(0)$, all of the same age. For convenience, let us classify this age as the zero age. At future time t, there are $n_1(t)$ of these members still in the population. Therefore, $n_1(t)$ and $n(0)$ are connected through a **survival function**, $f(t)$, as follows

$$n_1(t) = n(0)f(t)$$

Let us assume that at time t_1 we placed m of zero age. At any future time $t > t_1$ the number of these individuals will be

$$m(t) = m(t_1)f(t - t_1) \qquad t > t_1$$

Consider that there is a **replacement rate** $r(t)$ of the zero-age individuals. In the time interval from τ to $\tau + \Delta\tau$, $r(\tau_1)\Delta\tau$ individuals are placed in the population. τ_1 must be in the range $\tau \leq \tau_1 \leq \tau + \Delta\tau$. The survival law dictates that $r(\tau_1)\Delta\tau f(t - \tau_1)$ individuals will still be present in the population at time t. We can therefore think that we split the interval $[0, t]$ into subintervals of length $\Delta\tau$ and add up the number of survivors for each interval. At the limit, as $\Delta\tau$ approaches zero, the summation becomes an integral. Hence, the number of species at time t is

$$n(t) = n(0)f(t) + \int_0^t r(\tau)f(t-\tau)d\tau \tag{8.49}$$

Let us assume that in 2005 the population of deer was 55,000 and their survival function was the exponential function $\exp(-t)$. Suppose that we must determine a rate function r(t) so that the population is a linear function of time, that is,

$$n(t) = n(0) + at$$

where a is a constant.

Solution: Here n(0) = 55,000 and r(t) must satisfy the equation

$$n(0) + at = n(0)e^{-t} + \int_0^t r(x)e^{-(t-x)}dx \qquad t \geq 0 \qquad (8.50)$$

Taking the LT of both sides of the above equation, we find

$$\frac{n(0)}{s} + \frac{a}{s^2} = \left[\frac{n(0)}{s+1} + R(s)\frac{1}{s+1}\right] \quad \text{or} \quad R(s) = \frac{(n(0)+a)s + a}{s^2} = \frac{n(0)+a}{s} + \frac{a}{s^2}$$

Therefore, the rate function is $r(t) = n(0) + a + at$ ∎

Note: *Observe that the Laplace transform was used in (8.49), even if it was not a differential equation.*

*Example 8.5.8: Systems of differential equations. Find the voltage $v_2(t)$ for a delta function (impulse response, $h(t)$) and step function (step response) inputs to the system shown in Figure 8.5.7. Assume zero initial conditions, which indicates we find the zero-state response for the step function.

Solution: Using the node equation law, we write

$$\frac{v_1(t)}{R} + C\frac{dv_1(t)}{dt} + i(t) = i_i(t); \quad \frac{v_2(t)}{R} + C\frac{dv_2(t)}{dt} - i(t) = 0;$$

$$v_1(t) - v_2(t) = L\frac{di(t)}{dt} \qquad (8.51)$$

Taking transforms of both sides and setting $G = 1/R$, we obtain

$$GV_1(s) + CsV_1(s) + I(s) = I_i(s)$$

$$GV_2(s) + CsV_2(s) - I(s) = 0 \qquad (8.52)$$

$$V_1(s) - V_2(s) - LsI(s) = 0$$

The desired output is

$$V_2(s) = \frac{I_i(s)}{LC^2s^3 + 2LCGs^2 + (LG^2 + 2C)s + 2G} \qquad (8.53)$$

Chapter 8: Laplace transform

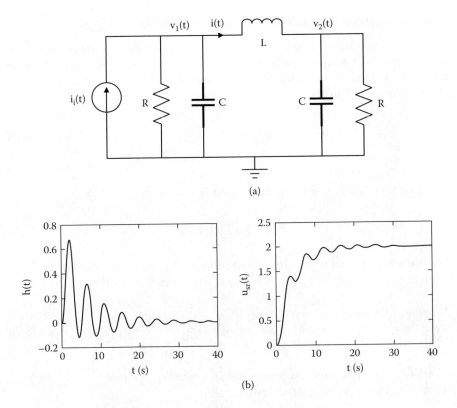

Figure 8.5.7 Illustration of Example 8.5.8.

Let us assume the following values: $L = 1$, $R = 4$, and $C = 1$. Then (8.53) becomes

$$V_2(s) = I_i(s) \frac{1}{s^3 + 0.5s^2 + 2.0625s + 0.5}$$

For the impulse response, $I_i(s) = 1$, and for step response, $I_i(s) = 1/s$. To obtain the time domain of these two responses we used the following MATLAB programs:

```
>>t=0:0.05:40;
>>h=impulse([0 0 0 1],[1 0.5 2.0625 0.5],t]);
>>subplot(2,2,1);plot(t,h);xlabel('t s');ylabel('h(t)');
```

For the step response we used the commands:

```
>>t=0:0.05:40;
>>sr=step([0 0 0 1],[1 0.5 2.0625 0.5],t]);
```

The outputs are shown in Figure 8.5.7b. ∎

8.6 Frequency response of LTI systems

The frequency response of systems is defined when the initial conditions are zero; hence, we address only the zero-state response. We know that the output of an analog system is given by the convolution of the input and the impulse response of the system:

$$\underbrace{g(t)}_{\text{output}} = \underbrace{f(t)}_{\text{input}} * \underbrace{h(t)}_{\text{impulse response}} \tag{8.54}$$

The LT of both sides of (8.54) gives (see Section 8.2)

$$G(s) = F(s)H(s) \tag{8.55}$$

The time and transformed representation of a system are shown diagrammatically in Figure 8.6.1. Because $h(t)$ is the inverse LT of $H(s)$,

$$h(t) = \mathcal{L}^{-1}\{H(s)\} = \mathcal{L}^{-1}\left\{\frac{G(s)}{F(s)}\right\} \tag{8.56}$$

When two systems are connected in cascade, as shown in Figure 8.6.2, we obtain the following equations:

$$g_1(t) = f(t) * h_1(t) \quad G_1(s) = F(s)H_1(s); \quad g(t) = g_1(t) * h_2(t) \quad G(s) = G_1(s)H_2(s)$$

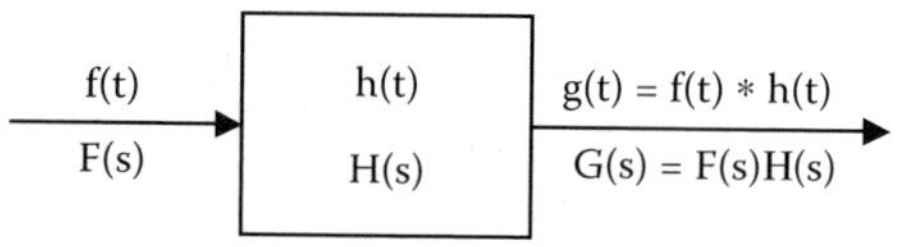

```
  f(t)  ┌──────────┐  g(t) = f(t) * h(t)
──────▶ │   h(t)   │ ──────────────────▶
  F(s)  │   H(s)   │  G(s) = F(s)H(s)
        └──────────┘
```

Figure 8.6.1 Diagrammatic representation of an analog system.

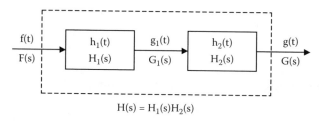

$$H(s) = H_1(s)H_2(s)$$

Figure 8.6.2 Representation of two systems in cascade.

Chapter 8: Laplace transform

Eliminating $G_1(s)$ from the transformed equations above, we find

$$G(s) = H_1(s)H_2(s)F(s)$$

which shows that the combined transfer function is

$$\boxed{H(s) = H_1(s)H_2(s)} \qquad (8.57)$$

Therefore, if n systems are connected in series, their total transfer function is

$$H(s) = H_1(s)H_2(s)H_3(s)\cdots H_n(s) \qquad (8.58)$$

If two systems are connected in parallel, as shown in Figure 8.6.3, we have

$$g(t) = g_1(t) + g_2(t) = f(t) * h_1(t) + f(t) * h_2(t)$$

from which

$$G(s) = F(s)H_1(s) + F(s)H_2(s) = F(s)[H_1(s) + H_2(s)] \qquad (8.59)$$

This equation shows that the transfer function of two systems connected in parallel is given by

$$\boxed{H(s) = H_1(s) + H_2(s)} \qquad (8.60)$$

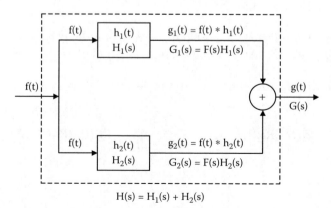

Figure 8.6.3 Representation of two systems in parallel.

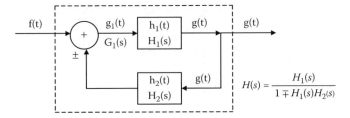

Figure 8.6.4 Representation of systems in feedback form.

For systems with feedback connection, as shown in Figure 8.6.4, we write

$$g_1(t) = f(t) \pm g(t) * h_2(t) \quad \text{and} \quad g(t) = g_1(t) * h_1(t) = f(t) * h_1(t) \pm g(t) * h_2(t) * h_1(t)$$

The LT of the second equation above gives

$$G(s) = F(s)H_1(s) \pm G(s)H_1(s)H_2(s)$$

from which

$$\boxed{G(s) = F(s)\frac{H_1(s)}{1 \mp H_1(s)H_2(s)}} \tag{8.61}$$

and

$$\boxed{H(s) = \frac{G(s)}{F(s)} = \frac{H_1(s)}{1 \mp H_1(s)H_2(s)}} \tag{8.62}$$

These results are consistent with those discussed for block diagram configurations (see Figure 2.5.4).

Example 8.6.1: Determine the transfer function, $I_2(s)/V(s)$, for the system shown in Figure 8.6.5a. Also, find the magnitude and phase response functions.

Solution: The differential equations describing the system are

$$Ri_1(t) - Ri_2(t) = v(t) \qquad L\frac{di_2(t)}{dt} + Ri_2(t) - Ri_1(t) = 0$$

from which

$$I_1(s) = \frac{1}{R}V(s) + I_2(s) \qquad I_2(s) = I_1(s)\frac{R}{Ls + R}$$

Chapter 8: Laplace transform

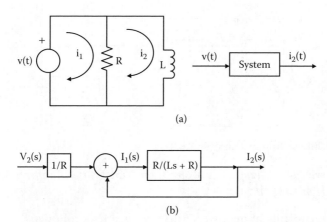

Figure 8.6.5 Illustration of Example 8.6.1.

These two equations are shown in block diagram representation in Figure 8.6.5b. By block diagram simplification (see Figure 2.5.4) or by eliminating $I_1(s)$ from these equations we obtain

$$H(s) = \frac{I_2(s)}{V(s)} = \frac{1}{Ls} \quad \text{or} \quad H(j\omega) = \frac{1}{j\omega L}$$

As a consequence,

$$|H(j\omega)| = \frac{1}{\omega L} \quad \phi(\omega) = -\frac{\pi}{2} = \text{constant}$$

∎

Example 8.6.2: Deduce an expression for the current (voltage input–current output) in the circuit shown in Figure 8.6.6 if a cosine is applied with $\omega = 5$.

Solution: Taking the LT of the KVL equation that describes the system, we find the transfer function to be equal to

$$H(s) = \frac{1}{s+2}$$

The output of the system, if the input is a complex exponential function, is

$$i(t) = \text{Re}\{e^{j\omega t} * h(t)\} = \text{Re}\{\int_{-\infty}^{\infty} h(x)e^{j\omega(t-x)}dx\} = \text{Re}\{e^{j\omega t}\int_{-\infty}^{\infty} h(x)e^{-j\omega x}dx\}$$

$$= \text{Re}\{e^{j\omega t}H(j\omega)\}$$

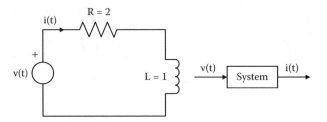

Figure 8.6.6 Illustration of Example 8.6.2.

Note: *We can exchange linear operations (integrations, differentiations, etc.) with the Re{} or Im{} operations.*

Based on our results above, we write the output as follows:

$$i(t) = \text{Re}\left\{\frac{1}{j5+2}e^{j5t}\right\} = \text{Re}\left\{\frac{1}{\sqrt{5^2+2^2}e^{j\tan^{-1}(5/2)}}e^{j5t}\right\} = \frac{1}{\sqrt{29}}\cos[5t - \tan^{-1}(5/2)]$$

When there is a sum of exponential function inputs

$$f(t) = a_1 e^{j\omega_1 t} + a_2 e^{j\omega_2 t} + \cdots + e^{j\omega_n t} a_n \tag{8.63}$$

the output function becomes

$$g(t) = a_1 H(j\omega_1) e^{j\omega_1 t} + a_2 H(j\omega_2) e^{j\omega_2 t} + \cdots + a_n H(j\omega_n) e^{j\omega_n t} \tag{8.64}$$

■

Example 8.6.3: Repeat Example 8.6.2 for an input voltage $v(t) = 2 + 5\cos 3t + 6\sin 6t$.

Solution: For the given circuit,

$$H(j\omega) = \frac{1}{j\omega + 2}$$

and, thus, we write

$$i(t) = \text{Re}\left\{2\frac{1}{j0+2}e^{j0t}\right\} + \text{Re}\left\{5\frac{1}{j3+2}e^{j3t}\right\} + \text{Im}\left\{6\frac{1}{j6+2}e^{j6t}\right\}$$

$$= 1 + \frac{5}{\sqrt{13}}\cos\left(3t - \tan^{-1}\frac{3}{2}\right) + \frac{6}{\sqrt{40}}\cos\left(6t - \tan^{-1}\frac{6}{2}\right)$$

A basic and very important property of any system is its **filtering properties**, which define how different frequencies are attenuated and phase shifted as they pass through the system. Further, the energy of a signal is associated with the amplitude of each harmonic. Thus, the filtering properties of a system will dictate the amount of energy that is to be transferred by the system and the percentage for each particular frequency component. Therefore, we generally prefer to plot the magnitude $|H(j\omega)|$ and the phase $\phi(\omega) = \tan^{-1}[H_i(\omega)/H_r(\omega)]$ vs. frequency ω, where $H_i(\omega)$ and $H_r(\omega)$ are the imaginary and real components of $H(j\omega)$ and are real functions of ω. ∎

Example 8.6.4: Deduce and plot the frequency characteristics (filtering) of the circuit shown in Figure 8.6.7a.

Solution: Applying the KCL and taking the LT of the differential equation, we obtain

$$C\frac{dv(t)}{dt} + \frac{v(t)}{R} = i(t) \qquad H(s) = \frac{V(s)}{I(s)} = \frac{1}{C}\frac{1}{s+(1/RC)}$$

The magnitude and phase spectrums are obtained by setting $s = j\omega$. These are

$$|H(j\omega)| = \frac{1}{C}\frac{1}{\sqrt{\omega^2 + (1/RC)^2}} \qquad \phi(\omega) = -\tan^{-1}(\omega RC)$$

The magnitude and phase characteristics are graphed in Figure 8.6.7b. ∎

Example 8.6.5: Deduce and plot the frequency characteristics of the system shown in Figure 8.6.8.

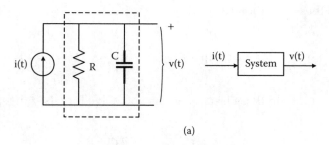

(a)

Figure 8.6.7 Illustration of Example 8.6.4.

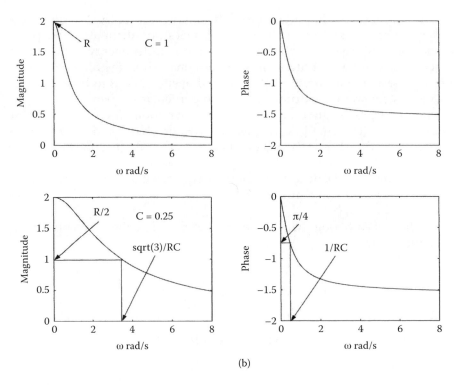

(b)

Figure 8.6.7 (continued).

Solution: From an inspection of the circuit, we write the two KCL equations:

$$C\frac{dv_c(t)}{dt} + \frac{1}{L}\int[v_c(t) - v_o(t)]dt = i(t)$$

$$\frac{1}{L}\int[v_o(t) - v_c(t)]dt + \frac{v_o(t)}{R} = 0$$

The LTs of these equations are

$$CsV_c(s) + \frac{1}{Ls}[V_c(s) - V_o(s)] = I(s); \quad \frac{1}{Ls}[V_o(s) - V_c(s)] + \frac{V_o(s)}{R} = 0$$

Eliminate $V_C(s)$ from these last equations and solve for $H(s) \triangleq V_o(s)/I(s)$. The result is

$$H(s) \triangleq \frac{V_o(s)}{I(s)} = \frac{R/LC}{s^2 + (R/L)s + 1/LC}$$

Chapter 8: Laplace transform

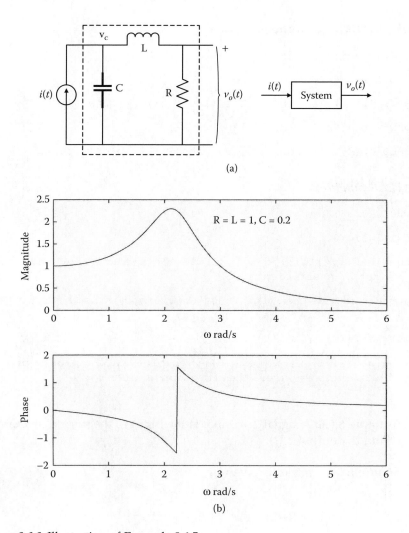

Figure 8.6.8 Illustration of Example 8.6.5.

Therefore,

$$|H(j\omega)| = \frac{1/LC}{\left[(-\omega^2 + 1/LC)^2 + \left(\frac{R}{L}\right)^2 \omega^2\right]^{1/2}}; \quad \phi(\omega) = -\tan^{-1}\left(\frac{(R/L)\omega}{1/LC - \omega^2}\right)$$

The above characteristics are plotted in Figure 8.6.8b for $L = R = 1$ and $C = 0.2$. These plots can be done by the following two approaches.

Book MATLAB calculations

```
>>w=0:0.02:6;
>>r=1;l=1;c=0.2;
>>H=(r/(l*c))./sqrt((-w.^2+1/(l*c))^2+(r/l)^2*w.^2);
>>phw=-atan((r/l)*w./(1/(l*c)-w.^2));
>>subplot(2,1,1);plot(w,H);subplot(2,1,2);plot(w,phw);
```

MATLAB functions

```
>>w=0:0.02:6;
>>r=1;l=1;c=0.2;
>>num=[0 0 r/(l*c)]; den[1 r/l 1/(l*c)];%numerator and denom-
    %inator coefficients;
>>H=freqs(num,den,w);%the function freqs() is used for Laplace
    %transform rational functions;
>>phw=angle(H);
>>subplot(2,1,1);plot(w,abs(H)); subplot(2,1,2);plot(w,phw);
    %note that we used the function abs();
```

■

Example 8.6.6: A signal $2 + 3\sin 5t$ is the input to the system of Example 8.6.5. Find the output $v_o(t)$.

Solution: The output is given by

$$v_o(t) = \text{Re}\{2H(j0)e^{j0t}\} + \text{Im}\{3H(j5)e^{j5t}\}$$
$$= \text{Re}\{2|H(j0)|e^{j\phi(0)}e^{j0t}\} + \text{Im}\{3|H(j5)|e^{j\phi(5)}e^{j5t}\}$$

Therefore, we obtain

$$v_o(t) = 2\frac{R/LC}{[(1/LC)^2 + (R/L)^2 0]^{1/2}}$$

$$+ 3\frac{R/LC}{[(-25+1/LC)^2 + (R/L)25]^{1/2}} \sin\left(5t - \tan^{-1}\left[\frac{5R/L}{1/LC - 25}\right]\right)$$

■

8.7 Pole location and the stability of LTI systems

The physical idea of stability is closely related to a bounded system response to a sudden disturbance or input. If the system is displaced slightly from its equilibrium state, several different behaviors are possible. If the system remains near the equilibrium state, the system is said to be **stable**. If the system tends to return to the equilibrium state or tends to a bounded or limited state, it is said to be **asymptotically stable**. Here, it should be noted that the stability can be examined by studying a system either through it impulse response $h(t)$ or through its Lapalace-transformed system $H(s)$.

Let us assume that we can expand the transfer function in terms of its roots as follows:

$$H(s) = \frac{A_1}{s - s_1} + \frac{A_2}{s - s_2} + \cdots + \frac{A_k}{s - s_k} + \cdots + \frac{A_n}{s - s_n} \qquad (8.65)$$

The time response for an applied impulse due to the kth pole will be of the form $A_k e^{s_k t}$. Thus, the nature of the response will depend on the location of the roots s_k in the complex s-plane. Because the controlling differential equation that describes the system has real coefficients, the roots either are real or, if complex, will occur in complex conjugate pairs. Three general cases exist that depend intimately on the **order** and **location** of the poles s_k in the s-plane. These are:

1. The point representing s_k lies to the left of the imaginary axis in the s-plane.
2. The point representing s_k lies on the $j\omega$-axis.
3. The point representing s_k lies to the right of the imaginary axis in the s-plane.

We examine each of these alternatives for both simple-order and higher-order poles.

Simple-order poles

Case I

The root is a real number $s_k = \sigma_k$, and it is located on the negative real axis of the s-plane. The response due to this root will be of the form

$$\text{response} = A_k e^{\sigma_k t} \qquad \sigma_k < 0 \qquad (8.66)$$

This expression indicates that after a lapse time, the response will be vanishingly small.

For the case when a pair of complex conjugate roots exists, the response is given by

$$\text{response} = A_k e^{s_k t} + A_k^* e^{s_k^* t} \qquad (8.67)$$

The response terms can be combined, noting that $A_k = a + jb$ and $s_k = \sigma_k + j\omega_k$

$$\text{response} = (a+jb)e^{(\sigma_k+j\omega_k)t} + (a-jb)e^{(\sigma_k-j\omega_k)t}$$

or

$$\text{response} = 2\sqrt{a^2+b^2}\, e^{\sigma_k t} \cos(\omega_k t + \phi_k), \quad \phi_k = \tan^{-1}(b/a), \quad \sigma_k < 0 \qquad (8.68)$$

This response is a damped sinusoid, and it ultimately decays to zero.

Case II

The point representing s_k lies in the imaginary axis. This condition is a special case of Case I, but now $\sigma_k = 0$. The response for complex conjugate poles (see (8.68)) is

$$\text{response} = 2\sqrt{a^2+b^2}\, \cos(\omega_k t + \phi_k), \quad \phi_k = \tan^{-1}(b/a), \quad \sigma_k = 0 \qquad (8.69)$$

Observe that there is no damping, and the response is thus a sustained oscillatory function. Such a system has a bounded response to a bounded input, and the system is defined as **stable** even though it is oscillatory.

Case III

The point representing s_k lies in the right half of the s-plane. The response function will be of the form

$$\text{response} = A_k e^{\sigma_k t} \qquad \sigma_k > 0 \qquad (8.70)$$

for real roots and

$$\text{response} = 2\sqrt{a^2+b^2}\, e^{\sigma_k t} \cos(\omega_k t + \phi_k), \quad \phi_k = \tan^{-1}(b/a), \quad \sigma_k > 0 \qquad (8.71)$$

for complex conjugate roots. Because both functions increase with time without limit, even for bounded inputs, the system for these functions is said to be **unstable**.

Chapter 8: Laplace transform

Multiple-order poles

We now examine the situation when multiple-order poles exist. The following cases are examined.

Case I

Multiple real poles exist in the left half of the s-plane. As previously discussed (see Table 8.3.1), a second-order real pole (two repeated roots) gives rise to the response function

$$\text{response} = (A_{k1} + A_{k2}t)e^{\sigma_k t} \qquad \sigma_k < 0 \qquad (8.72)$$

Because σ_k is negative and because the exponential decreases faster than the linearly increasing time, the response eventually becomes zero. The system with such poles is stable.

Case II

Multiple poles exist on the imaginary axis. The response function is made up of the responses due to each pair of poles, and it is

$$\text{response} = (A_{k1} + A_{k2}t)e^{j\omega_k t} + (A_{k1}^* + A_{k2}^*t)e^{-j\omega_k t} \qquad (8.73)$$

Following the procedure discussed above for the simple complex poles, this result can be written

$$\text{response} = 2\sqrt{a^2 + b^2}\cos(\omega_k t + \phi_k) + 2\sqrt{c^2 + d^2}\,t\cos(\omega_k t + \theta_k) \qquad (8.74)$$

$$\phi_k = \tan^{-1}(b/a) \qquad \theta_k = \tan^{-1}(d/c)$$

Because the second term is oscillatory and increases linearly with time, the system is unstable.

Case III

Multiple roots exist in the right hand of the s-plane. For a double real root, for example, the solution in this case will be

$$\text{response} = (A_{k1} + A_{k2}t)e^{\sigma_k t} \qquad \sigma_k > 0 \qquad (8.75)$$

For complex roots, the solution will be

$$\text{response} = e^{\sigma_k t}[2\sqrt{a^2 + b^2}\cos(\omega_k t + \phi_k) + 2\sqrt{c^2 + d^2}\,t\cos(\omega_k t + \theta_k)] \qquad (8.76)$$

In both cases, owning to the exponential factor, the response increases with time and the system is unstable.

Note:

1. *A system with simple poles is unstable if one or more poles of its transfer function appear in the right half of the s-plane.*
2. *A system whose transfer function has simple poles is stable when all of the poles are in the left half of the s-plane and on its boundary (imaginary axis).*
3. *A system with multiple poles is unstable if one or more of its poles appear on the imaginary axis or in the right half of the s-plane.*
4. *When all of the multiple poles of the system are confined to the left-hand s-plane, the system is stable.*

The impulse response of a system and the location of its poles are shown in Figure 8.7.1.

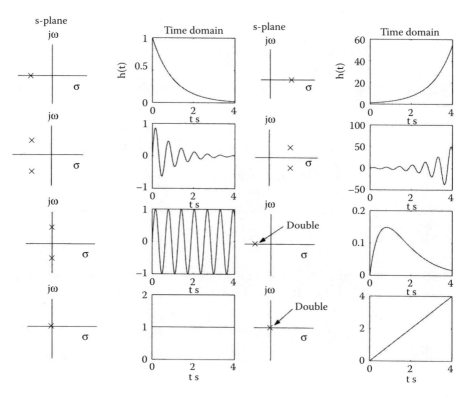

Figure 8.7.1 Location of poles and corresponding impulse response of the system.

*8.8 Feedback for linear systems

The feedback concept is one of the most important engineering discoveries of the last century. In general, the introduction of feedback is to adjust the output so that it coincides with the desired one. The following are some of the features that can be achieved by using feedback.

1. Stabilization of the system is possible.
2. The system output sensitivity may be reduced due to the system parameter variations.
3. Reduction of input disturbance to the system is possible.
4. Feedback improves the system transient response.
5. Feedback improves the steady-state output.

Cascade stabilization of systems

For cascaded systems, it is possible, by selecting the appropriate transfer functions, to cancel zeros or poles, thus producing stabilization. It was found in Chapter 2 (see Figure 2.5.4) that the combined transfer function of two systems in cascade is given by $H(s) = H_1(s)H_2(s)$. If the first system transfer function is of the form $H_1(s) = s/(s - 1.2)$, we observe that the pole is on the right side of the s-plane, which indicates a nonstable system. Hence, if we add in series a second system with a transfer function of the form $H_2(s) = (s - 1.2)/s$, the combined transfer function will be $H(s) = 1$. Thus, we succeeded in creating a stable system because the ROC is the whole s-plane. In practice, however, most of the times we cannot create a system with a pole having the exact value that is needed for stabilization. If the first system, for example, is of the form $H_1(s) = s/(s - 1.2001)$, then the total transfer function is

$$H(s) = H_1(s)H_2(s) = \frac{s}{s-1.2001} \frac{s-1.2}{s} = \frac{s-1.2}{s-1.2001}$$

and indicates an unstable combined system.

In communications, for example, a signal leaves a transmitter (antenna), propagates through space, and reaches the receiver (cell phone). During transmissions, the information-carrying signal is distorted by the transfer function of the space it traveled, known as **channel**. One remedy would be to create a system in front of the receiver that would be the reciprocal of the space transfer function. This system is known as an **equalizer** and acts as a **compensator** that reduces the distortion or completely eliminates it. For example, let the channel transfer function be of the form

$$H_1(s) = \frac{s}{s+0.5}$$

The equalizer must have a transfer function of the form

$$H_2(s) = \{H_1(s)\}^{-1} = \frac{s+0.5}{s} = 1 + 0.5\frac{1}{s}$$

The impulse response of the above filter is $h_2(t) = \delta(t) + 0.5u(t)$. This filter is stable and can serve as an equalizer.

Most of the time, the functional form of the communication channel is not known. One way to circumvent this difficulty, especially in slow-changing channels, is to transmit a known signal and then design the appropriate inverse filter that eliminates the channel disturbance. The sequence of the known signals is called the **training sequence**.

Parallel composition

The transfer function of two systems in parallel is given by (see also Figure 2.5.4)

$$H(s) = H_1(s) + H_2(s)$$

The transfer function of the system shown in Figure 8.8.1 is

$$Y(s) = D(s) + [H_1(s) + H_2(s)]V(s)$$

where $D(s)$ is the transformed desired signal and $V(s)$ is the transformed noise that must be eliminated. If we can create a filter such that $H_2(s) = -H_1(s)$, then the desired signal is detected. This type of process is known as **noise cancellation**. One of the important tasks of engineers is to build **noise cancellers** that suppress or remove unwanted signals (noise). One common case is when a person speaks in a microphone, as a person in a convention center or a pilot in the cockpit, and noise is also introduced (the noise of the attendees or the engine noise). Figure 8.8.2a shows a physical setup of

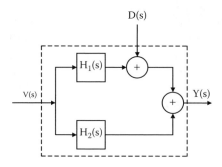

Figure 8.8.1 A noise-canceling system.

Chapter 8: Laplace transform

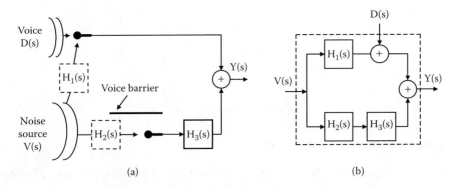

Figure 8.8.2 Noise cancellation: (a) physical setup and (b) diagrammatic representation.

the situation discussed, and Figure 8.8.2b shows the diagrammatic representation of the system. From the figure we obtain

$$V(s)H_1(s) + D(s) + V(s)[H_2(s)H_3(s)] = Y(s)$$

or we can write

$$Y(s) = V(s)H_1(s) + D(s) + V(s)H_2(s)\left(-\frac{H_1(s)}{H_2(s)}\right) = D(s)$$

From the above results we observe that if we introduce in the second parallel branch a system ($H_3(s)$) with the transfer function $-H_1(s)/H_2(s)$, we are able to eliminate the noise and receive the desired signal. A common use of such a scheme is in telephone systems. At the receiving and sending ends, the two-wire transmission changes into a four-wire transmission by a device known as the **hybrid**. The imperfection of the hybrid devices creates echoes (we hear our own voice) that we sometimes encounter during conversation.

Feedback stabilization

Figure 8.8.3 shows a simple negative feedback system with unity return (see also Figure 2.5.2 and Figure 2.5.4). Let the **open-loop transfer function** be of the form

$$G(s) = G_1(s)G_2(s) = \frac{1}{(s-0.6)(s+1.2)}$$

This system is unstable due to the pole at +0.6. This pole creates in the time domain an exponential function of the form $e^{0.6t}$, which becomes unbounded as $t \to \infty$. If the open system is modified to a negative feedback system, as

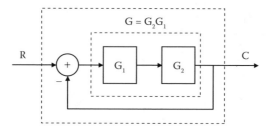

Figure 8.8.3 Unity feedback system.

shown in Figure 8.8.3, the total transfer function (**closed-loop transfer function**) is

$$T = \frac{C}{R} = \frac{1}{s^2 + 0.6s + 0.28}$$

The poles of this new system are $-0.3000 + j0.4359$, $-0.3000 - j0.4359$. Because the real part of both poles is negative, this indicates a sinusoidal decaying function in the time domain. Observe also that the closed-loop poles are different than those of the open loop.

Note: *An unstable system may become stable by introducing a negative feedback path.*

Sensitivity in feedback

The input–output relation for a unit feedback system (see Figure 8.8.3) is

$$C(s) = \frac{G(s)}{1 + G(s)} R(s)$$

If we assume that there is an incremental change of the open-loop transfer function by $\Delta G(s)$, there will be an incremental change of the output. Therefore, we will have the following relation:

$$C(s) + \Delta C(s) = \frac{G(s) + \Delta G(s)}{1 + G(s) + \Delta G(s)} R(s)$$

Proceeding to rearrange the above equation, we find

$$\Delta C(s) = \frac{G(s) + \Delta G(s)}{1 + G(s) + \Delta G(s)} R(s) - C(s) = \frac{G(s) + \Delta G(s)}{1 + G(s) + \Delta G(s)} R(s) - \frac{G(s)}{1 + G(s)} R(s)$$

$$= \frac{\Delta G(s)}{[1 + G(s) + \Delta G(s)][1 + G(s)]} R(s)$$

Chapter 8: Laplace transform

Since, in general, $|\Delta G(s)| \ll |G(s)|$, the last expression can be approximated as follows

$$\Delta C(s) \cong \frac{\Delta G(s)}{[1+G(s)][1+G(s)]} R(s) = \frac{G(s)\Delta G(s)}{G(s)[1+G(s)][1+G(s)]} R(s)$$

$$= \frac{\Delta G(s)}{G(s)} \frac{1}{1+G(s)} C(s)$$

We can now write the above equation in the form

$$\frac{\Delta C(s)}{C(s)} = \frac{1}{1+G(s)} \frac{\Delta G(s)}{G(s)} = S(s)\frac{\Delta G(s)}{G(s)} \quad (8.77)$$

where $S(s)$ is called the system **output sensitivity function**. Since $|S(s)| = |1/[1 + G(s)]| < 1$, the variation of the open-loop transfer function due to the perturbation of its parameter will reduce the perturbation of the output by the factor of $|S(s)| < 1$.

Rejection of disturbance using feedback

Figure 8.8.4 shows a feedback configuration that completely eliminates the disturbance $V(s)$. Based on the design shown and referring to Figure 2.5.2, we obtain

$$C(s) = V(s) + G(s)[R(s) - C(s)]$$

The above equation can be written in the form

$$C(s) = \frac{G(s)}{1+G(s)} R(s) + \frac{1}{1+G(s)} V(s) = \frac{G(s)}{1+G(s)} R(s) + S(s)V(s) \quad (8.78)$$

It is apparent that for $|S(s)V(s)| \cong 0$ in the frequency range of $V(s)$, the feedback configuration is able to diminish and eliminate the disturbance.

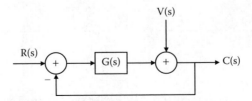

Figure 8.8.4 Feedback system with an external disturbance.

Step response

The voltage across the capacitor of an RLC series circuit to a step input is given by (8.43) and is also given below with the following substitutions:

$$E = 1, \omega_r^2 = K, 2\alpha = 6$$

$$V_{zs} = \frac{1}{s}\frac{K}{s^2 + 6s + K}$$

The second factor of the above equation is produced by the feedback system shown in Figure 8.8.5a. The voltage responses for $K = 5, 15$, and 60 are shown in Figure 8.8.5b. We observe that we can shorten the rise time with the drawback in creating overshoots. In this case, **the percent overshoot** is about 25%. The **rise time** is defined as the time required for the response to rise

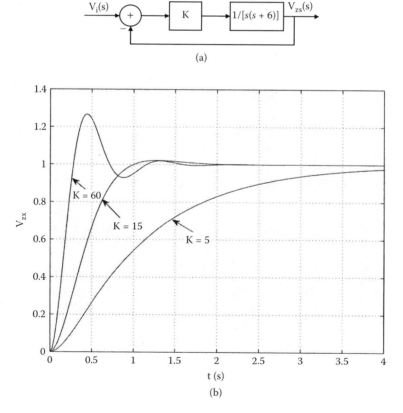

Figure 8.8.5 (a) Block diagram of a second-order system. (b) The unit step function.

Chapter 8: Laplace transform

from 10 to 90% of the steady-state value. From Figure 8.8.5b we can categorize the step responses in three main categories. For this case, we have (a) $K < 5$ is said to be an **overdamped** case, (b) $K = 15$ is said to be **critically damped**, and c) $K > 15$ is said to be **underdamped**.

Book MATLAB m-file: step8_8_1

```
%step8_8_1 is an m file to produce one of the
%curves of Figure 8.8.5b;n=numerator coefficient vector;
%d=denominator coefficient vector;H(s)=15/(s^2+6s+15);
n=15;d=[1 6 15];
sys=tf(n,d);
[y,t]=step(sys,4);%4 is the desired end of time axis;
plot(t,y);
text(3,1.2,'a)');%this sets the text 'a)' at point (3,1.2);
```

Proportional controllers

Let a voltage source be attached to an inductor. The relationship is well known to be equal to

$$L\frac{di(t)}{dt} = v(t) \quad \text{or} \quad \frac{di(t)}{dt} = \frac{1}{L}v(t) \quad \text{also } I(s) = \frac{1}{Ls}V(s) \quad \text{or} \quad H_2(s) = \frac{I(s)}{V(s)} = \frac{1}{Ls}$$

where $i(t)$ is the current in the circuit at time t, $v(t)$ is the voltage source, and L is the inductance. Based on the above relationships, we can create a unity feedback system made of the **plant** and **controller**. The feedback configuration is shown in Figure 8.8.6, where G_2 is called the plant and G_1 is called the controller. This particular feedback controller is known as the **proportional controller**. Since the feedback loop transfers the output I to the summation

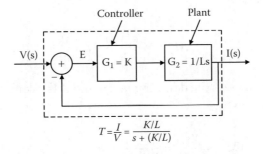

Figure 8.8.6 Feedback proportional controller with gain K.

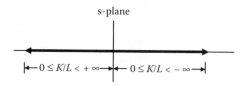

Figure 8.8.7 Root locus of a proportional controller.

point, it is apparent that the **error signal** E will be zero if the input is equal to the output. The total transfer function for the feedback system is

$$T(s) = \frac{KG_2(s)}{1+KG_2(s)} = \frac{K/L}{s+(K/L)} \qquad (8.79)$$

with a pole at $-K/L$. To obtain the range of values of the gain K must take to create a bounded feedback system, we can vary K and plot the values of the root in the s-plane. For the present case, as K varies from $-\infty$ to $+\infty$ the pole ranges from $+\infty$ to $-\infty$. The trace of the values of the roots is known as the **root locus**. Figure 8.8.7 shows the root locus for this case. If the system had additional roots, each root would have produced its own root locus (**branch**).

If the input is a step function, the output in LT form and in the time domain is

$$I(s) = \frac{K/L}{s+(K/L)} \frac{1}{s} = -\frac{1}{s+(K/L)} + \frac{1}{s} \quad \text{or} \quad i(t) = u(t) - e^{-(K/L)t}u(t)$$

We observe that at time infinity the current is equal to its input. The first term is known as the **steady-state response** $i_{ss}(t)$. Hence, the **error signal** (**tracking error**), $i(t) - u(t) = e(t) = -\exp[-(K/L)t]$, decreases faster the larger the gain K values are.

Example 8.8.1: The position of the mass shown in Figure 8.8.8a is found by solving the equation

$$M\frac{dv(t)}{dt} + Dv(t) = M\frac{d^2x(t)}{dt^2} + D\frac{dx(t)}{dt} = Af(t)$$

where M is the mass D is the damping factor and A is a constant. The transfer function of this system is

$$G_2(s) \triangleq \frac{X(s)}{F(s)} = \frac{A}{Ms^2+Ds} = \frac{A/M}{s[s+(D/M)]}$$

Chapter 8: Laplace transform

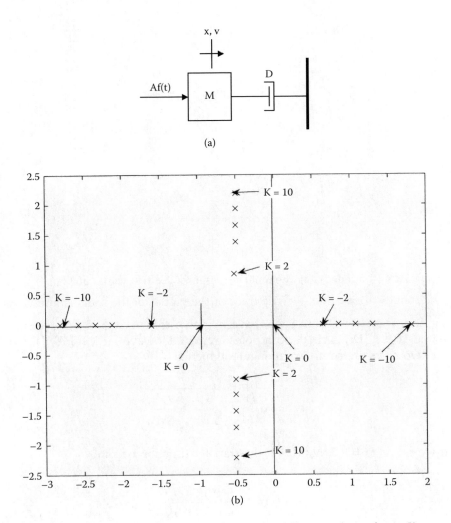

Figure 8.8.8 (a) A mechanical system. (b) Roots for different values of gain K.

The step response of this system is unstable due to the double root at $s = 0$. If we use a proportional controller with unit feedback loop, as shown in Figure 8.8.6, the total transfer function is

$$T(s) \triangleq \frac{X(s)}{F(s)} = \frac{KG_2(s)}{1 + KG_2(s)} = \frac{KA}{Ms^2 + Ds + KA}$$

$$s_{1,2} = -\frac{D}{2M} \pm \sqrt{\frac{D^2}{4M^2} - \frac{KA}{M}} \equiv \text{roots of the characteristic equation } Ms^2 + Ds + KA$$

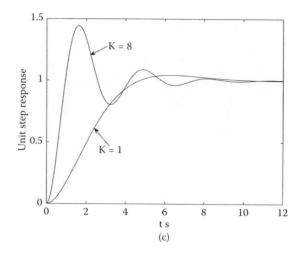

Figure 8.8.8 (continued). (c) The unit step response for two values of K.

If K is positive, both roots have negative real parts and the system is stable. If $K < [D^2/(4MA)]$, both poles are real. However, if $K > [D^2/(4MA)]$, the two roots are complex conjugate, which is

$$s_{1,2} = -\frac{D}{2M} \pm j\sqrt{\frac{KA}{M} - \frac{D^2}{4M^2}}$$

If we set $A/M = 0.5$, $D/M = 1$, the transfer function becomes

$$T(s) = \frac{0.5K}{s^2 + 0.2s + 0.5K}$$

Figure 8.8.8b shows the roots for different values of the gain K and Figure 8.8.8c the unit step response of the feedback system.

Book MATLAB m-file: ex8_8_1

```
%m-file for plotting the roots of the total
%transfer function of the Ex 8.8.1; name:ex8_8_1;
for k=1:5
    z(:,k)=roots([1 1 0.5*2*k]);
end;
plot(real(z(1,:)),imag(z(1,:)),'xk')%'xk'=plot black x's;
```

Chapter 8: Laplace transform

```
hold on;
plot(real(z(2,:)),imag(z(2,:)),'xk');
for m=1:5
    z1(:,k)=roots([1 1 -0.5*2*k]);
end;
plot(real(z1(1,:)),imag(z1(1,:)),'xk');
hold on;
plot(real(z1(2,:)),imag(z1(2,:)),'xk');■
```

We must mention that the proportional controller does not always produce a closed-loop stability.

Proportional integral differential (PID) controllers

Before we proceed with an example, we must mention that a proportional controller is just a constant, as it was developed above. If the controller has in integrator, its transform domain representation is proportional to 1/s. And similarly, the differentiator is proportional to s. This type of controller is basically a generalization of a proportional controller.

Example 8.8.2: The KVL of a series *RLC* relaxed circuit and the equivalent differential equation with respect to charge are

$$L\frac{di(t)}{dt} + Ri(t) + \frac{1}{C}\int i(t)dt = v(t) \quad \text{or} \quad L\frac{d^2q(t)}{dt^2} + R\frac{dq(t)}{dt} + \frac{1}{C}q(t) = v(t) \quad (8.80)$$

The LT of the second equation and the transfer function are

$$Ls^2Q(s) + RsQ(s) + (1/C)Q(s) = V(s); \quad G_2(s) \triangleq \frac{Q(s)}{V(s)} = \frac{1}{Ls^2 + Rs + (1/C)} \quad (8.81)$$

From the results of Problem 8.8.7 we observe that the transient disappears to about 10 units of time, which may be very slow for some applications. In addition, the overshoot may be also objectionable.

To alleviate these objections, we can use a **PID** controller, which is of the form

$$G_1(s) = K_1 + \frac{K_2}{s} + K_3 s = \frac{K_3 s^2 + K_1 s + K_2}{s} \quad (8.82)$$

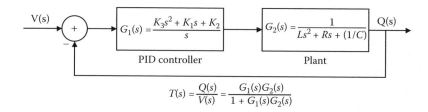

Figure 8.8.9 The RLC series circuit with a PID controller in a unit feedback form.

The feedback controller is shown in Figure 8.8.9. The configuration indicates that by selecting the values of K_i's, we can create the following individual controllers: (a) **proportional** controller, $K_3 = K_2 = 0$, (b) **integral** controller, $K_3 = K_1 = 0$, and (c) **derivative** controller, $K_1 = K_2 = 0$. ∎

Example 8.8.3: To evaluate the PID controller, we set $L = R = 1$ and $1/C = 1.30$. The total transfer function (closed-loop transfer function) is

$$T(s) = \frac{G_1(s)G_2(s)}{1+G_1(s)G_2(s)} = \frac{K_3 s^2 + K_1 s + K_2}{s^3 + (1+K_3)s^2 + (1.3+K_1)s + K_2} \tag{8.83}$$

Setting $K_1 = 8$ and $K_2 = K_3 = 0$, the step response of the P controller is given by

$$Q_P(s) = \frac{1}{s}\frac{8}{s^2 + s + 9.3} \tag{8.84}$$

Setting $K_1 = K_3 = 8$ and $K_2 = 0$, the step response of the PD controller is given by

$$Q_{PD}(s) = \frac{1}{s}\frac{8s+8}{s^2 + 9s + 9.3} \tag{8.85}$$

Finally, with $K_1 = K_3 = 8$ and $K_2 = 4$, the step response of the PID controller is given by

$$Q_{PID}(s) = \frac{1}{s}\frac{8s^2 + 8s + 4}{s^3 + 9s^2 + 9.3s + 4} \tag{8.86}$$

Figure 8.8.10 shows the step responses for the above three cases and the open-loop response with $L = R = 1$ and $(1/C) = 1.3$. ∎

Chapter 8: Laplace transform

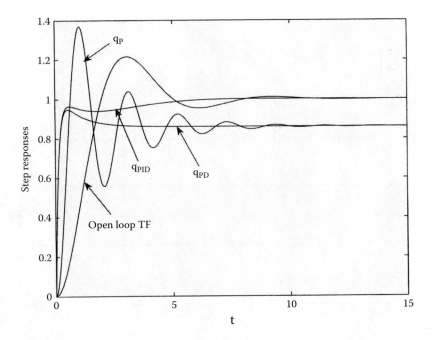

Figure 8.8.10 Step responses illustrating Example 8.8.3.

*8.9 Bode plots

Bode plots represent the magnitude and phase vs. frequency based on 10 base log-log scales. The data are plotted as follows:

$$\text{decibels (dB)} = 20\log|H(j\omega)| \text{ vs. } \log\omega, \quad \log(.) \triangleq \log_{10}(.) \tag{8.87}$$

$$\theta(\omega) \text{ vs. } \log(\omega) \tag{8.88}$$

Bode plots are extensively used in feedback control studies since they are one several techniques used to specify whether a system is stable.

Bode plots of constants

Based on (8.87), and due to the fact that $C = Ce^{j0}$ and $-C = Ce^{j\pi} = Ce^{-j\pi}$, we obtain

$$dB = 20\log|C| = \begin{cases} \text{positive number} & |C| > 1 \\ \text{negative number} & |C| < 1 \end{cases}$$

$$\theta = \begin{cases} 0 & C > 1 \\ \pi = -\pi & C < 1 \end{cases} \tag{8.89}$$

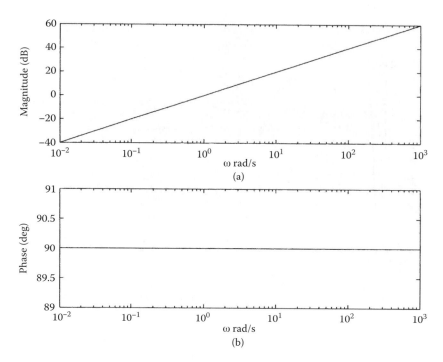

Figure 8.9.1 Magnitude and phase Bode plots for a differentiator.

For example, the number −0.5 gives $20\log(0.5) = -6.0206$ and $\theta = \pi$ or $-\pi$. This means that the plot will show a straight line at height −6.0206.

Bode diagram for differentiator

The frequency transfer function, amplitude, and phase of a differentiator are

$$H(j\omega) = j\omega = \omega\, e^{j\frac{\pi}{2}}; \; |H(j\omega)|_{dB} = 10\log\sqrt{(j\omega)(-j\omega)} = 20\log\omega; \; \theta(\omega) = \frac{\pi}{2} \quad (8.90)$$

Since $s = j\omega$, the transfer function of a differentiator is $H(s) = s/1$. Based on the Book MATLAB m-file given below, Figure 8.9.1 was found.

Book MATLAB m-file: bode_differentiator

```
%m-file: bode_differentiator
num=[1 0];den=[0 1];
sys=tf(num,den);
w=logspace(-2,3,100);%(-2,3,100)=create 100 points
  %along x-axis from 10^{-2}=0.01 to 10^{3}=1000
  %in a log scale;
```

Chapter 8: Laplace transform

```
[ma,ph]=bode(sys,w);
mag=reshape(ma,[100,1]);%since ma=1x1x100 it reshapes
  %to 100 by 1 vector;
phg=reshape(ph,[100,1]);
subplot(2,1,1);semilogx(w,20*log10(mag),'k');
xlabel('\omega rad/s');ylabel('Magnitude (dB)');
subplot(2,1,2);semilogx(w,phg,'k');
xlabel('\omega rad/s');ylabel('Phase (deg)');
```

Bode diagram for an integrator

The frequency transfer function, amplitude. and phase of an integrator are $(\log(1) = 0)$:

$$H(j\omega) = \frac{1}{j\omega}, \quad |H(j\omega)|_{dB} = 20\log\frac{1}{\omega} = -20\log\omega, \quad \theta(\omega) = -\frac{\pi}{2} \quad (8.91)$$

If we plot the Bode diagram of an integrator, the magnitude line has the opposite inclination to that of the differentiator.

Bode diagram for a real pole

Let the transfer function of a system be given by

$$H(s) = \frac{p}{p+s} \quad \text{or} \quad H(j\omega) = \frac{p}{p+j\omega} = \frac{1}{1+j\frac{\omega}{p}} = \frac{1}{\sqrt{1+\left(\frac{\omega}{p}\right)^2}} e^{-j\tan^{-1}\left(\frac{\omega}{p}\right)} \quad (8.92)$$

$$= |H(j\omega)| e^{j\theta(\omega)}$$

The magnitude and phase of the above transfer function for a Bode plot are

$$|H(j\omega)| = 20\log\frac{1}{\sqrt{1+\left(\frac{\omega}{p}\right)^2}} = 20\log(1) - 20\log\sqrt{1+\left(\frac{\omega}{p}\right)^2} = -20\log\sqrt{1+\left(\frac{\omega}{p}\right)^2}$$

$$\theta(\omega) = -\tan^{-1}\left(\frac{\omega}{p}\right) \quad (8.93)$$

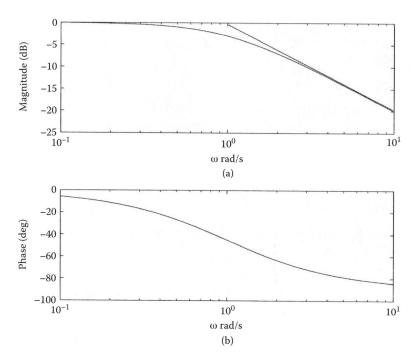

Figure 8.9.2 Magnitude and phase Bode plots for $H(j\omega) = 1/(1+j\omega)$.

For $\omega \ll p$, the magnitude is about equal to $-20\log(\sim 1) = 0$, a constant. For the same case, the angle is equal to $-\tan^{-1}(\sim 0) = \sim 0$. On the other hand, if $\omega \gg p$, the magnitude is about equal to $-20\log(\omega/p) = -20\log(\omega) + 20\log(p)$, which is a straight line with a negative slope of -20 dB/decade. The Bode plots are shown in Figure 8.9.2 as curves, and the approximations are shown as straight lines for the particular case when $p = 1$.

For the case of a transfer function having a real zero (see Problem 8.9.1), we expect the curves to be reflected with respect to the frequency axis since the zero is the reciprocal of the pole.

Example 8.9.1: Find the transfer function for the system shown in Figure 8.9.3. Plot the s-plane of zeros and poles, the locus, and the Bode plots.

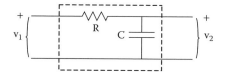

Figure 8.9.3 The integrating circuit.

Chapter 8: Laplace transform

Solution: The voltage ratio system function is

$$\frac{V_2(s)}{V_1(s)} \triangleq H(s) = \frac{\dfrac{1}{Cs}}{R+\dfrac{1}{Cs}} = \frac{1}{1+RCs} = \frac{\dfrac{1}{RC}}{s+\dfrac{1}{RC}}, \qquad H(j\omega) = \frac{\dfrac{1}{RC}}{j\omega+\dfrac{1}{RC}}$$

There are three critical ranges to be examined:

1. For low frequencies, $H(j\omega) \to 1$, with phase angle approximately zero.
2. For high frequencies, $H(j\omega) \to 1/(j\omega RC)$, with angle approximately $-90°$.
3. For frequency $\omega = 1/RC$, $H(j\omega) = 1/(1+j1) = 0.707 \angle -45°$.

The appropriate figures are given in Figures 8.9.4a to c. The following Book MATLAB m-files can be used to produce Figure 8.9.4c.

Book MATLAB m-file: ex8_9_1

```
%m-file for Ex8.9.1:ex8_9_1
num=[0 1];den=[1 1];
sys=tf(num,den);
bode(sys,'k',{0.1 100});
```

Book MATLAB m-file: ex8_9_1a

```
%another way to find the Body plots;
%m file: ex8_9_1a
num=[0 1]; den=[1 1];
w=0.1:0.01:100;
[mag,pha,w]=bode(num,den,w);
subplot(2,1,1);semilogx(w,20*log10(mag),'k');
xlabel('\omega rad/s');ylabel('Magnitude (dB)');
subplot(2,1,2);semilogx(w,pha,'k');
xlabel('\omega rad/s');ylabel('Phase (deg)');
```

∎

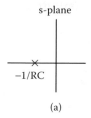

(a)

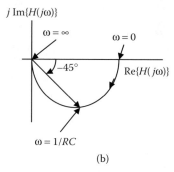

(b)

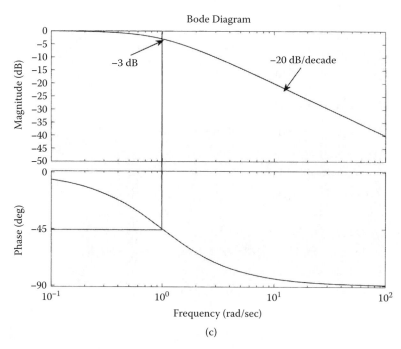

(c)

Figure 8.9.4 Illustration of Example 8.9.1.

Chapter 8: Laplace transform

Important definitions and concepts

1. One-sided Laplace transform
2. Laplace transform properties
3. Transfer function of systems
4. Inverse Laplace transform
5. Impulse response of systems using Laplace transform
6. Zero-state and zero-input solutions of differential equations
7. Frequency response of linear time-invariant systems
8. Filtering of signals by systems
9. Stability and location and order of poles
10. Cascade stabilization
11. Equalizer
12. Channel
13. Compensators
14. Parallel stabilization
15. Feedback stabilization
16. Sensitivity in feedback systems
17. Negative and positive feedback systems
18. Output sensitivity function in feedback systems
19. Step response
20. Percent overshoot
21. Rise time of a response
22. Overdamped response
23. Critically damped response
24. Proportional controller
25. Error signal in feedback control
26. Steady-state response
27. Proportional integral differential controllers
28. Bode plots

Chapter 8 Problems

Section 8.1

1. Deduce the LT of the following functions:

    a. Entries 5, 6, 9, and 14 in Table 8.3.1.
    b. $t^2 + 2t + 1$, $(1 + \cos 4t)/2$, $t \sin \omega t$ for $t \geq 0$.

2. Find the LT and the region of convergence of the following functions for $t \geq 0$:

$$2 + 3t, \quad 4e^{-0.1t}, \quad \sinh 2t, \quad 1 + \sin t, \quad e^{2t}$$

3. Find the LT of the following functions for $t \geq 0$:

$$2 - 8t^3$$

$$t \cos 2t$$

$$e^t \cos t$$

$$e^{-t} \cosh t$$

4. Find the LT of the following functions for $t \geq 0$:

$$1 + e^{-2t}, \quad u(t) - u(t-1)$$

$$\cos 2t - e^t \sin 3t$$

Section 8.2

1. a. Find the LT of the following functions for $t \geq 0$:

$$\frac{de^{-3t}}{dt}, \quad e^{-2t}u(t), \quad e^{-(t/2)}$$

b. Verify your results in a by using the LT properties.
2. Find the LT of the functions shown in Figure P8.2.2.

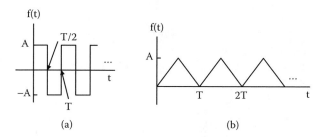

Figure P8.2.2

3. Apply the initial value and final value properties to the following functions:

$$\frac{s}{s^2+3} \qquad \frac{s^2+s+3}{s^2+3} \qquad \frac{s}{s^2-b^2} \qquad \frac{s+a}{(s+a)^2+b^2}$$

4. Find the LT of the following functions for $t \geq 0$:

$$e^{-t}u(t-2) + \frac{dg(t)}{dt} \qquad g(t) + \int_0^t e^{-|x|} dx$$

5. If $f(t)$ is an even function, find the relationship between $F(s)$ and $F(-s)$. Repeat for the case when $f(t)$ is an odd function.
6. Determine the LT of the functions shown in Figure P8.2.6.

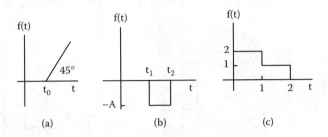

Figure P8.2.6

7. Find the LT of the functions shown in Figure P8.2.7 and then deduce the LT of the related $h(t)$ using the appropriate Laplace property for each case.

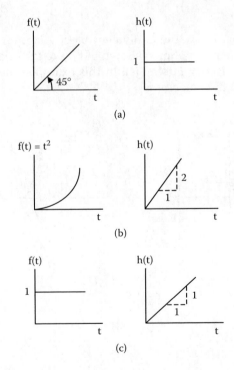

Figure P8.2.7

8. Find the LT of the following functions for $t \geq 0$:

$$e^{-at} \cos bt, \quad te^{-at}, \quad t \cos bt$$

Section 8.3

1. Draw the equivalent block diagram representation of the systems shown in Figure P8.3.1 and find the transfer function, as indicated, for each.

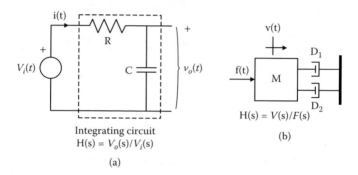

Figure P8.3.1

2. **Microphone.** Draw the equivalent block diagram representation of the system (a microphone) shown in Figure P8.3.2 and find the transfer function $H(s) = E_o(s)/F_a(s)$. In this example, the voltage is designated e and the velocity v.

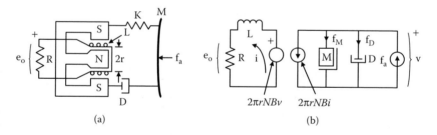

Figure P8.3.2

3. Find the equivalent block diagram of the systems shown in Figure P8.3.3 and determine the transfer function, as indicated, in each case.

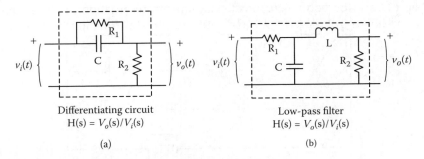

Differentiating circuit
$H(s) = V_o(s)/V_i(s)$

(a)

Low-pass filter
$H(s) = V_o(s)/V_i(s)$

(b)

Figure P8.3.3

4. Determine the transfer function $H(s) = C(s)/R(s)$ for the systems shown in Figure P8.3.4.

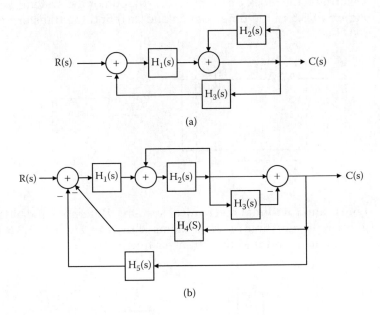

Figure P8.3.4

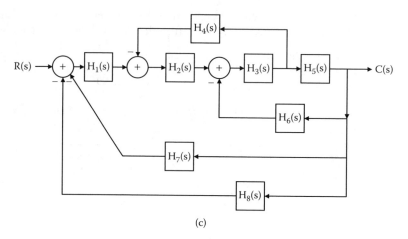

Figure P8.3.4 (continued).

5. Determine the block diagram representation of the system shown in Figure P8.3.5 in the time domain and also find the transfer function $I(s)/V(s)$.

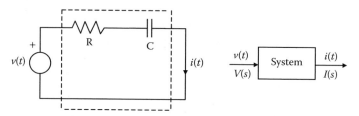

Figure P8.3.5

6. **Linear and rotational mechanical systems**. Determine the block diagram representation of the systems shown in Figure P8.3.6 in the time domain and find their transfer functions.

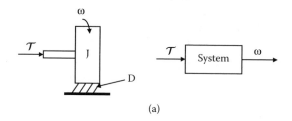

(a)

Figure P8.3.6

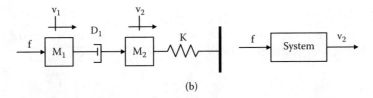

(b)

Figure P8.3.6 (continued).

7. Find the transfer function for the systems shown in Figure P8.3.7.

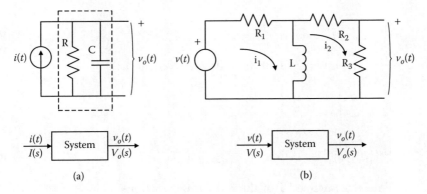

Figure P8.3.7

8. **Linear mechanical system**. Find the transfer function for the system shown in Figure P8.3.8.

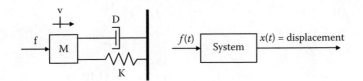

Figure P8.3.8

Section 8.4

1. Find the inverse LT of the following functions by means of partial fraction expansions. Check you result using MATLAB.

   a. $\dfrac{1}{s} - \dfrac{1}{s}e^{-2s}$
   b. $\dfrac{2s+1}{s^2+4s+7}$
   c. $\dfrac{1}{(s+2)^4}$

   d. $\dfrac{s+1}{(s+2)(s+3)^2}$
   e. $\dfrac{s^2+2s+1}{s^2(s+1)}$
   f. $\dfrac{s^2+2s+1}{s^2+3s+5}$

   g. $\dfrac{e^{-2s}}{(s+1)^2(s^2+3)}$

2. Find the inverse LT of the following functions:

   a. $\dfrac{a}{s^2+b^2}$
   b. $\dfrac{2}{s^2-5s+4}$
   c. $\dfrac{2}{(s^2+1)^2}$

   d. $\dfrac{2s-3}{(s+1)^2+25}$
   e. $\dfrac{4}{(s-1)(s^2+1)}$
   f. $\dfrac{s}{s^2-a^2}$

3. Find the inverse LT of the following functions:

   a. $\dfrac{6}{s(s+3)}$
   b. $\dfrac{s^2+2s+5}{(s+3)(s+5)^2}$

   c. $\dfrac{5s+13}{s(s^2+4s+13)}$
   d. $\dfrac{s}{(s^2+9)(s+2)}$

Section 8.5

Note: We must have in mind that $y(0)$ means $y(0+)$ to avoid any inconsistencies when input functions are discontinuous at $t = 0$, such as the step function.

1. Solve the following differential equations by LT methods for $t \geq 0$:

   a. $\dfrac{d^2y(t)}{dt^2} + 3\dfrac{dy(t)}{dt} + 2y(t) = 0$ $y(0) = 5, \dfrac{dy(0)}{dt} = 0$

   b. $\dfrac{d^2y(t)}{dt^2} + 3\dfrac{dy(t)}{dt} + 2y(t) = \delta(t)$ initially relaxed

c. $\dfrac{d^2y(t)}{dt^2} + 5y(t) = \sin 2t + e^{-3t}$ initially relaxed

d. $\dfrac{d^2y(t)}{dt^2} + 3\dfrac{dy(t)}{dt} + 2y(t) = t^2 + 3t$ $y(0) = 2,\ \dfrac{dy(0)}{dt} = -8$

2. Determine the driving point current in the circuits of Figure P8.5.2 for $t > 0$. Assume that these circuits are initially relaxed.

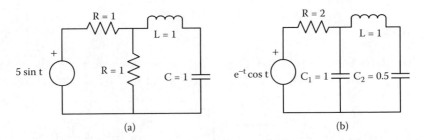

Figure P8.5.2

3. Determine the currents $i_1(t)$ and $i_2(t)$ for $t > 0$ in the network shown in Figure P8.5.3.

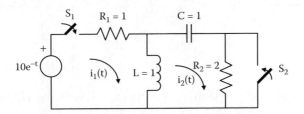

Figure P8.5.3

4. Find the velocity v for the system shown in Figure P8.5.4 for $t > 0$ if $f(0) = V_0$ and $f(t) = \sin(\omega t + \phi)$.

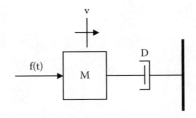

Figure P8.5.4

5. Determine the LT of the output $v_o(t)$ for $t > 0$ of the system shown in Figure P8.5.5 when the input is $v(t) = te^{-2t}$. Use the convolution property.

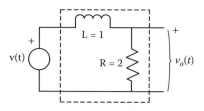

Figure P8.5.5

6. Refer to Figure P8.5.6a. Prove that the portion external to the rectangle can be replaced by the circuit of Figure P8.5.6b for an initially relaxed system.

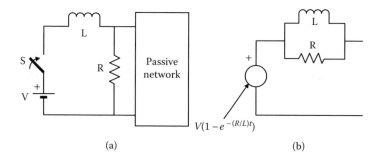

Figure P8.5.6

Hint: Consider the similarity of this result with the Thevenin theorem for the steady state.

7. Determine the impulse response of a series RLC circuit for the following circuit constants:

   a. $R = 4, C = 1, L = 1$
   b. $R = 1, C = 4, L = 1$
   c. $R = 2, C = 0.1, L = 1$

8. Determine the impulse response of the systems shown in Figure P8.5.8.

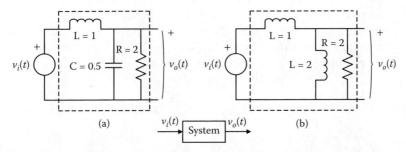

Figure P8.5.8

9. Determine the impulse response of the system shown in Figure P8.5.9. The input is $f(t)$ and the output is $v_2(t)$.

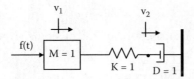

Figure P8.5.9

10. Use LT techniques to find the output voltage of the relaxed circuit shown in Figure P8.5.10 for an input voltage $v_i(t) = e^{-t}$ $t > 0$. Verify your results using the convolution integral method.

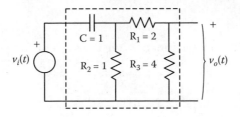

Figure P8.5.10

11. **Ecology; environmental engineering.** Consider two species that co-exist in a given region and that interact in a specific way. Denote the number of each species at time t by $N_1(t)$ and $N_2(t)$, respectively. Assume that the rate of change of their numbers is proportional to the number present. Find the change in their numbers if the proportionality constants are $a = -1$, $b = 0$, $c = 2$, and $d = -1$. Based on these suggestions, the governing equations are:

$$\frac{dN_1(t)}{dt} = aN_1(t) + bN_2(t) \qquad \frac{dN_2(t)}{dt} = cN_1(t) + dN_2(t)$$

$b = 0$ means that the second species does not compete with the first species; $c = 2$ implies that the first species competes with the second. Next, assume that initial numbers were $N_1(0) = 2000$ and $N_2(0) = 1000$. Using the LT method, find the number of each species vs. time.

Section 8.6

1. Find the transfer function for the systems shown in Figure P8.6.1.

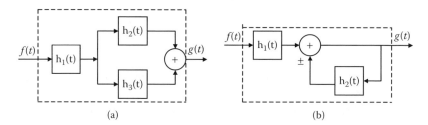

Figure P8.6.1

2. Deduce the transfer function for the system shown in Figure P8.6.2.

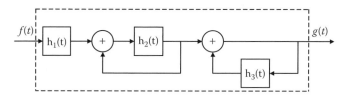

Figure P8.6.2

3. Determine the transfer functions of the systems shown in Figure P8.6.3.

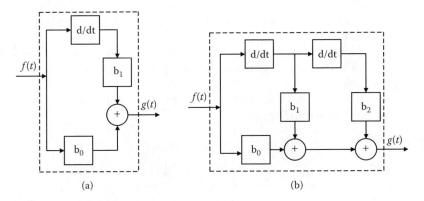

Figure P8.6.3

4. Determine the output of the circuits shown in Figure P8.6.4.

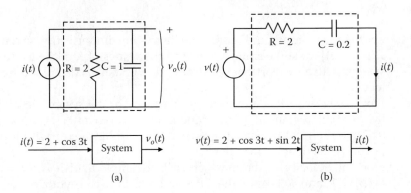

Figure P8.6.4

5. Determine and plot the magnitude filtering characteristics of the system shown in Figure P8.6.5.

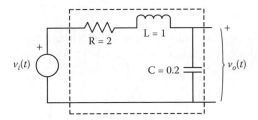

Figure P8.6.5

Section 8.8

1. If the transfer function of a communication channel is $H_1(s) = s^2/[(s-a)(s-a^*)]$, can the inverse function $H_1(s)^{-1}$ be considered an equilazer?
2. Let the open-loop transfer function be $G(s) = 1/[(s-1.1)(s+1.9)]$. Find whether the negative closed-loop feedback system is stable or unstable.
3. Find the condition for eliminating the disturbance for the feedback system shown in Figure P8.8.3.

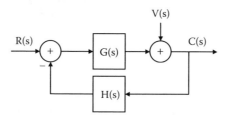

Figure P8.8.3

4. Let the plant shown in Figure 8.8.6 have some imperfection given by $1/[L(s-\varepsilon)]$ for some small $\varepsilon > 0$. Find the values of K of the feedback system with a proportional controller that is stable. Also plot the root locus.
5. Find the steady state and transient response of the system described by (8.79) if the input is the function $v(t) = A\cos\omega_0 t\, u(t)$. Equation (8.79) is the total transfer function of a simple inductor with a proportional feedback controller.
6. Find the velocity of the mass given in Example 8.8.1 if $A/M = 0.5$, $D/M = 1$, and K takes values 0.8 and 0.6. Compare your results to those given in the example. Use the method of partial fraction expansion.
7. Set $L = 1$, $R = 1$, and $(1/C) = 1.3$ in (8.81) and find the unit step function response.

Section 8.9

1. Plot the Bode diagram of the transfer function $H(s) = 1 + s$.
2. Plot the Bode diagrams for the transfer function

$$H(s)\bigg|_{s=j\omega} = \frac{\omega_r^2}{s^2 + 2\varsigma\omega_r s + \omega_r^2}\bigg|_{s=j\omega}$$

Use $\omega_r = 1$ and $\varsigma = 0.1, 0.5$ and 1.5.

Chapter 8: Laplace transform

3. Draw the Bode diagrams for the transfer function and its asymptotes. The transfer function is

$$H(s)\big|_{s=j\omega} = \frac{s-2}{s}$$

Appendix 8.1: Proofs of Laplace transform properties
Linearity
This property is the result of the linear operation of an integral.

Time derivative

$$\mathcal{L}\left\{\frac{df(t)}{dt}\right\} = \int_0^\infty \frac{df(t)}{dt} e^{-st} dt = f(t)e^{-st}\Big|_0^\infty - \int_0^\infty f(t)\frac{de^{-st}}{dt} = -f(0) + s\int_0^\infty f(t)e^{-st} dt$$

$$= sF(s) - f(0); \text{ only for functions that } \lim_{t\to\infty} f(t)e^{-st} = 0$$

Integral

$$\mathcal{L}\left\{\int_{-\infty}^t f(x)dx\right\} = \int_0^\infty \left[\int_{-\infty}^t f(x)dx\right] e^{-st} dt = \int_0^\infty \left[\int_{-\infty}^t f(x)dx\right] \frac{d}{dt}\left[-\frac{e^{-st}}{s}\right]$$

$$= \left[-\frac{e^{-st}}{s}\int_{-\infty}^t f(x)dx\right]_0^\infty + \int_0^\infty \frac{e^{-st}}{s}\frac{d}{dt}\left[\int_{-\infty}^t f(x)dx\right] dt = -\frac{e^{-s\infty}}{s}\int_{-\infty}^\infty f(x)dx$$

$$+\frac{e^{-s0}}{s}\int_{-\infty}^0 f(x)dx + \frac{1}{s}\int_0^\infty f(t)e^{-st}dt = \frac{F(s)}{s} + \frac{1}{s}\int_{-\infty}^0 f(t)dt;\quad \int_{-\infty}^0 f(t)dt = \text{initial}$$

value of the integral of $f(t)$ at $t = 0$; the value of $\int_{-\infty}^\infty f(x)dx$ must be finite

Note that the term

$$(1/s)\int_{-\infty}^0 f(t)dt$$

is the LT of a step function of amplitude

$$\int_{-\infty}^0 f(t)dt.$$

This factor is a very important result in network problems since it shows that initial conditions associated with integral functions are automatically step functions in the LT development.

Multiplication by exponential

$$\mathcal{L}\{e^{-at}f(t)\} = \int_0^\infty e^{-(s+a)t}dt = \frac{1}{-(s+a)}\left[e^{-(s+a)t}\Big|_0^\infty\right] = \frac{1}{s+a}$$

provided that $e^{-(s+a)\infty}$ has zero value.

Multiplication by t

$$\mathcal{L}\{tf(t)\} = \int_0^\infty tf(t)e^{-st}dt = -\int_0^\infty f(t)\frac{de^{-st}}{ds}dt = -\frac{d}{ds}\left[\int_0^\infty f(t)e^{-st}dt\right] = -\frac{d}{ds}F(s)$$

Time shifting

$$\mathcal{L}\{f(t-a)u(t-a)\} = \int_0^\infty f(t-a)u(t-a)e^{-st}dt$$

By setting $x = t - a$, we obtain for $t = 0$, $x = -a$; for $t = \infty$, $x = \infty$; $dx = dt$ and $t = x + a$. Therefore, the above equation becomes

$$\mathcal{L}\{f(x)u(x)\} = \int_{-a}^\infty f(x)u(x)e^{-s(x+a)}dx = e^{-sa}\int_0^\infty f(x)e^{-sx}dx = e^{-sa}F(s)$$

where a is a positive constant.

Complex frequency shift

$$\mathcal{L}\{e^{s_0 t}f(t)\} = \int_0^\infty e^{-(s-s_0)t}f(t)dt = \int_0^\infty e^{-s't}f(t)dt = F(s') = F(s-s_0)$$

Scaling

$$\mathcal{L}\{f(at)\} = \int_0^\infty f(at)e^{-st}dt = \int_0^\infty f(x)e^{-(s/a)x}d(x/a) = \frac{1}{a}F\left(\frac{s}{a}\right) \quad a>0$$

Time convolution

$$\mathcal{L}\{f(t) * h(t)\} = \mathcal{L}\left\{\int_0^\infty f(x)h(t-x)dx\right\} = \int_0^\infty \left[\int_0^\infty f(x)h(t-x)dx\right]e^{-st}dt$$

$$= \int_0^\infty f(x)dx \int_0^\infty h(t-x)e^{-st}dt = \int_0^\infty f(x)dx \int_0^\infty h(y)e^{-s(y+x)}dy$$

$$= \int_0^\infty f(x)e^{-sx}dx \int_0^\infty h(y)e^{-sy}dy = F(s)H(s)$$

where we set $t - x = y$, and for $t = 0$, $y = -x$ (but $h(y) = 0$ for $y < 0$ and the second integral starts from 0); for $t = \infty$, $y = \infty$. We also find that $dt = dy$.

Initial value

$$\mathcal{L}\left\{\frac{df(t)}{dt}\right\} = \int_0^\infty \frac{df(t)}{dt}e^{-st}dt = sF(s) - f(0) \quad \text{or} \quad sF(s) = \int_0^\infty \frac{df(t)}{dt}e^{-st}dt + f(0)$$

Taking the limit of the above equation, we find

$$\lim_{s\to\infty} sF(s) = \int_0^\infty f(t)\left(\lim_{s\to\infty} e^{-st}\right)dt + f(0) = f(0)$$

where the exchange of integral and limit operations was performed due to the linearity of the two operators. The initial value theorem does not apply for functions that have an impulse (delta function) at $t = 0$.

Final value

$$\int_0^\infty \frac{df(t)}{dt}e^{-st}dt = sF(s) - x(0) \quad \text{or} \quad \lim_{s\to 0}\int_0^\infty \frac{df(t)}{dt}e^{-st}dt = \int_0^\infty \frac{df(t)}{dt}\lim_{s\to 0}(e^{-st})dt$$

$$= \int_0^\infty \frac{df(t)}{dt}dt = f(\infty) - f(0) = \lim_{s\to 0}[sF(s) - x(0)] = \lim_{s\to 0}sF(s) - x(0) \quad \text{or}$$

$$\lim_{t\to\infty} f(t) = \lim_{s\to 0} sF(s)$$

provided that $f(t)$ exists as the value of t approaches infinity.

chapter 9

The z-transform, difference equations, and discrete systems

The z-transform method provides a powerful tool for solving difference equations of any order, and hence plays a very important role in digital systems analysis. This chapter includes a study of the z-transform, its properties, and its applications.

The z-transform method provides a technique for transforming a difference equation into an algebraic equation. Specifically, the z-transform converts a sequence of numbers $\{y(n)\}$ into a function of complex variable $Y(z)$, thereby allowing algebraic process and well-defined mathematical procedures to be applied in the solution process. In this sense, the z-transform plays the same general role in the solution of difference equations that the Laplace transform plays in the solution of differential equations, and in a roughly parallel way. Inversion procedures that parallel one another also exist.

9.1 The z-transform

To understand the essential features of the z-transform, consider a **one-sided sequence** of numbers $\{y(n)\}$ taken at uniform time intervals. This sequence might be the values of a continuous function that has been sampled at uniform time intervals; it could, of course, be a number sequence, e.g., the value of the amount that is present in a bank account at the beginning of each month, including the interest. This number sequence is written

$$\{y(n)\} = \{y(0), y(1), y(2), \ldots, y(n), \ldots\} \tag{9.1}$$

We now create the series

$$Y(z) = \frac{y(0)}{z^0} + \frac{y(1)}{z} + \frac{y(2)}{z^2} + \cdots = y(0) + y(1)z^{-1} + y(2)z^{-2} + \cdots \tag{9.2}$$

In this expression, z denotes the general complex variable and $Y(z)$ denotes the z-transform of the sequence $\{y(n)\}$. In this more general form, the one-sided z-transform of a sequence $\{y(n)\}$ is written

$$Y(z) \triangleq \mathcal{Z}\{y(n)\} = \sum_{n=0}^{\infty} y(n)z^{-n} \qquad (9.3)$$

This expression can be taken as the definition of the one-sided z-transform.

Since the exponent of the z's is equal to the distance from the sequence element $y(0)$, we identify the negative exponents as the amount of delay. This interpretation is not explicit in the mathematical form of (9.3), but it is implied when the shifting properties of functions are considered. This same concept will occur when we apply the z-transform to the solution of difference equations. Initially, however, we study the mathematics of the z-transform.

When the sequence of numbers is obtained by sampling a function $y(t)$ every T seconds — for example, by using an analog-to-digital (A/D) converter — the numbers represent sample values $y(nT)$ for $n = 0, 1, 2, \ldots$. This suggests that there is a relationship between the Laplace transform of a continuous function and the z-transform of a sequence of samples of the function at the time constants $\cdots -nT, -(n-1)T, \cdots -T, 0, T, 2T, \cdots, nT, (n+1)T, \cdots$. To show that there is such a relationship, let $y(t)$ be a function sampled at time constants T seconds apart. The sampled function is (see (6.2))

$$y_s(t) = y(t)\operatorname{comb}_T(t) = \sum_{n=-\infty}^{\infty} y(nT)\delta(t-nT) \qquad (9.4)$$

The Laplace transform of this equation is

$$Y_s(s) \triangleq \mathcal{L}\{y_s(t)\} = \mathcal{L}\left\{\sum_{n=-\infty}^{\infty} y(nT)\delta(t-nT)\right\} = \sum_{n=-\infty}^{\infty} y(nT)\mathcal{L}\{\delta(t-nT)\}$$
$$= \sum_{n=-\infty}^{\infty} y(nT)e^{-nTs} \qquad (9.5)$$

noting that the Laplace operator operates on time t. If we make the substitution $z = e^{sT}$, then

$$Y_s(s)\big|_{z=e^{sT}} = \sum_{n=-\infty}^{\infty} y(nT)z^{-n} = Y(z) \qquad (9.6)$$

Hence, $Y(z)$ is the z-transform of the sequence of samples of $y(t)$, namely, $y(nT)$, with $n = 0, 1, 2, \ldots$. From this discussion, we observe that the z-transform may be viewed as the Laplace transform of the sampled time function $y_s(t)$, with an appropriate change of variables. This interpretation is in addition to that given in (9.3), which, as already noted, specifies that $Y(z)$ is the z-transform of the number sequence $\{y(nT)\}$ for $n = \ldots -2, -1, 0, 1, 2, \ldots$ The above equation is the two-sided z-transform.

Example 9.1.1: Deduce the z-transform of the discrete function

$$y(n) = \begin{cases} 0 & n \leq 0 \\ 1 & n = 1 \\ 2 & n = 2 \\ 0 & n \geq 3 \end{cases}$$

Solution: From the defining equation (9.3), we write

$$Y(z) = \frac{1}{z} + \frac{2}{z^2} = z^{-1} + z^{-2} = \frac{z+2}{z^2}$$

Observe that this function possesses a second-order pole at the origin and a zero at −2. Observe that $y(n)$ could be written as $y(n) = \delta(n-1) + 2\delta(n-2)$. The definition of the discrete delta function is

$$\delta(n) = \begin{cases} 1 & n = 0 \\ 0 & n \neq 0 \end{cases} \tag{9.7}$$

■

Example 9.1.2: Deduce the z-transform of the function

$$f(t) = Ae^{-at} \qquad t \geq 0$$

which is sampled every T seconds, that is, at times $t = nT$.

Solution: The sampled values are

$$\{f(nT)\} = A, Ae^{-aT}, Ae^{-2aT}, \ldots$$

The z-transform of this sequence is written

$$F(z) = A\left[1 + \frac{e^{-aT}}{z} + \left(\frac{e^{-aT}}{z}\right)^2 + \left(\frac{e^{-aT}}{z}\right)^3 + \cdots\right]$$

The series can be written in closed form, recalling that

$$\frac{1}{1-x} = 1 + x^1 + x^2 + x^3 + \cdots \qquad |x| < 1$$

Thus, we have that

$$\mathcal{Z}\{e^{-anT}u(nT)\} \triangleq F(z) = \frac{A}{1 - \frac{e^{-aT}}{z}} = \frac{A}{1 - e^{-aT}z^{-1}} = \frac{Az}{z - e^{-aT}} \qquad (9.8)$$

The convergence is satisfied if

$$|e^{-aT}z^{-1}| = e^{-aT}|z^{-1}| < 1 \text{ or } |z| > e^{-aT}$$

∎

Example 9.1.3: Deduce the z-transform of the function

$$f(t) = e^{-t} + 2e^{-2t} \qquad t \geq 0$$

which is sampled at time intervals $T = 0.1$ s.

Solution: We use the results of Example 9.1.2 to write

$$F(z) = \frac{z}{z - e^{-0.1}} + \frac{2z}{z - e^{-0.2}} = \frac{z}{z - 0.9048} + \frac{2z}{z - 0.8187} = \frac{3z^2 - 2.6283}{z^2 - 1.7235z + 0.7408}$$

∎

Example 9.1.4: Find the z-transform of the given functions when sampled every T seconds:

a. $f(t) = u(t)$
b. $f(t) = tu(t)$
c. $f(t) = e^{-bt}u(t) \quad b > 0$
d. $f(t) = \sin \omega t \, u(t)$

Solution:

a. $\mathcal{Z}\{f(nT)\} = \mathcal{Z}\{u(nT)\} = \sum_{n=0}^{\infty} u(nT)z^{-n} = (1 + z^{-1} + z^{-2} + \cdots) \qquad (9.9)$

$$= \frac{1}{1 - z^{-1}} = \frac{z}{z - 1} \qquad |z| > 1$$

b. $\mathcal{Z}\{f(nT)\} = \mathcal{Z}\{nTu(nT)\} = \sum_{n=0}^{\infty} nTu(nT)z^{-n} = Tz^{-1} + 2Tz^{-2} + 3Tz^{-3} + \cdots$ (9.10)

$$= -Tz\frac{d}{dz}(z^{-1} + z^{-2} + z^{-3} + \cdots) = -Tz\frac{d}{dz}[z^{-1}(1 + z^{-1} + z^{-2} + z^{-3} + \cdots)]$$

$$= -Tz\frac{d}{dz}\left[z^{-1}\frac{z}{z-1}\right] = \frac{Tz}{(z-1)^2} \qquad |z| > 1$$

c. $\mathcal{Z}\{f(nT)\} = \mathcal{Z}\{e^{-bnT}u(nT)\} = \mathcal{Z}\{u(nT)c^{-n}\}; \quad c = e^{bT};$ (9.11)

$$= \sum_{n=0}^{\infty} u(nT)c^{-n}z^{-n} = 1 + c^{-1}z^{-1} + c^{-2}z^{-2} + \cdots = \frac{1}{1 - c^{-1}z^{-1}}$$

$$= \frac{cz}{cz - 1} = \frac{ze^{bT}}{ze^{bT} - 1} = \frac{z}{z - e^{-bT}} \qquad |z| > e^{-bT}$$

d. $\mathcal{Z}\{f(nT)\} = \mathcal{Z}\{u(nT)\sin\omega nT\} = \mathcal{Z}\left\{u(nT)\frac{e^{j\omega nT} - e^{-j\omega nT}}{2j}\right\}$ (9.12)

$$= \sum_{n=0}^{\infty}\frac{u(nT)}{2j}c_1^{-n}z^{-n} - \sum_{n=0}^{\infty}\frac{u(nT)}{2j}c_2^{-n}z^{-n}; \; c_1 = e^{-j\omega T}, \; c_2 = e^{j\omega T}$$

$$= \frac{1}{2j}\left[\frac{c_1 z}{c_1 z - 1} - \frac{c_2 z}{c_2 z - 1}\right] = \frac{z\sin\omega T}{z^2 - 2z\cos\omega T + 1} \qquad |z| > 1$$

The inequalities that appear in the solutions are found from the fact that the summations to converge the factor of the geometric series must have an absolute value of less than 1 (see Section 9.2). ∎

9.2 Convergence of the z-transform

The function $F(z)$ for a specified value of z may be either finite or infinite. The **region of convergence** (ROC) is the set of values of z in the complex z-plane for which the magnitude of $F(z)$ is finite, whereas the set of values of z for which the magnitude of $F(z)$ is infinite is the **region of divergence**. The region of convergence is determined by considering the defining expression (9.3) and examining the complex values of z for which

$$\sum_{n=0}^{\infty} f(nT)z^{-n}$$

has finite values. If we write z in polar form, $z = re^{j\theta}$, then

$$\sum_{n=0}^{\infty}\left|f(nT)z^{-n}\right| = \sum_{n=0}^{\infty}\left|f(nT)(re^{j\theta})^{-n}\right| = \sum_{n=0}^{\infty}\left|f(nT)r^{-n}\right|\left|e^{-j\theta n}\right| = \sum_{n=0}^{\infty}\left|f(nT)r^{-n}\right| \quad (9.13)$$

since $|e^{-j\theta n}| = [\cos^2 n\theta + \sin^2 n\theta]^{1/2} = 1$. For this sum to be finite, we find numbers M and R such that $|f(nT)| \le MR^n$ for $n \ge 0$. Thus,

$$\sum_{n=0}^{\infty}\left|f(nT)z^{-n}\right| \le M\sum_{n=0}^{\infty} R^n r^{-n} = M\sum_{n=0}^{\infty}\left(\frac{R}{r}\right)^n \quad (9.14)$$

For this sum to be finite, it is required that $R/r < 1$ or $r > R$. That is, $F(z)$ is absolutely convergent for all z in the region where $|z| = r > R$. A separate test is required to establish whether the boundary belongs in the ROC.

Example 9.2.1: Find the z-transform of the signal specified and discuss its properties.

$$f(n) = \begin{cases} c^n & n = 0, 1, 2, \ldots \\ 0 & n = -1, -2, \ldots \end{cases} \quad (9.15)$$

The real constant c takes the values (a) $0 < c < 1$ and (b) $c > 1$.

Solution: The time sequences for the two cases are shown in Figures 9.2.1a and b. The z-transform is given by

$$F(z) = \sum_{n=0}^{\infty} c^n z^{-n} = \sum_{n=0}^{\infty} (c^{-1}z)^{-n} = 1 + c^{-1}z + (c^{-1}z)^2 + \cdots + (c^{-1}z)^n + \cdots \quad (9.16)$$

Initially, we consider the sum of the first n terms (up to the $(c^{-1}z)^{n-1}$ term) of this geometric series, which is given by

$$F_n = \frac{1-(c^{-1}z)^{(n-1)+1}}{1-c^{-1}z} = \frac{1-(c^{-1}z)^n}{1-c^{-1}z} \quad (9.17)$$

Next, we set $c^{-1}z = c^{-1}|z|e^{j\theta}$, where θ is the argument of the complex number $c^{-1}z$. Hence, we write $(c^{-1}z)^{-n} = (cz^{-1})^n = (cr^{-1}e^{-j\theta})^n = (c|z|^{-1})^n e^{-j\theta n}$.

Chapter 9: The z-transform, difference equations, and discrete systems 407

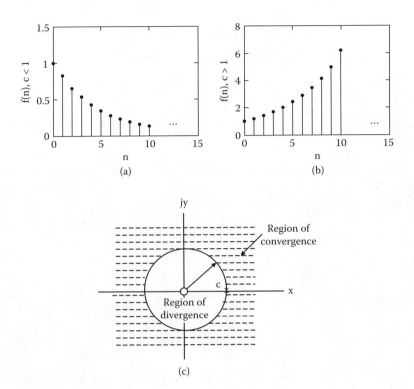

Figure 9.2.1 Discrete signal c^n.

Case 1

For values of z for which $|cz^{-1}| < 1$ ($|cz^{-1}| = |c| \, |z^{-1}| = c|z^{-1}|$ for real c), the magnitude of the complex number $(cz^{-1})^n$ approaches zero as $n \to \infty$. As a consequence,

$$F(z) = \lim_{n \to \infty} F_n(z) = \frac{1}{1 - cz^{-1}} = \frac{z}{z - c} \qquad |cz^{-1}| < 1 \text{ or } |z| > |c| \qquad (9.18)$$

Case 2

For the general case where c is a complex number, the inequality $|cz^{-1}| < 1$ leads to $|c| < |z|$, which implies that the series converges when $|z| > |c|$ and diverges when $|z| < |c|$. Thus, we see that the ROC and divergence in the complex z-plane for $F(z)$ are those shown in Figure 9.2.1c.

To establish whether the boundary of the circle in Figure 9.2.1c belongs to the ROC, we apply L'Hospital's rule to (9.17). Thus,

$$\lim_{z \to c} F_n(z) = \lim_{z \to c} \frac{\frac{d}{d(cz^{-1})}\left(1-(cz^{-1})^n\right)}{\frac{d}{d(cz^{-1})}\left(1-(cz^{-1})\right)} = \lim_{z \to c} \frac{-n(cz^{-1})^{n-1}}{-1} = n$$

and hence $\lim_{z \to c} F_n(z) \to \infty$. Clearly, the boundary belongs to the region of divergence. ∎

Example 9.2.2: Deduce the z-transform and discuss the properties of the impulse functions

$$y(n) = \delta(n) = \begin{cases} 1 & n = 0 \\ 0 & n \neq 0 \end{cases} \tag{9.19}$$

$$y(n) = \delta(n-N) = \begin{cases} 1 & n = N \\ 0 & n \neq N \end{cases} \tag{9.20}$$

Solution: From the basic definition of the z-transform we write that

$$Y(z) = \delta(0)z^{-0} + \delta(1)z^{-1} + \cdots = 1 + 0 + 0 + \cdots = 1$$

Since $Y(z)$ is independent of z, the entire plane is the ROC. By an application of the basic definition, (9.20) becomes

$$Y(z) = \sum_{n=0}^{\infty} \delta(n-N)z^{-n} = 0z^{-0} + 0z^{-1} + \cdots + 1z^{-N} + 0z^{-(N+1)} + \cdots = z^{-N}$$

Since $Y(z) \to \infty$ only for $z = 0$, the entire z-plane is the ROC except for an infinitesimal part of the region. ∎

Example 9.2.3: Deduce the z-transform of the function

$$y(n) = \begin{cases} a^n \sin n\omega & n \geq 0 \quad a > 0 \\ 0 & n < 0 \end{cases} \tag{9.21}$$

Indicate the ROC, the region of divergence, and the poles and zeros in the z-plane.

Solution: The given functions are shown in Figures 9.2.2a and b for two different values of a. Clearly, for $a = 1$, the function is a sinusoidal discrete signal. The z-transform is given by

Chapter 9: The z-transform, difference equations, and discrete systems

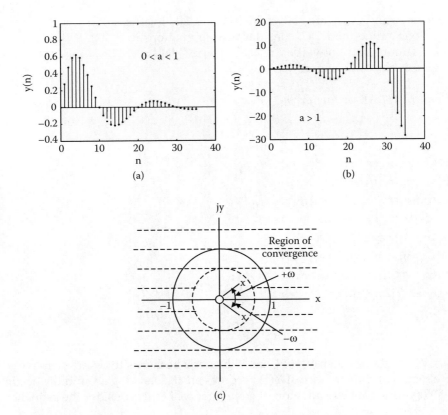

Figure 9.2.2 Discrete signal $a^n \sin n\omega$, its poles, zeros, and ROC.

$$Y(z) = \sum_{n=0}^{\infty} a^n \sin n\omega \, z^{-n} = \sum_{n=0}^{\infty} a^n \frac{e^{jn\omega} - e^{-jn\omega}}{2j} z^{-n} \qquad (9.22)$$

$$= \frac{1}{2j} \sum_{n=0}^{\infty} (ae^{j\omega} z^{-1})^n - \frac{1}{2j} \sum_{n=0}^{\infty} (ae^{-j\omega} z^{-1})^n = \frac{1}{2j} \left[\frac{1}{1 - ae^{j\omega} z^{-1}} - \frac{1}{1 - ae^{-j\omega} z^{-1}} \right]$$

$$= \frac{z^{-1} a \sin \omega}{1 - 2az^{-1} \cos \omega + a^2 z^{-2}} \qquad |z| > a$$

Next, multiply the numerator and denominator of this expression by z^2 to find

$$Y(z) = \frac{za \sin \omega}{z^2 - 2a(\cos \omega)z + a^2} = \frac{za \sin \omega}{[z - a\cos \omega - ja\sin \omega][z - a\cos \omega + ja\sin \omega]}$$

$$= \frac{za \sin \omega}{(z - ae^{j\omega})(z - ae^{-j\omega})} \qquad (9.23)$$

The zeros and poles are shown in Figure 9.2.2c for the case $a < 1$.

To examine the ROC, consider the summations in (9.22). It is seen that each summation converges if $|ae^{j\omega}z^{-1}| = |az^{-1}| \, |e^{j\omega}| = |az^{-1}| < 1$ or $|z| > a$. This region is shown in Figure 9.2.2c.

Book MATLAB m-file: ex_9_2_3

```
%m-file:ex9_2_3
n=0:35;
y1=0.9.^n.*sin(0.1*pi*n);
y2=1.1.^n.*sin(0.1*pi*n);
subplot(2,2,1);stem(n,y1,'k.');
xlabel('n');ylabel('y(n)');
hold on;text(25,0.6,'0<a<1');
hold on;text(35,0.8,'a)');
subplot(2,2,2);stem(n,y2,'k.');
xlabel('n');ylabel('y(n)');
hold on;text(10,-20,'a>1')
hold on;text(35,10,'b)');
```
∎

When the sequence $\{y(k)\}$ has values for both positive and negative k, the region of convergence of $Y(z)$ becomes an annular ring around the origin or $Y(z)$ does not have a region of convergence. Let us consider the sequence

$$y(n) = \begin{cases} 3^n & n \geq 0 \\ 4^n & n < 0 \end{cases}$$

This is a bilateral function, and the definition of the bilateral z-transform is given by

$$\boxed{\mathcal{Z}\{y(n)\} = \sum_{n=-\infty}^{\infty} y(n)z^{-n}} \qquad (9.24)$$

For our specific function,

$$Y(z) = \sum_{n=0}^{\infty} 3^n z^{-n} + \sum_{n=-\infty}^{-1} 4^n z^{-n} = \sum_{n=0}^{\infty} 3^n z^{-n} + \sum_{n=\infty}^{1} 4^{-n} z^n = \sum_{n=0}^{\infty} 3^n z^{-n} + \sum_{n=1}^{\infty} 4^{-n} z^n$$

The first summation converges as $n \to \infty$ provided that $|3z^{-1}| < 1$ or $|z| > 3$. If we set $R^+ = 3$ for positive n's, we see that the region of convergence for

Chapter 9: The z-transform, difference equations, and discrete systems

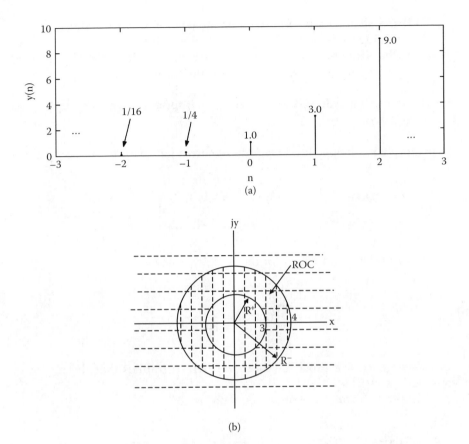

Figure 9.2.3 ROC for a two-sided sequence.

positive n's is $|z| > R^+$. Similarly, the second summation will converge if $|4^{-1}z| < 1$ or $|z| < 4$ and the region of convergence for negative n's is $|z| < R$ with $R^- = 4$. The sequence $y(n)$ and the region of convergence depicted by the double-line region are shown in Figures 9.2.3a and b, respectively.

The reader can easily show, following parallel steps to the above, that the sequence

$$y(n) = \begin{cases} 4^n & n \geq 0 \\ 3^n & n < 0 \end{cases}$$

has no region of convergence.

From the foregoing discussion, we conclude that:

1. The region of convergence of a two-sided sequence is a ring in the z-plane centered at the origin.
2. The region of convergence of a sequence of finite duration is the entire z-plane, except possibly the points $z = 0$ or $z = \infty$.

3. If the sequence is right-handed, that is, $n \geq 0$, then the region of convergence is beyond a circle of finite radius.
4. If the sequence is left-handed, that is, $n < 0$, then the region of convergence is within a circle of finite radius.

The following example shows why it is important to specify the region of convergence.

Example 9.2.4: Specify the regions of convergence for the two sequences

$$x(n) = a^n \qquad a > 1, \qquad n = 0, 1, 2, \cdots \qquad \text{a)}$$
$$y(n) = -a^n \qquad a > 1, \qquad n = -1, -2, -3, \cdots \qquad \text{b)} \qquad (9.25)$$

Solution: The z-transform of the first sequence is

$$X(z) = \sum_{n=0}^{\infty} (az^{-1})^n = \frac{1}{1 - az^{-1}} = \frac{z}{z - a} \qquad (9.26)$$

For convergence, we must have $|az^{-1}| < 1$, which implies that the region of convergence is $|z| > a$.

The z-transform of the other sequence is

$$Y(z) = -\sum_{n=-1}^{-\infty} a^n z^{-n} = -\sum_{n=1}^{\infty} a^{-n} z^n = -\sum_{n=0}^{\infty} a^{-n} z^n + 1 = -\frac{1}{1 - a^{-1}z} + 1 = \frac{z}{z - a} \qquad (9.27)$$

The region of convergence for this function is found as follows: $|a^{-1}z| < 1$ or $|z| < a$. We observe that although the two sequences are completely different in the time domain (one increases positively with positive time and the other decreases absolutely with negative time), they have the same function in their z presentation. However, each one has its own ROC, and this distinguishes one sequence from the other. The inverse z-transform is an integral where the integration path is inside the ROC. Therefore, to find the right time function, the integration must be conducted inside the corresponding ROC. ∎

9.3 Properties of the z-transform

The most important properties of the z-transform for one-sided sequences are given next. The properties are summarized and accompanied by examples to elucidate their use. At the end of this chapter, in Appendix 9.1, we provide the proofs. The one-sided sequences are of great importance because

Chapter 9: The z-transform, difference equations, and discrete systems

all detected signals are of finite extent and their starting points can always be referenced at $t = 0$ ($n = 0$).

Summary of z-transform properties

1. *Linearity* $\qquad \mathcal{Z}\{ax(n)+by(n)\} = a\mathcal{Z}\{x(n)\}+b\mathcal{Z}\{y(n)\}$
2. *Right-shifting property*
   a. $\mathcal{Z}\{y(n-m)\} = z^{-m}\mathcal{Z}\{y(n)\} = z^{-m}Y(z)$ $\qquad$ zero intial conditions
   b. $\mathcal{Z}\{y(n-m)\} = z^{-m}Y(z) + \sum_{i=0}^{m-1} y(i-m)z^{-i}$ $\qquad$ initial conditions present
3. *Left-shifting property* $\qquad \mathcal{Z}\{y(n+m)\} = z^{m}Y(z) - \sum_{i=0}^{m-1} y(i)z^{m-i}$
4. *Time scaling* $\qquad \mathcal{Z}\{a^{n}y(n)\} = Y(a^{-1}z) = \sum_{n=0}^{\infty}(a^{-1}z)^{-n}$
5. *Periodic sequence* $\qquad \mathcal{Z}\{y(n)\} = \dfrac{z^{N}}{z^{N}-1}\mathcal{Z}\{y_{(1)}(n)\},$
   N = number of time units in a period
   $y_{(1)}(n)$ = first period
6. *Multiplication by n* $\qquad \mathcal{Z}\{ny(n)\} = -z\dfrac{dY(z)}{dz}$
7. *Initial value* $\qquad y(n_0) = z^{n_0} Y(z)\big|_{z\to\infty} \qquad y(n) = 0$ for $n < n_0$
8. *Final value* $\qquad \lim_{n\to\infty} y(n) = \lim_{n\to 1}(1-z^{-1})Y(z)$ provided (∞) exists
9. *Convolution* $\qquad \mathcal{Z}\{y(n)\} = \mathcal{Z}\{h(n)*x(n)\} = \mathcal{Z}\left\{\sum_{m=0}^{\infty} h(n-m)x(m)\right\} = H(z)X(z)$
10. *Bilateral convolution*

$$\mathcal{Z}\{y(n)\} = \mathcal{Z}\{h(n)*x(n)\} = \mathcal{Z}\left\{\sum_{m=-\infty}^{\infty} h(n-m)x(m)\right\} = H(z)X(z)$$

Example 9.3.1: Shifting property. Find the z-transform of the functions shown in Figures 9.3.1a and b using the right-shifting property.

Solution: First, consider the z-transform of the unit step function u(n) shown in Figure 9.3.1b, which is given by

$$U(z) = \sum_{n=0}^{\infty} u(n)z^{-n} = 1 + z^{-1} + z^{-2} + z^{-3} + \cdots = \dfrac{1}{1-z^{-1}} = \dfrac{z}{z-1}$$

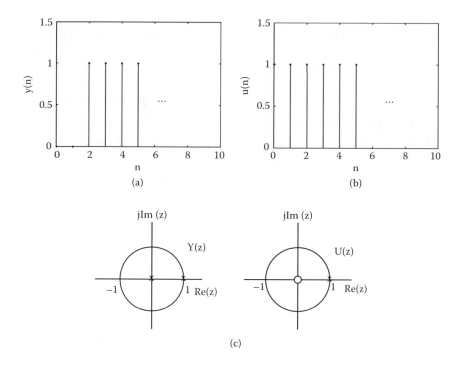

Figure 9.3.1 Shifted discrete signal.

The discrete-time function in Figure 9.3.1a is $y(n) = u(n - 2)$. The z-transform of this equation is

$$\mathcal{Z}\{y(n)\} = \mathcal{Z}\{u(n-2)\} = z^{-2}\mathcal{Z}\{u(n)\} = \frac{1}{z^2}\frac{z}{z-1} = \frac{1}{z(z-1)}$$

The pole-zero configurations are shown in Figure 9.3.1c. We observe from this figure that $U(z)$ does not have poles at zero, whereas the combination of its shifted version $y(n)$ does possess poles (single) at the origin for this particular shifting. ∎

Example 9.3.2: Shifting property. Find the z-transform of an RL series circuit, with a voltage input and the current as the output. The initial condition is $i(0) = 2$. Discretize the analog system.

Solution: The Kirchhoff voltage law (KVL) is

$$L\frac{di(t)}{dt} + Ri(t) = v(t)$$

Chapter 9: The z-transform, difference equations, and discrete systems 415

The discretized form of the above equation is given by

$$L\frac{i(nT)-i(nT-T)}{T}+Ri(nT)=v(nT)$$

or

$$i(nT)-\frac{1}{1+T\frac{R}{L}}i(nT-T)=\frac{T}{L}\frac{1}{1+T\frac{R}{L}}v(nT)$$

Taking into consideration the linearity and shifting property, we obtain

$$I(z)-a[z^{-1}I(z)+i(0-0)z^{-0}]=\frac{T}{L}aV(z) \qquad a=\frac{1}{1+T\frac{R}{L}}$$

Finally, we obtain the algebraic relation

$$I(z)=\frac{2a}{1-az^{-1}}+\frac{T}{L}\frac{a}{1-az^{-1}}V(z)$$

■

Example 9.3.3: Time scaling property. The z-transform of $y(n) = \sin n\omega$ for $n = 0, 1, 2, \ldots$ is equal to $z\sin\omega/(z^2-2z\cos\omega+1)$ (see (9.12) with $T = 1$). By application of the scaling property, we can write the z-transform of the function $a^n y(n) = a^n \sin n\omega$ by inserting $a^{-1}z$ for z. This substitution leads to

$$\mathcal{Z}\{a^n \sin n\omega\} = \frac{a^{-1}z\sin\omega}{a^{-2}z^2-2a^{-1}z\cos\omega+1}$$

The result is the same as (9.23), which was deduced by a different approach.

■

Example 9.3.4: Left-shifting property. Find the z-transform of the output of the system shown in Figure 9.3.2.

Solution: From the diagram, the difference equation that describes the system is

$$\delta(n+1)-3y(n)=y(n+1) \quad \text{or} \quad y(n+1)+3y(n)=\delta(n+1)$$

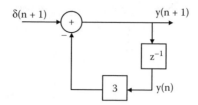

Figure 9.3.2 First-order discrete system.

Take the z-transform of both sides of the equation, recalling the linearity and the left-shifting property, with the result

$$\mathcal{Z}\{y(n+1)\} + 3\mathcal{Z}\{y(n)\} = \mathcal{Z}\{\delta(n+1)\}$$

which, by applying the left-shifting property, yields

$$zY(z) - zy(0) + 3Y(z) = z \quad \text{or} \quad Y(z) = \frac{z[1+y(0)]}{z+3}$$

■

Example 9.3.5: Periodic sequence property. Find the z-transform of the sequence shown in Figure 9.3.3.

Solution: Use periodic sequence property No 5 with $N = 4$ to find

$$Y(z) = \frac{z^4}{z^4 - 1}\mathcal{Z}\{y_{(1)}(n)\} = \frac{z^4}{z^4 - 1}(1 + z^{-1} + z^{-2}) = \frac{z^2(z^2 + z + 1)}{z^4 - 1}$$

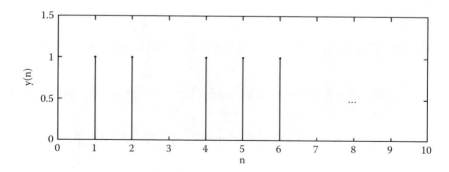

Figure 9.3.3 Periodic discrete function.

■

Chapter 9: The z-transform, difference equations, and discrete systems

Example 9.3.6: Multiplication by n property. Deduce the z-transform of the functions $nu(n), n^2 u(n), n(n+1)u(n), n(n-1)u(n)$.

Solution: Since $\mathcal{Z}\{u(n)\} = z/(z-1)$, then multiplication by n property No 6, we have

$$\mathcal{Z}\{nu(n)\} = -z\frac{d}{dz}\left(\frac{z}{z-1}\right) = \frac{z}{(z-1)^2}$$

We can continue this procedure to find

$$\mathcal{Z}\{n^2 u(n)\} = \mathcal{Z}\{n[nu(n)]\} = -z\frac{d}{dz}\left(\frac{z}{(z-1)^2}\right) = \frac{z(z+1)}{(z-1)^3}$$

Because $n(n+1) = n^2 + n$, we use the linearity property and, adding the results above, find $\mathcal{Z}\{n(n+1)u(n)\} = \mathcal{Z}\{n^2 u(n)\} + \mathcal{Z}\{nu(n)\} = 2z^2/(z-1)^3$. Similarly, for the last case we find $\mathcal{Z}\{n(n-1)u(n)\} = \mathcal{Z}\{n^2 u(n)\} - \mathcal{Z}\{nu(n)\} = 2z/(z-1)^3$.

Example 9.3.7: Initial value property. To find the initial value of the function given by (9.23), we proceed as follows:

$$y(0) = z^0 Y(z)\Big|_{z\to\infty} = \frac{z^{-1}a\sin\omega}{1-2az^{-1}\cos\omega + a^2 z^{-2}}\Big|_{z\to\infty} = 0$$

which agrees with the value deduced from (9.21). ∎

Example 9.3.8: Final value property. We know that the z-transform of the function $y(n) = 0.9^n u(n)$ is $Y(z) = z/(z-0.9)$, and therefore the final value property is given by

$$\lim_{z\to 1}(1-z^{-1})\frac{z}{z-0.9} = \lim_{z\to 1}\frac{z-1}{z-0.9} = \frac{1-1}{1-0.9} = 0$$

as it should be. ∎

Example 9.3.9: Convolution property. The input signal sequence $x(n)$ and the impulse response $h(n)$ of a system are shown in Figures 9.3.4a and b. Deduce the output of the system $w(n)$.

Solution: Figures 9.3.4a and b show the input function $x(n)$ and the impulse response function $h(n)$ of the system. Figure 9.3.4c shows the reflected

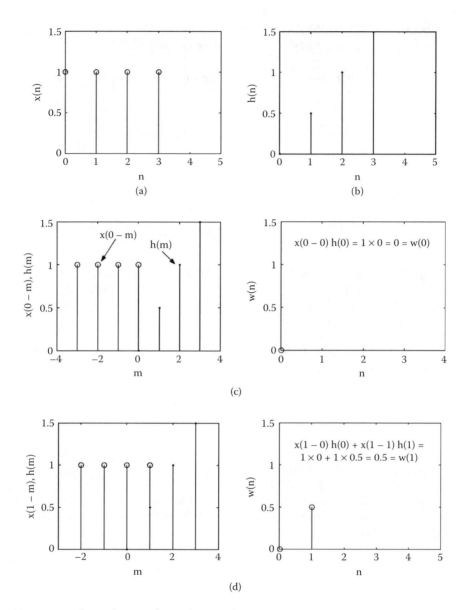

Figure 9.3.4 Convolution of two discrete functions.

function $x(0 - m)$ in the m-domain with zero shift. In the same figure we have plotted the impulse function in the m-domain. From the figure we see that there is an overlap at 0, and their multiplication gives us a zero result. Hence, the first value of the output at $n = 0$ is $w(0) = 0$. The output is plotted on the right-hand side of the same figure as a function of n. The left part of Figure 9.3.4d shows the input function $x(m)$ shifted by one to the right (a positive number, +1, was introduced) and was plotted in the m-domain. The impulse

Chapter 9: The z-transform, difference equations, and discrete systems 419

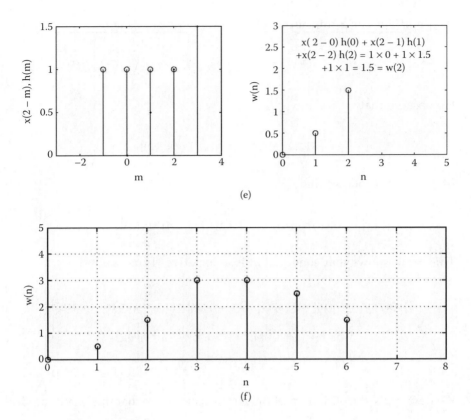

Figure 9.3.4 (continued).

response was also plotted in the m-domain. Next, we multiplied point by point the two functions and added the result. The value found is shown as the number 0.5 at $n = 1$ in the right-hand figure. However, we have plotted the output function $w(n)$ for both shifts $n = 0$ and $n = 1$. Similarly, the result for shift 2 gives the output to be equal to 1.5. The right-hand figure of Figure 9.3.4d gives the output $w(n)$ for shifts 0, 1, and 2. Proceed th same way to obtain the final output function, which is given in Figure 9.3.4e.

The reader can use the following MATALAB algorithm to obtain the output function $w(n)$:

```
x=[1 1 1 1]; h=[0 0.5 1 1.5];w=conv(x,h);n=0:6;stem(n,w,'k');
    %conv is a MATLAB function;
```
∎

Example 9.3.10: Convolution property. Find the z-transform of the convolution of the following three functions:

$$x(n) = a_0\delta(n) + a_1\delta(n-1)$$

$$y(n) = b_0\delta(n) + b_1\delta(n-1)$$

$$q(n) = c_0\delta(n) + c_1\delta(n-1)$$

Solution: The z-transforms of these three functions are

$$X(z) = a_0 + a_1 z^{-1} \qquad Y(z) = b_0 + b_1 z^{-1} \qquad Q(z) = c_0 + c_1 z^{-1}$$

We observe that

$$[X(z)Y(z)]Q(z) = X(z)[Y(z)Q(z)]$$

from which it follows that

$$[x(n) * y(n)] * q(n) = x(n) * [y(n) * q(n)]$$

This result shows that convolution is **associative**. Thus, we write

$$\mathcal{Z}\{x(n) * y(n) * q(n)\} = X(z)Y(z)Q(z) = [a_0 b_0 + (a_1 b_0 + a_0 b_1)z^{-1} + a_1 b_1 z^{-2}]$$
$$(c_0 + c_1 z^{-1}) = a_0 b_0 c_0 + (a_1 b_0 c_0 + a_0 b_1 c_0 + a_1 b_0 c_1)z^{-1}$$
$$+ (a_1 b_0 c_1 + a_0 b_1 c_1 + a_1 b_1 c_0)z^{-2} + a_1 b_1 c_1 z^{-3}$$

We next apply the definition of the z-transform to see that this result implies

$$x(n) * y(n) * q(n) =$$

$$\left(\sum_{m=0}^{n} [a_0 \delta(n-m) + a_1 \delta(n-m-1)][a_0 \delta(n-m) + a_1 \delta(n-m-1)] \right)$$

$$* [c_0 \delta(n) + c_1 \delta(n-1)] = \left(\sum_{m=0}^{n} [a_0 b_0 \delta(n-m)\delta(m) + a_1 b_0 \delta(n-m-1)\delta(m) \right.$$

$$\left. + a_0 b_1 \delta(n-m)\delta(m-1) + a_1 b_1 \delta(n-m-1)\delta(m-1)] \right) * [c_0 \delta(n) + c_1 \delta(n-1)]$$

$$= [a_0 b_0 \delta(n) + a_1 b_0 \delta(n-1) + a_0 b_1 \delta(n-1) + a_1 b_1 \delta(n-2)] * [c_0 \delta(n) + c_1 \delta(n-1)]$$
$$= a_0 b_0 c_0 \delta(n) + (a_1 b_0 c_0 + a_0 b_1 c_0 + a_0 b_0 c_1)\delta(n-1)$$
$$+ (a_1 b_0 c_1 + a_0 b_1 c_1 + a_1 b_1 c_0)\delta(n-2) + a_1 b_1 c_1 \delta(n-3)$$

We have left as an exercise showing that the convolution is also **commutative**.

∎

Chapter 9: The z-transform, difference equations, and discrete systems

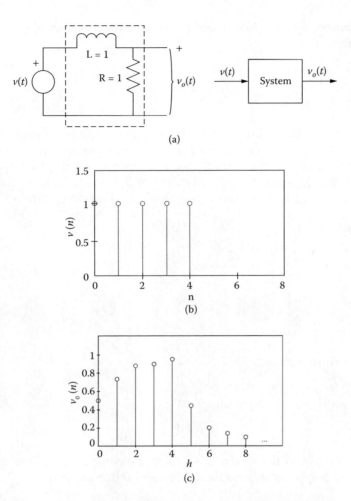

Figure 9.3.5 Impulse response of an RL circuit in discrete form with a discrete pulse as input.

Example 9.3.11: Convolution property. Find the output of the relaxed system shown in Figure 9.3.5a if the input is that shown in 9.3.5b. Express the system in its discrete form.

Solution: A direct application of the KVL yields the equation

$$\frac{di(t)}{dt} + i(t) = v(t), \qquad \frac{dv_o(t)}{dt} + v_o(t) = v(t)$$

The second equation follows from the first since $v_o(t) = Ri(t)$ with $R = 1$ ohm. If we assume the sampling time $T = 1$, then from Section 3.2 we find that $dv_o(t)/dt \cong v_o(n) - v_o(n-1)$, and thus the above equation takes the following discretized form:

$$v_o(n) = \frac{1}{2}v(n) + \frac{1}{2}v_o(n-1)$$

Next, we proceed to determine the impulse response of the system. We use the fact that the z-transform of the output of the system is equal to the system function $H(z)$ for a delta function input $v(n) = \delta(n)$ (the exact same approach as was used in Laplace transform (LT) studies). Also, the inverse transform of $H(z)$ is the impulse response $h(n)$. Therefore, $H(z)$ is given by

$$\mathcal{Z}\{v_o(z)\} = \frac{1}{2}\mathcal{Z}\{\delta(n)\} + \frac{1}{2}\mathcal{Z}\{v_o(n-1)\} \quad \text{or} \quad H(z) = \frac{V_0(z)}{V(z)} = \frac{1}{2}\frac{1}{1-\frac{1}{2}z^{-1}}$$

$$= \frac{1}{2}\left[1 + \left(\frac{1}{2}\right)z^{-1} + \left(\frac{1}{2}\right)^2 z^{-2} + \cdots\right], \quad V(z) = 1$$

It follows from (9.2) that

$$h(n) = \frac{1}{2}\left(\frac{1}{2}\right)^n \quad n \geq 0$$

Since the output of the system is equal to the convolution of the input and its impulse response, this result can be used in the expression

$$v_o(n) = \sum_{m=0}^{n} h(n-m)v(m)$$

The outputs at successive time steps are

$$v_o(0) = h(0)v(0) = \frac{1}{2} \times 1 = \frac{1}{2}$$

$$v_o(1) = h(1)v(0) + h(0)v(1) = \frac{1}{4} \times 1 + \frac{1}{2} \times 1 = \frac{3}{4}$$

$$v_o(2) = h(2)v(0) + h(1)v(1) + h(0)v(2) = \frac{1}{8} \times 1 + \frac{1}{4} \times 1 + \frac{1}{2} \times 1 = \frac{7}{8}$$

⋮

Chapter 9: The z-transform, difference equations, and discrete systems 423

The output is shown in Figure 9.3.5c. Observe that the resulting shape of the output is the same as that for the corresponding continuous system. It is expected that if we had selected the sampling time interval to be shorter, the output voltage would be closer to its corresponding analog case, and the shorter the sampling time, the closer to the analog case the output voltage would have resulted. ∎

9.4 z-Transform pairs

Just as with other transforms, the z-transform of a discrete function and its inverse are given as

$$\mathcal{Z}\{f(n)\} \triangleq F(z) = \sum_{n=0}^{\infty} f(n)z^{-n} \quad \text{or} \quad F(z) = \sum_{n=0}^{\infty} f(nT)z^{-n}$$

$$\mathcal{Z}^{-1}\{F(z)\} \triangleq \{f(n)\} = \frac{1}{2\pi j} \oint_C F(z)z^{n-1}dz$$

(9.28)

where $\mathcal{Z}^{-1}$ denotes the inverse z-transformation. Since the inverse z-transform includes integration in the complex z-plane, we will not study it. It is beyond the scope of this course. However, we would like to point out that the integration path must be inside the ROC for the integral to give us the right answer. In our studies in this text we will rely on tables to obtain the inverse z-transforms. Table 9.4.1 provides some common z-transform pairs.

9.5 Inverse z-transform

As already discussed in our studies, we assume that an $F(z)$ corresponds to a sequence $\{f(n)\}$ that is bounded as $n \to \infty$. To find the inverse z-transform, we cast the transform function into a form that is amenable to simple tabular lookup using Table 9.4.2. The approach parallels that followed in performing similar operations using the Laplace transform. The functions that we will be concerned with are rational functions of z, that is, they are the ratios of two polynomials. Ordinarily, these are **proper fractions** since the degree of the numerator polynomial is less than the degree of the denominator polynomial. If the functions are not proper functions, we perform long division until the degree of the numerator is less than that of the denominator. This results is a power series and a proper fraction.

Example 9.5.1: Determine the inverse z-transform of the function

$$F(z) = \frac{1}{1 - 0.1z^{-1}} \qquad (9.29)$$

Table 9.4.1 Properties of the z-Transform ($n \geq 0$)

1.	$\mathcal{Z}\{ax(n)+by(n)\}$	$aX(z)+bY(z)$
2.	$\mathcal{Z}\{y(n-m)\}$	$z^{-m}Y(z)+\sum_{i=1}^{m}y(-i)z^{-(m-i)}$
3.	$\mathcal{Z}\{y(n+m)\}$	$z^{m}Y(z)-\sum_{n=0}^{m-1}y(n)z^{m-n}$
4.	$\mathcal{Z}\{a^{n}y(n)\}$	$Y\left(\dfrac{z}{a}\right)$
5.	$\mathcal{Z}\{y(n)\}$	$\dfrac{z^{N}}{z^{N}-1}Y_{1}(z)$; $y_{(1)}$ is the first period of a periodic sequence $y(n)=y(n+N)$
6.	$\mathcal{Z}\{y(n)*h(n)\}$	$Y(z)H(z)$
7.	$\mathcal{Z}\{ny(n)\}$	$-z\dfrac{d}{dz}Y(z)$
8.	$\mathcal{Z}\{n^{m}y(n)\}$	$\left(-z\dfrac{d}{dz}\right)^{m}Y(z)$
9.	$y(n_{0})$	$z^{n_{0}}Y(z)\big|_{z\to\infty}$; n_{0} is the initial value of the sequence and $Y(z)=\sum_{n=0}^{\infty}y(n_{0}+n)z^{-n}$
10.	$\lim_{z\to\infty}y(n)$	$\lim_{z\to 1}(1-z^{-1})Y(z)$

Solution: The function possesses a simple pole at $z = 0.1$ and a zero at $z = 0$. The ROC is a circle in the complex z-plane with radius $|z| > 0.1$. We can proceed by dividing the numerator by the denominator, which results in the following infinite series in powers of z^{-1}:

$$F(z) = 1 + 0.1z^{-1} + (0.1)^{2}z^{-2} + (0.1)^{3}z^{-3} + \cdots$$

Thus, we have that

$$F(z) = 1 + 0.1z^{-1} + (0.1)^{2}z^{-2} + (0.1)^{3}z^{-3} + \cdots$$

Chapter 9: The z-transform, difference equations, and discrete systems   425

Table 9.4.2 Common z-Transform Pairs

Entry Number	$f(n), f(nT)$ for $n \geq 0$	$F(z) = \sum_{n=0}^{\infty} f(n) z^{-n}$	Radius of Convergence $\|z\| > R$
1.	$\delta(n)$	1	0
2.	$\delta(n-m)$	z^{-m}	0
3.	1	$\dfrac{z}{z-1}$	1
4.	n	$\dfrac{z}{(z-1)^2}$	1
5.	n^2	$\dfrac{z(z+1)}{(z-1)^3}$	1
6.	n^3	$\dfrac{z(z^2+4z+1)}{(z-1)^4}$	1
7.	a^n	$\dfrac{z}{z-a}$	$\|a\|$
8.	na^n	$\dfrac{az}{(z-a)^2}$	$\|a\|$
9.	$n^2 a^n$	$\dfrac{az(z+a)}{(z-a)^3}$	$\|a\|$
10.	$\dfrac{a^n}{n!}$	$e^{a/z}$	0
11.	$(n+1)a^n$	$\dfrac{z^2}{(z-a)^2}$	$\|a\|$
12.	$\dfrac{(n+1)(n+2)a^n}{2!}$	$\dfrac{z^3}{(z-a)^3}$	$\|a\|$
13.	$\dfrac{(n+1)(n+2)\cdots(n+m)a^n}{m!}$	$\dfrac{z^{m+1}}{(z-a)^{m+1}}$	$\|a\|$
14.	$\sin n\omega T$	$\dfrac{z \sin \omega T}{z^2 - 2z \cos \omega T + 1}$	1
15.	$\cos n\omega T$	$\dfrac{z(z - \cos \omega T)}{z^2 - 2z \cos \omega T + 1}$	1
16.	$a^n \sin n\omega T$	$\dfrac{az \sin \omega T}{z^2 - 2az \cos \omega T + a^2}$	$\|a\|^{-1}$

Table 9.4.2 (continued) Common z-Transform Pairs

Entry Number	$f(n), f(nT)$ for $n \geq 0$	$F(z) = \sum_{n=0}^{\infty} f(n)z^{-n}$	Radius of Convergence $\|z\| > R$
17.	$a^{nT} \sin n\omega T$	$\dfrac{a^T z \sin \omega T}{z^2 - 2a^T z \cos \omega T + a^{2T}}$	$\|a\|^{-T}$
18.	$a^n \cos n\omega T$	$\dfrac{z(z - a \cos \omega T)}{z^2 - 2az \cos \omega T + a^2}$	$\|a\|^{-1}$
19.	$e^{-anT} \sin n\omega T$	$\dfrac{ze^{-aT} \sin \omega T}{z^2 - 2e^{-aT} z \cos \omega T + e^{-2aT}}$	$\|z\| > \|e^{aT}\|$
20.	$e^{-anT} \cos n\omega T$	$\dfrac{z(z - e^{-aT} \cos \omega T)}{z^2 - 2e^{-aT} z \cos \omega T + e^{-2aT}}$	$\|z\| > \|e^{aT}\|$
21.	$\dfrac{n(n-1)}{2!}$	$\dfrac{z}{(z-1)^3}$	1
22.	$\dfrac{n(n-1)(n-2)}{3!}$	$\dfrac{z}{(z-1)^4}$	1
23.	$\dfrac{n(n-1)(n-2)\cdots(n-m+1)}{m!} a^{n-m}$	$\dfrac{z}{(z-a)^{m+1}}$	1
24.	e^{-anT}	$\dfrac{z}{z - e^{-aT}}$	$\|e^{aT}\|$
25.	ne^{-anT}	$\dfrac{ze^{-aT}}{(z - e^{-aT})^2}$	$\|e^{aT}\|$

It follows from (9.2) that the corresponding sequence is

$$f(n) = \begin{cases} 1, 0.1, (0.1)^2, (0.1)^3, \ldots & n \geq 0 \\ 0 & n < 0 \end{cases} \quad (9.30)$$

which is the sequence

$$f(n) = (0.1)^n \qquad n \geq 0 \qquad (9.31)$$

∎

Chapter 9: The z-transform, difference equations, and discrete systems 427

Example 9.5.2: Find the inverse z-transform of the function

$$F(z) = \frac{1}{(1-0.2z^{-1})(1+0.2z^{-1})} = \frac{1}{1-0.04z^{-2}} \qquad (9.32)$$

Solution: One approach is to proceed as in the foregoing example. By long division, the following polynomial results in

$$F(z) = 1 + 0.04z^{-2} + (0.04)^2 z^{-4} + \cdots = (0.2)^{2n}(z^{-1})^{2n}$$

with region of convergence $|z| > 0.2$. This series corresponds to the sequence

$$f(m) = \begin{cases} (0.2)^m & m = 2n \quad n \geq 0 \\ 0 & m = 2n+1 \quad n > 0 \\ 0 & n < 0 \end{cases} \qquad (9.33)$$

A different approach calls for separating the function $F(z)$ into partial fraction form. Now, two factors must be considered: (1) the resulting function must be proper function, and (2) many entries in Table 9.4.2 involve z in the numerator of the resulting expression for $F(z)$. This need is achieved by considering $F(z)/z$. Thus, we modify $F(z)$ to $F(z)/z$:

$$\frac{F(z)}{z} = \frac{z}{(z-0.2)(z+0.2)} = \frac{A}{z-0.2} + \frac{B}{z+0.2}$$

where

$$A = \frac{z}{z+0.2}\bigg|_{z=0.2} = 0.5, \quad B = \frac{z}{z-0.2}\bigg|_{z=-0.2} = \frac{-0.2}{-0.4} = 0.5$$

$$F(z) = \frac{1}{2}\left[\frac{z}{z-0.2} + \frac{z}{z+0.2}\right]$$

From appropriate entries of Table 9.4.2, the inverse transform is

$$f(n) = \begin{cases} \frac{1}{2}[(0.2)^n + (0.2)^{-n}] & n \geq 0 \\ 0 & n < 0 \end{cases} \qquad (9.34)$$

The reader can easily verify that (9.33) and (9.34) yield identical results. ∎

Example 9.5.3: Find the inverse z-transform of the function

$$F(z) = \frac{1}{(1-0.2z^{-1})z^{-2}} = \frac{z^3}{z-0.2} \qquad (9.35)$$

Solution: By long division, we obtain

$$F(z) = z^2 + 0.2z + (0.2)^2 \frac{z}{z-0.2}$$

From Table 9.4.2, the inverse transform is

$$f(n) = \delta(n+2) + 0.2\delta(n+1) + (0.2)^2(0.2)^n$$

where the last term is applicable for $n \geq 0$. Therefore, this equation is equivalent to

$$f(n) = \begin{cases} 0.2^{n+2} & n \geq -2 \\ 0 & n < -2 \end{cases} \qquad (9.36)$$

It is recalled that (9.35) could be expanded into the form

$$F(z) = z^2 + 0.2z + (0.2)^2 z^0 + (0.2)^3 z^{-1} + (0.2)^4 z^{-2} + \cdots$$

$$= z^2 \left[1 + \frac{0.2}{z} + \frac{0.2^2}{z^2} + \cdots \right]$$

The inverse transform of the bracketed term is 0.2^n, and the factor z^2 indicates a shift to the left of two sample periods. Thus, (9.36) is realized.

From the above, we note that to find the inverse z-transform, we must:

- Initially ignore any factor of the form z^n, where n is an integer
- Expand the remaining part into partial fraction form
- Use z-transform tables or properties to obtain the inverse z-transform of each term in the expansion
- Combine the results and perform the necessary shifting required by z^n omitted in the first step ∎

Example 9.5.4: Find the inverse z-transform of the function

$$F(z) = \frac{z^2 - 3z + 8}{(z-2)(z+2)(z+3)}$$

Chapter 9: The z-transform, difference equations, and discrete systems 429

Solution: Expand $F(z)/z$ in partial fraction form for reasons already discussed:

$$\frac{F(z)}{z} = \frac{z^2 - 3z + 8}{z(z-2)(z+2)(z+3)} = \frac{A}{z} + \frac{B}{z-2} + \frac{C}{z+2} + \frac{D}{z+3}$$

where

$$A = z\left.\frac{F(z)}{z}\right|_{z=0} = -\frac{2}{3}$$

$$B = (z-2)\left.\frac{F(z)}{z}\right|_{z=2} = \left.\frac{z^2 - 3z + 8}{z(z+2)(z+3)}\right|_{z=2} = \frac{3}{20}$$

$$C = (z+2)\left.\frac{F(z)}{z}\right|_{z=-2} = \left.\frac{z^2 - 3z + 8}{z(z-2)(z+3)}\right|_{z=-2} = \frac{9}{4}$$

$$D = (z+3)\left.\frac{F(z)}{z}\right|_{z=-3} = \left.\frac{z^2 - 3z + 8}{z(z-2)(z+2)}\right|_{z=-3} = -\frac{26}{15}$$

Therefore,

$$F(z) = -\frac{2}{3} + \frac{3}{20}\frac{z}{z-2} + \frac{9}{4}\frac{z}{z+2} - \frac{26}{15}\frac{z}{z+3}$$

This leads to the following values for $\{f(n)\}$ using Table 9.4.2:

$$f(n) = -\frac{2}{3}\delta(n) + \frac{3}{20}2^n + \frac{9}{4}(-2)^n - \frac{26}{15}(-3)^n$$

If we set $n = 0$, then $f(0) = 0$ and MATLAB below ignores this point. However, if we divide the numerator by the denominator, we obtain the sequence $z^{-1}(1 - 6z^{-1} + 30z^{-2} - 102z^{-3} + \cdots)$, which indicates a shifted sequence to the right and MATLAB ignores the shifting.

Book MATLAB m-file: ex_9_5_4

```
%m-file: ex_9_5_4
num=[0 1 -3 8];den=[conv(conv([1 -2],[1 2]),[1 3])];
[r,p,k]=residue(num,den);

r =[5.2000 -4.5000 0.3000]; p=[-3.0000 -2.0000 2.0000]; k=[]
```

Based on the MATLAB results we write

$$F(z) = 5.2\frac{z}{z+3} - 4.5\frac{z}{z+2} + 0.3\frac{z}{z-2}$$

Note the corresponding numbers between r and p vectors. The inverse z-transform of the above equation is

$$f(n) = 5.2(-3)^n - 4.5(-2)^n + 0.3(2)^n$$

Although the two results look different the sequence $\{f(n)\}$, the two cases are numerically identical. ∎

Example 9.5.5: Find the inverse z-transform of the function

$$F(z) = \frac{z^2 - 9}{(z-1)(z-2)^2}$$

Observe that the function has a single- and a second-order pole.

Solution: The function $F(z)/z$ is expanded in partial fraction form as follows:

$$\frac{F(z)}{z} = \frac{z^2 - 9}{z(z-1)(z-2)^2} = \frac{A}{z} + \frac{B}{(z-1)} + \frac{C_1}{(z-2)} + \frac{C_2}{(z-2)^2} \quad (9.37)$$

We can find three of the unknown constants using the relations

$$A = \left.\frac{z^2 - 9}{(z-1)(z-2)^2}\right|_{z=0} = \frac{-9}{(-1)(4)} = \frac{9}{4}$$

$$B = \left.\frac{z^2 - 9}{z(z-2)^2}\right|_{z=1} = \frac{1-9}{1\times(-1)^2} = -8$$

$$C_2 = \left.\frac{z^2 - 9}{z(z-1)}\right|_{z=2} = \frac{4-9}{2(1)} = -\frac{5}{2}$$

These constants are introduced in (9.37), leaving a relation involving the one remaining constant C_1. One procedure for finding the constant is to select

Chapter 9: The z-transform, difference equations, and discrete systems

any two appropriate values for z, avoiding the roots of the rational polynomial, thereby creating an equation with the unknown. In particular, if we choose $z = 3$, we obtain the following expression:

$$\frac{3^2-9}{3(3-1)(3-2)^2} = \frac{9}{4}\frac{1}{3} - 8\frac{1}{3-1} + C_1\frac{1}{3-2} - \frac{5}{2}\frac{1}{(3-2)^2} \quad \text{or} \quad C_1 = \frac{23}{4}$$

First we introduce the values of all the constants found in (9.37), next we multiply both sides by z, and finally we take the inverse z-transform using Table 9.4.2. The result is

$$f(n) = \frac{9}{4}\delta(n) - 8u(n) + \frac{23}{4}2^n - \frac{5}{2}n2^{n-1}$$

For $n = 0, 1, 2, 3$, and 4, the value of the sequence is $\{f(n)\} = \{0\ 1\ 5\ 8\ 4\}$.

Book MATLAB m-file: ex_9_5_5

```
%m-file: ex_9_5_5
num=[0 1 0 -9];
den=[1 -5 8 -4];
f=dimpulse(num,den,5);%the number 5 indicates the number
                      %of values desired of f(n);
```

The values found using the above MATLAB m-file are identical to those found above. ∎

9.6 Transfer function

The z-transform provides a very important technique in the solution of the difference equation. As part of this process, the transfer function plays an important role.

Example 9.6.1: Deduce an expression for the impulse response of the circuit shown in Figure 9.6.1a using the z-transform technique.

Solution: The controlling equation of the circuit is

$$L\frac{di(t)}{dt} + Ri(t) = v(t)$$

The appropriate discrete form of this equation for sampling time $T_s = 1$ is (see (3.7) and (3.8))

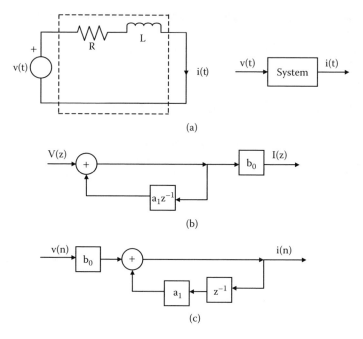

Figure 9.6.1 Series RL circuit.

$$L[i(n) - i(n-1)] + Ri(n) = v(n)$$

from which we obtain

$$i(n) = \frac{1}{L+R}v(n) + \frac{L}{L+R}i(n-1) \triangleq b_0 v(n) + a_1 i(n-1)$$

Taking the z-transform of both sides of the above equation and using the appropriate properties in Section 9.3, the above equation becomes

$$I(z) = b_0 V(z) + a_1 z^{-1} I(z) \quad \text{or} \quad I(z) = \frac{b_0}{1 - a_1 z^{-1}} V(z) = H(z)V(z) \quad (9.38)$$

This equation relates the input–output relation explicitly in the transformed domain of a discrete system. The quantity $H(z) = I(z)/V(z)$, or equivalently the ratio of the system output to its input, is the **system transfer function** for the discrete system. Further, the inverse transform of $H(z)$ is the impulse response $h(n)$ of the system. Thus, for our circuit with a delta function excitation, ($\mathcal{Z}\{\delta(n)\} = 1$), we have

$$I(z) \triangleq H(z) = \frac{b_0}{1 - a_1 z^{-1}} \quad (9.39)$$

Chapter 9: The z-transform, difference equations, and discrete systems   433

The inverse transform is easily found to be

$$h(n) = b_0(a_1)^n \qquad n \geq 0 \qquad (9.40)$$

Figures 9.6.1b and c shows the feedback configuration of the discrete system in the transformed and time domains. ∎

Example 9.6.2: Determine the response of the first-order system specified by (9.38) to a unit step response by z-transform and convolution methods.

Solution: The unit step sequence, which is written

$$u(n) = \begin{cases} 1 & n = 0, 1, 2, \cdots \\ 0 & n < 0 \end{cases}$$

has the z-transformed value

$$U(z) = \frac{z}{z-1} = \frac{1}{1-z^{-1}}$$

The response, by writing y for i, is given by (see (9.38))

$$Y(z) = b_0 \frac{z}{z-a_1} \frac{z}{z-1} = b_0 \left[\frac{Az}{z-a_1} + \frac{Bz}{z-1} \right]$$

where

$$A = \frac{z}{z-1}\bigg|_{z=a_1} = \frac{a_1}{a_1-1}, \qquad B = \frac{z}{z-a_1}\bigg|_{z=1} = \frac{1}{1-a_1}$$

Thus,

$$Y(z) = \frac{b_0}{1-a_1} \left(\frac{-a_1}{1-a_1 z^1} + \frac{1}{1-z^{-1}} \right)$$

The inverse transform is

$$y(n) = \frac{b_0}{1-a_1}[-a_1(a_1)^n + (1)^n] = \frac{b_0}{1-a_1}(1 - a_1^{n+1}) \qquad n \geq 0$$

It is recalled that the derivative of the step function response of a system is its impulse response. To show that the result is consistent with the result of Example 9.6.1, we consider the derivative of $y(n)$ in its discrete form representation. Ignoring at first the constant factor $b_0/(1 - a_1)$, we obtain

$$\frac{y(n) - y(n-1)}{1} = 1 - a_1^{n+1} - 1 + a^n = (1 - a_1)a_1^n$$

Therefore, the impulse response is given by

$$h(n) = \frac{b_0}{1 - a_1} a_1^n (1 - a_1) = b_0 a_1^n \qquad n \geq 0$$

We can proceed to find the output in the forgoing example by using the convolution equation. Here we write, using the results of Example 9.6.1,

$$y(n) = \sum_{m=0}^{n} h(n-m)u(n) = \sum_{m=0}^{n} b_0 a_1^{n-m} u(m)$$

The output at successive time steps is

$$y(0) = b_0(a_1)^0 = b_0$$

$$y(1) = b_0(a_1 + 1)$$

$$y(2) = b_0(a_1^2 + a_1 + 1)$$

$$\vdots$$

$$y(n) = b_0(a_1^n + a_1^{n-1} + \cdots + 1) = b_0 \left(\frac{1 - a_1^{n+1}}{1 - a_1} \right)$$

where the formula for the finite geometric series was used in the last summation. This result is identical with that above using the z-transform method. ∎

When discrete systems are interconnected, the rules that apply to continuous systems are also applied for discrete systems. For example, if the impulse responses of two systems connected in cascade are known, the impulse response of the total system is

$$h(n) = h_1(n) * h_2(n) \tag{9.41}$$

and in the z-domain

$$\boxed{H(z) = H_1(z)H_2(z) = \mathcal{Z}\{h_1(n) * h_2(n)\}} \tag{9.42}$$

Chapter 9: The z-transform, difference equations, and discrete systems

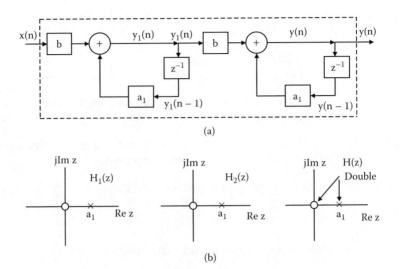

Figure 9.6.3 Two first-order discrete systems in cascades.

Example 9.6.3: Deduce the transfer function for the system shown in Figure 9.6.3a.

Solution: Consider initially the portion of the system (subsystem) shown between $x(n)$ and $y_1(n)$ described by the difference equation

$$y_1(n) = bx(n) + a_1 y_1(n-1)$$

The z-transform of this expression is

$$Y_1(z) = bX(z) + a_1 z^{-1} Y_1(z) \quad \text{or} \quad Y_1(z) = \frac{b}{1 - a_1 z^{-1}} X(z) = H_1(z) X(z)$$

The portion of the system between $y_1(n)$ and $y(n)$ (a cascaded subsystem with the first subsystem) is described by a similar expression whose z-transform is

$$Y(z) = \frac{b}{1 - a_1 z^{-1}} Y_1(z) = H_2(z) Y_1(z)$$

Substitute the known expression for $Y_1(z)$ into this final expression to obtain

$$Y(z) = \left(\frac{b}{1 - a_1 z^{-1}} \right) \left(\frac{b}{1 - a_1 z^{-1}} \right) X(z) = H_1(z) H_2(z) X(z) = H(z) X(z)$$

where

$$H(z) = H_1(z)H_2(z) = \left(\frac{b}{1-a_1z^{-1}}\right)^2$$

The pole-zero configurations for each subsystem and for the combined systems are shown in Figure 9.6.3b. ∎

Example 9.6.4: Find the transfer function for the first-order system, shown in Figure 9.6.4a, and sketch the pole-zero configuration.

Solution: The difference equation describing the system is

$$y(n) = b_0 x(n) + b_1 x(n-1)$$

The z-transform of this equation and the transfer function of the system are

$$Y(z) = b_0 X(z) + b_1 z^{-1} X(z) \quad \text{or} \quad H(z) \triangleq \frac{Y(z)}{X(z)} = b_0 + b_1 z^{-1} = b_0 \frac{z + \frac{b_1}{b_0}}{z}$$

The pole-zero configuration is shown in Figure 9.6.4b. ∎

Note: *In Example 9.6.1, b_0 as well as a_0 (1 in this case) and a_1 are different than zero. This type of system is known as the first-order **infinite impulse response (IIR)** system. Observe that an analog first-order RL series circuit results in an IIR digital system. In Example 9.6.4, a_0 (= 1), b_0, and b_1 are different than zero. This type of first-order digital system is known as the **finite impulse response (FIR)** system.*

Figure 9.6.5 shows a combined FIR and IIR system. The difference equation describing the total system is found by the following two equations, which are obtained by inspection of Figure 9.6.5. These are

$$x_1(n) - a_1 y(n-1) = y(n), \qquad x_1(n) = b_0 x(n) + b_1 x(n-1)$$

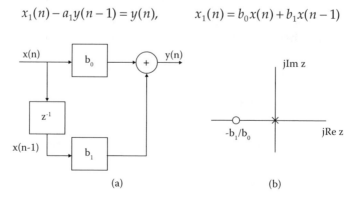

(a) (b)

Figure 9.6.4 First-order discrete system.

Chapter 9: The z-transform, difference equations, and discrete systems

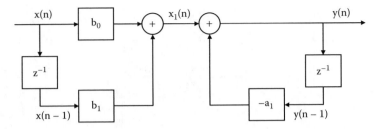

Figure 9.6.5 A first-order combined FIR and IIR digital system.

Therefore, the difference equation describing the system is

$$y(n) + a_1 y(n-1) = b_0 x(n) + b_1 x(n-1) \quad (9.43)$$

The a's and b's can take either positive or negative values. Taking the z-transform of both sides of (9.43), we obtain the transfer function of the system, which is

$$\boxed{H(z) \triangleq \frac{Y(z)}{X(z)} = \frac{b_0 + b_1 z^{-1}}{1 + a_1 z^{-1}}} \quad (9.44)$$

*Higher-order transfer functions

The general case of a system is described by the following difference equation:

$$y(n) + a_1 y(n-1) + a_2 y(n-2) + \cdots + a_p y(n-p) = b_0 x(n) + b_1 x(n-1) + b_2 x(n-2)$$
$$+ \cdots + b_q x(n-q) \quad (9.45)$$

Taking the z-transform and solving for the ratio $Y(z)/X(z)$, we obtain the transfer function

$$\boxed{H(z) \triangleq \frac{Y(z)}{X(z)} = \frac{b_0 + b_1 z^{-1} + \cdots + b_q z^{-q}}{1 + a_1 z^{-1} + \cdots + a_p z^{-p}} = \frac{\sum_{n=0}^{q} b_n z^{-n}}{1 + \sum_{n=1}^{p} a_n z^{-n}}} \quad (9.46)$$

This equation indicates that if we know the transfer function $H(z)$, then the output $Y(z)$ to any input $X(z)$ (or equivalently, $x(n)$) can be determined.

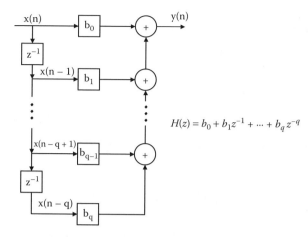

Figure 9.6.6 A qth-order FIR digital system.

If we set $a_1, a_2, \cdots, a_p$ equal to zero, (9.45) becomes

$$y(n) = b_0 x(n) + b_1 x(n-1) + b_2 x(n-2) + \cdots + b_q x(n-q) \qquad (9.47)$$

This expression defines a **qth-order FIR filter**, and such filters are examined in a later chapter. The block diagram of a FIR filter is shown in Figure 9.6.6.

For the case when $b_0 = 1$ and the rest of the b_i's are zero, the difference equation (9.45) becomes

$$y(n) + a_1 y(n-1) + a_2 y(n-2) + \cdots + a_p y(n-p) = x(n) \qquad (9.48)$$

This equation defines a **pth-order IIR filter**. A block diagram representation of this equation is shown in Figure 9.6.7.

Finally, if none of the constants is zero in (9.45), the block diagram representation of the combined FIR and IIR system is that shown in Figure 9.6.8.

9.7 Frequency response of first-order discrete systems

Suppose that the input to the system is the function z^n. Then, using the convolution property of system response, the output is given by

$$y(n) = z^n * h(n) = \sum_{m=0}^{\infty} h(m) z^{n-m} = z^n \sum_{m=0}^{\infty} h(m) z^{-m} = z^n H(z) \qquad (9.49)$$

If we set $z = e^{j\omega}$ in this expression, we have

$$y(n) = e^{j\omega n} H(e^{j\omega}) \qquad (9.50)$$

Chapter 9: The z-transform, difference equations, and discrete systems

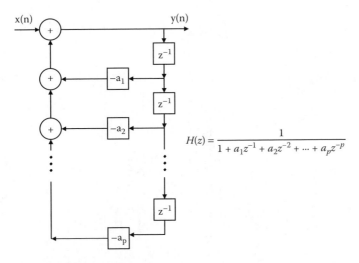

Figure 9.6.7 A pth-order IIR digital system.

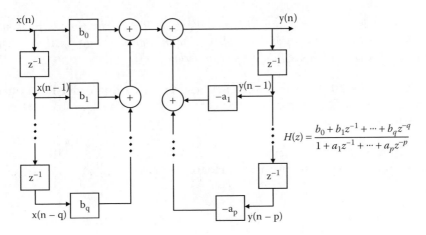

Figure 9.6.8 A combined FIR and IIR digital system.

Therefore, the transfer function for the first-order systems is

$$H(e^{j\omega}) = \frac{b_0 + b_1 e^{-j\omega}}{1 + a_1 e^{-j\omega}} \qquad (9.51)$$

This is the **frequency response function**.

If we set $\omega = \omega + 2\pi$ in $H(.)$ we find that $H(e^{j(\omega+2\pi)}) = H(e^{j\omega}e^{j2\pi}) = H(e^{j\omega})$, which indicates that the frequency response function is periodic with period 2π. If, on the other hand, we had introduced $z = e^{j\omega T}$ (T = sampling time), then the frequency response function would be of the form

$$H(e^{j\omega T}) = \frac{b_0 + b_1 e^{-j\omega T}}{1 + a_1 e^{-j\omega T}} \qquad (9.52)$$

If we set

$$\omega = \omega + \frac{2\pi}{T}$$

in $H(.)$, we find that

$$H\left(e^{j(\omega + \frac{2\pi}{T})T}\right) = H(e^{j\omega T} e^{j2\pi}) = H(e^{j\omega T}),$$

which indicates that the frequency response function is periodic with period $2\pi/T$.

Note: *Discrete systems with unit sampling time ($T = 1$) have periodic frequency response functions with period 2π, and those with time sampling equal to T have periodic frequency response functions with period $2\pi/T$.*

Example 9.7.1: Find the frequency response of a first-order FIR system and plot its amplitude and phase spectra for $b_0 = 1$ and $b_1 = 0.5$. Plot both cases using $T = 1$ and $T = 2$.

Solution: The frequency response function of a first-order FIR system is found from (9.51) by setting $a_1 = 0$. Hence, we have

$$H(e^{j\omega T}) = \frac{b_0 + b_1 e^{-j\omega T}}{1} = b_0 + b_1 e^{-j\omega T} = b_0 + b_1 \cos\omega T - jb_1 \sin\omega T$$

From the above equation, the amplitude and phase spectra are

$$\sqrt{H(e^{j\omega T}) H^*(e^{j\omega T})} = \sqrt{(b_0 + b_1 e^{j\omega T})(b_0 + b_1 e^{-j\omega T})} = \sqrt{b_0^2 + 2b_0 b_1 \cos\omega T + b_1^2}$$

$$ph\{H(e^{j\omega T})\} = -\tan^{-1}\frac{b_1 \sin\omega T}{b_0 + b_1 \cos\omega T}$$

The plots are shown in Figure 9.7.1. Note that for $T = 1$, the periodicity of the spectra is $2\pi/1 = 2\pi$, and for $T = 2$, the periodicity is $2\pi/2 = \pi$. ∎

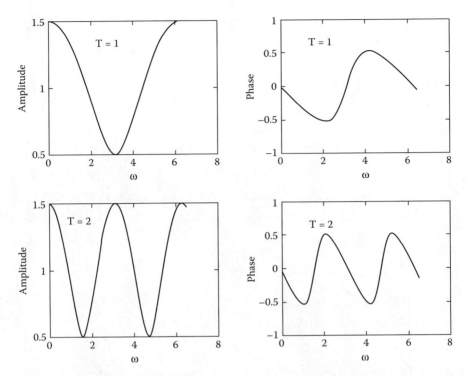

Figure 9.7.1 Spectra of a first-order FIR system for $T = 1$ and $T = 2$.

Example 9.7.2: Find the frequency characteristics of the system shown in Figure 9.7.2, which is made up of two FIR systems in cascade.

Solution: The difference equation describing the system is

$$y(n) = 4x(n) + 4x(n-1) + x(n-2)$$

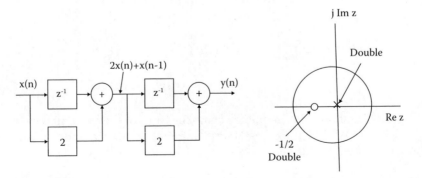

Figure 9.7.2 Two first-order FIR systems (identical) in cascade.

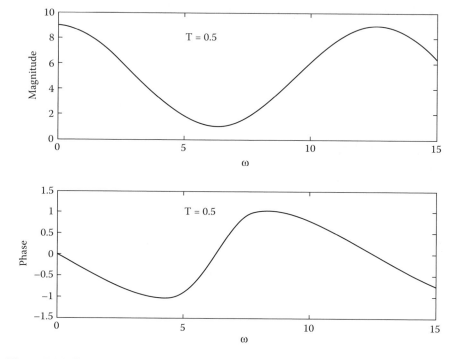

Figure 9.7.3 Frequency characteristics of two FIR systems in cascade.

The system function for this system is

$$H(z) \triangleq \frac{Y(z)}{X(z)} = 4 + 4e^{-j\omega T} + e^{-j2\omega T}$$

The frequency responses are obtained from the relationships

$$\left[\left[H(z)H^*(z)\right]_{z=e^{j\omega T}}\right]^{1/2} = \left[H(z)H(z^{-1})\Big|_{z=e^{j\omega T}}\right]^{1/2}$$

$$= (33 + +40\cos\omega T + 8\cos 2\omega T)^{1/2}$$

Figure 9.7.3 shows the frequency characteristics of the combined system using the MATLAB functions **abs(H)** and **angle(H)**. Note that the periodicity in this case is equal to $2\pi/0.5 = 4\pi$.

Note: *The amplitude functions are even functions of the frequency and the phase functions are odd functions of the frequency.* ∎

Chapter 9: The z-transform, difference equations, and discrete systems

Phase shift in discrete systems

Let us assume that a discrete system is described by the difference equation

$$y(n) = x(nT) + 0.6x(nT - T)$$

The system function for this is

$$H(z) \triangleq \frac{Y(z)}{X(z)} = 1 + 0.6z^{-1}$$

and the frequency response function is then given by

$$\left[H(z)H(z^{-1}) \Big|_{z=e^{j\omega T}} \right]^{1/2} = (1.36 + 1.2 \cos \omega T)^{1/2}$$

and the phase is

$$\theta(\omega) = \arg H(e^{j\omega T}) = -\tan^{-1} \frac{0.6 \sin \omega T}{1 + 0.6 \cos \omega T}$$

Since we have shown above that when the input is the complex function $e^{j\omega T}$ the output is $H(e^{j\omega T})e^{j\omega nT} = |H(e^{j\omega T})| e^{j\theta(\omega)} e^{j\omega T}$, when the input is $\cos \omega nT$, the output is the real part of the output, which is

$$y(nT) = |H(e^{j\omega T})| \cos[\omega nT + \theta(\omega)]$$

If we set $T = 0.5$ and $\omega = 0.4\pi$ in the above equations, then the input and output are shown in Figure 9.7.4. Observe the phase shift to the left and the decrease of the amplitude of the output signal. The general input–output relations of a discrete system to sinusoidal inputs are shown in Figure 9.7.5.

*9.8 Frequency response of higher-order digital systems

If we set $z = e^{j\omega T}$ in (9.46) we obtain the general frequency response function

$$\boxed{H(e^{j\omega T}) = \frac{b_0 + b_1 e^{-j\omega T} + b_2 e^{-j2\omega T} + \cdots + b_q e^{-jq\omega T}}{a_0 + a_1 e^{-j\omega T} + a_2 e^{-j2\omega T} + \cdots + a_p e^{-jp\omega T}} = H(z) \Big|_{z=e^{j\omega T}}} \quad (9.53)$$

of a combined system of qth-order FIR and a pth-order IIR. If, for example, we want to study a third-order system, then the transfer function is

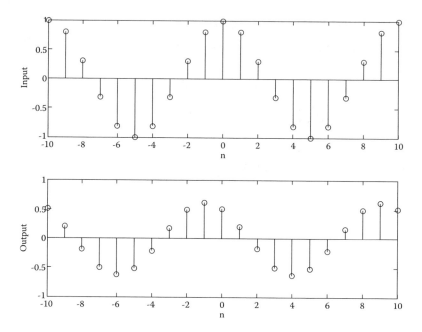

Figure 9.7.4 The input–output results of a discrete system.

$$\begin{array}{c} e^{j\omega T} \\ \xrightarrow{} \\ x(nT) = \cos\omega nT \\ x(nT) = \sin\omega nT \end{array} \boxed{H(z)} \begin{array}{l} y(nT) = H(e^{j\omega T})e^{j\omega T} = \left|H(e^{j\omega T})\right|e^{j[\omega T + \theta(\omega)]} \\ y(nT) = \left|H(e^{j\omega T})\right|\cos[\omega nT + \theta(\omega)] \\ y(nT) = \left|H(e^{j\omega T})\right|\sin[\omega nT + \theta(\omega)] \end{array}$$

Figure 9.7.5 Input–output relationship for discrete sinusoidal signals.

$$H(z) = \frac{b_0 + b_1 e^{-j\omega T} + b_2 e^{-j2\omega T}}{a_0 + a_1 e^{-j\omega T} + a_2 e^{-j2\omega T}} \tag{9.54}$$

The amplitude squared is given by

$$|H(z)|^2 \Big|_{z=e^{j\omega T}} = H(z)H(z^{-1}) \Big|_{z=e^{j\omega T}} = \frac{d_2 z^2 + d_1 z + d_0 + d_1 z^{-1} + d_2 z^{-2}}{c_2 z^2 + c_1 z + c_0 + c_1 z^{-1} + c_2 z^{-2}} \Big|_{z=e^{j\omega T}}$$

$$= \frac{d_0 + \sum_{n=1}^{2} 2d_n \cos n\omega T}{c_0 + \sum_{n=1}^{2} 2c_n \cos n\omega T} \tag{9.55}$$

Chapter 9: The z-transform, difference equations, and discrete systems

where

$$c_n = \sum_{k=0}^{k-n} a_k a_{n+k} \qquad d_n = \sum_{k=0}^{k-n} b_k b_{n+k} \qquad (9.56)$$

For example,

$$c_0 = \sum_{k=0}^{2-0} a_k a_{0+k} = a_0 a_0 + a_1 a_1 + a_2 a_2$$

$$c_1 = \sum_{k=0}^{2-1} a_k a_{1+k} = a_0 a_1 + a_1 a_2$$

$$c_2 = \sum_{k=0}^{2-2} a_k a_{2+k} = a_0 a_2$$

Example 9.8.1: Find and plot the frequency characteristics of a general third-order discrete system with the following constants: $b_0 = 1$, $b_1 = 0.8$, $b_2 = 0.1$, $a_0 = 1$, $a_1 = -0.6$, $a_2 = 0.8$, $T = 2$.

Solution: The frequency characteristics are plotted in Figure 9.8.1 using the MATLAB functions **abs**(H) and **angle**(H). The transfer function with the given data is

$$H(e^{j2\omega}) = \frac{1 + 0.8 e^{-j2\omega} + 0.1 e^{-j4\omega}}{1 - 0.6 e^{-j2\omega} + 0.8 e^{-j4\omega}}$$

■

Example 9.8.2: Find the frequency response function of the system shown in Figure 9.8.2.

Solution: We can solve this problem in two different ways. From the figure, the difference equations describing the first and second subsystems are respectively

$$x_1(n-1) + 2x_1(n) = y(n)$$

$$-[y(n-1) + 2y(n)] + x(n) = x_1(n)$$

Substituting $x_1(n)$ from the second equation into the first, we obtain

$$5y(n) + 4y(n-1) + y(n-2) = 2x(n) + x(n-1)$$

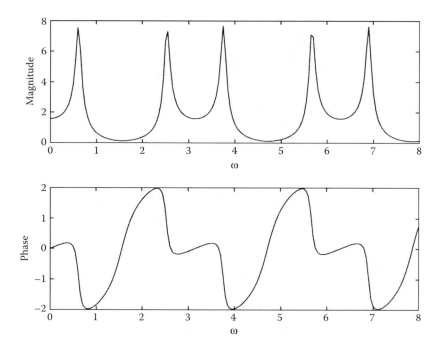

Figure 9.8.1 Second-order combined FIR and IIR discrete systems.

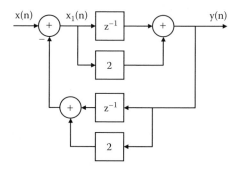

Figure 9.8.2 A discrete interconnected system.

By taking the z-transform of the above equation and solving for the ratio of output over input, we obtain the transfer function

$$H(z) = \frac{2+z^{-1}}{5+4z^{-1}+z^{-2}}$$

Since the transfer function for each subsystem is the same, $H_s(z) = 2 + z^{-1}$, the transfer function of the total system is found using the feedback transfer function given in Figure 2.5.4. ∎

9.9 z-Transform solution of first-order difference equations

We now study the use of z-transform methods in the solution of linear difference equations with constant coefficients. The following examples will explain how to use the z-transform method for the solution of difference equations.

Example 9.9.1: Solve the discrete-time problem defined by the equation

$$y(n) + 2y(n-1) = 3.5u(n) \qquad (9.57)$$

with $y(-1) = 0$ and $u(n)$ is the discrete unit step function.

Solution: Begin by taking the z-transform of both sides of (9.57). This is

$$\mathcal{Z}\{y(n)\} + 2\mathcal{Z}\{y(n-1)\} = 3.5\mathcal{Z}\{u(n)\}$$

By Section 9.3 and Table 9.4.1, we write

$$Y(z) + 2z^{-1}Y(z) = 3.5\frac{z}{z-1} \quad \text{or} \quad Y(z) = 3.5\frac{z}{z-1}\frac{z}{z+2} = \frac{6}{7}\frac{z}{z-1} + \frac{7}{3}\frac{z}{z+2}$$

The inverse z-transform of this equation is

$$y(n) = \frac{7}{6}u(n) + \frac{7}{3}(-2)^n \qquad n = 0, 1, 2, \cdots \qquad \blacksquare$$

If you want to use the MATLAB function **residue(.)**, you must solve for the $Y(z)/z$ function and then multiply both sides by z.

Example 9.9.2: Repeat Example 9.9.1, but now with the initial condition $y(-1) = 4$.

Solution: We again begin by taking the z-transform of both sides and use the right-shift property (see Section 9.3):

$$\mathcal{Z}\{y(n)\} + 2\mathcal{Z}\{y(n-1)\} = 3.5\mathcal{Z}\{u(n)\} \quad \text{or} \quad Y(z) + 2z^{-1}Y(z) + 2y(-1) = 3.5U(z)$$

Upon solving for Y(z), we obtain

$$Y(z) = \underbrace{\frac{3.5}{1+2z^{-1}}U(z)}_{\substack{\text{zero-state}\\\text{response}}} - \underbrace{\frac{2y(-1)}{1+2z^{-1}}}_{\substack{\text{zero-input}\\\text{response}}}$$

The inverse transform of the *zero-input response* is the solution of the homogeneous difference equation $y(n)+2y(n-1)=0$, a result that can be readily verified by setting consecutively $n = 0, 1, 2, \ldots$ in the equation. Specifically, the results are

$$\text{zero-input response} = \mathcal{Z}^{-1}\left\{-\frac{2y(-1)}{1+2z^{-1}}\right\} = -2y(-1)(-2)^n u(n)$$

$$\text{zero-state response} = \mathcal{Z}^{-1}\left\{\frac{3.5}{1+2z^{-1}}\frac{1}{1-z^{-1}}\right\} = \mathcal{Z}^{-1}\left\{\frac{3.5z^2}{(z+2)(z+1)}\right\}$$

$$= \mathcal{Z}^{-1}\left\{7\frac{z}{z+2} - 3.5\frac{z}{z+1}\right\}$$

The complete solution is the sum of these two responses. Hence, we write

$$y(n) = \underbrace{4(-2)^{n+1}}_{\substack{\text{zero-input}\\\text{response}}} + \underbrace{7(-2)^n - 3.5u(n)}_{\substack{\text{zero-state}\\\text{response}}} \qquad n = 0, 1, 2, \cdots$$

and

$$y(n) = \underbrace{-(-2)^n}_{\text{transient}} - \underbrace{3.5u(n)}_{\text{steady state}} \qquad n = 0, 1, 2, \cdots$$

■

Example 9.9.3: Determine the output of the discrete approximation of the system shown in Figure 9.9.1a for a sampling time T. The outputs for $T = 0.2$ and $T = 1$ are to be plotted, and the results compared with the output of the continuous system. The input is a unit step current source $i(t) = u(t)$, and an initial condition $v_o(0) = 2$ V.

Solution: The differential equation describing the system is

$$\frac{dv_o(t)}{dt} + \frac{v_o(t)}{0.5} = i(t)$$

Chapter 9: The z-transform, difference equations, and discrete systems

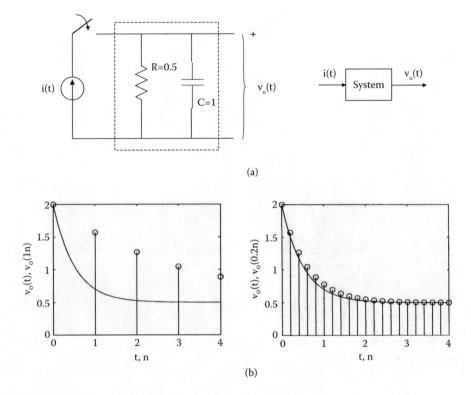

Figure 9.9.1 Illustration of Example 9.9.3.

The analogous discrete form of this equation is

$$\frac{v_o(nT) - v_o(nT-T)}{T} + 2v_o(nT) = i(nT)$$

From this,

$$v_o(nT) = \frac{T}{1+2T} i(nT) + \frac{1}{1+2T} v_o(nT-T) = b_0 i(nT) + a_1 v_o(nT-T)$$

The z-transform of this equation gives (see Section 9.3)

$$V_o(z) = \underbrace{\frac{b_0}{1-a_1 z^{-1}} I(z)}_{\text{zero-state response}} + \underbrace{\frac{a_1}{1-a_1 z^{-1}} v_o(-T)}_{\text{zero-input response}}$$

Since the continuous case $v_o(0) = 2$, we must refer back to the difference equation and set the appropriate values to find the value of $v_o(-T)$. Hence, we find

$$v_o(0) = b_0 i(0) + a_1 v_0(-T) \quad n = 0, \text{ or } v_o(-T) = (2-b_0)/a_1$$

Thus, we obtain

$$\text{zero-input response} = \mathcal{Z}^{-1}\left\{b_0 \frac{z^2}{(z-1)(z-a_1)}\right\} = \mathcal{Z}^{-1}\left\{\frac{b_0}{1-a_1}\frac{z}{z-1} + \frac{b_0 a_1}{a_1-1}\frac{z}{z-a_1}\right\}$$

$$= \frac{b_0}{1-a_1} u(n) + \frac{b_0 a_1}{a_1-1} a_1^n u(n)$$

$$\text{zero-state response} = \mathcal{Z}^{-1}\left\{\frac{(2-b_0)z}{(z-a_1)}\right\} = (2-b_0) a_1^n u(n)$$

The solution is the sum of the *zero-input* and *zero-state* solutions. The solution of the continuous case is easily found to be equal to

$$v_o(t) = 0.5 + 1.5 e^{-2t} \quad t \geq 0$$

The results for the continuous and discrete cases $T = 1$ and $T = 0.2$ are shown in Figure 9.9.1b. ∎

*9.10 Higher-order difference equations

The class of linear discrete systems under discussion is described by the general difference equation

$$y(n) + a_1 y(n-1) + a_2 y(n-2) + \cdots + a_p y(n-p) = b_0 x(n) \quad (9.58)$$

The constant in front of $y(n)$ does not appear since we can always divide both sides of the equation by that number. We have seen that such an equation can arise when a differential equation is transformed into an equivalent difference equation using approximations of the derivatives. Hence,

$$\frac{dy(t)}{dt} \cong \frac{y(nT) - y(nT-T)}{T} \quad \text{a)}$$

$$\frac{d^2 y(t)}{dt^2} \cong \frac{y(nT) - 2y(nT-T) + y(nT-2T)}{T^2} \quad \text{b)} \quad (9.59)$$

$$\frac{d^3 y(t)}{dt^3} \cong \frac{y(nT) - 3y(nT-T) + 3Y(nT-2T) - y(nT-3T)}{T^3} \quad \text{c)}$$

Chapter 9: The z-transform, difference equations, and discrete systems

We would like to mention again that when the coefficients a_i are independent of n, the system is time invariant; otherwise, it is time varying.

For $T = 1$, the second-order difference equation can always be written in the form

$$\boxed{y(n) + a_1 y(n-1) + a_2 y(n-2) = b_0 x(n)} \qquad (9.60)$$

We can assert that a complete and unique solution to the above equation can be found if the initial conditions are known, say,

$$y(-1) = A \qquad y(0) = B \qquad (9.61)$$

where A and B are constants. In this connection, we state certain theorems without proof (see Finizio and Ladas*).

Definition 9.10.1: If $\{a(n)\}$ and $\{b(n)\}$ denote two sequences, the determinant

$$C[a(n), b(n)] = \begin{vmatrix} a(n) & b(n) \\ a(n-1) & b(n-1) \end{vmatrix} \qquad (9.62)$$

is known as the **Casoratian** or **Wroskian** determinant. ∎

Theorem 9.10.1: Two solutions, $y_1(n)$ and $y_2(n)$, of the linear homogeneous difference equation (the input $b_0 x(n)$ is set equal to zero) are linearly independent if and only if their Casoratian

$$C[y_1(n), y_2(n)] = \begin{vmatrix} y_1(n) & y_2(n) \\ y_1(n-1) & y_2(n-1) \end{vmatrix} \qquad (9.63)$$

is different from zero for all values of $n = 0, 1, 2, \ldots$ ∎

Theorem 9.10.2: If $y_1(n)$ and $y_2(n)$ are two linear independent solutions of the homogeneous equation and $y_p(n)$ is the particular solution to the nonhomogeneous equation (9.60), then the general solution is

$$y(n) = y_h(n) + y_p(n) = C_1 y_1(n) + C_2 y_2(n) + y_p(n) \qquad (9.64)$$

where C_1 and C_2 are arbitrary constants and can be determined from the initial conditions. ∎

---

* N. Finizio and G. Ladas, *An Introduction to Differential Equations*, Belmont, CA: Wadsworth Publishing Co., 1982.

Table 9.10.1 Solutions to Homogeneous Difference Equations

Difference Equation	Characteristic Equation
$y(n) + a_1 y(n-1) + a_2 y(n-2) = 0$	$\lambda^2 + a_1 \lambda + a_2 = 0$
$\lambda_1 \neq \lambda_2$	$y(n) = c_1 \lambda_1^n + c_2 \lambda_2^n$
$\lambda_1 = \lambda_2 = \lambda$	$y(n) = c_1 \lambda^n + c_2 n \lambda^n$
$\lambda_1 = a + jb;\ \lambda_2 = a - jb$	$y(n) = C_1 e^{an} \cos bn + C_2 e^{an} \sin bn$

Note: All the c_i's and C_i's are constants.

Theorem 9.10.3: The homogeneous difference equation

$$y(n) + a_1 y(n-1) + a_2 y(n-2) = 0 \tag{9.65}$$

with constant and real coefficients and with the roots of the characteristic equation

$$\lambda^2 + a_1 \lambda + a_2 = 0 \tag{9.66}$$

denoted λ_1 and λ_2, has the possible solutions shown in Table 9.10.1.

Solution: Assume that $y(n) = c\lambda^n$ is a solution of (9.65). Then

$$c\lambda^n + a_1 c \lambda^{n-1} + a_2 c \lambda^{n-2} = 0 \quad \text{or} \quad \lambda^2 + a_1 \lambda + a_2 = 0$$

which is (9.66). ∎

Example 9.10.1: Solve the difference equation

$$y(n) - by(n-1) + ay(n-2) = u(n) \quad a, b > 0$$

The constants are to be selected so that the second-order system is (a) **critically damped**, (b) **underdamped**, or (c) **overdamped**. The starting conditions are zero: $y(-1) = 0$, $y(-2) = 0$.

Solution: Taking the z-transform of both sides of the difference equation and solving for $Y(z)$, we obtain

$$Y(z) = \frac{z^2}{z^2 - bz + a} \frac{z}{z - 1} \tag{9.67}$$

The denominator of the first factor, which is the characteristic equation of the difference equation, has two roots that are specified by

Chapter 9: The z-transform, difference equations, and discrete systems

$$z_{1,2} = \frac{b \pm \sqrt{b^2 - 4a}}{2} \qquad (9.68)$$

Critically damped case ($b^2 = 4a$)

We set $b = 0.8$, and hence $a = 0.8$. The two roots are 0.4 and 0.4. Thus, (9.67) becomes

$$Y(z) = \frac{z^2}{(z-0.4)^2} \frac{z}{z-1} = \frac{Az}{z-1} + \frac{Bz}{z-0.4} + \frac{Cz^2}{(z-0.4)^2}$$

By straightforward methods we find $A = 25/9$, $B = -1.1111$, and $C = -2/3$. The inverse transform of this equation is

$$y(n) = \frac{25}{9} - 1.1111(0.4)^n - \frac{2}{3}(n+1)(0.4)^n$$

The response to a step function can also be found using the MATLAB function `[y,t]=stepz(num,den)`. For this particular case and for $n = 0, 1, 2, \ldots 30$ we write

`[y,t]=stepz([1 0 0],[1 -0.8 0.16],30);`

Overdamped case ($4a < b^2$)

If we select $b = 0.9$ and $a = 0.1$, the inequality holds. The two roots are 0.1298 and 0.7702. The z-transform of the output is given by

$$Y(z) = 5\frac{z}{z-1} + 0.0303\frac{z}{z-0.1298} - 4.0303\frac{z}{z-0.7702}$$

and its inverse is

$$y(n) = 5 + 0.0303(0.1298)^n - 4.0303(0.7702)^n \qquad n = 0,1,2,\cdots$$

Underdamped case ($b^2 < 4a$)

In this case, two conjugate roots exist, and these roots are poles of $Y(z)$. To proceed, we write the denominator of $Y(z)$ in the form

$$z^2 - bz + a = (z - ce^{j\theta})(z - ce^{-j\theta}) \qquad (9.69)$$

By expanding the right-hand side and equating like powers of z, we find that

$$a = c^2 \quad \text{and} \quad b = 2c\cos\theta$$

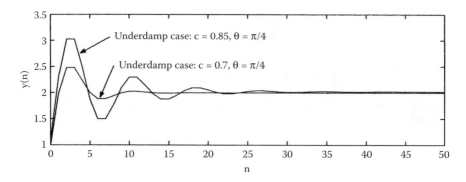

Figure 9.10.1 Step response of a second-order system.

Thus, if a and b are known, then c and θ are readily obtained. By combining (9.69) with (9.67), we write

$$Y(z) = \frac{z^2}{(z - ce^{j\theta})(z - ce^{-j\theta})} \frac{z}{z-1}$$

This fraction is now expanded into fractional form, which is

$$Y(z) = \frac{1}{1 - 2c\cos\theta + c^2} \frac{z}{z-1} - \frac{ce^{j2\theta}}{j2\sin\theta(1 - ce^{j\theta})} \frac{z}{z - ce^{j\theta}}$$

$$+ \frac{ce^{-j2\theta}}{j2\sin\theta(1 - ce^{-j\theta})} \frac{z}{z - ce^{-j\theta}}$$

The inverse z-transform of this equation is

$$y(n) = \frac{1}{1 - 2c\cos\theta + c^2} u(n) - \frac{ce^{j2\theta}}{j2\sin\theta(1 - ce^{j\theta})} c^n e^{jn\theta} + \frac{ce^{-j2\theta}}{j2\sin\theta(1 - ce^{-j\theta})} c^n e^{-jn\theta}$$

This expression can be written in a more convenient form by writing $1 - ce^{j\theta} = re^{-j\phi}$; then, $1 - ce^{-j\theta} = re^{j\phi}$. Figure 9.10.1 shows the underdamped cases for $c = 0.85$ and $\theta = \pi/4$ and for $c = 0.7$ and $\theta = \pi/4$. ∎

Method of undetermined coefficients

The particular solution of a nonhomogeneous equation is the method of **undetermined coefficients**. The method is particularly efficient for input functions that are linear combinations of the following functions:

Chapter 9: The z-transform, difference equations, and discrete systems   455

1. n^k, where n is a positive integer or zero
2. a^n, where a is a nonzero constant
3. cos an, where a is a nonzero constant
4. sin an, where a is a nonzero constant
5. A product (finite) of two or more functions of type 1 to 4

This method works because any derivative of the input function $x(n)$ is also possible as a linear combination of functions of the five types above. For example, the function $2n^2$ or any derivative of $2n^2$ is a linear combination of the sequences n^2, n and 1, all of which are of type 1. Hence, what is required in any case is the appropriate sequences for which any derivative of the input function $x(n)$ can be constructed by a linear combination of these sequences. Clearly, if $x(n) = \cos 3n$, the appropriate sequences are cos $3n$ and sin $3n$. The following examples clarify these assertions.

Example 9.10.2: Consider the system shown in Figure 9.10.2. Find the general solution if the input is $x(n) = 3^n u(n)$ and initial conditions are $y(-1) = 0$ and $y(-2) = 1$.

Solution: From Figure 9.10.2, and taking into consideration the time shift of the output function, the controlling difference equation of this system is

$$y(n) - 5y(n-1) + 6y(n-2) = 3^n \qquad n = 0, 1, 2, \cdots \qquad (9.70)$$

The characteristic equation obtained from the corresponding homogeneous equation is

$$\lambda^2 - 5\lambda + 6 = 0$$

with roots $\lambda_1 = 2$ and $\lambda_2 = 3$. The two solutions are

$$y_1(n) = 2^n \qquad y_2(n) = 3^n \qquad (9.71)$$

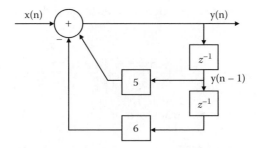

Figure 9.10.2 A discrete-time system.

Since one of the homogeneous solutions is proportional to the input function, the particular solution is of the form $y_p(n) = Bn3^n$, where B is a constant. Introducing the particular solution in (9.70), we obtain

$$Bn3^n - 5b(n-1)3^{n-1} + 6B(n-2)3^{n-2} = 3^n$$

Solving for the unknown, we find that $B = 3$. Hence, the total solution is

$$y(n) = C_1 y_1(n) + C_2 y_2(n) + y_p(n) = C_1 2^n + C_2 3^n + 3n(3^n) \qquad (9.72)$$

Subjecting this solution to the given initial conditions yields

$$\frac{1}{2}C_1 + \frac{1}{3}C_2 = 1$$

$$\frac{1}{4}C_1 + \frac{1}{9}C_2 = \frac{5}{3}$$

From this system we find $C_1 = 16$ and $C_2 = -21$. The complete solution is

$$y(n) = 16(2^n) - 21(3^n) + 3n(3^n) \qquad (9.73)$$

We now solve this problem by the z-transform method. The z-transform of (9.70) is

$$Y(z) - 5[z^{-1}Y(z) + y(-1)] + 6[z^{-2}Y(z) + y(-1)z^{-1} + y(-2)z^{-0}] = \frac{z}{z-3}$$

which can be written as follows:

$$Y(z) = -\frac{6z^2}{(z-2)(z-3)} + \frac{z^3}{(z-2)(z-3)^2}$$

The terms to the right are expanded (two terms and three terms, respectively) as follows:

$$Y(z) = 12\frac{z}{z-2} - 18\frac{z}{z-3} + 4\frac{z}{z-2} - 6\frac{z}{z-3} + 3\frac{z^2}{(z-3)^2}$$

By Table 9.4.2, the inverse z-transform is

$$y(n) = 12(2^n) - 18(3^n) + 4(2^n) - 6(3^n) + 3(n+1)3^n = 16(2^n) - 21(3^n) + 3n(3^n)$$

which is (9.73) as anticipated. ∎

Chapter 9: The z-transform, difference equations, and discrete systems

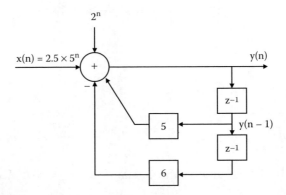

Figure 9.10.3 Second-order discrete system.

Example 9.10.3: Find the solution for the system shown in Figure 9.10.3. The initial conditions are $y(-1) = 1$, $y(-2) = 0$.

Solution: The difference equation describing the system of Figure 9.10.3 is

$$y(n) - 5y(n-1) + 6y(n-2) = 2.5 \times 5^n + 2^n \qquad (9.74)$$

As found in the example above, the roots are the same, and hence the general solution of the homogeneous equation is

$$y_h(n) = C_1 2^n + C_2 3^n \qquad (9.75)$$

We observe that the input 2^n is also a solution to the homogeneous equation. This suggests that we try the following particular solution:

$$y_p(n) = A5^n + Bn2^n \qquad (9.76)$$

where the constants A and B are undetermined coefficients. These constants can be found by substituting (9.76) into (9.74). When this is done, we obtain

$$A5^n + Bn2^n - 5[A5^{n-1} + B(n-1)2^{n-1}]$$
$$+ 6[A5^{n-2} + B(n-2)2^{n-2}] = 2.5 \times 5^n + 2^n$$

Rearranging terms, we have

$$\left(A - 5A\frac{1}{5} + 6A\frac{1}{25}\right)5^n + \left(Bn + \frac{5B - 5Bn}{2} + \frac{6Bn - 12B}{4}\right)2^n = 2.5 \times 5^n + 2^n$$

By equating coefficients of similar terms, we find that

$$A = \frac{62.5}{6} \qquad B = -2$$

and therefore, the total solution is

$$y(n) = y_h(n) + y_p(n) = C_1 2^n + C_2 3^n + \frac{62.5}{6} 5^n - 2n2^n$$

Applying the initial conditions, the constants are $C_1 = -8.667$ and $C_2 = 6.750$. ∎

Example 9.10.4: Find the particular solution of the nonhomogeneous equation

$$y(n) - y(n-2) = 5n^2 \qquad (9.77)$$

Solution: The roots of the characteristic equation are readily found to be $\lambda_1 = 1$ and $\lambda_2 = -1$. Since the characteristic equation is $\lambda^2 - 1 = 0$, we can use the MATLAB function **roots([1 0 -1])** to find the roots. Thus, the solution to the homogeneous equation is

$$y_h(n) = C_1(1)^n + C_2(-1)^n \qquad n = 0,1,2,\cdots$$

We observe that the function $5n^2$ and its derivatives can be found by the linear combination of sequences n^2, n and 1. However, 1 is a solution of the homogeneous equation, so we choose as a trial particular solution n times the sequence n^2, n and 1:

$$y_p(n) = An^3 + Bn^2 + Cn \qquad (9.78)$$

Substitute this trial solution in (9.77) and equate coefficients of equal power terms. The coefficients are found to be $A = 5/6$, $B = 5/2$, and $C = 5/3$. ∎

Table 9.10.2 gives the suggested forms for the particular solutions for a specified $x(n)$.

Table 9.10.2 Method of Undetermined Coefficients

$x(n)$	$y_p(n)$
n^m	$A_1 n^m + A_2 n^{m-1} + \cdots + A_m n + A_{m+1}$
a^n	Aa^n
$\cos \theta n$ or $\sin \theta n$	$A_1 \cos \theta n + A_2 \sin \theta n$
$n^m a^n$	$a^n (A_1 n^m + A_2 n^{m-1} + \cdots + A_m n + A_{m+1})$
$a^n \cos \theta n$ or $a^n \sin \theta n$	$a^n (A_1 \cos \theta n + A_2 \sin \theta n)$

Chapter 9: The z-transform, difference equations, and discrete systems

Important definitions and concepts

1. One-sided z-transform
2. Convergence of z-transform
3. Region of convergence
4. Two-sided z-transform and its region of convergence
5. The z-transform properties: linearity, right-shifting property, left-shifting property, time scaling, periodic sequence, multiplication by n, initial value, final value, convolution, and bilateral convolution
6. Inverse z-transform
7. System transfer function
8. Infinite impulse response (IIR) system
9. Finite impulse system (FIR)
10. Higher-order transfer functions
11. Frequency response of discrete systems
12. z-transform solution of difference equations
13. Zero-state and zero-input solution of discrete systems
14. Casoratian or Wroskian determinant
15. Critically damped, underdamped, and overdamped systems
16. Method of undetermined coefficients

Chapter 9 Problems

Section 9.1

1. Find the z-transform for the sequences shown in Figure P9.1.1.

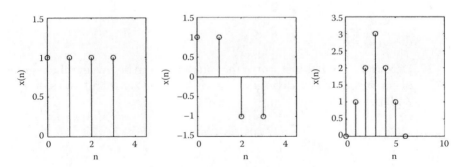

Figure P9.1.1

2. Determine the z-transform of the following sequences:

   a. $f(n) = \begin{cases} (1/2)^n & n = 0,1,2,\cdots \\ 0 & n < 0 \end{cases}$

b. $f(n) = \begin{cases} 0 & n \leq 0 \\ -1 & n = 1 \\ a^n & n = 2, 3, \cdots \end{cases}$

3. Find the z-transform of the following functions:

   a. $f(n) = \delta(n)$
   b. $f(n) = \delta(n - m)$
   c. $f(n) = 2\delta(n) - 2.5\delta(n - 2)$

4. Determine the z-transform of of the following functions:

   a. $f(n) = 2\delta(n) + 3\delta(n - 1) + 4\delta(n - 2)$
   b. $f(n) = \{1, -a, a^2, -a^3, a^4, -a^5, \cdots\}$

Section 9.2

1. Find the z-transforms and their regions of convergence and plot the pole-zero configurations of the sequences given for $n \geq 0$.

   a. $y(n) = 1 + n$
   b. $y(n) = n^2$
   c. $y(n) = a^n + a^{-n}$ $a > 1$
   d. $y(n) = e^{jn\theta}$

2. Determine the z-transform of the function given below and the region of convergence and divergence, and show the zeros and poles on the z-plane.

$$y(n) = \begin{cases} a^n \cos nb & n \geq 0, \, a > 0 \\ 0 & n < 0 \end{cases}$$

3. Determine the z-transforms and their regions of convergence and plot the pole configuration of the following sequences:

   a. $y(n) = \begin{cases} 2 & n = 0, 1, 2, 3, 4 \\ 3^n & n = 5, 6, \cdots \end{cases}$
   b. $y(n) = (1/3)^n + (1/4)^n$ $n = 0, 1, 2, \cdots$

4. Determine the z-transform and the region of convergence and plot the pole-zero configuration of the following sequence:

$$y(n) = \begin{cases} 2^n & 0 \le n \le 5 \\ 0 & \text{otherwise} \end{cases}$$

5. Show that the region of convergence of the sequence $y(n) = a(b)^n$ for $n = 0, 1, 2, \cdots$ depends only on b.
6. Determine the z-transform of the following functions when sampled every T seconds:

   a. $y(n) = \cos \omega t \, u(t)$

   b. $y(n) = a^t \sin \omega t \, u(t)$

   c. $y(n) = a^t \cos \omega t \, u(t)$

   d. $y(n) = e^{-at} \sin \omega t \, u(t)$

   e. $y(n) = e^{-at} \cos \omega t \, u(t)$

   f. $y(n) = a^t \, u(t)$

7. Determine the z-transform and the region of convergence of the following functions that are sampled every T seconds:

   a. $y(n) = e^{-at} u(t)$, a is a real number

   b. $y(n) = u(t) - u(t - 4T)$

Section 9.3

1. Generalize the z-transform property $\mathcal{Z}\{ny(n)\} = -z\dfrac{d}{dz}Y(z)$ and show that

$$\mathcal{Z}\{n^m y(n)\} = \left(-z\dfrac{d}{dz}\right)^m Y(z)$$

2. Two sequences are given. Determine their z-transforms, their regions of convergence, and their pole-zero configurations. Compare the two results and state your observations.

   a. $y(n) = 2e^{-n} \quad n = 0, 1, 2, \cdots$

   b. $y(n) = 2e^{-(n-2)} u(n-2) \quad n = 0, 1, 2, \cdots$

3. Determine the z-transforms and the regions of convergence of the following sequences:

   a. $y(n) = e^{\pm j\omega_0 n} x(n) \quad n = 0, 1, 2, \cdots$

   b. $y(n) = z_0^n x(n) \quad n = 0, 1, 2, \cdots$

4. Deduce the z-transform of the sequence shown in Figure P9.3.4 and compare the pole-zero configurations between $y(n)$ and $y_{(1)}(n)$.

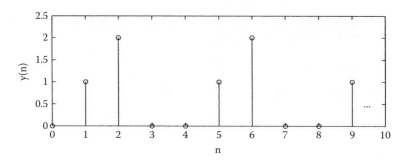

Figure P9.3.4

5. Given the relationship $\mathcal{Z}\{u(n)\} = 1/(1 - z^{-1})$, use the multiplication by n property to obtain the inverse transform of $Y(z) = z^{-1}/(1 - z^{-1})^2$.

6. Prove that the convolution is commutative; hence, we have the following:

$$y(n) = \sum_{m=0}^{n} h(n-m)x(m) = \sum_{m=0}^{n} h(m)x(n-m)$$

7. Deduce the z-transform of the output $y(n)$ of the system shown in Figure P9.3.7 if the input is a unit step function $u(n)$ and $h_3(n) = 0.5^n u(n)$.

8. Use the convolution theorem to find the voltage output of the system shown in Figure P9.3.8 for an input current $i(n) = e^{-0.2n} \quad n = 0, 1, 2, \cdots$. Compare these results to those found in the corresponding continuous-time case.

9. Show that the z-transform of the shifted function $y(t)u(t)$ sampled every T seconds and shifted to the right by mT seconds is $z^{-m}Y(z)$.

10. Deduce the z-transform of the shifted function $(nT - 2T)[u(nT - 2T) - u(nT - 5T)]$.

11. Show that the z-transform of the function $y(t)u(t)$ sampled every T seconds and shifted to the left by mT seconds is equal to

$$z^m [Y(z) - \sum_{n=0}^{m-1} y(nT) z^{-n}].$$

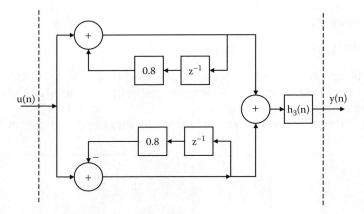

Figure P9.3.7

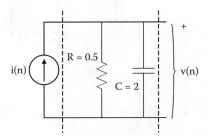

Figure P9.3.8

12. Determine the z-transform of the output of the system shown in Figure P9.3.12.

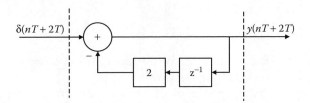

Figure P9.3.12

13. Deduce the z-transform of the following functions using the time-scaling property:

   a. $y(n) = a^n e^{-bn} u(n)$

   b. $y(n) = a^n \cos n\omega\, u(n)$

c. $y(n) = na^n u(n)$

d. $y(n) = a^n e^{-bn} \sin n\omega \, u(n)$

14. Show that the z-transform of the function $f(t) = a^t y(t)u(t)$ sampled every T seconds (time-scaling property of sampled signals) is $F(z) = Y(a^{-T}z)$.

15. Determine the z-transform of the sequences shown in Figure P9.3.15.

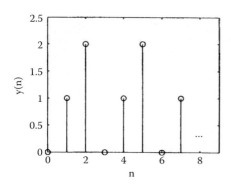
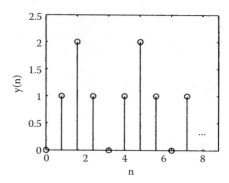

Figure P9.3.15

16. Show that the z-transform of the function $f(t) = ty(t)u(t)$ sampled every T seconds is $F(z) = -Tz[dY(z)/dz]$.

17. Use the results of Problem 9.3.16 to deduce the z-transforms of the following functions:

    a. $y(nT) = nTu(nT)$

    b. $y(nT) = (nT)^2 u(nT)$

    c. $y(nT) = nT(nT + T)u(nT)$

    d. $y(nT) = nT(nT - T)u(nT)$

18. Show that the z-transform of the convolution of two sampled functions is given by $\mathcal{Z}\{h(nT) * x(nT)\} = H(z)X(z)$.

19. Determine the response of the system shown in Figure P9.3.19 in discrete form. Assume $T = 0.5$. Use the convolution theorem of the input–output relationship of LTI systems.

Section 9.5

1. Deduce the function $f(n)$ specified by the following one-sided z-transforms. Employ fraction expansion.

Chapter 9: The z-transform, difference equations, and discrete systems   465

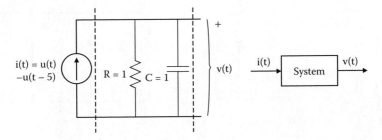

Figure P9.3.19

    a. $F(z) = 1/[(1-z^{-1})(1-2z^{-1})]$

    b. $F(z) = z^{-1}/[(2-z^{-1})^2]$

    c. $F(z) = [z^2(z+2)]/(z^2+4z+3)$

2. Determine the inverse functions for the following z-transforms:

    a. $z/[(z-1)(z-2)]$ $1 < |z| < 2$

    b. $z^2/[(z-1)^2(z-2)]$ $1 < |z| < 2$

    c. $z^3/[(z-1)(z-2)^2]$ $1 < |z| < 2$

    d. Repeat part (c) for the entire region $|z| > 1$.

3. Find the inverse z-transform of the function $F(z) = z^2 - 3z + 8)/[(z-2)(z+2)(z+3)]$ for $3 < |z| < 2$ and $2 < |z| < 3$.
4. Determine the inverse z-transform of the function $F(z) = (z\sin a)/[z^2 - (2\cos a)z + 1]$ for $|z| > 1$.
5. Determine the inverse z-transforms of the following functions:

    a. $(z+2)/[z^2(z-1)]$

    b. $(z^3 + 2z^2 + z + 1)/(z^3 + z^2 - 5z + 3)$

6. Derive the z-transforms of the functions shown in Figure P9.5.6 and then deduce the inverse transforms.
7. Deduce the inverse z-transform of the following functions:

    a. $(18z^2 - 12z)/[(z-3)(z-1)]$

    b. $z^2/[(z-2)(z-1)^2]$

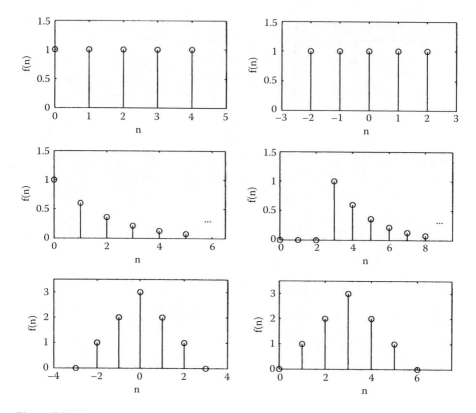

Figure P9.5.6

Section 9.6

1. A discrete-time system function is

$$H(z) = \frac{2(3-z^{-1})}{1-2z^{-1}+z^{-2}}$$

   The input is the signal specified by $x(0) = 0$, $x(1) = 1$, $x(2) = 2$, $x(3) = 0$. Determine $y(n)$.

2. A discrete-time system function is

$$H(z) = \frac{2+z^{-1}}{1+3z^{-1}+z^{-2}}$$

   The input is the unit step function $u(n)$. Determine the system response (output).

3. A system is described by its system function

$$H(z) = \frac{az}{z-b} + \frac{cz}{z-d}$$

b and d are real and have magnitudes less than unity. Show different block diagram configurations that have this $H(z)$.
4. Find the transfer function of the systems shown in Figure P9.6.4.

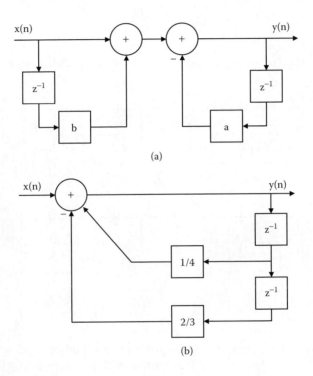

Figure P9.6.4

Section 9.7

1. Find the frequency response of the first-order IIR system and plot its amplitude and phase spectra for $b_0 = 0.5$, $a_1 = -0.8$.
2. Find the frequency response of the system shown in Figure P9.7.2. Observe that the system is made up of two first-order IIR systems in cascade.

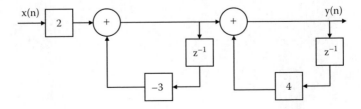

Figure P9.7.2

3. Sketch the general shape of the magnitude of $H(e^{j\omega})$ as a function of the frequency omega. The system function is

$$H(z) = \frac{(z^2+1)(z^2-1)}{z^4+0.8}$$

4. Find the frequency response of the systems shown in Figure P9.7.4.

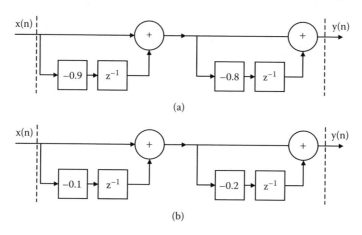

Figure P9.7.4

5. A system is defined by the difference equation $y(n) = ay(n-1) + (1-a)x(n)$. Determine the response of the system to the input signal

$$x(n) = \begin{cases} \sin \omega nT & n = 0, 1, 2, \cdots \\ 0 & n < 0 \end{cases}$$

Section 9.9

1. Discretize the systems shown in Figure P9.9.1 and find their solutions using a direct solution of the difference equations. Next, verify your result using the z-transform method.

Chapter 9: The z-transform, difference equations, and discrete systems

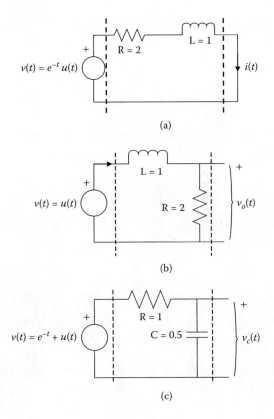

Figure P9.9.1

2. Refer to the system shown in Figure P9.9.2. Deduce the zero-state response and the zero-input response. Also, find its transfer function. Use the appropriate substitutions to create the analogous discrete system. With the sampling time $T = 0.2$, an input function $f(t) = e^{-t}u(t)$, and initial condition $v(0) = 1$, determine the output $v(0.2n)$. Determine the impulse response function $h(n)$.

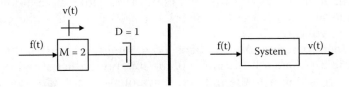

Figure P9.9.2

3. Let us approximate the first-order differential in a forward-type format as as follows:

$$\frac{dy(t)}{dt} \cong \frac{y(nT+T) - y(nT)}{T}$$

Using the above formula, find the solution to the differential equation

$$\frac{dy(t)}{dt} + 2y(t) = u(t) \quad y(0) = 1$$

4. Use the z-transform technique to determine the solution to the following difference equations:

  a. $3y(n) + 2y(n-1) = 5u(n-1)$ $y(-1) = 0$

  b. $3y(n+1) + 2y(n) = 5u(n)$ $y(0) = 0$

  Observe that one equation can be derived from the other by one unit time shift.

5. **Radioactive decay.** A radioactive substance decays at a rate proportional to the amount of the substance that is present. Therefore, the differential equation describing the decay is

$$\frac{dN(t)}{dt} = -kN(t).$$

The number of atoms present at time t is $N(t)$, and k is a positive proportionality constant. If the initial number of atoms present are N_0, and knowing that the half-time is $T_{1/2}$, the proportionality constant can be found from the solution and is equal to $\ln 2 / T_{1/2}$. If the radioactive isotope I_{125} is injected in a cancerous prostate gland, with the initial amount of atoms 10^{20}, how many radioactive atoms remain after 4 years if the half-life of I_{125} is 60 days? Use the equivalent discrete case.

6. **Bioengineering.** Assumptions: D_0 = milligrams (mg) of dye injected into a patient's heart at $t = 0$; the mixture of blood and dye in the heart is uniform; the mixture flows out at a constant rate r l/min; the heart is a container with constant volume V liters. Since $dD(t)/dt$ is the rate change of the dye in the heart at time t, we obtain $dD(t)/dt$ = rate dye in-rate dye out. Here, rate is zero and

$$\text{rate out} = \frac{D(t)}{V} r,$$

where $D(t)/V$ is the concentration of the dye in the heart. Hence, the differential equation is

$$\frac{dD(t)}{dt} = -\frac{r}{V}D(t)$$

Discretize the above equation and find the quantity $D(t)/V$, a useful measure of cardiac output, if $r = 5$ l/min, $V = 0.1$ l, and $D_0 = 100$ mg.

7. Use the z-transform techniques to find the solution to the following difference equations ($n > 0$) with zero initial conditions:

   a. $y(n+1) + y(n) = -2$

   b. $y(n+1) + y(n) = 2\sin(n\pi/4)$

Section 9.10

1. Solve the homogeneous equations:

   a. $y(n) - 3y(n-1) + 2y(n-2) = 0$ $y(-2) = 2$, $y(-1) = 4$

   b. $y(n) - 8y(n-1) + 15y(n) = 0$ $y(-2) = 0$, $y(-1) = 1$

2. Solve the homogeneous equations:

   a. $y(n+2) - 3y(n+1) + 2y(n) = 0$ $y(0) = 2$, $y(1) = 4$

   b. $y(n+2) - 8y(n+1) + 15y(n) = 0$ $y(0) = 0$, $y(1) = 1$

3. Solve the homogeneous equations:

   a. $y(n) - 7y(n-1) + 10y(n-2) = 0$ $y(-2) = 0$, $y(-1) = 1$

   b. $y(n) + 6y(n-1) + 18y(n-2) = 0$ $y(-2) = 0$, $y(-1) = 1$

4. Solve the equations given in Problem 9.10.1 using the z-transform methods.

5. **Fibonacci numbers.** The sequence 1, 1, 2, 3, 5, 8, 13, ... is called the **Fibonacci sequence**. Observe that after the second element, each term is the sum of the two preceding terms. Therefore, if $y(n)$ denotes the nth term in the sequence, the unique solution is the initial value problem $y(n + 2) = y(n + 1) + y(n)$, where $n = 1, 2, 3, 4, \cdots$, $y(0) = 1$, and $y(1) = 1$. Show that the solution is

$$y(n) = \frac{1}{\sqrt{5}}\left[\left(\frac{1+\sqrt{5}}{2}\right)^n - \left(\frac{1-\sqrt{5}}{2}\right)^n\right]$$

6. Determine the distance traveled by the system shown in Figure P9.10.6 if $x(-1) = 0$ and $x(-2) = 1$ with the sampling time $T = 1$. Use the z-transform method.

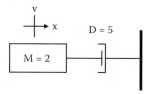

Figure P9.10.6

7. **Population model.** A pair of adult rabbits is assumed to produce a pair of rabbits every month. Every new pair of rabbits is assumed to follow the same sequence. How many pairs will there be after n months if we begin (initial condition) with a pair of newborn rabbits? (This is the Fibonacci rabbit problem.)

8. Use the z-transform methods to solve the following difference equations:

   a. $y(n+1) - y(n) = 2^n$ $y(0) = 1$
   b. $y(n+1) - 3y(n) = 2$ $y(0) = 0$
   c. $y(n+2) - 8y(n+1) + 15y(n) = 5 \times 2^n$ $y(0) = y(1) = 0$

9. **National income (engineering economics).** A national income model is given by the equation $v(n) = c(n) + i(n) + g(n)$, where $c(n)$ = consumer expenditure, $i(n)$ = induced private investment, and $g(n)$ = government expenditures. It is specified* that $c(n) = av(n-1)$ and $i(n) = c[c(n) - c(n-1)] = ab[v(n-1) - v(n-2)]$, where a and b are constants (a is referred to as the marginal propensity to consume). Thus, at any counting period — for example, say any four months — the national income is given by

$$v(n) = a(1+b)v(n-1) - abv(n-2) + g(n)$$

Determine the national income if $a = 1$, $b = 1$, $g(n) = u(n)$, and $v(-1) = v(-2) = 0$. Use the z-transform methods.

   1. The national income is the sum of consumer expenditure, induced private investment, and government expenditures.
   2. Consumer expenditure in each period is directly proportional to the national income in that period.

---

* A.P. Samuelson, Interactions between the multiplier analysis and the principle of acceleration, *Rev. Econ. Stat.*, 21, 75, 1939.

Chapter 9: The z-transform, difference equations, and discrete systems

3. Induced private investment in any period is directly proportional to the increase in the national income for that period above the national income for the preceding period.
4. The government expenditure stays constant from one period to the next.

10. **Loan repayments (engineering economics).** Compound interest is equal to ap, where a is the fraction of a year and $p\%$ is the interest per year. It is supposed that compound interest at 8% per year is charged on the outstanding debt, with conversion period $a = 1/12$ of the year as the period between payments. Find the debt 20 years later if the beginning debt was 200,000.00 and the payments were 2000.00 per month.

*Appendix 9.1: Proofs of the z-transform properties

Linearity

This property is due to the fact that the summation is a linear operation. Furthermore, if we have the sum of m sequences and take their z-transform, we have m regions of convergence (ROCs) or, equivalently, m radius. The ROC of the sum is equal to the greatest radius found.

Right shifting

Proof: Begin with the definition of the z-transform:

$$Y(z) = \sum_{n=0}^{\infty} y(n) z^{-n}$$

Multiply by z^{-m} and then substitute $-q$ for $-m - n$. The result is

$$z^{-m} Y(z) = z^{-m} \sum_{q=m}^{\infty} y(q-m) z^{-q} = \sum_{q=0}^{\infty} y(q-m) z^{-q} = \sum_{n=0}^{\infty} y(n-m) z^{-n}$$

where, since m is a dummy variable, it has been changed to the letter n. The third term in the expression is obtained by invoking the one-sided character of $y(n)$ with y(negative argument) = 0. For the case when we find that $y(-m)$ (initial condition) has values other than zero, we must add the quantity $y(-m) + y(1-m)z^{-1} + \cdots + y(-1)z^{-(m-1)}$ to the left-hand member of the above equation. The multiplication by z^{-m} ($m > 0$) creates a pole at $z = 0$ and deletes a pole at infinity. Therefore, the ROC is the same as the ROC of $y(n)$, with possible exclusion of the origin.

Left-shift property

Proof: From the basic definition,

$$\mathcal{Z}\{y(n+1)\} = \sum_{n=0}^{\infty} y(n+1)z^{-n}$$

Now, set $n + 1 = m$ to find

$$\mathcal{Z}\{y(m)\} = \sum_{m=1}^{\infty} y(m)z^{-m+1} = z\sum_{m=1}^{\infty} y(m)z^{-m} = z\sum_{m=0}^{\infty} y(m)z^{-m} - zy(0) = zY(z) - zy(0)$$

By a similar procedure, we can show that

$$\mathcal{Z}\{y(n+m)\} = z^m Y(z) - z^m y(0) - z^{m-1} y(1) - \cdots - zy(m-1)$$

$$= z^m Y(z) - \sum_{n=0}^{m-1} y(n) z^{m-n}$$

Time scaling

Proof: From the definition of the z-transform,

$$\mathcal{Z}\{a^n y(n)\} = \sum_{n=0}^{\infty} a^n y(n) z^{-n} = \sum_{n=0}^{\infty} y(n)(a^{-1}z)^{-n} = Y(a^{-1}z)$$

which indicates that in the z-transform of $y(n)$, wherever we see z we substitute it with $a^{-1}z$.

Periodic sequences

Proof: The z-transform of the first period is

$$\mathcal{Z}\{y_{(1)}(n)\} = \sum_{n=0}^{N-1} y_{(1)}(n) z^{-n} = Y_{(1)}(z)$$

Because the period is repeated every N discrete time units, we can use the right-shift property to write

$$\mathcal{Z}\{y(n)\} = \mathcal{Z}\{y_{(1)}(n)\} + \mathcal{Z}\{y_{(1)}(n-N)\} + \mathcal{Z}\{y_{(1)}(n-2N)\} + \cdots$$

$$= Y_{(1)}(z) + z^{-N} Y_{(1)}(z) + z^{-2N} Y_{(1)}(z) + \cdots = Y_{(1)}(z)(1 + z^{-N} + z^{-2N} + \cdots)$$

$$= \frac{1}{1-z^{-N}} Y_{(1)}(z) = \frac{z^N}{z^N - 1} Y_{(1)}(z)$$

Chapter 9: The z-transform, difference equations, and discrete systems

where the geometric series formula was used.

Multiplication by n

Proof: From the basic definition,

$$\mathcal{Z}\{ny(n)\} = \sum_{n=0}^{\infty} y(n)nz^{-n} = z\sum_{n=0}^{\infty} y(n)(nz^{-n-1}) = z\sum_{n=0}^{\infty} y(n)\left[-\frac{d}{dz}z^{-n}\right]$$

$$= -z\frac{d}{dz}\sum_{n=0}^{\infty} y(n)z^{-n} = -z\frac{dY(z)}{dz}$$

Initial value

Proof: Let the sequence be zero for $n < m$. Then,

$$Y(z) = \sum_{n=m}^{\infty} y(n)z^{-n} = y(m)z^{-m} + y(m+1)z^{-m-1} + \cdots$$

Multiply both sides by z^m to obtain

$$y(m) + y(m+1)z^{-1} + y(m+2)z^{-2} + \cdots = z^m Y(z)$$

As $z \to \infty$, all terms to the left of the above expression go to zero besides the first.

Final value

Proof: Begin with $y(n) - y(n-1)$ and consider

$$\mathcal{Z}\{y(n) - y(n-1)\} = Y(z) - z^{-1}Y(z) = \sum_{n=0}^{\infty}[y(n) - y(n-1)]z^{-n}$$

by the right-shifting property. This is written

$$(1-z^{-1})Y(z) = \lim_{N\to\infty}\sum_{n=0}^{N}[y(n) - y(n-1)]z^{-n}$$

Take the limit as $z \to 1$, or

$$\lim_{z\to 1}(1-z^{-1})Y(z) = \lim_{z\to 1}\lim_{N\to\infty}\sum_{n=0}^{N}[y(n) - y(n-1)]z^{-n}$$

Interchange the summation on the right:

$$= \lim_{N \to \infty} \lim_{z \to 1} \sum_{n=0}^{N} [y(n) - y(n-1)]z^{-n} = \lim_{N \to \infty} \sum_{n=0}^{N} [y(n) - y(n-1)]$$

$$= \lim_{N \to \infty} [y(0) - y(-1) + y(1) - y(0) + y(2) - y(1) + \cdots] = \lim_{N \to \infty} y(N)$$

since $y(-1) = 0$. The limit $z \to 1$ will give meaningful results only when the point $z = 1$ is located within the ROC of $Y(z)$.

Convolution

Proof: The z-transform of the convolution summation is

$$\mathcal{Z}\{y(n)\} \triangleq Y(z) = \sum_{n=0}^{\infty} z^{-n} \sum_{m=0}^{n} h(n-m)x(m) = \sum_{n=0}^{\infty} z^{-n} \sum_{m=0}^{\infty} h(n-m)x(m)$$

We substituted $n = \infty$ in the second summation since h (negative argument) is zero. Next, write $n - m = q$ and invert the order of summation:

$$Y(z) = \sum_{m=0}^{\infty} x(m) \sum_{n=0}^{\infty} h(n-m)z^{-n} = \sum_{m=0}^{\infty} x(m)z^{-m} \sum_{q=-m}^{\infty} h(q)z^{-q}$$

$$= \sum_{m=0}^{\infty} x(m)z^{-m} \sum_{q=0}^{\infty} h(q)z^{-q} = X(z)H(z)$$

since $h(q) = 0$ for $q < 0$.

chapter 10

Analog filter design

The term **filter**, as used in this text, is a frequency-selective network design to operate on an input signal to produce a desired output signal. That is, a filter passes signals of certain frequencies and blocks signals of other frequencies. The signals may be continuous-time entities that may be stated in time or frequency terms. The signals may also be discrete-time entities, and these signals may also be stated in time or frequency terms.

Filters are usually categorized according to their behavior in the frequency domain and are specified in terms of their magnitude and phase characteristics. Based on their magnitude or transfer response, filters are classified as low-pass, high-pass, band-pass, and band-stop. Based on phase characteristics, filters are often linear phase devises.

Modern filter design techniques, which date from the 1930s, employ a two-step procedure. The first step is to find an analytic approximation to the specified filter characteristic as a transfer function. The second step is a physical realization of this transfer function by passive or active networks. In this text we will discuss the first step because we will need the analog form of filters to develop the corresponding digital ones. The second step is beyond our studies in this text. Our intent is not to provide an in-depth presentation of analog filters in electrical engineering, but to give the fundamentals of two types of filters that will be used to develop digital ones.

10.1 General aspects of filters

Ideal low-pass, high-pass, band-pass, and band-stop analog filters are shown in Figure 10.1.1. Well-developed procedures exist for approximating these response characteristics and involve such functions as Butterworth, Chebyshev, elliptic, and others. The use of these functions has the advantage that the formulas for them are well established and design tables are available.

As a step in meeting our interests in digital filter design, we first study the approximation problem in the analog domain and then the means of converting from the s-plane, in which the $H(s)$ approximation exists, to the z-plane and the corresponding $H(z)$. The developed system function $H(z)$

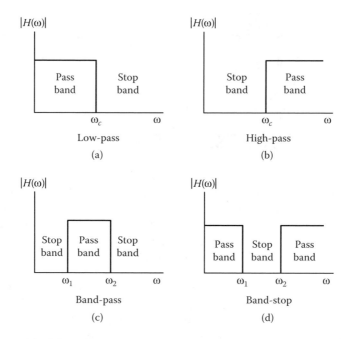

Figure 10.1.1 Ideal frequency response characteristics of analog filters.

can then be realized by discrete systems, either by transformation to difference equation form, which can be adapted to computer calculations, or by direct hardware implementation. The resulting difference equation can be considered to denote a digital filter approximation to the analog filter.

From a practical point of view, the implementation of the filter on a digital computer places accuracy constraints on the realization of the transfer function. This problem arises because the registers can contain only a finite number of bits at any one instance. Therefore, when we are dealing with devices having registers that can contain a limited number of bits, we must be aware that the filters to implement may differ considerably from the desired ones. These sources of error within a digital filter are referred to as **round-off noise**, and methods have been developed to reduce their effect.

It is important to understand that when we talk about filters, we actually mean systems. Filters (systems) can be of a diverse nature. For example, a painting is a two-dimensional optical signal (input) to our eyes. The filter (system) that processes and interprets the optical signal is our eye–brain combination, and the output is the painting that we actually see. Observe Figure 10.1.2 at close range and at arm's length. At the first position the black-and-white changes are sharp and well defined. At the second position the sharpness has disappeared and the picture looks smoother. Because the high frequencies (in this case, the number of black-and-white variations of parallel black lines per 1° angle from the center of our retina; it must not be confused with the frequencies that make up the light) in the optical signal do not pass through the eye–brain filter (actually a band-pass filter with its

Figure 10.1.2 Observe the effect of filtering of the visual system by studying the figure when held several feet away. (Harmon, L. et al., Masking in visual recognition: Effects of two-dimensional filtered noise, *Science*, vol. 180, June 15, 1973, p. 1194. © AAAS.)

lower end at 0.5 cycles/degree and its high end at 35 cycles/degree; the eye sees best at about 26 cycles/degree), smoothing of the picture occurs. The same phenomenon was discussed in Chapter 4 when we tried to reproduce a discontinuous periodic signal by retaining only the low-frequency components.

Other examples of nonelectrical systems that possess filtering characteristics include the economy of the country (input = goods and services, output = gross national product), the structure of a building (input = earthquake vibrations, output = building oscillations), the heartbeat (input = message for a girl or boy, output = number of heartbeats per minute), the airplane (input = direction and speed of wind, output = direction of flaps), etc. From this brief discussion, it can be seen that the study of filters — electrical, biological, chemical, and so on — can be far ranging and is most important.

10.2 Butterworth filter

We now examine the use of the Butterworth function to approximate the ideal low-pass filter shown in Figure 10.1.1a. The **amplitude response** of the nth-order Butterworth filter is given by

$$\left|H_n(j\omega)\right| = \frac{1}{\sqrt{1+(\omega/\omega_c)^{2n}}} \quad (10.1)$$

where ω_c is the cutoff frequency and ω is the frequency in rad/s. If we had set the cutoff frequency equal to 1, the above formula would represent the normalized amplitude response.

At ω equal to 0 the magnitude has the maximum value of 1. At the cutoff frequency, the so-called cutoff point, the magnitude is equal to

$$\left|H(j\omega_c)\right| = \frac{1}{\sqrt{2}}\left\|H(j\omega)\right\|_{max} = 0.707\left\|H(j\omega)\right\|_{max}$$

Because the function of all orders approaches the value 1 smoothly without overshoots, the function is called **maximally flat**. It is observed that that the approximation to the ideal filter improves as n increases.

To obtain the transfer function form of the Butterworth filter, we make use of the fact that the normalized formula can be written in the form $H_n(j\omega) = H_n(s)|_{s=j\omega}$ and rewrite (10.1) in squared form:

$$H_n(s)H_n(-s) = \frac{1}{1+[(j\omega)^2/j^2]^n} = \frac{1}{1+(-s^2)^n} \qquad (10.2)$$

We write the denominator polynomial in the form

$$D(s)D(-s) = 1+(-s^2)^n \qquad (10.3)$$

The roots of the function are deduced from

$$1+(-s^2)^n = 0$$

from which it follows that

$$(-1)^n s^{2n} = -1 = e^{j(2k-1)\pi} \qquad k=1, 2, 3, \cdots, 2n \qquad (10.4)$$

so that

$$s^{2n} = e^{j(2k-1)\pi} e^{jn\pi}$$

The kth root is

$$s_k = \sigma_k + j\omega_k = e^{j(2k+n-1)\pi/2n} = je^{j(2k-1)\pi/2n} \qquad (10.5)$$

In expanded form, this expression is

$$\boxed{s_k = -\sin\left[(2k-1)\frac{\pi}{2n}\right] + j\cos\left[(2k-1)\frac{\pi}{2n}\right] \qquad 1 \le k \le 2n} \qquad (10.6)$$

It can be seen from (10.5) that the roots of s_k are on a unit circle and spaced $\pi/2n$ radians apart. Moreover, no s_k can occur on the $j\omega$-axis since $2k-1$ cannot be an even integer. We thus see that there are n left-half plane roots and n right-half plane roots. The left-half plane roots are associated with $H(s)$ since σ_k is negative for these roots, and the right-half plane roots

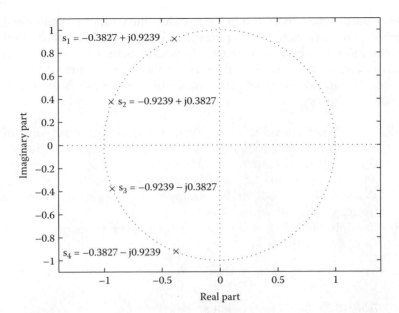

Figure 10.2.1 The left-hand-side poles of the fourth-order Butterworth filter.

are associated with $H(-s)$. These results are shown in Figure 10.2.1 for the case $n = 4$:

$$s_1 = e^{j5\pi/8} \quad s_2 = e^{j7\pi/8} \quad s_3 = e^{j9\pi/8} \quad s_4 = e^{j11\pi/8}$$

The poles due to $H(-s)$ located on the right side of the s-plane are not shown. They are symmetrically located with respect to the imaginary axis. The polynomial constructed from these poles is

$$D(s) = (s-s_1)(s-s_2)(s-s_3)(s-s_4) = 1 + 2.6131s + 3.4142s^2 + 2.6131s^3 + s^4$$

We can use the following MATLAB function to plot the roots: **zplane**([], roots([1 2.6131 3.4142 2.6131 1])). If there are also zeros, we substitute [] with roots([b]), where b is a vector containing the polynomial coefficients.

Note: *The expression for $D(s)$ is for the normalized condition $\omega_c = 1$. For $\omega_c \neq 1$, we must substitute s/ω_c for s in the expression.*

The normalized $D(s)$ of the Butterworth function of order n is the polynomial

$$D(s) = 1 + a_1 s + a_2 s^2 + \cdots + a_{n-1} s^{n-1} + s^n \qquad (10.7)$$

with the coefficients computed in the manner shown above. Note that a_0 and a_n are always unity because the poles are all on the unit circle. Table 10.2.1 gives the values of the coefficients for these functions for $n = 2$ to $n = 8$, and Table 10.2.2 gives the Butterworth polynomials (10.7) in factored form.

The steps to find the transfer function of a Butterworth filter are summarized as follows:

1. Use the normalized form of (10.1), $\omega_c = 1$, to obtain the value of n or $-10\log_{10}|H(j\omega)|^2 = 10\log_{10}(1+\omega^{2n}) \geq $ dB attenuation for frequencies above $\omega = k\omega_c = k \times 1$, where k is some given positive number.
2. If the deduced value of n is nonintegral, select the next higher integer as the order of the filter. For example, if in a specific case $n = 3.325$, we select a fourth-order filter.
3. Use Table 10.2.2 to obtain the the normalized transfer function

$$H(s) = \frac{1}{D(s)}$$

Table 10.2.1 Coefficients of Butterworth Polynomial

n	a_1	a_2	a_3	a_4	a_5	a_6	a_7
2	1.4142						
3	2.0000	2.0000					
4	2.6131	3.4142	2.6131				
5	3.2361	5.2361	5.2361	3.23.61			
6	3.8637	7.4641	9.1416	7.4641	3.8637		
7	4.4940	10.0978	14.5918	14.5918	10.0978	4.4940	
8	5.1528	13.1371	21.8462	25.6884	21.8462	13.1371	5.1258

Table 10.2.2 Factors of Butterworth Polynomials

n	
1	$s+1$
2	$s^2 + 1.4142s + 1$
3	$(s+1)(s^2+s+1)$
4	$(s^2+0.7654s+1)(s^2+1.8478s+1)$
5	$(s+1)(s^2+0.6180s+1)(s^2+1.6180s+1)$
6	$(s^2+0.5176s+1)(s^2+1.4142s+1)(s^2+1.9319s+1)$
7	$(s+1)(s^2+0.4450s+1)(s^2+1.2470s+1)(s^2+1.8019s+1)$
8	$(s^2+0.3902s+1)(s^2+1.1111s+1)(s^2+1.6639s+1)(s^2+1.9616s+1)$
9	$(s+1)(s^2+0.3473s+1)(s^2+s+1)(s^2+1.5321s+1)(s^2+1.8794s+1)$

Chapter 10: Analog filter design

4. Substitute s/ω_c for ω in step 3 for the nonnormalized form.
5. If we wish to normalize $|H(s/\omega_c)|$ with amplitude starting at unity, multiply $|H(s/\omega_c)|$ by the value of the constant of the polynomial $D(s/\omega_c)$.

Example 10.2.1: Deduce the transfer function of a Butterworth filter that has an attenuation of at least 10 dB at twice the cutoff frequency, where $\omega_c = 2.5 \times 10^3$ rad/s.

Solution: We initially find the normalized Butterworth filter. At $\omega = 2 \times \omega_c = 2 \times 1$ ($\omega_c=1$, normalized),

$$|H_n(j2)|^2 = \frac{1}{1+2^{2n}} \tag{10.8}$$

The dB attenuation is given by

$$-10\log_{10}\left(|H_n(j2)|^2\right) = 10\log_{10}(1+2^{2n}) \geq 10 \text{ or } \log_{10}(1+2^{2n}) = \log_{10}(10^1)$$

Therefore,

$$1 + 2^{2n} = 10 \text{ or } 2^{2n} \geq 9$$

The order of the filter must then be $n = (\log_2 9)/2 = 1.5850$. Instead of logarithm transformation, the reader can use the MATLAB function **log2**(9)/2 to obtain the above value. Hence, a second-order Butterworth satisfies this requirement. The corresponding transfer function of the normalized filter is (see Table 10.2.2)

$$H_2(s) = \frac{1}{s^2 + 1.4142s + 1} \tag{10.9}$$

For a cutoff frequency $\omega_c = 2.5 \times 10^3$ rad/s, the transfer function is

$$H_2(s/\omega_c) = \frac{1}{1.6 \times 10^{-7} s^2 + 5.6568 \times 10^{-4} s + 1} \tag{10.10}$$

Figure 10.2.2 shows the amplitude and phase characteristics of the normalized second-order filter. If we multiply the axis of the normalized filter by $\omega_c = 2.5 \times 10^3$ rad/s, we obtain the unnormalized filter characteristics for this case.

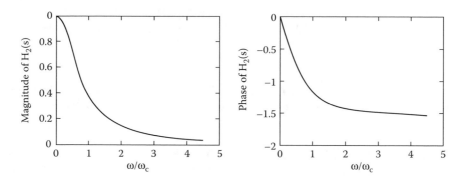

Figure 10.2.2 Phase and amplitude characteristics of the second-order normalized Butterworth filter.

The denominator polynomial has conjugate roots with negative real parts. We split the function $D(s)$ of (10.10) into two polynomials, separating the even and odd powers of s:

$$D(s) = e(s) + o(s) = \underbrace{1.6 \times 10^{-7} s^2 + 1}_{e} + \underbrace{5.6569 \times 10^{-4} s}_{o} \quad (10.11)$$

The roots of each component are purely imaginary (zero included), and additionally, the roots alternate; thus, $-2.5 \times 10^3 < s < 2.5 \times 10^3$. Polynomials that possess these properties are known as **Hurwitz polynomials**.

Observe also that (10.10) is of the general form

$$H(s) = \frac{1}{a_0 + a_1 s + a_2 s^2 + \cdots + a_{n-1} s^{n-1} + a_n s^n} = \frac{k}{D(s)} \quad (10.12)$$

It can be shown that a transfer function of this form can be realized with passive elements if and only if $D(s)$ is a Hurwitz polynomial.

Suppose that we wish to find a two-port network of the terminated form shown in Figure 10.2.3a appropriate to (10.10). Since $D(s)$ is of the second order, the lossless two-port network shown in Figure 10.2.3b appears to be an appropriate form. The transfer function is

$$\frac{V_o(s)}{V_i(s)} = H(s) = \frac{1}{LCs^2 + (L+C)s + 2} \quad (10.13)$$

This form differs from (10.10) by 2 in the denominator. If we multiply the numerator and denominator of (10.10) by 2, we obtain an equivalent transfer function:

Chapter 10: Analog filter design

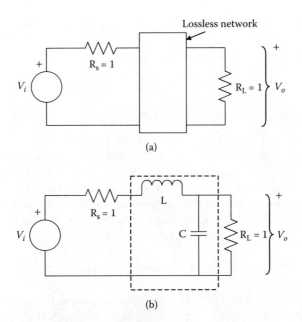

Figure 10.2.3 Second-order Butterworth filter.

$$\frac{H(s)}{2} = \frac{1}{3.2 \times 10^{-7} s^2 + 1.1314 \times 10^{-3} s + 2} \tag{10.14}$$

By comparing these two equations, we obtain

$$CL = 3.2 \times 10^{-7} \qquad C + L = 1.1314 \times 10^{-3}$$

These expressions can be solved to yield $C = 5.6164 \times 10^{-4}$ F and $L = 5.6976 \times 10^{-4}$ H. ∎

Example 10.2.2: A low-pass normalized Butterworth filter must satisfy the following: (a) $|H_n(j0.5)|^2 > 0.9$ in pass-band, $0 \leq \omega \leq 1$, and (b) $|H_n(j0.5)|^2 < 0.05$ in the stop-band, $1 \leq \omega$. Find its order and plot its amplitude response.

Solution: From (a) we have $1/[1 + (0.5)^{2n}] > 0.9$ or $n > 1.5850$, and from (b), $1/[1 + (1.5)^{2n}] < 0.05$ or $n > 3.6309$, which implies that $n = 4$ will suffice and satisfies both inequalities.

Book MATLAB m-file: ex_10_2_2

```
%m-file: ex_10_2_2
w=0:0.01:4;%frequency range from 0 to 4 rad/s;
b=[0 0 0 1];
```

```
a=[1 2.613 3.414 2.613 1];
H=freqs(b,a,w);
plot(w,abs(H),'k');xlabel('\omega rad/s');
ylabel('Magn. 4th-order Butt. filter');
```
∎

10.3 Chebyshev low-pass filter

An examination of Figure 10.2.2 shows that the Butterworth low-pass amplitude response approaches the ideal in the region of low frequencies and also in the region of high-value frequencies. However, it does not produce very good approximation in the neighborhood of the cutoff frequency ($\omega = 1$). The Chebyshev low-pass filter possesses a sharper cutoff response than the Butterworth filter, but it also possesses amplitude variations within the pass-band. A number of general properties are:

1. The oscillations in the pass-band have equal amplitudes for a given value of ε.
2. The curves for n even always start from the trough of the ripple, whereas the curves for n odd always start from the peak.
3. At the normalized cutoff frequency of 1, all curves pass through the same point.

The amplitude response of the Chebyshev low-pass filter is specified by

$$\boxed{|H_n(j\omega)| = \frac{1}{\sqrt{1+\varepsilon^2 C_n^2(\omega)}} = \frac{1}{\sqrt{1+\varepsilon^2 C_n^2(s/j)}}} \qquad n = 1, 2, 3, \cdots \qquad (10.15)$$

where ε is a constant and $C_n(\omega)$ is the Chebyshev polynomials given by the equations

$$C_n(\omega) = \cos(n\cos^{-1}\omega) \qquad |\omega| \le 1 \qquad \text{a)}$$
$$= \cosh(n\cosh^{-1}\omega) \qquad |\omega| > 1 \qquad \text{b)} \qquad (10.16)$$

The analytic form of Chebyshev polynomials for order 0 to 10 is tabulated in Table 10.3.1.

By (10.16) and the recurrence relationship

$$C_{n+1}(\omega) = 2\omega C_n(\omega) - C_{n-1}(\omega) \qquad n = 1, 2, 3, \cdots \qquad \text{a)}$$
$$C_1(\omega) = \omega C_0(\omega) \qquad \text{b)} \qquad (10.17)$$

the nth-order Chebyshev polynomial possesses the properties shown in Table 10.3.1.

Table 10.3.1 Chebyshev Polynomials

n	$C_n(\omega)$
0	1
1	ω
2	$2\omega^2 - 1$
3	$4\omega^3 - 3\omega$
4	$8\omega^4 - 8\omega^2 + 1$
5	$16\omega^5 - 20\omega^3 + 5\omega$
6	$32\omega^6 - 48\omega^4 + 18\omega^2 - 1$
7	$64\omega^7 - 112\omega^5 + 56\omega^3 - 7\omega$
8	$128\omega^8 - 256\omega^6 + 160\omega^4 - 32\omega^2 + 1$
9	$256\omega^9 - 476\omega^7 + 432\omega^5 - 120\omega^3 + 9\omega$
10	$512\omega^{10} - 1280\omega^8 + 1120\omega^6 - 400\omega^4 + 50\omega^2 - 1$

1. For any n,

$$0 \le |C_n(\omega)| \le 1 \qquad 0 \le |\omega| \le 1$$
$$|C_n(\omega)| > 1 \qquad |\omega| > 1$$

2. $C_n(\omega)$ is monotonically increasing for $\omega > 1$ for all n.
3. $C_n(\omega)$ is an odd polynomial if n is odd. $C_n(\omega)$ is an even polynomial if n is even.
4. $C_n(0) = 0$ for n odd.
5. $|C_n(0)| = 1$ for n even.

The function $|H(j\omega)|$ attains its maximum value 1 at the zeros of $C_n(\omega)$ and its minimum value $1/\sqrt{1+\varepsilon^2}$ at the points where $C_n(\omega)$ attains its maximum value of 1. Thus, the ripples in the pass-band $0 \le \omega \le 1$ have a peak-to-peak value of

$$r = 1 - 1/\sqrt{1+\varepsilon^2} \tag{10.18}$$

The ripple in decibels is given by

$$\boxed{r_{dB} = -20 \log_{10} \frac{1}{\sqrt{1+\varepsilon^2}} = 10 \log_{10}(1+\varepsilon^2)} \tag{10.19}$$

Outside of the pass-band $\omega > 1$, $|H(j\omega)|$ decreases monotonically.

In the unnormalized case, when $\omega_c \neq 1$, the Chebyshev low-pass filter is defined by

$$|H(j\omega)| = \frac{1}{\sqrt{1+\varepsilon^2 C_n^2(\omega/\omega_c)}} = \frac{1}{\sqrt{1+\varepsilon^2 C_n^2(s/j\omega_c)}} \quad n=1,2,3,\cdots \quad (10.20)$$

To find the pole location of $H_n(s)$, where $s = j\omega$, we study the denominator of the normalized function ($\omega_c = 1$)

$$H_n(s)H_n(-s) = \frac{1}{1+\varepsilon^2 C_n^2(s/j)} \quad (10.21)$$

More specifically, the poles of the function occur when $C_n(s/j) = \pm\sqrt{-1/\varepsilon^2}$ or when (see (10.16))

$$\cos[n\cos^{-1}(s/j)] = \pm j/\varepsilon \quad (10.22)$$

To proceed we define

$$\cos^{-1}(s/j) = \alpha - j\beta \quad (10.23)$$

Combine this with (10.22), from which

$$\cos n\alpha \cosh n\beta + j(\sin n\alpha \sinh n\beta) = \pm \frac{j}{\varepsilon} \quad (10.24)$$

because $\cos jx = \cosh x$ and $\sin jx = j\sinh x$. Equate real and imaginary terms on each side of the equation:

$$\begin{aligned} \cos n\alpha \cosh n\beta &= 0 & \text{a)} \\ \sin n\alpha \sinh n\beta &= \pm \frac{1}{\varepsilon} & \text{b)} \end{aligned} \quad (10.25)$$

It follows from (10.25a), since $\cos n\beta \neq 0$, that

$$\alpha = (2k-1)\frac{\pi}{2n} \quad k = 1, 2, 3, \cdots, 2n \quad (10.26)$$

Because $\sin n\alpha = \pm 1$, (10.25b) together with (10.26) yields for β:

Chapter 10: Analog filter design

$$\beta = \pm \frac{1}{n}\sinh^{-1}\left(\frac{1}{\varepsilon}\right) \tag{10.27}$$

Equation (10.23) can be used to specify the poles:

$$\begin{aligned} s_k = \sigma_k + j\omega_k = j\cos(\alpha - j\beta) &= -\sin\left[(2k-1)\frac{\pi}{2n}\right]\sinh\left[\frac{1}{n}\sinh^{-1}\left(\frac{1}{\varepsilon}\right)\right] \\ &\quad - j\cos\left[(2k-1)\frac{\pi}{2n}\right]\cosh\left[\frac{1}{n}\sinh^{-1}\left(\frac{1}{\varepsilon}\right)\right] \end{aligned} \tag{10.28}$$

These points are located on an ellipse in the s-plane. The reader can write one of the following simple MATLAB programs (in this case for the sixth-order Chebyshev filter) and see that indeed the roots are located on an ellipse.

Book MATLAB program

```
y=roots([1 1.1592 2.1718 1.5898 1.1719 0.4324 0.0948]);
>> plot(real(y),imag(y),'xk');
>> axis equal;
```

Book MATLAB program

```
zplane([],[1 1.1592 2.1718 1.5898 1.1719 0.4324 0.0948]);
```

To prove that the locus is an ellipse, let

$$\sigma_k = -\sin\left[(2k-1)\frac{\pi}{2n}\right]\sinh\left[\frac{1}{n}\sinh^{-1}\left(\frac{1}{\varepsilon}\right)\right] \tag{10.29}$$

$$\omega_k = \cos\left[(2k-1)\frac{\pi}{2n}\right]\cosh\left[\frac{1}{n}\sinh^{-1}\left(\frac{1}{\varepsilon}\right)\right] \tag{10.30}$$

It follows from (10.29) and (10.30) that

$$\frac{\sigma_k^2}{\left(\sinh\left[\frac{1}{n}\sinh^{-1}\left(\frac{1}{\varepsilon}\right)\right]\right)^2} + \frac{\omega_k^2}{\left(\cosh\left[\frac{1}{n}\sinh^{-1}\left(\frac{1}{\varepsilon}\right)\right]\right)^2} = 1 \tag{10.31}$$

This equation is of an ellipse with the minor axis (x-axis) equal to the square root of the denominator of σ_k and the major axis (y-axis) equal to the square root of the denominator of ω_k.

The desired response, using only the roots with negative real parts, is given by the following transfer function:

$$H(s) = \frac{K}{a_0 + a_1 s + a_2 s^2 + \cdots + a_{n-1} s^{n-1} + s^n} \tag{10.32}$$

The steps to find the Chebyshev filter are summarized as follows:

1. For a specified dB ripple, ε is deduced from (10.19).
2. The attenuation in dB at a specified frequency that is a multiple of the critical frequency must satisfy the equation (see (10.20))

$$-10 \log_{10} |H(j\omega)|^2 = 10 \log_{10} \left[1 + \varepsilon^2 C_n^2 \left(\frac{\omega}{\omega_c} \right) \right] \geq dB$$

3. From step 2, n is determined. Next, use (10.28) to obtain the roots.
4. The constant K of (10.32) is selected to meet the direct current (dc) gain level dictated by the problem at hand.
5. The normalized transfer function is then given by

$$\frac{H(s)}{K} = \frac{1}{(-1)^n \prod_{k=1}^{n} \left(\frac{s}{s_k} - 1 \right)} \tag{10.33}$$

6. For the case that the cutoff frequency is different than unity, the transfer function is

$$\frac{H(s)}{K} = \frac{1}{(-1)^n \prod_{k=1}^{n} \left(\frac{(s/\omega_c)}{s_k} - 1 \right)} \tag{10.34}$$

Table 10.3.2 gives the coefficients in the denominator of (10.32).

Example 10.3.1: Design a Chebyshev filter with a 1-dB ripple in the pass-band and an attenuation of at least 20 dB at twice the cutoff frequency, specified as 3 kHz.

Chapter 10: Analog filter design

Table 10.3.2 Coefficients of the Polynomial in (10.32)

n	a_0	a_1	a_2	a_3	a_4	a_5	a_6	a_7
				r = 0.5 dB		ε = 0.3493		
1	2.8628							
2	1.5162	1.4256						
3	0.7157	1.5349	1.2529					
4	0.3791	1.0255	1.7169	1.1974				
5	0.1789	0.7525	1.3096	1.9374	1.1725			
6	0.0948	0.4324	1.1719	1.5898	2.1718	1.1592		
7	0.0447	0.2821	0.7557	1.6479	1.8694	2.4127	1.1512	
8	0.0237	0.1525	0.5736	1.1486	2.1840	2.1492	2.6567	1.1461
				r = 1.0 dB		ε = 0.5088		
1	1.9652							
2	1.1025	1.0977						
3	0.4913	1.2384	0.9883					
4	0.2756	0.7426	1.4539	0.9528				
5	0.1228	0.5805	0.9744	1.6888	0.9368			
6	0.0689	0.3071	0.9393	1.2021	1.9308	0.9283		
7	0.0307	0.2137	0.5486	1.3575	1.4288	2.1761	0.9231	
8	0.0172	0.1073	0.4478	0.8468	1.8369	1.6552	2.4230	0.9198

Solution: Follow the steps outlined above to obtain:

1. Since $C_n(1) = 1$ $n = 0, 1, 2, \cdots$, then $1 = 10 \log_{10}(1 + \varepsilon^2)$ from which $10^{0.1} = 1 + \varepsilon^2$. Thus, $\varepsilon = \sqrt{1.2589 - 1} = 0.5088$.
2. $10 \log_{10}[1 + 0.5088^2 \, C_n^2(2)] \geq 20$ dB from which $1 + 0.5088^2 \, C_n^2(2) \leq 10^2$. Thus, we find that $C_n^2(2) \geq 99/(0.5088^2) = 382.42$. From Table 10.3.1 we find that $C_3^2(2) = (4 \times 2^3 - 3 \times 2)^2 = 676$ and all others less $n = 3$ have values of less than 382.42. Hence, we choose $n = 3$.
3. From (10.28), we obtain

$$s_1 = \sigma_1 + j\omega_1 = -\sin\left(\frac{\pi}{6}\right)\sinh\left[\frac{1}{3}\sinh^{-1}\left(\frac{1}{0.5088}\right)\right] + j\cos\left(\frac{\pi}{6}\right)$$
$$\times \cosh\left[\frac{1}{3}\sinh^{-1}\left(\frac{1}{0.5088}\right)\right] = -0.2471 + j0.9660$$

In a similar manner, we find

$$s_2 = \sigma_2 + j\omega_2 = -0.4942 \qquad s_3 = \sigma_3 + j\omega_3 = -0.2471 - j0.9660$$

4. Assume $K = 1$.

5. The normalized transfer function is

$$H_3(s) = \frac{s_1 s_2 s_3}{(-1)^3 (s-s_1)(s-s_2)(s-s_3)}$$

$$= \frac{-0.4913}{(-1)^3 [s^3 - (s_1+s_2+s_3)s^2 + (s_1 s_2 + s_1 s_3 + s_2 s_3)s - s_1 s_2 s_3]}$$

$$= \frac{0.4913}{s^3 + 0.9883 s^2 + 1.2384 s + 0.4913}$$

6. In this example, $\omega_c = 2\pi \times 3 \times 10^3 = 6\pi \times 10^3$ and the final function is

$$H_3\left(\frac{s}{6\pi \times 10^3}\right) =$$

$$\frac{0.4913}{\left(\dfrac{s}{6\pi \times 10^3}\right)^3 + 0.9883 \left(\dfrac{s}{6\pi \times 10^3}\right)^2 + 1.2384 \left(\dfrac{s}{6\pi \times 10^3}\right) + 0.4913}$$

Figure 10.3.1a shows the pole location on the s-plane, and Figure 10.3.1b shows the frequency response of the filter.

Book MATLAB m-file: ex_10_3_1a

```
%m-file for Ex 10.3.1: ex_10_3_1a
r=roots([1/(6*pi*10^3)^3  0.9883*(1/((6*pi*10^3)^2))...
    1.2384*(1/(6*pi*10^3))  0.4913]);
cr=-r;%conjugate roots to the right of the s-plane;
p=[r'  cr'];
plot(real(p),imag(p),'x');
axis equal;
grid on;
```

Book MATLAB m-file: ex_10_3_1b

```
%m-file for Ex_10.3.1:ex_10_3_1b
w=0:500:80000;
h=0.4913./((j*w/(6*pi*10^3)).^3+0.9883*(j*w/...
    (6*pi*10^3)).^2+1.2384*(j*w/(6*pi*10^3))+0.4913);
plot(w,abs(h));
```

∎

Chapter 10: Analog filter design

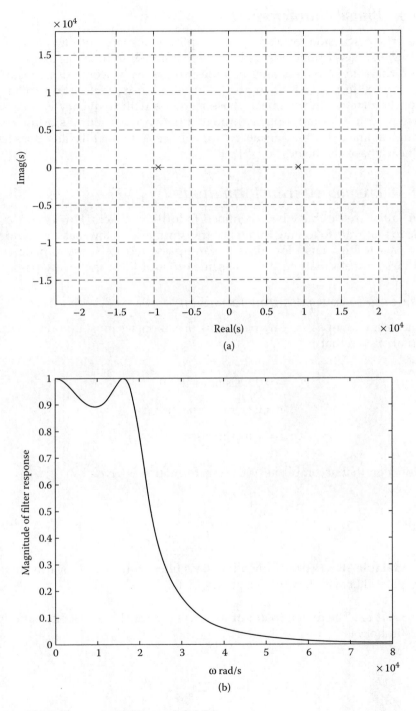

Figure 10.3.1 Illustration of Example 10.3.1.

10.4 Phase characteristics

The prior discussion focused on the amplitude response of low-pass filters. However, phase characteristics have been ignored in these discussions, and these characteristics generally become progressively worse (less linear) as the amplitude response is improved. The phase characteristic is often an important factor since a linear phase response with frequency is necessary if we wish to transmit a pulse through a network without distortion, even though a time delay may ensue. As we will see, a digital FIR filter does have a linear phase characteristic.

10.5 Frequency transformations

Our initial studies have been confined to low-pass filters. However, we are able to make the following frequency transformations: low-pass to low-pass, low-pass to high-pass, low-pass to band-pass, and low-pass to band-stop. All these transformations give the same order of filter as the low-pass one.

Low-pass-to-low-pass transformation

To achieve a low-pass-to-low-pass transformation, we must introduce a function of ω such that

$$\omega = 0 \text{ transforms } w = 0$$
$$\omega = \pm\infty \text{ transforms } w = \pm\infty$$
$$\omega = 1 \text{ transforms } w = w_c$$

A function that accomplishes this transformation is given by

$$s = \frac{w}{w_c} \tag{10.35}$$

Example 10.5.1: Transform a low-pass filter with a cutoff of 1 to another low-pass filter with a cutoff frequency of 1.8.

Solution: The denominator of the second-order low-pass filter, after the substitution, is

$$D(w) = 1 + 1.4142\frac{w}{1.8} + \left(\frac{w}{1.8}\right)^2$$

Chapter 10: Analog filter design

And hence the transformed filter is

$$H_2(w) = \frac{1}{1+1.4142\dfrac{w}{1.8}+\left(\dfrac{w}{1.8}\right)^2} = \frac{1.8^2}{1.8^2+1.4142\times 1.8w+w^2}$$

The reader can easily verify the transformation. ∎

We can also use the MATLAB function **lp2lp()**. To solve Example 10.5.1, we write **[nt,dt]=lp2lp([0 0 1],[1 1.4142 1]);** (note that the vectors nt and dt start from the highest order and go to the zero order of these polynomials).

Low-pass-to-high-pass transformation

To effect this transformation we set

$$s = \frac{w_c}{w} \qquad (10.36)$$

Example 10.5.2: Design a second-order Chebyshev high-pass filter with a cutoff frequency of 300 rad/s. The ripple factor is defined at 1 dB. Use the MATLAB m-file.

Solution: The following m-file produces the desired output.

Book MATLAB m-file: ex_10_5_2

```
%m-file for the Ex 10.5.2:ex_10_5_2
w=0:10:1000;
n=[0 0 1];d=[1 1.0977 1.5162];
[nt,dt]=lp2hp(n,d,300);
h=freqs(nt,dt,w);
plot(w,abs(h))
```
∎

Low-pass-to-band-pass transformation

The required function must effect the following correspondence:

$\omega = 0$ transforms to $w = \pm w_0$, the center frequency

$\omega > 0$ transforms to $w > w_0$, and $w < -w_0$

$\omega < 0$ transforms to $w_0 > w > -w_0$

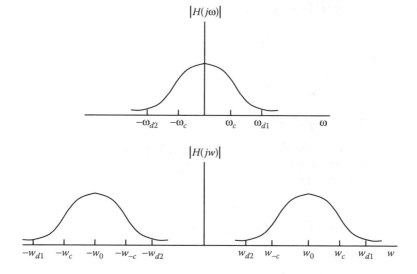

Figure 10.5.1 Transformation from a low-pass to a band-pass filter.

Figure 10.5.1 shows these relationships in graphical form.

The transformation that accomplishes the required correspondence between the low-pass and the band-pass case is given by

$$s = \frac{s_n^2 + w_0^2}{s_n(w_c - w_{-c})} = \frac{s_n^2 + w_c w_{-c}}{s_n W} \quad (10.37)$$

where $w_0^2 = w_c w_{-c}$ and $W = w_c - w_{-c}$ =bandwidth. From the factor $w_c w_{-c}$, it is evident that only a geometric symmetry exists. Hence, to obtain the band-pass $H(s)$ from the low-pass $H(s)$, substitute $(s_n^2 + w_0^2)/s_n W$ for the variable s.

Example 10.5.3: Design a band-pass filter that meets the following specifications:

1. The 3-dB attenuation occurs at 8000 and 12,000 rad/s.
2. The attenuation must be at least 20 dB for frequencies lower than 4000 rad/s and higher than 20,000 rad/s.

Solution: The requisite transformation is given by ($w_c = 12 \times 10^3$ rad/s $w_{-c} = 8 \times 10^3$ rad/s)

$$s = \frac{s_n^2 + 9.6 \times 10^7}{s_n \times 4 \times 10^3}$$

Chapter 10: Analog filter design

or

$$j\omega = \frac{-w^2 + 8\times 10^3 \times 12 \times 10^3}{jw \times 4 \times 10^3}$$

or

$$\omega = \frac{w^2 - 9.6 \times 10^7}{w \times 4 \times 10^3} \tag{10.38}$$

Now, substitute $w_{d2} = 4 \times 10^3$ rad/s in this equation, from which

$$\omega_{d2} = -\frac{1.6 \times 10^7 - 9.6 \times 10^7}{4 \times 10^3 \times 4 \times 10^3} = 5$$

Substitute $w_{d1} = 2 \times 10^4$ in the equation to find

$$\omega_{d1} = \frac{1.6 \times 10^8 - 9.6 \times 10^7}{2 \times 10^4 \times 4 \times 10^3} = 3.8$$

Since 3.8 is smaller than 5, we must design a normalized filter with $\omega_c = 1$ and $\omega_{d1} = 3.8$. However, from Figure 10.5.2a, we find that the Butterworth filter at $\omega = 3.8$ meets line $n = 2$ just below 20 dB, which satisfies the specifications, with its normalized function (see Table 10.2.2)

$$H_2(s) = \frac{1}{s^2 + 1.4142s + 1} \tag{10.39}$$

Combine (10.38) and (10.39) to find the desired band-pass filter transfer function:

$$H_2(s) = \frac{1}{\left(\dfrac{s_n^2 + 9.6 \times 10^7}{s_n \times 4 \times 10^3}\right)^2 + 1.4142 \dfrac{s_n^2 + 9.6 \times 10^7}{s_n \times 4 \times 10^3} + 1} \tag{10.40}$$

Introduce $s_n = jw$ into this equation to obtain a function of w rad/s. A plot of $20 \log_{10}|H(jw)|$ vs. w rad/s is shown in Figure 10.5.2b. Observe that the –3 dB points are at 8×10^3 and 12×10^3 rad/s, as originally specified. In addition, the attenuation at 4×10^3 and 20×10^3 rad/s exceeds –20 dB, as required. ∎

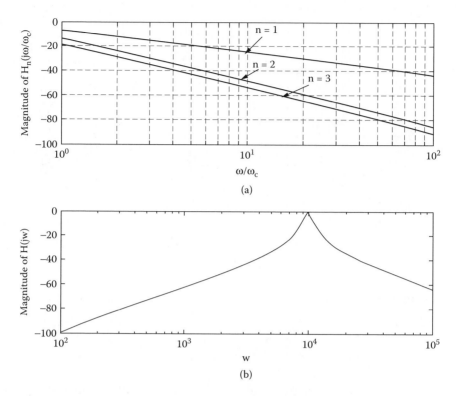

Figure 10.5.2 Illustrating Example 10.5.3.

We can also use the MATLAB function as follows:

`[nt,dt]=lp2bp(num,den,w0,bw);`%w0=center frequency of the
　　　　　　　　　　　　　　　　　%band-pass filter;
　　　　　　　　　　　　　　　　　%bw=bandwidth of the band-pass
　　　　　　　　　　　　　　　　　%filter;

Therefore, for our example above, we write the following.

Book MATLAB program

```
[nt,dt]=lp2bp([0 0 1],[1 1.4142 1],10^4,4*10^3);
x=100:10:1000;y=1050:50:10000;z=10500:500:100000; w=[x y z];
H=freqs(nt,dt,w);
semilogx(w,-abs(20*log10(H)));
```

Low-pass-to-band-stop transformation

This transformation is roughly the inverse of that for the low-pass to band-pass filter. It is (see Figure 10.5.1)

Chapter 10: Analog filter design

$$s = \frac{s_n(w_{d1} - w_{d2})\omega_{d1}}{s_n^2 + w_{d1}w_{d2}} = \frac{jw(w_{d1} - w_{d2})\omega_{d1}}{-w^2 + w_{d1}w_{d2}} \quad (10.41)$$

Example 10.5 4: Design a band-stop filter to meet the following specifications:

1. The attenuation between 1500 and 2200 rad/s must be at least 30 dB.
2. The attenuation for ω less than 500 rad/s and larger than 2800 rad/s must be less than −3 dB.

Solution: For the data given, (10.41) yields

$$f(s_n) = s = \frac{s_n(2200 - 1500)\omega_{d1}}{s_n^2 + 2200 \times 1500} \quad \text{or} \quad f(\omega) = \omega = \frac{w700\omega_{d1}}{s_n^2 + 3.3 \times 10^6} \quad (10.42)$$

The values of $f(w)$ for $f(2800)$ and $f(500)$ are

$$f(2800) = \frac{2800 \times 700\omega_{d1}}{-(2800)^2 + 3.3 \times 10^6} = -0.4317\omega_{d1}$$

$$f(500) = \frac{500 \times 700\omega_{d1}}{-(500)^2 + 3.3 \times 10^6} = 0.1147\omega_{d1}$$

Since we require that the attenuation be less than −3 dB for $\omega \geq -0.4317\omega_{d1}$ and $\omega \leq 0.4317\omega_{d1}$, the pass-band cutoff frequency of the original low-pass filter is chosen to be $\omega_c = 0.4317\omega_{d1}$. Begin with the normalized $\omega_c = 1$, then $\omega_{d1} = 1/0.4317 = 2.32$ rad/s. Thus, the low-pass filter that we must first design has a 3-dB pass-band cutoff frequency at 1 rad/s and an attenuation of at least 30 dB at frequencies larger than 2.32 rad/s. From Figure 10.5.2a, we see that we require a third-order Butterworth filter. The transform function is $H(s) = 1/[(s+1)(s^2 + s + 1)]$. Hence, we find

$$H_n(s) = \frac{1}{\left(\dfrac{s_n 700 \times 2.32}{s_n^2 + 3.3 \times 10^6} + 1\right)\left[\left(\dfrac{s_n 700 \times 2.32}{s_n^2 + 3.3 \times 10^6}\right)^2 + \dfrac{s_n 700 \times 2.32}{s_n^2 + 3.3 \times 10^6} + 1\right]} \quad (10.43)$$

This transfer function can be found by setting $s_n = jw$ and then plotting the $20 \log_{10} |H(jw)|$ vs. w.

We can plot the filter response by writing the following.

Book MATLAB program

```
w=200:10:4000;
x=((j*w*700*2.32)./(-w.^2+3.3*10^6)+1);
y=x.^2+x+1;
H=1./(x.*y);
plot(w,20*log10(abs(H)))
```
■

We can also use the following MATLAB function:

[nt,dt]=lp2bs(num,den,w0,bw);%w0=center frequency of the
%band-stop filter;

%bw=bandwidth of the band-stop
%filter;

For Example 10.5.4 we write the following:

Book MATLAB program

```
[nt,dt]=lp2bs([0 0 0 1],[1 2 2 1],(1500+2200)/2,(2800-500));
w=200:10:4000;
H=freqs(nt,dt,w);
plot(w,20*log10(abs((H))));
```

10.6 Analog filter design using MATLAB functions

Butterworth filter design

The m-file function **[z,p,k]=buttap(n)** provides the zeros and the poles in the vectors **z** and **p**, and the gain factor k for an nth-order filter with a 3-dB cutoff frequency.

To find the **numerator** and **denominator** coefficients, we use the m-file function **[num,den]=zp2tf(z,p,k)**. Having this information, we find the transfer function using the m-file function **[H,w]=freqs(num,den)**, where **w** is the frequency range vector, a set of 200 frequencies. If we provide the vector **w**, we use the following form of the m-file: **[H]=freqs(num,den,w)**.

The order of low-pass filter and the 3-dB cutoff frequency is produced by the m-file function **[n,wn]=buttord(wp,ws,rp,rs,'s')**, where n = order, wn = 3-dB cutoff frequency, wp = pass-band edge frequency, ws = stop-band edge frequency, rp = maximum pass-band attenuation in dB, and rs = minimum stop-band attenuation in dB. The edge frequencies are as follows for different types of filters:

Low-pass: $wp = 0.1$, $ws = 0.2$
High-pass: $wp = 0.2$, $ws = 0.1$
Band-pass: $wp = [0.2\ \ 0.7]$, $ws = [0.1\ \ 0.8]$
Band-stop: $wp = [0.1\ \ 0.8]$, $ws = [0.2\ \ 0.7]$

Different **types** of filters can be found using the following m-file function:

`[num,den]=butter(n,wn,'filter type','s');`%filter type=high
%for high pass); filter type=
%stop (for band stop); filter type=pass (for pass band);

The m-files for the Chebyshev filter are:

`[z,p,k]=cheb1ap(n,rp)`
`[num,den]=cheby1(n,rp,wn,'s')`
`[num,den]=cheby1(n,rp,wn,'filter type','s')`
`[n,wn]=cheb1ord(wp,ws,rp,'s')`

Important definitions and concepts

1. Butterworth filter
2. Maximally flat filter
3. Hurwitz polynomials
4. Normalized filters
5. Chebyshev filter
6. Chebyshev polynomials
7. Ripple constant of a Chebyshev filter
8. Phase characteristics of filters
9. Frequency transformations
10. MATLAB functions for frequency transformations of analog filters

Chapter 10 Problems

Section 10.2

1. A Butterworth filter must have an attenuation of at least 20 dB at 6 kHz, twice the cutoff frequency. Deduce the transfer function of the appropriate low-pass filter.
2. Show that the square of (10.1), with $\omega_c = 1$, is equal to (10.2).
3. First deduce $D(s)$ for the third-order Butterworth filter and then demonstrate that $D(s)D(-s) = 1 + \omega^6$.
4. Deduce a low-pass Butterworth filter that has an attenuation of at least 20 dB at three times the frequency $\omega_c = 2 \times 10^4$ rad/s.
5. Deduce the transfer function of a Butterworth filter that has an attenuation of at least 10 dB at twice the cutoff frequency $\omega_c = 10^5$ rad/s. Find the values of L and C for the filter shown in Figure 10.2.3.
6. Determine the impulse response functions of the first-order and third-order Butterworth filters.
7. A low-pass normalized filter ($\omega_c = 1$) filter must satisfy the following:
   a. $|H_n(j0.5)|^2 < 0.9$ in the pass-band.
   b. $|H_n(j1.5)|^2 < 0.05$ in the stop-band.
   Find its order.

8. Show that the high-frequency roll-off of an nth-order Butterworth filter is $2n$ dB/decade. Also, show that the first five $(2n - 1)$ derivatives of a third-order Butterworth filter are zero at $\omega = 0$.

Section 10.3

1. Use (10.16) to find $C_n(\omega)$ for $n = 0, 1, 2, 3$ and for both $|\omega| \leq 1$ and $\omega > 1$.
2. The ripple of a Chebyshev filter in decibels is given by (a) $r_{dB} = 0.05$, (b) $r_{dB} = 0.1$, and (c) $r_{dB} = 1.5$. Find the corresponding ε's.
3. Write the Chebyshev functions $|H_2(j\omega)|$ and $|H_3(j\omega)|$ if $\omega_c = 4$ rad/s and $\varepsilon = 1$.
4. For $n = 3$ and $\varepsilon = 0.8$, determine the poles of a low-pass Chebyshev filter. Plot the poles on the complex plane. Find the major and minor axes of the ellipse that they define.
5. Plot the Chebyshev functions of several orders. The following Book MATLAB Function will plot the first four polynomials:

```
function[w,w1,w2,tn1,tn2,Tn]=sschebfunct(wmax,dw)
            %introduce small values of wmax e.g 1.5;
w=0:dw:wmax;%wmax=maximum frequency in rad/s;
n=[1 2 3 4];%this function will produce the first
            %four polynomials;
w1=0:dw:1;
w2=1+dw:dw:wmax;
tn1=cos(n'*acos(w1));
tn2=cosh(n'*acosh(w2));
Tn=[tn1 tn2];%an nx(wmax/dw) matrix;
plot(w,Tn(1,:),w,Tn(2,:),'g',w,Tn(3,:),'r',w,Tn(4,:),'k');
            %will plot the third order polynimial since
            %Tn(3,:)=vector of the third row and all the
            %columns of Tn;
```

6. Design a low-pass Chebyshev filter with a 1.2-dB ripple in the band-pass and an attenuation of at least 30 dB at three times the cutoff frequency of 8 rad/s.
7. Determine the transfer function of a second-order Chebyshev low-pass filter with a cutoff frequency of $100/2\pi$ Hz, a 0.5-dB ripple in the pass-band, and a unit magnitude at $\omega = 0$.
8. Design a Chebyshev low-pass filter to meet the following specifications:
   a. Bandwidth, 10^3 rad/s
   b. Ripple, 0.1 dB
   c. Attenuation at least 30 dB for $\omega \geq 6 \times 10^3$ rad/s

   Find the transfer function of this filter.

Section 10.5

1. The design of Problem 10.3.6 is to provide the basis for a high-pass filter with cutoff at 3 rad/s. Specify the transfer function.
2. Design a pass-band filter to meet the following specifications:
   a. The 3-dB attenuations are at $1500 \times 2\pi$ and $2500 \times 2\pi$ rad/s.
   b. The attenuation should be at least 30 dB for frequencies lower than $700 \times 2\pi$ rad/s and higher than $5000 \times 2\pi$ rad/s.
3. Design a band-stop filter with an attenuation of at least 30 dB between 10^4 and 1.3×10^4 Hz. The attenuation for ω less than 4×10^3 Hz and more than 16×10^3 Hz must be less than -3 dB.

chapter 11

Finite Impulse Response (FIR) filters

FIR filters (nonrecursive) are filters whose present output is terminated from the present and past inputs, but is independent of its previous outputs. Because no feedback is present, FIR filters are stable. Furthermore, such filters are associated with zero phase or linear phase characteristics, and so no phase distortion occurs in the output.

Linear phase FIR digital filters have many advantages, such as guaranteed stability, freedom from phase distortion, and low coefficient sensitivity. Such filters are used where frequency dispersion is harmful. The design of such filters is well established. However, the design problem has the shortcoming of complexity, particularly in sharp filters. Several methods have been proposed in the literature for reducing the complexity of sharp filters. In this text we shall present the two fundamental approaches, the use of discrete Fourier series and discrete Fourier transform.

11.1 Properties of FIR filters

Causality

An ideal filter has an infinite number of terms of its impulse response. To alleviate this problem, we select $2N$ terms symmetrically with respect to $h(0)$ and shift the resulting sequence N steps to the right, producing a sequence $\{h(n)\}$ for $0 \leq n \leq 2N$. The result of the truncation produces the Gibbs' phenomenon.

Frequency normalization

Since the desired frequencies are given in Hz, we must transform them to the normalized digital ones. Therefore, we write

$$\omega_p = \Omega_p T = 2\pi F_p T = \frac{\Omega_p}{F_s} = \frac{2\pi \Omega_p}{\Omega_s} \qquad F_s = \frac{1}{T} \qquad \text{a)}$$

$$\omega_a = \Omega_a T = 2\pi F_a T = \frac{\Omega_a}{F_s} = \frac{2\pi \Omega_a}{\Omega_s} \qquad F_s = \frac{1}{T} \qquad \text{b)} \qquad (11.1)$$

The lowercase ω's indicate digital frequencies (rad/unit time) and the capital Ω's indicate analog frequencies (rad/s). T is the sampling time, F_s is the sampling frequency in Hz, Ω_p is the pass-band frequency, and Ω_a is the stop-band frequency (attenuation frequency). For example, if the pass-band is 5 kHz and the sampling frequency is 15 kHz, then we find that

$$\omega_p = 2\pi \times 5 \times 10^3 [1/(15 \times 10^3)] = 2\pi/3$$

Phase consideration

The transfer function $H(z)$ of a causal FIR filter of length N is given by

$$H(z) = \sum_{n=0}^{N-1} h(n) z^{-n} \qquad (11.2)$$

The corresponding frequency response is given by the discrete-time Fourier transform

$$H(e^{j\omega}) = \sum_{n=0}^{N-1} h(n) e^{-j\omega n} = H(z)\Big|_{z=e^{j\omega}} = |H(e^{j\omega})| e^{j\theta(\omega)} \qquad (11.3)$$

To ensure the filter with **linear phase**, we must impose the conditions

$$h(n) = h(N-1-n) \qquad \text{a)}$$
$$h(n) = -h(N-1-n) \qquad \text{b)} \qquad (11.4)$$

The first filter is called **symmetric** and the second **antisymmetric**.

The phase and group delays of a filter are defined respectively by

$$\tau_p = -\frac{\theta(\omega)}{\omega} \qquad \tau_{gp} = -\frac{d\theta(\omega)}{d\omega} \qquad (11.5)$$

If we substitute, for example, $N = 3$ in (11.3) we obtain the relations

Chapter 11: Finite Impulse Response (FIR) filters

$$\theta(\omega) = -\tau_p \omega = \tan^{-1} \frac{-\sum_{n=0}^{3-1} h(n) \sin n\omega}{\sum_{n=0}^{3-1} h(n) \cos n\omega} \quad \text{or} \quad \tan \omega \tau_p = \frac{\sum_{n=0}^{3-1} h(n) \sin n\omega}{\sum_{n=0}^{3-1} h(n) \cos n\omega} = \frac{\sin \omega \tau_p}{\cos \omega \tau_p}$$

The above equality holds if

$$\sum_{n=0}^{3-1} h(n) \sin n\omega \cos \tau_p \omega - \sum_{n=0}^{3-1} h(n) \cos n\omega \sin \tau_p \omega =$$

$$\sum_{n=0}^{3-1} h(n) \sin(\tau_p \omega - n\omega) = 0 \tag{11.6}$$

Therefore, by expanding the summation, we find

$$h(0) \sin \omega \tau_p + h(1) \sin(\omega \tau_p - \omega) + h(2) \sin(\omega \tau_p - 2\omega) = 0$$

The above equation is satisfied if $\tau_p = (3-1)/2$ and $h(0) = h(3-1-0)$, and hence, for N = odd, the general solution for the symmetric case is

$$\tau_p = \frac{N-1}{2} \quad h(n) = h(N-1-n) \quad n = 0, 1, 2, \cdots, N-1 \tag{11.7}$$

We therefore observe that FIR filters can have constant phase and group delays over the entire baseband. For this to happen, the impulse response must be **symmetrical** about the midpoint, that is, between $(N-2)/2$ and $N/2$ for **even** N and about samples $(N-1)/2$ for **odd** N.

For the antisymmetrical case we have the relations

$$\theta(\omega) = \theta_0 - \tau_p \omega = \pm \pi - \tau_p \omega \qquad \tau_p = (N-1)/2 \tag{11.8}$$

$$h(n) = -h(N-1-n) \qquad n = 0, 1, 2, \cdots, N-1 \tag{11.9}$$

Scaling the digital transfer function

To scale the magnitude of the filter, we multiply the transfer function by a scaling constant K such that the scaled transfer function has a maximum gain of zero dB. In the case of a low-pass filter, we usually set $K = 1/H(1)$, a direct current (dc) gain of 0 dB, and for a high-pass filter, we set $K = 1/H(-1)$, yielding a gain of 0 dB at $\omega = \pi$. For the band-pass transfer function we set $K = 1/|H(e^{j\omega_c})|$, where ω_c is the center frequency. For the band-stop transfer function K is reciprocal to the maximum of the two values $H(-1)$ and $H(1)$.

Symmetric FIR low-pass filters

A FIR filter of length N is characterized by the difference equation

$$y(n) = b_0 x(n) + b_1 x(n-1) + \cdots + b_{N-1} x(n-N+1) = \sum_{m=0}^{N-1} b_m x(n-m) = b_n * x(n) \quad (11.10)$$

If we set $h(n) = b_n$ and recalling the convolution property of the z-transform, we obtain

$$H(z) = \frac{Y(z)}{X(z)} = \sum_{n=0}^{N-1} h(n) z^{-n} \quad (11.11)$$

If we set $N = 5$ in (11.11) and consider a symmetric sequence (see (11.7)), we find

$$H(e^{j\omega}) = h(0) + h(1)e^{-j\omega} + h(2)e^{-j2\omega} + h(3)e^{-j3\omega} + h(4)e^{-j4\omega}$$

$$= \sum_{n=0}^{(5-3)/2} h(n) e^{-j\omega n} + h\left(\frac{5-1}{2}\right) e^{-j\omega(5-1)/2} + \sum_{n=(5+1)/2}^{5-1} h(n) e^{-j\omega n} \quad (11.12)$$

By using (11.7) ($h(n) = h(N - 1 - n)$) and then letting $5 - 1 - n = m$, the last summation of (11.12) becomes

$$\sum_{n=(5+1)/2}^{5-1} h(n) e^{-j\omega n} = \sum_{m=0}^{(5-3)/2} h(5-1-m) e^{-j\omega(5-1-m)} = \sum_{n=0}^{(5-3)/2} h(n) e^{-j\omega(5-1-n)} \quad (11.13)$$

where $h(5 - 1 - m) = h(m)$ and m was renamed n. Combining next (11.13) and (11.12) we obtain the expression

$$H(e^{j\omega}) = e^{-j\omega(5-1)/2} \left[h\left(\frac{5-1}{2}\right) + \sum_{n=0}^{(5-3)/2} h(n) \cos\left[\omega\left(\frac{5-1}{2} - n\right)\right] \right]$$

Chapter 11: Finite Impulse Response (FIR) filters

If we then set

$$\frac{N-1}{2} - n = k,$$

where N is an **odd** integer in place of 5 and rename $k = n$, we end up with the general equation

$$H(e^{j\omega}) = e^{-j\omega(N-1)/2} \sum_{n=0}^{(N-1)/2} a(n)\cos\omega n \qquad (11.14)$$

$$a(0) = h\left(\frac{N-1}{2}\right) \qquad a(n) = 2h\left[\left(\frac{N-1}{2} - n\right)\right]$$

In a similar approach the antisymmetric case can be developed.

Book MATLAB function m-file: ssfir1(h)

```
function[H,w]=ssfir1(h)
%this function finds the frequency response of
%a FIR filter with an odd N(=length of filter)
%and symmetric form starting from n=1 to N+1, the
%middle point is (N+1)/2;w is the frequency range
%from 0 to pi; H is real and it is the FIR response;
N=length(h);
n=2:(N+1)/2;
w=0:pi/512:pi-(pi/512);
H1=cos(w'*(n-1));
H2=H1*(2*h(((N+1)/2)-n+1))';
H=h((N+1)/2)+H2;
```

Figure 11.1.1 shows the response of a FIR low-pass filter with $N = 61$ in dB scale form.

11.2 FIR filters using the Fourier series approach

Since $H(e^{j\omega})$ is periodic with period 2π, $[H(e^{j(\omega+2\pi)}) = H(e^{j\omega})]$, it can be expanded in discrete-time Fourier transform (DTFT) in the form

$$H(e^{j\omega}) = \sum_{n=-\infty}^{\infty} h(n)e^{-j\omega n} \qquad (11.15)$$

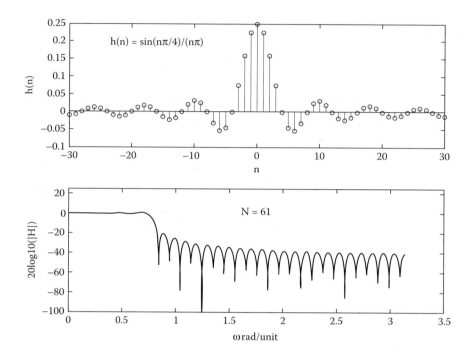

Figure 11.1.1 FIR low-pass filter with $N = 61$ and symmetric.

The inverse DTFT is (see also Chapter 7)

$$h(n) = \frac{1}{2\pi} \int_{-\pi}^{\pi} H(e^{j\omega}) e^{jn\omega} d\omega \tag{11.16}$$

If we set $z = e^{j\omega}$ in (11.15), we obtain

$$H(z) = \sum_{n=-\infty}^{\infty} h(n) z^{-n} \tag{11.17}$$

The above transfer function is **noncausal** and **infinite order**. Therefore, we must truncate it and shift it to the right such that $h(n)$ has values for positive n's only. Next we set

$$h(n) = 0 \qquad |n| > \frac{N-1}{2} \qquad N = \text{odd integer} \tag{11.18}$$

Chapter 11: Finite Impulse Response (FIR) filters

to obtain

$$H(z) = h(0) + \sum_{n=1}^{(N-1)/2} [h(-n)z^n + h(n)z^{-n}] \quad (11.19)$$

The shifted equivalent of (11.19) is the causal filter

$$H_{cs}(z) = z^{-(N-1)/2} H(z) \quad (11.20)$$

which is the desired transfer function.

Example 11.2.1: Design a low-pass FIR filter with $\Omega_p = 2$ rad/s and sampling frequency $\Omega_s = 12$ rad/s. Use $N = 21$ and $N = 91$.

Solution: From the given sampling frequency we obtain

$$\Omega_s = 2\pi F_s = 12, \quad F_s = 12/2\pi = 6/\pi$$

The normalized pass-band edge discrete frequency is

$$\omega_p = \Omega_p T = \Omega_p / F_s = 2/(6/\pi) = \pi/3$$

The desired low-pass filter has a magnitude of 1 in the range of $0 \le \omega \le \pi/3$ and zero otherwise. Hence, from (11.16) we obtain the impulse response

$$h_d(n) = \frac{1}{2\pi} \int_{-\pi/3}^{\pi/3} e^{jn\omega} d\omega = \frac{\sin \frac{n\pi}{3}}{n\pi}$$

For a causal transfer function with odd N and symmetrical $h(n)$, its frequency response is

$$H_{cs}(e^{j\omega}) = e^{-j\omega(N-1)/2} \left[h_d\left(\frac{N-1}{2}\right) + \sum_{n=1}^{(N-1)/2} 2h_d\left(\frac{N-1}{2} - n\right) \cos n\omega \right]$$

where $h_d(n)$ was found first in the range $-(N-1)/2 \le n \le (N-1)/2$ and then shifted to the right by $(N-1)/2$ to recreate $h(n)$. The amplitude responses for $N = 21$ and $N = 91$ are shown in Figure 11.2.1.

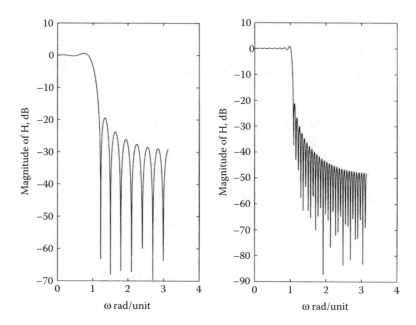

Figure 11.2.1 FIR filter using Fourier series approach.

Book MATLAB function m-file: ex_11_2_1

```
function[ampH,w]=ex_11_2_1(hd)
%this function solves the Ex 11.2.1 for low-pass FIR filter;
N=length(hd);
M=(N-1)/2;
a=[hd(M+1) 2*hd(M:-1:1)];%middle point of hd is at M+1;
n=[0:M];
w=[0:511]'*pi/511;
ampH=cos(w*n)*a';
%w*n is a (512x1)x(1x(M+1))=512x(M+1) matrix and the value
%of each element e.g. (k,m) is equal to cos(w(k)*n(m));
%ampH is a 512x1 vector where each element is the sum
%over n for each row at each w (omega);
```

To execute the above m-function we write:

```
N=10; %this plots the left-hand side of the figure;
n=[-N:N];
hd=sin((n*pi/3)+eps)./(n*pi+eps);
hd(N+1)=1/3;%observe the correction, MATLAB gives zero;
[ampH,w]=ex_11_2_1;
subplot(1,2,1);plot(w,20*log10(ampH),'k');                    ∎
```

11.3 FIR filters using windows

Windows

To reduce Gibbs' phenomenon (oscillations), we must multiply the impulse response, $h(n)$, with special finite length function $w(n)$, known as **window function**. Therefore, the widowed impulse response is

$$h_w(n) = w(n)h(n) \qquad (11.21)$$

The two basic characteristics of a window are the width of the **main lobe** and the **ripple ratio**. The ripple ratio is defined by the relation

$$r = \frac{100(\text{maximum side-lobe amplitude})}{\text{main-lobe amplitude}}\% \qquad (11.22)$$

For example, the rectangular window has a ripple ratio of 21.70 and a relative side-lobe level of −13.27 dB, and the Blackman window has a ripple factor of 0.12 and a relative side-lobe level of −58.42. Table 11.3.1 gives the most useful windows.

Table 11.3.1 Window Functions for FIR Filter Design

Name of Window	$w(n), -N \le n \le N$
1. Bartlett (triangular)	$1 - \dfrac{\|n\|}{N}$
2. Blackman	$0.42 + 0.5\cos\left(\dfrac{2\pi n}{2N+1}\right) + 0.08\cos\left(\dfrac{4\pi n}{2N+1}\right)$
3. Hamming	$0.54 + 0.46\cos\left(\dfrac{2\pi n}{2N+1}\right)$
4. Hann	$\dfrac{1}{2}\left[1 + \cos\left(\dfrac{2\pi n}{2N+1}\right)\right]$
5. Kaiser	$I_0\left[\beta\sqrt{1-\left(\dfrac{n}{N}\right)^2}\right]/I_0(\beta)$

β = adjustable parameter
$I_0(x)$ = zero-order modified Bessel function

$$I_0(x) = 1 + \sum_{k=1}^{\infty}\left[\frac{1}{k!}\left(\frac{x}{2}\right)^k\right]^2$$

Note: In practice, 20 to 25 terms of $I_0(x)$ are sufficient for accurate results.

Example 11.3.1: Design a low-pass FIR filter using the Fourier series approach and the windowing approach. The constants are pass-band frequency 20 rad/s and sampling frequency 60 rad/s.

Solution: From (11.1) we find the normalized pass-band frequency to be $\omega_p = 2\pi\Omega_p/\Omega_s = 2\pi 20/60 = 2\pi/3$. From (11.16) we obtain

$$h(n) = \frac{1}{2\pi}\int_{-2\pi/3}^{2\pi/3} 1 \times e^{jn\omega}d\omega = \frac{\sin\left(\frac{2\pi}{3}n\right)}{n\pi}$$

The following Book MATLAB program gives the requested results:

Book MATLAB m-file: ex_11_3_1

```
%m-file for the Ex 11.3.1: ex_11_3_1
n=-50:50;hn=sin(2*pi*n/3)./(n*pi+eps);hn(51)=2/3;
[ah,w]=ex_11_2_1(hn);
hw=hn.*hamming(101)';%hamming(M) is a MATLAB function
        %that gives a column vector of length M;
[ahw,w]=ex_11_2_1(hw);
plot(w,20*log10(ah),'k');hold on;
plot(w,20*log10(ahw),'k');
xlabel('\omega rad/unit');ylabel('Magnitude, dB');
```

Figure 11.3.1 shows a Fourier series FIR low-pass filter and a windowed one using the Hamming window and the same impulse response function.

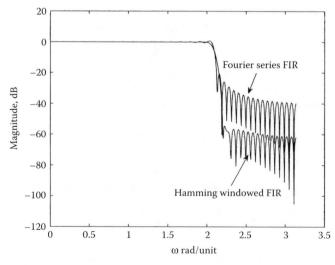

Figure 11.3.1 Fourier series and windowed FIR filter.

Chapter 11: Finite Impulse Response (FIR) filters

Note: *The sampling time is $T = 2\pi/60 = \pi/30$, and hence the frequency range is from 0 to $\pi/(\pi/30) = 30$ rad/s.* ∎

If the pass-band frequency and attenuation-band frequency are given, then we generally use the cutoff frequency

$$\omega_c = \frac{\omega_p + \omega_a}{2} \tag{11.23}$$

High-pass FIR filters

The only new information needed is (see Figure 11.3.2b)

$$H(e^{j\omega}) = \begin{cases} 0 & |\omega| < \omega_c \\ 1 & \omega_c \le |\omega| \le \pi, \end{cases} \quad \omega_c = \frac{\omega_p + \omega_a}{2} \tag{11.24}$$

Hence, the impulse response is found as follows (see (11.16))

$$h(n) = \frac{1}{2\pi} \int_{-\pi}^{-\omega_c} 1 \times e^{j\omega n} d\omega + \frac{1}{2\pi} \int_{\omega_c}^{\pi} 1 \times e^{j\omega n} d\omega$$

$$= \begin{cases} 1 - \dfrac{\omega_c}{\pi} & n = 0 \\ -\dfrac{\sin(\omega_c n)}{n\pi} & |n| > 0 \end{cases} \tag{11.25}$$

The procedure is identical to that given above for low-pass filters (see Figure 11.3.2a).

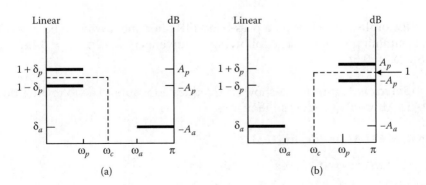

Figure 11.3.2 Ideal frequency response for low-pass and high-pass filters.

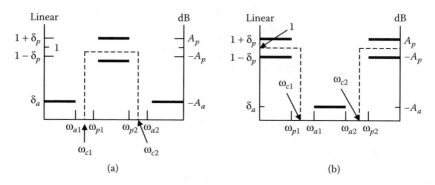

Figure 11.3.3 Band-pass and band-stop FIR filter specifications.

Band-pass FIR filters

For this case, the impulse response is (see Figure 11.3.3a)

$$h(n) = \frac{\sin(\omega_{c2} n) - \sin(\omega_{c1} n)}{n\pi} \qquad |n| \geq 0 \qquad (11.26)$$

$$\omega_{c1} = (\omega_{a1} + \omega_{p1})/2 \qquad \omega_{c2} = (\omega_{a2} + \omega_{p2})/2$$

Band-stop FIR filters

The impulse response for this case is (see Figure 11.3.3b)

$$h(n) = \begin{cases} 1 - \dfrac{\omega_{c2} - \omega_{c1}}{\pi} & n = 0 \\ \dfrac{\sin(\omega_{c1} n) - \sin(\omega_{c2} n)}{n\pi} & |n| > 0 \end{cases} \qquad (11.27)$$

$$\omega_{c1} = (\omega_{a1} + \omega_{p1})/2 \qquad \omega_{c2} = (\omega_{a2} + \omega_{p2})/2$$

Example 11.3.2: Design a pass-band FIR filter and a windowed one with the Hamming window. The following are given: $\omega_{a1} = 0.2\pi$, $\omega_{p1} = 0.4\pi$, $\omega_{a2} = 0.8\pi$, $\omega_{p1} = 0.6\pi$.

Solution: Figure 11.3.4 shows the requested results. The following Book MATLAB m-file produces the results.

Book MATLAB m-file: ex_11_3_2

```
%Book MATLAB m-file:ex_11_3_2;
n=-30:30;wc1=0.3*pi;wc2=0.7*pi;
hn=(sin(wc2*n)-sin(wc1*n))./(n*pi+eps);
```

Chapter 11: Finite Impulse Response (FIR) filters

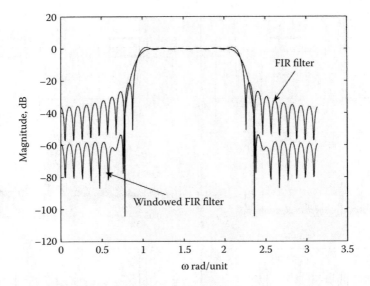

Figure 11.3.4 Band-pass FIR and windowed FIR filter.

```
hn(31)=0.4;
[ah,w]=ex_11_2_1(hn);
hnw=hn.*hamming(61)';%MATLAB gives the window in column form;
[ahw,w]=ex_11_2_1(hnw);
plot(w,20*log10(ah),'k');hold on;plot(w,20*log10(ahw),'k');
xlabel('\omega rad/unit');ylabel('Magnitude, dB');
```
■

Example 11.3.3: Find a low-pass FIR such that it eliminates the upper frequency of the signal

$$s(t) = \sin 20t + \sin 40t$$

which is contaminated with noise (a signal that is nonpredictable from one time to the next). To produce such a signal, we will use the MATLAB function **randn(1,M)**. More will be said about random signals in Chapter 13.

Solution: The Book MATLAB m-file given below produces the desired results. From Figure 11.3.5 we observe the following: The first figure shows the signal equal to the sum of two sine waves and the second the same signal but added noise. The third figure shows the DFT of the noisy signal. We observe that the spectrum is made up of two approximately delta functions, which indicates that two sinusoidal signals are present. However, we also see frequencies present in the whole spectral range. These frequencies came from the added noise. The fourth figure shows a 101-term impulse response of a low-pass FIR filter. The fifth figure shows the output of the filter that

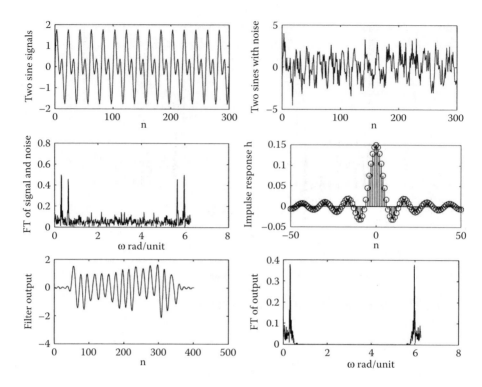

Figure 11.3.5 Low-pass filtering.

was produced by taking the convolution between the input and the impulse response of the filter. Observe that the signal is not pure sinusoid. This is more evident from the sixth figure, where the delta function is not alone but is accompanied with noise. This is true because the low-pass filter passes all the frequencies from 0 to 0.15π, including the noise. However, we are able to eliminate the second sine signal completely.

Book MATLAB m-file: ex_11_3_3

```
%Book m-file for Ex 11.3.3; ex_11_3_3
ws=400;%ws=sampling frequency rad/s;
m=0:300;
T=2*pi/ws;%T=time sampling;
s1=sin(20*m*T)+sin(40*m*T);
s=sin(20*m*T)+sin(40*m*T)+randn(1,301);%signal plus noise;
w=0:2*pi/512:2*pi-(2*pi/512);
fs=fft(s,512)/301;
n=-50:50;%102 impulse response of the filter;
hn=sin(0.15*pi*n)./(n*pi+eps);%hn=impulse response;
hn(51)=0.15;%corrects the MATLAB value that gives 0;
```

Chapter 11: Finite Impulse Response (FIR) filters

```
y=conv(s,hn);%y=the output of the system (filter);
%conv(.,.) is a MATLAB function giving the convolution
%of two sequences;
fy=fft(y,512)/402;
subplot(3,2,1);plot(m,s1,'k');ylabel('Two sine signals');
subplot(3,2,2);plot(m,s,'k');ylabel('Two sines with noise');
subplot(3,2,3);plot(w,abs(fs),'k');ylabel('FT of signal and...
    noise');
subplot(3,2,4);stem(n,hn,'k');ylabel('Impulse response h');
subplot(3,2,5);plot(y,'k');ylabel('Filter output');
subplot(3,2,6);plot(w,abs(fy),'k');ylabel('FT of output'); ■
```

*11.4 Prescribed filter specifications using a Kaiser window

Consider the low-pass filter specifications shown in Figure 11.3.2a. The pass-band ripple and minimum stop band attenuation in decibels are given by

$$A_p = 20\log\frac{1+\delta}{1-\delta} \quad \text{a)} \quad A_a = -20\log\delta \quad \text{b)} \tag{11.28}$$

The **transition bandwidth** is given by

$$B_t = \omega_a - \omega_p \tag{11.29}$$

The procedure to find a FIR filter obeying specific restrictions is:

1. Perform transformation from the frequency domain (rad/s) to the digital frequency domain (rad/unit) using (11.1).
2. Determine $h(n)$ by assuming the ideal low-pass filter

$$H(e^{j\omega}) = \begin{cases} 1 & |\omega| \leq \omega_c \\ 0 & \omega_c < |\omega| \leq \pi \end{cases} \qquad \omega_c = \frac{\omega_a + \omega_p}{2} \tag{11.30}$$

from which we find (see (11.16))

$$h(n) = \frac{1}{2\pi}\int_{-\omega_c}^{\omega_c} 1 e^{j\omega n} d\omega \tag{11.31}$$

However, for a high-pass, band-pass, or stop-band filter we can use (11.25), (11.26), or (11.27), respectively.

3. Choose δ in (11.28), whichever of the two equalities gives the smaller one. That is,

$$\delta = \min(\delta_1, \delta_2), \quad \delta_1 = 10^{-0.05A_a}, \quad \delta_2 = \frac{10^{0.05A_p} - 1}{10^{0.05A_p} + 1} \quad (11.32)$$

4. Calculate A_a using (11.28b).
5. Choose parameter β as follows:

$$\beta = \begin{cases} 0 & A_a \leq 21 \\ 0.5842(A_a - 21)^{0.4} + 0.07886(A_a - 21) & 21 < A_a \leq 50 \\ 0.1102(A_a - 8.7) & A_a > 50 \end{cases} \quad (11.33)$$

6. Choose parameter D as follows:

$$D = \begin{cases} 0.9222 & A_a \leq 21 \\ \dfrac{(A_a - 7.95)}{14.36} & A_a > 21 \end{cases} \quad (11.34)$$

Then select the lowest odd value of N satisfying the inequality

$$N \geq \frac{2\pi D}{B_t} + 1$$

7. Use the desired window.
8. From

$$H'_w(z) = z^{-(N-1)/2} H_w(z), \quad H_w(z) = \mathcal{Z}[w(n)h(n)] \quad (11.35)$$

Note: *The parameter β is used in the Kaiser window formula (see Table 11.3.1).*

Example 11.4.1: Design a FIR filter satisfying the following specifications:

$$\Omega_{a1} = 20\,\text{rad/s}, \quad \Omega_{p1} = 30\,\text{rad/s}, \quad \Omega_{p2} = 50\,\text{rad/s}, \quad \Omega_{a2} = 60\,\text{rad/s}$$

Minimum attenuation	$0 \leq \Omega \leq 20$ rad/s:	40 dB
Maximum pass-band ripple	$30 < \Omega < 50$ rad/s:	0.2 dB
Maximum attenuation	$60 < \Omega < 80$ rad/s:	40 dB
Sampling frequency	160 rad/s	

Chapter 11: Finite Impulse Response (FIR) filters

Solution: From (11.1) we obtain

$$\omega_{a1} = \Omega_{a1}T = \frac{\Omega_{a1} \times 2\pi}{160} = \frac{20 \times 2\pi}{160} = 0.25\pi$$

and similarly

$$\omega_{p1} = 3\pi/8, \ \omega_{p2} = 5\pi/8, \ \omega_{a2} = 3\pi/4, \ \omega_{c1} = 5\pi/16, \ \omega_{c2} = 11\pi/16$$

From step 2 and (11.26) we find

$$h(n) = \frac{\sin(\omega_{c2}n) - \sin(\omega_{c1}n)}{n\pi} = \frac{\sin(11\pi n/16) - \sin(5\pi n/16)}{n\pi}$$

From step 3 we obtain

$$\delta_1 = 10^{-0.05 \times 40} = 0.01, \quad \delta_2 = \frac{10^{0.05 \times 0.2} - 1}{10^{0.05 \times 0.2} + 1} = 0.0115$$

and hence from step 4,

$$A_p = 20 \log_{10} \frac{1 + 0.01}{1 - 0.01} = 0.1737, \quad A_a = -20 \log_{10}(0.01) = 40$$

From step 5 we obtain

$$\beta = 0.5842(40 - 21)^{0.4} + 0.07886(40 - 21) = 3.3953$$

From step 6 we find

$$D = \frac{40 - 7.95}{14.36} = 2.2319, \ N > \frac{2\pi 2.2319}{(11\pi/16) - (5\pi/16)} = 11.9035, \ N = 13$$

The following Book MATLAB m-file gives the appropriate results.

Book MATLAB m-file function

```
function[ahnw,w]=ssbandpass_kaiser_fir(N,wc1,wc2,beta);
%Book MATLAB function
n=-N:N;
hn=(sin(wc2*n)-sin(wc1*n))./(n*pi+eps);
```

```
hn(N+1)=(wc2-wc1)/pi;
hnw=hn.*kaiser(2*N+1,beta)';
[ahnw,w]=ex_11_2_1(hnw);
```

To plot the response we write at the command window: `plot(w,20*log10(ahnw))`. The above function can be easily modified to introduce a different window. ∎

11.5 MATLAB FIR filter design

The MATLAB function **fir1** can be used to design low-pass, high-pass, band-pass, and band-stop linear phase FIR filters.

Low-pass FIR filter

```
b=fir1(N,wc);   %N=order of the filter(odd); wc=normalized cutoff
                %frequency (0<wc<1);
                %b=vector of the impulse response coefficients
                %arranged in ascending
                %powers of z⁻¹; the sampling frequency is assumed
                %2 Hz.
[h,w]=freqz(b,a,n);  %a=1, n=frequency response is calculated
                %at n points equally spaced
                %around the upper half of the unit circle
                %(usually n=512); w=contains n
                %points between 0 and π; b is the vector
                %containing the coefficients
                %of the denominator polynomial;
plot(w,20*log10(abs(h)));
```

High-pass FIR filter

```
b=fir1(N,wc,'high');  %N=order of the filter (even only);
                      %wc=filter stop band;
[h,w]=freqz(b,a,n);%a=1,n=usually 512;
plot(w,20*log10(abs(h)));
```

Band-pass FIR filter

```
b=fir1(N,[wc1wc2]);%band-pass=wc2-wc1;
[h,w]=freqz(b,1,512);
plot(w,20*log10(abs(h)));
```

In the command window use *help fir1* to obtain functions to produce multiple band-pass and band-stop designs.

Band-stop FIR filter

```
b=fir1(N, [wc1wc2],'stop');%wc1 and wc2 must be put in vector
                           %form;
[h,w]=freqz(b,1,512);
plot(w,20*log10(abs(h)));
```

Window use

If we want, for example, to create a band-stop filter with a Hamming window, we write:

```
b=fir1(N,[wc1 wc2],'stop',hamming(N+1));%N must be even;
```

Other windows are: blackman ($N+1$), hanning ($N+1$), chebwin ($N+1$), and kaiser ($N+1$,beta).

Important definitions and concepts

1. Causality of filter systems
2. Frequency normalization
3. Symmetric and antisymmetrical filters
4. Scaling of transfer functions
5. FIR design using the Fourier series
6. FIR design using windows
7. Main lobe of filter response
8. Ripple ratio
9. High-pass, band-pass, and band-stop FIR filters
10. Prescribed filter specifications
11. Transition bandwidth
12. MATLAB filter design

Chapter 11 Problems

Section 11.1

1. The band-pass frequency for a low-pass FIR filter is 120 rad/s and the attenuation frequency is 130 rad/s. Show the frequency axis in rad/s and in rad/unit.
2. For third-order filter prove (11.9).
3. Consider the filter $H(z) = [b_0 + b_1 z^{-1} + b_2 z^{-2}]/[1 + a_1 z^{-1} + a_2 z^{-2}]$ as a low-pass or high-pass filter. Find the attenuation factors K for these two cases.
4. Draw, by taking a small subset (10 to 15) of FIR impulse responses, the following causal cases: odd and antisymmetric, even and symmetric, even and antisymmetric.

Section 11.2

1. Design a low-pass filter with $\omega_p = 0.4$ and $N = 5$.
2. Design a low-pass filter with $\Omega_p = 2$ rad/s and sampling frequency $\Omega_s = 12$ rad/s. Use $N = 11$ and $N = 101$.
3. Plot the frequency responses of the rectangular, Hamming, and Blackman windows and estimate from your graphs the height of the first side lobe.

Section 11.3

1. Plot the low-pass FIR filter response if the band-pass frequency is 1000 rad/s and the sampling frequency is 2000 rad/s.
2. Repeat Problem 11.3.1, but using the Hamming window.
3. Repeat Problem 11.3.1, but using the Blackman window.
4. Find the response of a band-pass FIR given the following:

    a. Attenuation range from 0 to 30 rad/s
    b. Band-pass range from 40 to 80 rad/s
    c. Attenuation range from 90 to 200 rad/s
    d. Sampling frequency of 400 rad/s

5. Repeat Problem 11.3.4 but using Hamming and Blackman windows.
6. Find the response of a band-stop FIR filter given the following:

    a. Band-pass range from 0 to 100 rad/s
    b. Band-stop range from 120 to 150 rad/s
    c. Band-pass range from 170 to 300 rad/s
    d. Sampling frequency of 600 rad/s

7. Repeat Problem 11.3.6 but using Hamming and Blackman windows.

Section 11.4

1. Design a high-pass FIR filter given the following:

    a. Minimum attenuation band from 0 to 30 rad/s: 45 dB
    b. Maximum band-pass ripple from 40 to 100 rad/s: 0.1 dB
    c. Sampling frequency: 200 rad/s

2. Design a band-pass FIR filter given the following:

    a. Minimum attenuation from 0 to 50 rad/s: 50 dB
    b. Maximum band-pass ripple from 70 to 120 rad/s: 0.2 dB
    c. Minimum attenuation from 140 to 400 rad/s: 50 dB
    d. Sampling frequency: 800 rad/s

3. Repeat Problem 11.4.1 but using a Hamming window.
4. Repeat Problem 11.4.2 but using a Kaiser window.

*chapter 12

Infinite Impulse Response (IIR) filters

Infinite impulse response (IIR) filters, also known as recursive filters, take advantage of the continuous-time filters made up of networks with discrete elements. Both have impulse responses of infinite length and are described by rational fractional transfer functions in the frequency domain. We can start by designing the continuous-time filters and transforming them into digital format. The transformations, however, do not produce identical results since the output of the continuous case is due to the convolution of input and filter through integration, whereas the digital case is a convolution performed by summation. It can be said that the continuous-time and discrete-time systems converge more and more as the sampling time decreases.

12.1 The impulse-invariant method approximation in the time domain

In the impulse-invariant design method, the impulse response of the continuous method is sampled at equal instances of time. Hence, $h(t)$ becomes $h(nT)$. Suppose that the system function of a low-pass filter is

$$H(s) = \sum_{p=1}^{P} \frac{A_p}{s+s_p} \tag{12.1}$$

Assume that all poles are distinct; hence, the impulse response function is

$$h(t) = \mathcal{L}^{-1}\{H(s)\} = \sum_{p=1}^{P} A_p e^{-s_p t} \tag{12.2}$$

If $Th(nT)$ is the corresponding sampled version of $h(t)$, then we write

$$H(z) = T\sum_{n=0}^{\infty} h(nT)z^{-n} = T\sum_{n=0}^{\infty} z^{-n} \sum_{p=1}^{P} A_p e^{-s_p nT} = T\sum_{p=1}^{P} A_p \sum_{n=0}^{\infty} z^{-n} e^{-s_p nT} \qquad (12.3)$$

which, using the well-known formula of geometric series, is

$$H(z) = T\sum_{p=1}^{P} \frac{A_p}{1 - e^{-s_p T} z^{-1}} \qquad (12.4)$$

A comparison of (12.1) and (12.4) shows that a continuous low-pass filter specified by the system function $H(s)$ transforms into a digital filter specified by $H(z)$ via **impulse-invariant** techniques by setting

$$s + s_p = 1 - e^{-s_p T} z^{-1} \qquad (12.5)$$

Observe, therefore, that $H(z)$ can be obtained from $H(s)$ without evaluating $h(t)$ or $h(nT)$. As already noted, some degree of approximation exists in this transformation because the digital filter is necessarily band limited and periodic, whereas $H(s)$, a rational function of s, is not band limited. Because the frequency response of the digital filter comprises fewer frequency components than the analog filter, an added or folded difference due to the different number of terms of the series exists. This phenomenon is equivalent to that discussed in connection with the sampling theorem in Chapter 6. No error due to sampling occurs if $|H(j\Omega)| = 0$ for $|\Omega| > \Omega_s$. In this section, we introduce capital omega for the continuous case to facilitate our discussion.

The design of an IIR digital filter employs the following steps:

1. Begin with a given filter specification.
2. Create an analog transfer function $H(s)$ that meets the specifications of step 1.
3. Determine the impulse response of the analog filter by means of the Laplace inversion technique, $h(t) = \mathcal{L}^{-1}\{H(s)\}$.
4. Sample $h(t)$ at T-second intervals, thereby creating a sequence $\{h(nT)\}$.
5. Deduce $H(z)$ of the resulting digital filter by taking the z-transform of the discrete function $h(nT)$,

$$H(z) = T\sum_{n=0}^{\infty} h(nT) z^{-n}.$$

As an alternative procedure, proceed with steps 1 and 2 and then apply transformation (12.5).

Example 12.1.1: Determine the digital equivalent of the first-order Butterworth filter. The cutoff frequency Ω_c is 20 rad/s.

Chapter 12: Infinite Impulse Response (IIR) filters

Solution: The normalized analog transfer function is found from Table 10.2.2 to be

$$H(s) = \frac{1}{s+1}$$

The system transfer function is given by

$$H\left(\frac{s}{\Omega_c}\right) = H\left(\frac{s}{20}\right) = \frac{1}{(s/20)+1} = \frac{20}{s+20}$$

The impulse response of this filter is given by

$$h(t) = \mathcal{L}^{-1}\left\{H\left(\frac{s}{20}\right)\right\} = \mathcal{L}^{-1}\left\{\frac{20}{s+20}\right\} = 20e^{-20t} \qquad t \geq 0$$

The z-transform of the discrete function $Th(nT)$ is

$$H(z) = 20T \sum_{n=0}^{\infty} e^{-20nT} z^{-1} = 20T \sum_{n=0}^{\infty} \left(e^{-20T} z^{-1}\right)^n = \frac{20T}{1-e^{-20T}z^{-1}}$$

Note that we could have found $H(z)$ simply by setting $s + 20 = 1 - e^{-20T} z^{-1}$ in $H(s/\Omega_c)$, as shown in (12.4) and multiply by T. To proceed for the specified sampling frequency $\Omega_s = 80$ rad/s, we first deduce the absolute value of the analog transfer function:

$$\left|H\left(\frac{j\Omega}{20}\right)\right| = \frac{1}{|(j\Omega/20)+1|} = \frac{1}{\sqrt{1+(\Omega/20)^2}}$$

Next, we normalize the frequency using (11.1). Hence, we find

$$\omega_c = \Omega_c T = \Omega_c \frac{2\pi}{\Omega_s} = 20 \frac{2\pi}{80} = 0.5\pi$$

The equivalent expression becomes

$$H(z) = H(e^{j\omega}) = \frac{20T}{1-e^{-20T}e^{-j\Omega T}} = \frac{20T}{1-e^{-20T}e^{-j\omega}}$$

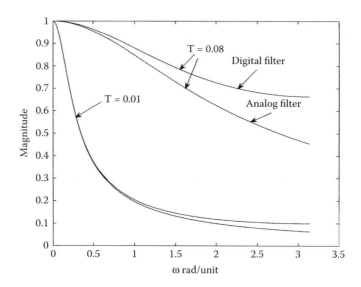

Figure 12.1.1 Comparison between the analog and its corresponding digital filter for two sampling times.

Since $\omega = \Omega T$, the analog case of the transfer function as a function of digital frequency is

$$\left| H\left(\frac{j\omega}{20}\right) \right| = \frac{1}{\left| \left(\frac{j\Omega T}{20T}\right) + 1 \right|} = \frac{1}{\sqrt{1 + \left(\frac{\omega}{20T}\right)^2}}$$

In Figure 12.1.1 we have plotted the above two equations for two sampling times: $T = 0.08$ and $T = 0.01$. From the figure and $T = 0.08$ we find that the cutoff requency is $\omega_c = \Omega_c T = 20 \times 0.08 = 1.6$. At that normalized frequency, the magnitude is 0.707, as it should be. However, the system frequency response is not as accurate at the same frequency value. This phenomenon is practically the same across all frequency ranges. For the case $T = 0.01$, we find that the cutoff frequency is $\omega_c = \Omega_c T = 20 \times 0.01 = 0.2$. At this frequency, both filters are identical and for the rest of the frequency range both have close values. The above tells us that the higher the sampling frequency is, the closer their responses are. If we desire to change the frequency axis to read in rad/s, we simply expand the axis, for example, by 1/0.08. For this sampling time, frequency 3.14 becomes 3.14/0.08 = 39.25 rad/s. ∎

Note: *If, for example, we had expanded the frequency range to more than π, the digital frequency response would repeat itself every 2π, whereas the analog filter response would be continuously decreasing.*

Chapter 12: Infinite Impulse Response (IIR) filters

Example 12.1.2: Determine the digital equivalent of a Butterworth third-order low-pass filter and find T if the sampling frequency is 15 times the normalized cutoff frequency. Repeat the procedure if the sampling frequency is 100 times the normalized cutoff frequency.

Solution: The normalized system function of this filter is deduced from Table 10.2.2:

$$H(s) = \frac{1}{s^3 + 2s^2 + 2s + 1} = \frac{1}{(s+1)(s^2+s+1)} = \frac{1}{(s+1)\left(s+\frac{1}{2}-j\frac{\sqrt{3}}{2}\right)\left(s+\frac{1}{2}+j\frac{\sqrt{3}}{2}\right)}$$

$$= \frac{A_1}{s+p_1} + \frac{A_2}{s+p_2} + \frac{A_3}{s+p_3}$$

The constant A_i's are easily found (see Chapter 8) to be

$$A_1 = 1 \quad A_2 = \frac{2}{-3+j\sqrt{3}} \quad A_3 = \frac{2}{-3-j\sqrt{3}}$$

The impulse response of this system is given by

$$h(t) = A_1 e^{-t} + A_2 e^{-p_2 t} + A_3 e^{-p_3 t}$$

Further, we note that $\Omega_s = 2\pi/T = 15\Omega_c = 15$, from which the sampling time $T = 0.4189$. The sampled impulse response function is

$$h(n0.4189) = 0.4189 \left[A_1 e^{-0.4189n} + A_2 e^{-0.4189 p_2 n} + A_3 e^{-0.4189 p_3 n} \right]$$

The z-transform of this function is

$$\frac{H(z)}{0.4189} = \frac{z}{z - e^{-0.4189}} + A_2 \frac{z}{z - e^{-0.4189 p_2}} + A_3 \frac{z}{z - e^{-0.4189 p_3}}$$

Using the following Book MATLAB program, we obtain the digital filter response shown in Figure 12.1.2.

Book MATLAB m-file: ex_12_1_2

```
%Book MATLAB m-file:ex_12_1_2
w=0:0.01:5;
a2=2/(-3+j*sqrt(3));a3=2/(-3-j*sqrt(3));
```

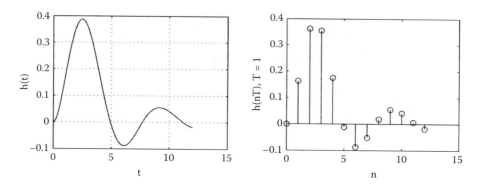

Figure 12.1.2 Impulse response $h(t)$ and $h(nT)$ for $T = 1$.

```
p2=.5-j*sqrt(3)*.5;p3=.5+j*sqrt(3)*.5;
T=.4189;
H=(exp(j*w*T)./(exp(j*w*T)-exp(-0.4189)))...
+(a2*exp(j*w*T)./(exp(j*w*T)-exp(-.4189*p2)))...
+(a3*exp(j*w*T)./(exp(j*w*T)-exp(-.4189*p3)));
subplot(2,1,1);plot(w,abs(H)*.4189);
Hs=freqs([0 0 0 1],[1 2 2 1],w);
subplot(2,1,2);plot(w,abs(Hs));
```

In general, a digital filter designed using the impulse-invariant method results in a transfer function in the form of the ratio of two polynomials. The difference equation written from this system function is a recursive expression, and the filter so realized is an infinite impulse response (IIR) filter. Clearly, the impulse-invariant response method is equivalent to analog filtering of an impulse-sampled input signal. The fact has already been discussed that for a sampled signal to approximate the analog signal, the sampling rate (Nyquist rate) must be at least twice the highest frequency component contained in the signal. However, a practical analog filter $H(s)$ is never strictly band limited. Therefore, an aliasing error occurs when this design method is used. As a practical matter, if the sampling frequency is five or more times the cutoff frequency of the low-pass analog filter, the aliasing effect on the frequency response is extremely small. ∎

Example 12.1.3: The normalized transfer function of the third-order Chebyshev filter with a 1-dB ripple is

$$H(s) = \frac{0.4913}{s^3 + 0.9883s^2 + 1.2384s + 0.4913} \quad (12.6)$$

Chapter 12: Infinite Impulse Response (IIR) filters

Determine the following:

1. The corresponding impulse-invariant digital filter
2. The amplitude characteristics of the digital filter
3. The impulse response $h(t)$ of $H(s)$
4. The sampled response $h(nT)$ of the digital filter for $T = 1$.

Solution:

1. Begin with the Chebyshev function in factored form (the roots can be found using the MATLAB function **roots([1 0.9883 1.2384 0.4913])**):

$$H(s) = \frac{0.4913}{(s+0.2471-j0.9660)(s+0.2471+j0.9660)(s+0.4942)}$$

This expression is written in partial fraction form, which is found using the MATLAM function **[r,p,k]=residue([0 0 0 0.4913],[1 0.9883 1.2384 0.4913])**. This expression is written in partial fraction form:

$$H(s) = \frac{-0.2471 - j0.0632j}{s+0.2471-j0.9660} + \frac{-0.2471+j0.0632}{s+0.2471+j0.9660} + \frac{0.4942}{s+0.4942} \quad (12.7)$$

Now, we set $s + p_i = 1 - e^{-p_i T} z^{-1}$ in (12.7), which specifies the impulse-invariant digital filter representation

$$\frac{H(z)}{T} = \frac{-0.2471-j0.0632j}{1-e^{-0.2471T+j0.9660T}z^{-1}} + \frac{-0.2471+j0.0632}{1-e^{-0.2471T-j0.9660T}z^{-1}} + \frac{0.4942}{1-e^{-0.4942T}z^{-1}} \quad (12.8)$$

2. The corresponding frequency response is obtained by substituting $z = e^{j\omega} = e^{j\Omega T}$, setting $T = 0.08$, and next, using the following m-file to produce the two responses.

Book MATLAB m-file: ex_12_1_3

```
%Book m-file: ex_12_1_3
w=0:0.05:8;T=0.08;
a1=-0.2471-j*0.0632;a2=conj(a1);ex=(-0.2471+j*0.966)*T;
hz=(a1./(1-exp(ex)*exp(-j*w*T)))+(a2./(1-exp(conj(ex))*...
    exp(-j*w*T)))...
+0.4942./(1-exp(-0.4942*T)*exp(-j*w*T));
```

```
hs=0.4913./(-j*w.^3-0.9883*w.^2+1.2384*j*w+0.4913);
subplot(2,1,1);plot(w,abs(hs));
subplot(2,1,2);plot(w,abs(hz)/max(abs(hz)));
```

3. Next, we apply the inverse Laplace transform to (12.7) and find $h(t)$ to be

$$h(t) = A_1 e^{-0.2471t+j0.9660t} + (A_1)^* e^{-0.2471t-j0.9660t} + 0.4942 e^{-0.4942t}$$
$$= e^{-0.2471t}[A_1 e^{j0.9660t} + (A_1 e^{+j0.9660t})^*] + 0.4942 e^{-0.4942t} \quad (12.9)$$
$$= e^{-0.2471t} \times 0.5102 \cos(0.9660t - 2.8912) + 0.4942 e^{-0.4942t}$$

4. The continuous and discrete forms of the impulse response are shown in Figure 12.1.2. ∎

12.2 Bilinear transformation

We start with a first-order differential equation and an identity relation:

$$\frac{dy(t)}{dt} + ay(t) = bx(t) \quad \text{a)} \qquad y(t) = \int_{t_0}^{t} y'(\tau)d\tau + y(t_0) \quad \text{b)} \qquad (12.10)$$

If we introduce $t = nT$ and $t_0 = nT - T$ in (12.10b) as $T \to 0$, we obtain the approximate expression where the integral is approximated by a rectangle whose height is the average value of the integrand $[y'(nT) + y'(nT-T)]/2$. Hence, we obtain

$$y(nT) = \frac{T}{2}[y'(nT) + y'(nT-T)] + y(nT-T) \quad (12.11)$$

From (12.10a) we also find that

$$y'(nT) = -ay(nT) + bx(nT) \quad (12.12)$$

Introducing (12.12) in (12.11) and simplifying, we find

$$\left(1 + \frac{aT}{2}\right)y(nT) - \left(1 - \frac{aT}{2}\right)y(nT-T) = \frac{bT}{2}[x(nT) + x(nT-T)] \quad (12.13)$$

Chapter 12: Infinite Impulse Response (IIR) filters

The z-transform of the above equation produces the transfer function, which is

$$\frac{Y(z)}{X(z)} = H(z) = \frac{(bT/2)(1+z^{-1})}{1+\frac{aT}{2}-\left(1-\frac{aT}{2}\right)z^{-1}} = \frac{b}{\frac{2}{T}\left(\frac{1-z^{-1}}{1+z^{-1}}\right)+a} \tag{12.14}$$

The transfer function of (12.10a) using the Laplace transform is

$$H(s) = \frac{b}{s+a} \tag{12.15}$$

Comparing the equations (12.15) and (12.14) we conclude that

$$\boxed{s = \frac{2}{T}\frac{z-1}{z+1} \quad \text{a)} \qquad z = \frac{1+Ts/2}{1-Ts/2} \quad \text{b)}} \tag{12.16}$$

In terms of the z-plane, this algebraic transformation uniquely maps the left-hand side of the s-plane, as shown in Figure 12.2.1a, into the interior of the unit circle in the z-plane, as shown in Figure 12.2.1b. Because no folding occurs, no folding error will arise. However, a shortcoming of the transformation is that the frequency response is nonlinear — that is, warped — in the digital domain.

The inverse transformation for $T = 2$ is given by

$$z = \frac{1+s}{1-s} \tag{12.17}$$

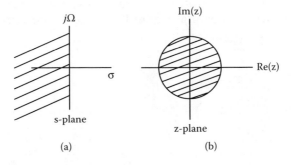

(a)   (b)

Figure 12.2.1 The bilinear transformation mapping.

In the general case, for $s = \sigma + j\Omega$

$$z = \frac{(1+\sigma) + j\Omega}{(1-\sigma) - j\Omega} \tag{12.18}$$

Therefore,

$$|z|^2 = \frac{(1+\sigma)^2 + \Omega^2}{(1-\sigma)^2 - \Omega^2} \tag{12.19}$$

Thus, a point on the $j\Omega$-axis in the s-plane ($\sigma = 0$) is mapped onto a point on the unit circle in the z-plane. A point on the left half of the s-plane with $\sigma < 0$ is mapped onto a point inside the unit circle in the z-plane. Any point on the right half of the s-plane, $\sigma > 0$, is mapped outside the unit circle of the z-plane. Any point in the s-plane is mapped onto a unique point in the z-plane and vice versa. Figure 12.2.1 shows this transformation.

If we insert $z = e^{j\omega}$ into (12.16a), we obtain a relationship between frequency Ω of the analog filter and the digital frequency. We find that

$$s = \sigma + j\Omega = \frac{2(e^{j\omega} - 1)}{T(e^{j\omega} + 1)} = \frac{2e^{j\omega/2}(e^{j\omega/2} - e^{-j\omega/2})}{Te^{j\omega/2}(e^{j\omega/2} + e^{-j\omega/2})} = j\frac{2}{T}\tan\frac{\omega}{2} \tag{12.20}$$

From this equation, by equating real and imaginary parts separately, we obtain

$$\boxed{\begin{array}{l} \sigma = 0 \\ \Omega = \dfrac{2}{T}\tan\dfrac{\omega}{2} \end{array}} \tag{12.21}$$

Observe that the relationship between the two frequencies is nonlinear. The effect is known as **warping**. Figure 12.2.2 shows the mapping between

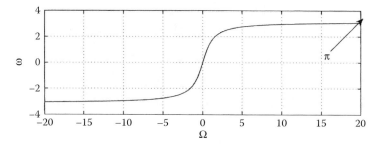

Figure 12.2.2 Mapping of Ω to ω via the bilinear transformation.

Chapter 12: Infinite Impulse Response (IIR) filters

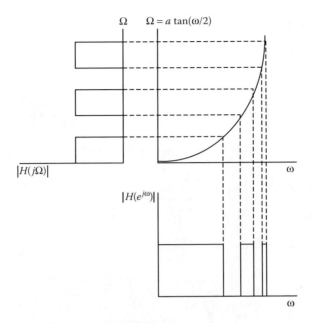

Figure 12.2.3 The frequency warping effect.

the two frequencies through (12.21). Figure 12.2.3 shows graphically the warping effect. Observe that the shapes of the filters are similar, but at high frequencies are disproportionately reduced.

Example 12.2.1: Use the bilinear transformation to determine the characteristic of a digital filter if the corresponding analog filter has the transfer function

$$H(s) = \frac{H_0}{s+p} \quad (12.22)$$

Solution: By the use of (12.16), we obtain

$$H(z) = H(s)\Big|_{s=2(z-1)/T(z+1)} = \frac{H_0}{\frac{2}{T}\frac{z-1}{z+1} + p} = H_0 T \frac{z+1}{2(z-1)+Tp(z+1)} \quad (12.23)$$

To illustrate the variations graphically, we choose (a) $H_0 = 0.8$, $p = 0.8$, and $T = 1$ and (b) $H_0 = 0.8$, $p = 0.8$, and $T = 0.2$. Figure 12.2.4 shows the three filter responses. The following Book MATLAB m-file produces the desired results. Note that when the sampling time is different than 1, the digital transfer function is of the form $H(e^{j\Omega T})$. This indicates that in (12.23) we must set $z = e^{j\Omega T}$ (see the Book m-file below).

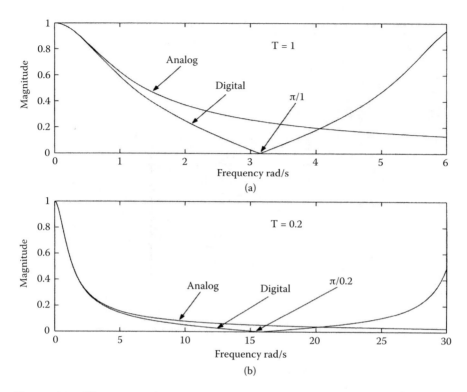

Figure 12.2.4 Illustration of Example 12.2.1.

Book MATLAB m-file: ex_12_2_1

```
%Book m-file:ex_12_2_1
w=0:0.05:6;
T=1;p=0.8;h0=0.8;
hs=h0./(j*w+p);
hz=h0*T*(exp(j*w)+1)./(2*(exp(j*w)-1)+p*T*(exp(j*w)+1));
T=0.2;
w1=0:0.05:30;
hs1=h0./(j*w1+p);
hz1=h0*T*(exp(j*w1*T)+1)./(2*(exp(j*w1*T)-1)+p*T*(exp(j*w1...
    *T)+1));
subplot(2,1,1);plot(w,abs(hs),'k',w,abs(hz)/max(abs(hz)),'k');
xlabel('Frequency rad/s');ylabel('Magnitude');
subplot(2,1,2);plot(w1,abs(hs1),'k',w1,abs(hz1)/max(abs...
    (hz1)),'k');
xlabel('Frequency rad/s');ylabel('Magnitude');                    ∎
```

Example 12.2.2: Design a digital Butterworth filter that meets the following specifications:

1. The 3-dB cutoff frequency ω_c is to occur at 0.4π rad/unit.
2. $T = 50$ μs.
3. At $2\omega_c$, the attenuation is 15 dB.

Solution: First, we find the analog-equivalent criteria for the required digital filter. Using (12.21), we have

$$\Omega_c = \frac{2}{T}\tan\frac{\omega_c}{2} = \frac{2}{50\times 10^{-6}}\tan\frac{0.4\pi}{2} = 29.062\times 10^3 \quad \text{rad/s}$$

$$\Omega\big|_{15dB} = \frac{2}{T}\tan\frac{2\omega_c}{2} = \frac{2}{50\times 10^{-6}}\tan\frac{0.8\pi}{2} = 123.107\times 10^3 \quad \text{rad/s}$$

By (10.1) we find

$$-10\log_{10}\left|H(j123.107\times 10^3)\right|^2$$

$$= -10\log_{10}\left|\frac{1}{1+[(123.107\times 10^3)/(29.062\times 10^3)]^{2n}}\right| \geq 15$$

from which

$$1+\left(\frac{123.107\times 10^3}{29.062\times 10^3}\right)^{2n} \geq 10^{1.5} = 31.6228$$

or

$$n \geq \frac{1}{2}\frac{\log_{10}(30.6228)}{\log_{10}\left(\dfrac{123.107\times 10^3}{29.062\times 10^3}\right)} = 1.1851$$

Hence, the minimum order of the Butterworth filter to meet the specifications is $n = 2$. The form of the filter is, from Table 10.2.2,

$$H(s) = \frac{1}{s^2 + 1.4142s + 1}$$

The analog normalized filter satisfying the specifications is

$$H_{un}(s) = H_{un}\left(\frac{s}{\Omega_c}\right) = H\left(\frac{s}{29.062 \times 10^3}\right)$$

$$= \frac{1}{\left(\frac{s}{29.062 \times 10^3}\right)^2 + 1.4142 \frac{s}{29.062 \times 10^3} + 1}$$

$$H_{un}(s) = \frac{(29.062 \times 10^3)^2}{s^2 + 41.0995 \times 10^3 s + (29.062 \times 10^3)^2}$$

Introduce (12.16) in this equation to write

$$H_{un}(z) = H_{un}(s) = \frac{(29.062 \times 10^3)^2}{\left(\frac{2(z-1)}{T(z+1)}\right)^2 + 41.0995 \times 10^3 \left(\frac{2(z-1)}{T(z+1)}\right) + (29.062 \times 10^3)^2}$$

which is the digital Butterworth design. ∎

12.3 Frequency transformation for digital filters

Frequency transformations are available, and these transforms permit a low-pass filter design to be adapted to another low-pass filter, a high-pass filter, a band-pass filter, or a band-stop filter. These transformations are considered below.

Low-pass-to-low-pass transformation

The transformation is specified by

$$z = \frac{z-c}{1-cz} \qquad c = \frac{\sin[(\omega_c - \overline{\omega_c})/2]}{\sin[(\omega_c + \overline{\omega_c})/2]},$$

$$\overline{\omega_c} = \text{cutoff frequency of the desired filter} \qquad (12.24)$$

$$\omega_c = \text{cutoff frequency of the given filter}$$

Example 12.3.1: Deduce a low-pass digital filter with $\overline{\Omega_c} = 30$ rad/s using the low-pass digital filter of Example 12.1.1 for the case $T = 0.05$ s.

Solution: The given digital filter has the system function

$$H(z) = \frac{20}{1 - e^{-20 \times 0.05} z^{-1}} = \frac{20z}{z - 0.3679} \qquad (12.25)$$

Chapter 12: Infinite Impulse Response (IIR) filters

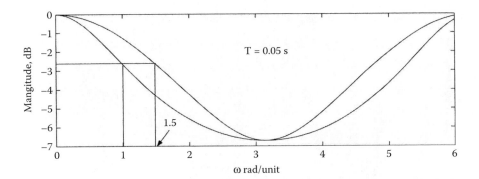

Figure 12.3.1 Low-pass-to-low-pass transformation of digital filter.

The original low-pass filter has a cutoff frequency of 20 rad/s. Working with normalization formats, we observe that if the cutoff frequency of the original is at 1, then the cutoff frequency of the desired one is at 30/20 = 1.5. Hence, the transformation gives the following expression:

$$H(z) = \frac{20\left(\frac{z-c}{1-cz}\right)}{\left(\frac{z-c}{1-cz}\right) - 0.3679} = \frac{20(z-c)}{(1+0.3679c)z - c - 0.3679}, \quad (12.26)$$

$$c = \frac{\sin[(1-1.5)/2]}{\sin[(1+1.5)/2]}$$

Figure 12.3.1 shows the plots of the two filters by substituting $z = e^{j\omega}$ in (12.25) and (12.26). Observe that the values of the filters at $\omega = \omega_c = 1$ (20 rad/s) and $\omega = \omega_c = 1.5$ (30 rad/s) are the same (−2.6636), a result that was expected in light of the requirement imposed on the problem. Since $T = 0.05$, we have $1/0.05 = 20$ rad/s and $1.5/0.05 = 30$ rad/s. ∎

Low-pass-to-high-pass transformation

The required transformation is given by

$$z = -\frac{z+c}{1+cz} \qquad c = -\frac{\cos[(\omega_c + \overline{\omega}_c)/2]}{\cos[(\omega_c - \overline{\omega}_c)/2]},$$

$$\overline{\omega}_c = \text{cutoff frequency of the desired filter} \qquad (12.27)$$

$$\omega_c = \text{cutoff frequency of the given filter}$$

Example 12.3.2: Using the digital filter of Example 12.1.1, determine the high-pass digital filter with $\overline{\Omega}_c = 30$ rad/s and $T = 0.05$ s.

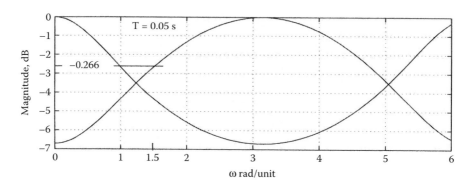

Figure 12.3.2 Low-pass-to-high-pass transformation of digital filter.

Solution: We begin with the given filter $H(z) = 20z/(z - 0.3679)$. Now apply the transformation by (12.27):

$$H(z) = \frac{-20\frac{z+c}{1+cz}}{-\frac{z+c}{1+cz} - 0.3679} = \frac{20(z+c)}{(1+0.3679c)z + 0.3679}$$

As in the previous example, we observe that the two filters have an identical value, -2.66, at normalized frequencies 1 for the low-pass filter and 1.5 for the high-pass filter. ∎

Low-pass-to-band-pass transformation

The transformation is given by

$$z = -\frac{z^2 - \frac{2ab}{b+1}z + \frac{b-1}{b+1}}{\frac{b-1}{b+1}z^2 - \frac{2ab}{b+1}z + 1} = -\frac{z^2 - a_1 z + b_1}{b_1 z^2 - a_1 z + 1} \qquad a_1 = \frac{2ab}{b+1} \qquad b_1 = \frac{b-1}{b+1}$$

$$a = \frac{\cos[(\overline{\omega}_u + \overline{\omega}_l)/2]}{\cos[(\overline{\omega}_u - \overline{\omega}_l)/2]} \qquad b = \cot\left(\frac{\overline{\omega}_u - \overline{\omega}_l}{2}\right)\tan\frac{\omega_c}{2} \qquad (12.28)$$

$\overline{\omega}_u$ = desired upper cutoff frequency

$\overline{\omega}_l$ = desired lower cutoff frequency

ω_c = low-pass filter cutoff frequency

Example 12.3.3: Determine a band-pass digital filter with an upper cutoff frequency of 0.8π rad/unit and a lower cutoff frequency of 0.4π rad/unit. Use the first-order digital filter of Example 2.1.1 with $T = 0.05$.

Chapter 12: Infinite Impulse Response (IIR) filters

Solution: The transformation is given by

$$H(z) = \frac{20(-)\dfrac{z^2 - a_1 z + b_1}{b_1 z^2 - a_1 z + 1}}{-\dfrac{z^2 - a_1 z + b_1}{b_1 z^2 - a_1 z + 1} - 0.3679}$$

$$= \frac{20(z^2 - a_1 z + b_1)}{(1 + 0.3679 b_1) z^2 - (a_1 + 0.3679 a_1) z + b_1 + 0.3679}$$

$$a_1 = \frac{2ab}{b+1} \quad b_1 = \frac{b-1}{b+1}$$

$$a = \frac{\cos[(0.8\pi + 0.4\pi)/2]}{\cos[(0.8\pi - 0.4\pi)/2]} \quad b = \cot\left(\frac{0.8\pi - 0.4\pi}{2}\right)\tan(1/2)$$

The normalized plot is given in Figure 12.3.3. Observe that the attenuation of -2.66 is at the normalized frequencies of 1, 0.4π, and 0.8π, as it should be. Since we have accepted a sampling time of 0.05, the corresponding unnormalized frequencies are 20, 8π, and 16π rad/s, respectively.

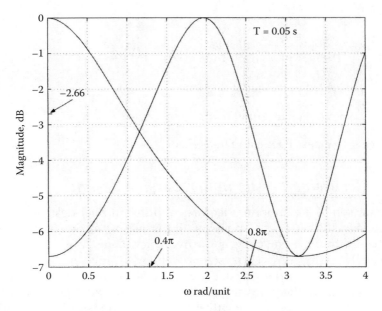

Figure 12.3.3 Low-pass-to-band-pass transformation of digital filter.

Book MATLAB m-file: ex_12_3_3

```
%Book m-file: ex_12_3_3
w=0:0.05:4;
a=cos(0.5*(0.8*pi+0.4*pi))/cos(0.5*(0.8*pi-0.4*pi));
b=cot(0.5*(0.8*pi-0.4*pi))*tan(1/2);
a1=2*a*b/(b+1);b1=(b-1)/(b+1);
num=20*(exp(j*2*w)-a1*exp(j*w)+b1);
den=(1+0.3679*b1)*exp(j*2*w)-(a1+0.3679*a1)*exp(j*w)+b1+0.3679;
hz=num./(den);
hzt=1./(1-exp(-1)*exp(-j*w));
plot(w,20*log10(abs(hzt)/max(abs(hzt))),'k',w,20*log10...
    (abs(hz)/max(abs(hz))),'k');
grid on;
xlabel('\omega rad/unit');ylabel('Magnitude, dB')
```
■

Low-pass-to-band-pass transformation

The transformation is given by

$$z = \frac{z^2 - \frac{2a}{1+b}z + \frac{1-b}{1+b}}{\frac{1-b}{1+b}z^2 - \frac{2a}{1+b}z + 1} = \frac{z^2 - a_1 z + b_1}{b_1 z^2 - a_1 z + 1} \qquad a_1 = \frac{2a}{1+b} \qquad b_1 = \frac{1-b}{1+b}$$

$$a = \frac{\cos[(\overline{\omega}_u + \overline{\omega}_l)/2]}{\cos[(\overline{\omega}_u - \overline{\omega}_l)/2]} \qquad b = \tan\left(\frac{\overline{\omega}_u - \overline{\omega}_l}{2}\right) \tan\frac{\omega_c}{2} \qquad (12.29)$$

$\overline{\omega}_u$ = desired upper cutoff frequency

$\overline{\omega}_l$ = desired lower cutoff frequency

ω_c = low-pass filter cutoff frequency

12.4 Recursive versus non-recursive design

A comparison of the important features of different filter designs is helpful. In recursive IIR filters, the poles of the transfer function can be placed anywhere within the unit circle in the frequency plane. As a result, IIR filters can usually match physical systems well. Also, high selectivity can be achieved using low-order transfer functions. With non-recursive FIR filters, on the other hand, the poles are fixed at the origin, and high selectivity can be achieved only by using a relatively high-order transfer function with its resulting computations. For the same filter specifications, the order of the non-recursive filter might be of the order of 5 to 10 times that of the recursive structure, with the consequent need for more electronic parts. Often, however, a recursive structure might not meet the specifications and, in such cases, the non-recursive filter can be used. An advantage in the design of the non-recursive filter is that FFT techniques can be used in the calculations.

Hardware implementation of filters requires that storage of input output data and also arithmetic operations be implemented, which requires using finite word length registers — for example, 8, 12, 16 bits. As a result, certain errors will occur. These errors are categorized as follows:

1. Optimization errors due to arithmetic operations, such as rounding off and truncation.
2. Quantization errors due to representing the input signal by a set of discrete values.
3. Quantization errors when the filter coefficients are represented by a finite number of bits.

It is left to the filter designer to decide on the various trade-offs between cost and precision in trying to reach a specified goal.

Important defintions and concepts

a. Impulse invariant method
b. Bilinear transformation
c. Warping effect from analog to digital transformation
d. Frequency transformations for digital filters

Chapter 12 Problems

Section 12.1

1. Fill in the steps of Example 12.1.2.
2. Determine the impulse-invariant digital filter corresponding to the second-order Butterworth filter.
3. Find $H(z)$ using the impulse-invariant method for a first-order normalized Chebyshev filter having a 0.5-dB ripple factor.

Section 12.2

1. Set in (12.26) $s = \sigma + j\Omega$ and $z = re^{j\omega}$; next, solve for σ and Ω to obtain them as two functions of r and ω. Based on these equations, show that the left-hand s-plane is mapped into the unit circle of the z-plane.
2. Create a strip in the s-plane with a width from $-j\Omega_s/2$ to $j\Omega_s/2$. Using the bilinear transformation and plotting several points, infer where all the points of this trip, which extends to $\pm j\infty$, are mapped in the z-plane.

3. Repeat Example 12.2.2 with the following difference: 3) At $2\omega_c$, the attenuation is to be 30 dB. Plot your result and compare with the results of Example 12.2.2.
4. Repeat Problem 12.2.3 but for a Chebyshev filter.

Section 12.3

1. Determine the digital equivalent of the first-order Butterworth filter using the bilinear transformation. The cutoff frequency is 25 rad/s. Next, deduce a low-pass digital filter using digital transformation with cutoff frequency of 40 rad/s. The sampling frequency is $T = \pi/40$ s.
2. Repeat Problem 12.3.1 but for a high-pass digital filter transformation.
3. Determine the digital equivalent of the second-order Butterworth filter using the bilinear transformation. The cutoff frequency is 20 rad/s. Next, deduce a band-pass digital filter using digital transformation. The lower frequency is 40 rad/s and the upper frequency is 60 rad/s. The sampling frequency is $\pi/200$ s.
4. Repeat Example 12.3.2 but for a band-stop digital filter with the same lower and upper frequencies.
5. Repeat Example 12.3.3 but for a band-stop digital filter with the same lower and upper frequencies.
6. Repeat Problem 12.3.3 but for a Chebyshev filter.

chapter 13

Random variables, sequences, and power spectra densities

13.1 Random signals and distributions

Most signals in practice are not deterministic but random. However, they can be described by precise mathematical analysis whose tools are contained in the theory statistical analysis. In this text we will be dealing only with discrete random signals. This can always be accomplished since we can sample the continuous signals at sampling rates at least at twice their highest frequency, thus avoiding aliasing. Remember that the signals we receive are band limited because all signals must be detected by a physical (not ideal) transducer, such as a voltmeter, one that measures the electrocardiogram, another that measures the earthquake, etc. All of these physical transducers cannot respond to a delta function excitation, and hence they pass only frequencies up to a specific value.

A **discrete random signal** $\{X(n)\}$ is a sequence of **indexed random variables** (rv's) assuming the values

$$\{x(0) \quad x(1) \quad x(2) \quad \cdots \} \tag{13.1}$$

The random sequence with values $\{x(n)\}$ is discrete with respect to sampling index n. In our case, we will assume that the random variable at any time n takes continuous values, and hence it is a continuous random variable at any time n. What we really say is that we can associate at each n an infinite set of values (continuous) of the random variable $X(n)$ within a specified range. This type of sequence is also known as **time series**. In case we study a continuous random signal, we will assume that we sample it at a high enough so that we construct a time series that is free of aliasing.

A particular rv $X(n)$ is characterized by its **probability density function** (pdf) $f(x(n))$:

$$f(x(n)) = \frac{\partial F(x(n))}{\partial x(n)} \qquad (13.2)$$

and its **cumulative density function** (cdf) $F(x(n))$ or distribution function:

$$F(x(n)) = p(X(n) \le x(n)) = \int_{-\infty}^{x(n)} f(y(n)) dy(n) \qquad (13.3)$$

The expression $p(X(n) \le x(n))$ is interpreted as the probability that the rv $X(n)$ will take values less than or equal to $x(n)$ at time n. As the value at time n approaches infinity, $F(x(n))$ approaches unity.

Similarly, the **multivariate** distributions are given by

$$F(x(n_1),\cdots,x(n_k)) = p(X(n_1) \le x(n_1),\cdots,X(n_k) \le x(n_k))$$

$$f(x(n_1),\cdots,x(n_k)) = \frac{\partial^k F(x(n_1),\cdots,x(n_k))}{\partial x(n_1)\cdots \partial x(n_k)} \qquad (13.4)$$

Note that we have used a capital letter to indicate rv. In general, we shall not keep this notation since it will be obvious from the context.

If, for example, we want to check the accuracy of reading a dial by a person, we will have two readings, one due to the person and another due to the instruments. A simultaneous plot of these two readings, each one associated with a different orthogonal axis, will produce a scattering diagram but with a linear dependence. The closer the points fall on a straight line, the more reliable the person's readings are. This example presents a case of a bivariate distribution.

To obtain a formal definition of a discrete-time stochastic process, we consider an experiment with a finite or infinite number of unpredictable outcomes from a sample space, $S(z_1,z_2, \ldots)$, each one occurring with a probability $p(z_i)$. Next, by some rule we assign a deterministic sequence $x(n, z_i)$, $-\infty < n < \infty$, to each element z_i of the sample space. The sample space, the probabilities of each outcome, and the sequences constitute a **discrete-time stochastic process** or **random sequence**. From this definition we obtain the following four interpretations:

- $x(n, z)$ is an rv if n is fixed and z is variable
- $x(n, z)$ is a sample sequence called realization if z is fixed and n is variable
- $x(n, z)$ is a number if both n and z are fixed
- $x(n, z)$ is a stochastic process if both n and z are variables

Chapter 13: Random variables, sequences, and power spectra densities   547

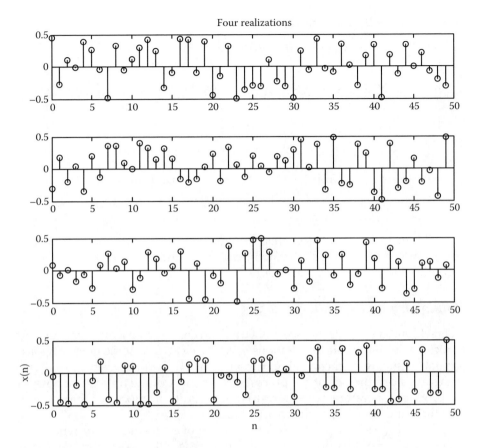

Figure 13.1.1 Ensemble of realizations.

Each time we run an experiment under identical conditions, we create a sequence of rv's $\{X(n)\}$ that is known as a **realization** and constitutes an **event**. A realization is one member of a set called the ensemble of all possible results from the repetition of an experiment. Figure 13.1.1 shows a typical ensemble of realizations.

Book MATLAB m-file

(To create a script file, we first create the file, as shown below, as a new MATLAB file. Then we save the file named "realizations.m," for example, in the directory c:\ap\samatlab. Be sure to add the .m at the end of the file name. When we are in the MATLAB command window we attach the above directory to the MATLAB path using the following command: >>path(path,'c:\ap\samatlab') or >>cd 'c:\ap\smatlab'. Then the only thing we have to do is to write realizations and automatically the MATLAB will produce the figure with four subplots.)

Book MATLAB m-file: realizations

```
%Book MATLAB m-file: realizations
for n=1:4
   x(n,:)=rand(1,50)-0.5;%x=4x50 matrix with each row having
                        %zero mean;
end;
m=0:49;
for i=1:4
   subplot(4,1,i);stem(m,x(i,:),'k');%plots four rows of matrix x
end;
xlabel('n');ylabel('x(n)')
```

Figure 13.1.1 shows four realizations of a stochastic process with zero mean value. With only slight modifications of the above script file we can produce any number of realizations.

Stationary and ergodic processes

It is seldom in practice that we will be able to create an ensemble of a random process with numerous realizations so that we can find some of its statistical characteristics, e.g., mean value, variance, etc. To find these statistical quantities, we need the pdf of the process, which most of the time is not possible to produce. Therefore, we will restrict our studies to processes that are easy to study and easy to handle mathematically.

The process that produces an ensemble of realizations and whose statistical characteristics do not change with time is called **stationary**. For example, the pdfs of the rv's $x(n)$ and $x(n + k)$ of the process $\{x(n)\}$ are the same independently of the values of n and k.

Since we will be unable to produce ensemble averages in practice, we are left with only one realization of the stochastic process. To overcome this difficulty, we assume that the process is **ergodic**. This characterization permits us to find the desired statistical characteristics of the process from only one realization at hand. We refer to those statistical values as **sample mean**, **sample variance**, etc. We must have in mind that these values are the approximate to the ensemble values. One of the main objectives of the statisticians is to find ways to give these sample values some confidence limits within which, with a high degree of probability, the ensemble value exists.

13.2 Averages

Mean value

The **mean** value or **expectation** value m_n at time n of a random variable $x(n)$ having pdf $f(x(n))$ is defined by

Chapter 13: Random variables, sequences, and power spectra densities

$$m_n = E\{x(n)\} = \int_{-\infty}^{\infty} x(n)f(x(n))dx(n) \qquad (13.5)$$

where $E\{\cdot\}$ stands for **expectation** operator. We can also use the ensemble of realizations to obtain the mean value using the **frequency interpretation** formula

$$m_n = \lim_{N \to \infty} \left\{ \frac{1}{N} \sum_{i=1}^{N} x_i(n) \right\} \qquad N = \text{number of realizations} \qquad (13.6)$$

where $x_i(n)$ is the ith outcome at sample index n (or time n) of the ith realization. Depending on the type of the rv, the mean value may or may not vary with time.

For an ergodic process, we find the sample mean (estimator of the mean) using the **time-average** formula

$$\hat{m} = \frac{1}{N} \sum_{n=0}^{N-1} x(n) \qquad (13.7)$$

It turns out that the ensample of sample mean $\hat{m}$ is equal to the population mean m, and therefore, we call the sample mean an **unbiased** estimator, in this case of the mean value.

Example 13.2.1: Show that the ensample mean of the sample mean is an unbiased estimator.

Solution:

$$E\{\hat{m}\} = E\{\frac{1}{N} \sum_{k=1}^{N} x(k)\} = \frac{1}{N} \sum_{k=1}^{N} E\{x(k)\} = \frac{1}{N} \sum_{k=1}^{N} m = \frac{Nm}{N} = m$$

Note that we were able to exchange the summation with the ensemble (an integral) because the integrant is not a function of k, as far as the integration is concerned, but a function of the values of x (see (13.5)). ∎

Correlation

The **cross-correlation** between two random variables is defined by

$$r_{xy}(m,n) = E\{x(m), y(n)\} = \iint x(m)y(n)f(x(m), y(n))dx(m)dy(n) \qquad (13.8)$$

where the integrals are from minus infinity to infinity. If $x(n) = y(n)$, the correlation is known as the **autocorrelation**. Having an ensemble of realizations, the frequency interpretation of the autocorrelation function is found using the formula

$$r_{xx}(m,n) = \lim_{N\to\infty} \left\{ \frac{1}{N} \sum_{i=1}^{N} x_i(m) x_i(n) \right\} \qquad (13.9)$$

Example 13.2.2: Using Figure 13.1.1, find the mean for $n = 10$ and the autocorrelation function for a time difference of 5: $n = 20$ and $n = 25$.

Solution: The desired values are

$$\mu_{10} = \frac{1}{4}\sum_{i=1}^{4} x_i(10) = \frac{1}{4}(0.3 - 0.1 - 0.4 + 0.45) = 0.06$$

$r_x(20, 25)$

$$= \frac{1}{4}\sum_{i=1}^{4} x_i(25)x_i(30) = \frac{1}{4}[(0.1)(-0.35) + (-0.15)(-0.4) + (0.35)(0.25) + (0.2)(-0.1)]$$

$$= 0.032$$

Because the number of realizations is very small, both values found above are not expected to be accurate. ∎

Figure 13.2.1 shows the mean value at 20 individual times and the autocorrelation function for 20 differences (from 0 to 49), known as **lags**. These results were found using the MALAB function given below. Note that as the number of realizations increases, the mean tends to zero and the autocorrelation tends to a delta function, as it should be. In this case, the random variables are independent, identically distributed (iid), and their pdf is Gaussian (white noise). If the rv's are independent, the autocorrelation is a delta function.

Book MATLAB function for mean and autocorrelation using the frequency interpretation approach

```
function[mx,rx]=ssmeanautocorensemble(M,N)
%function[mx,rx]=sameanautocorensemble(M,N);
%N=number of time instances;easily modified for
%other pdf's;
```

Chapter 13: Random variables, sequences, and power spectra densities 551

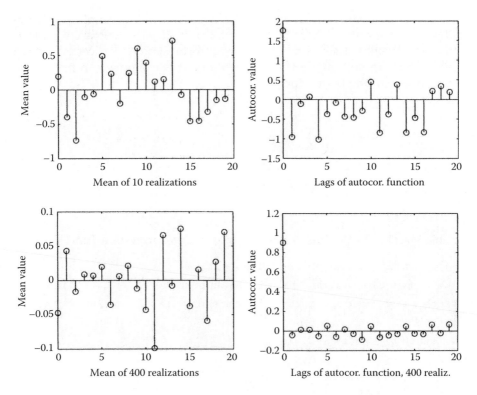

Figure 13.2.1 Means and autocorrelations based on frequency interpretations.

```
x=randn(M,N); %randn=MATLAB function producing zero mean
        %Gaussian distributed white noise; x=MxN matrix;
        %sum(x,1)=MATLAB function that sums all the rows;
        %sum(x,2)=MATLAB function that sums all the columns;
mx=sum(x,1)/M;
for i=1:N
    rx(i)=sum(x(:,1).*x(:,i))/M;
end;
```

Having the values of mx and rx, we can plot them as in Figure 13.2.1.

To find the **unbiased** sample autocorrelation function from one realization (ergotic process), we use the formula

$$\hat{r}(m) = \frac{1}{N-|m|} \sum_{n=0}^{N-|m|-1} x(n)x(n+|m|) \qquad m = 0, 1, \cdots, N-1 \qquad (13.10)$$

The absolute value of m ensures the symmetry of the sample autocorrelation sequency at $n = 0$. Although the formula gives an unbiased autocorrelation sequence, it sometimes produces autocorrelation matrices (discussed below) that do not have inverses. Therefore, it is customary in practice to use any one of the biased formulas

$$\hat{r}(m) = \frac{1}{N} \sum_{n=0}^{N-|m|-1} x(n)x(n+|m|) \qquad 0 \leq m \leq N-1$$

$$\hat{r}(m) = \frac{1}{N} \sum_{n=0}^{N-1} x(n)x(n-m) \qquad 0 \leq m \leq N-1$$

(13.11)

Book MATLAB function to find the biased autocorrelation function

```
function[r]=sssamplebiasedautoc(x,lg)
%this function finds the biased autocorrelation function
%with lag from 0 to lg; it is recommended that lg is 20-30 percent
%of N;
N=length(x);%x=data;lg=lag;
for m=1:lg
    for n=1:N+1-m
        xs(m,n)=x(n-1+m);
    end;
end;
r1=xs*x';
r=r1'./N;
```

We can also use MATLAB functions to obtain the biased or unbiased sample autocorrelation and cross-correlation sequences. The functions are

```
r=xcorr(x,y,'biased');
r=xcorr(x,y),'unbiased');
%x,y are N length vectors; r is a 2N-1 symmetric cross-
%correlation vector; in case the vectors are unequal
%the shorter one is padded with zeros;
```

Note: *If none of the options (i.e., biased or unbiased) is used, the default value is the biased, but the resulting vector is not divided by N.*

The reader is encouraged to find several interesting options in using the xcorr command by writing **help xcorr** or **doc xcorr** on the MATLAB command window.

Covariance

The **covariance** (autocovariance) of a random iid sequence is defined by

$$c_x(m,n) = E\{(x(m) - m_m)(x(n) - m_n)\} = E\{x(m)x(n)\} - m_m m_n$$
$$= r_x(m,n) - m_m m_n \qquad (13.12)$$

The **variance** is found by setting $m = n$ in (13.12). Thus,

$$c_x(n,n) = \sigma_n^2 = E\{(x(n) - m_n)^2\} = E\{x^2(n)\} - m_n^2 \qquad (13.13)$$

If the mean value is zero, then the variance and correlation function are identical:

$$c_x(n,n) = \sigma_n^2 = E\{x^2(n)\} = r_x(n,n) \qquad (13.14)$$

The estimator for the biased variance is given by

$$\boxed{\hat{\sigma}^2 = \frac{1}{N}\sum_{n=1}^{N}(x(n) - \hat{m})^2} \qquad (13.15)$$

and for the unbiased case we substitute N with $N - 1$. The variance above can be found from the Book MATLAB function **sssamplebiasedautoc (data_vector,lags)** at 0 lag. The reader can also use the MATLAB functions **std**(data_vector)=standard deviation=$\sqrt{\text{variance}}$ and **var**(data_vector)=variance.

Independent and uncorrelated rv's

If the joint pdf of two rv's can be separated into two pdfs, $f_{x,y}(m,n) = f_x(m)f_y(n)$, then the rv's are statistically **independent**. Hence,

$$E\{x(m)x(n)\} = E\{x(m)\}E\{x(n)\} = m_m m_n \qquad (13.16)$$

The above equation is a necessary and sufficient condition for the two random variables $x(m)$, $x(n)$ to be uncorrelated. Note that independent random variables are always uncorrelated. However, the converse is not necessarily true. If the mean value of any two uncorrelated rv's is zero, then the random variables are called **orthogonal**. In general, two rv's are called orthogonal if their correlation is zero.

13.3 Stationary processes

For a **wide-sense (or weakly) stationary** process (WSS), the cdf satisfies the relationship

$$F(x(m), x(n)) = F(x(m+k), x(n+k)) \qquad (13.17)$$

for any m, n, and k. This also applies for all statistical characteristics, such as mean value, variance, correlation, etc. If the above relationship is true for any number of rv's of the time series, then the process is known as **strictly stationary** process.

The basic **properties** of a wide-sense stationary process are (see Problem 13.3.1)

$$\begin{array}{ll} \text{a) } m_n(n) = m = \text{constant} & \text{c) } r_x(k) = r_x(-k) \\ \text{b) } r_x(m,n) = r_x(m-n) & \text{d) } r_x(0) \geq r_x(k) \end{array} \qquad (13.18)$$

Autocorrelation matrix

If $\mathbf{x} = [x(0)\ x(1)\ \ldots\ x(p)]^T$ is a vector representing a finite WSS random sequence, then the autocorrelation matrix is defined by

$$\begin{aligned}\mathbf{R}_x = E\{\mathbf{x}\mathbf{x}^T\} &= \begin{bmatrix} E\{x(0)x(0)\} & E\{x(0)x(1)\} & \cdots & E\{x(0)x(p)\} \\ E(x(1)x(0)) & E\{x(1)x(1)\} & \cdots & E\{x(1)x(p)\} \\ \vdots & \vdots & & \vdots \\ E\{x(p)x(0)\} & E\{x(p)x(1)\} & \cdots & E\{x(p)x(p)\} \end{bmatrix} \\ &= \begin{bmatrix} r_x(0) & r_x(-1) & \cdots & r_x(-p) \\ r_x(1) & r_x(0) & \cdots & r_x(-p+1) \\ & & \vdots & \\ r_x(p) & r_x(p-1) & \cdots & r_x(0) \end{bmatrix} \\ &= \begin{bmatrix} r_x(0) & r_x(1) & \cdots & r_x(p) \\ r_x(1) & r_x(0) & \cdots & r_x(p+1) \\ & & \vdots & \\ r_x(p) & r_x(p-1) & \cdots & r_x(0) \end{bmatrix}\end{aligned} \qquad (13.19)$$

since $r_x(k) = r_x(-k)$.

Chapter 13: Random variables, sequences, and power spectra densities

Example 13.3.1: (a) Find the biased autocorrelation function with a lag time of up to 20 of a sequence of 40 terms, which is a realization of rv's having a Gaussian distribution with zero mean value. (b) Create a 4×4 autocorrelation matrix.

Solution: The Book MATLAB m-function ex13_3_1 produces Figure 13.3.1 and is given below. To find the correlation matrix from the autocorrelation function, we use the following MATLAB function: **Rx=toeplitz(rx(1,1:4))**. In this example, a 4×4 matrix produced from the first four columns of correlation row vector rx is

$$\mathbf{R}_x = \begin{bmatrix} 2.0817 & 0.3909 & 0.1342 & -0.0994 \\ 0.3909 & 2.0817 & 0.3909 & 0.1342 \\ 0.1342 & 0.3909 & 2.0817 & 0.3909 \\ -0.0994 & 0.1342 & 0.3909 & 2.0817 \end{bmatrix}$$

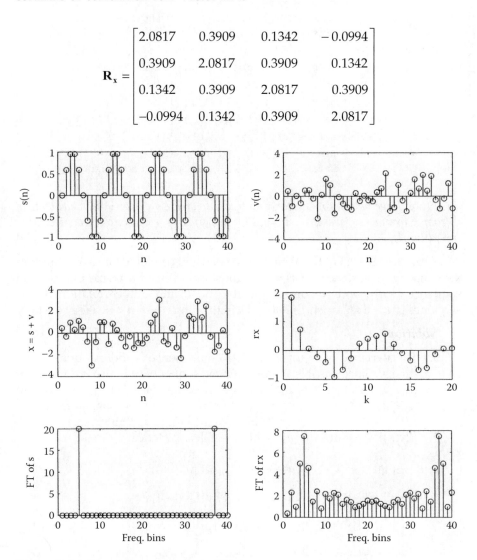

Figure 13.3.1 Illustration of Example 13.3.1.

Book MATLAB m file: ex13_3_1

```
%ex13_3_1 Book m-file;
n=0:39;
s=sin(.2*pi*n);
v=randn(1,40); %randn=MATLAB function producing white Gaussian
               %distributed rv's;
x=s+v;
rx=sssamplebiasedautoc(x,20);%Book MATLAB function creating the
               %autocorrelation of x;
fts=fft(s,40); %fft=MATLAB function executing the fast Fourier
               %transform;
ftrx=fft(rx,40);
subplot(3,2,1);stem(s,'k');xlabel('n');ylabel('s(n)');
subplot(3,2,2);stem(v,'k');xlabel('n');ylabel('v(n)');
subplot(3,2,3);stem(x,'k');xlabel('n');ylabel('x=s+v');
subplot(3,2,4);stem(rx,'k');xlabel('k');ylabel('rx');
subplot(3,2,5);stem(abs(fts),'k');xlabel('Freq. bins');
    ylabel('FT of s');
subplot(3,2,6);stem(abs(ftrx),'k');xlabel('Freq. bins');
    ylabel('FT rx');
```

Note: *If we have a row vector **x** and need to create a vector **y** with elements of x from k to m only, we write **y** = **x(1,k:m)**;. If **x** is a column vector, we write **y** = **x(k:m,1)**;.* ∎

Example 13.3.2: Let $\{v(n)\}$ be a zero mean, uncorrelated Gaussian random sequence with variance $\sigma_v^2(n) = \sigma^2$ = constant. (a) Characterize the random sequence $\{v(n)\}$, and (b) determine the mean and autocorrelation of the sequence $\{x(n)\}$ if $x(n) = v(n) + av(n-1)$, in the range $-\infty < n < \infty$. a is a constant.

Solution:

a. The variance of $\{v(n)\}$ is constant, and hence is independent of the time, n. Since $\{v(n)\}$ is an uncorrelated sequence, it is also independent due to the fact it is a Gaussian sequence. From (13.12) we obtain $c_v(l,n) = r_v(l,n) - m_l m_n = r_v(l,n)$ or $\sigma^2 = r_v(n,n)$ = constant. Hence, $r_v(l,n) = \sigma^2(l-n)$, which implies that $\{v(n)\}$ is a WSS process.
b. $E\{x(n)\} = 0$ since $E\{v(n)\} = E\{v(n-1)\} = 0$. Hence,

$$r_x(l,n) = E\{[v(l)+av(l-1)][v(n)+av(n-1)]\} = E\{v(l)v(n)\}$$

$$+ aE\{v(l-1)v(n)\} + aE\{v(l)v(n-1)\} + a^2 E\{v(l-1)v(n-1)\} = r_v(l,n)$$

$$+ ar_v(l-1,n) + ar_v(l,n-1) + a^2 r_v(l-1,n-1) = \sigma^2 \delta(l-n) + a\sigma^2 \delta(l-n+1)$$

$$+ a^2 \sigma^2 \delta(l-n) = (1+a^2)\sigma^2 \delta(r) + a\sigma^2 \delta(r+1) + a\sigma^2 \delta(r-1), \ l-n=r$$

Chapter 13: Random variables, sequences, and power spectra densities 557

Since the mean of $\{x(n)\}$ is zero, a constant, and its autocorrelation is a function of the lag factor $r = l - n$, it is a WSS process. ∎

Purely random process (white noise, WN)

A discrete process is purely random if the random variables $\{x(n)\}$ of a sequence are mutually independent and identically distributed variables. Since the mean and $cov(x(m), x(m + k))$ do not depend on time, the process is WSS. This process is also known as white noise and is given by

$$c_x(k) = \begin{cases} E\{(x(m) - m_x)(x(m+k) - m_x)\} & k = 0 \\ 0 & k = \pm 1, \pm 2, \cdots \end{cases} \quad (13.20)$$

Random walk (RW)

Let $\{x(n)\}$ be a purely random process with mean m_x and variance σ_x^2. A process $\{y(n)\}$ is a random walk if

$$y(n) = y(n-1) + x(n) \qquad y(0) = 0 \quad (13.21)$$

Solving the above equation iteratively, we find that

$$y(n) = \sum_{i=1}^{n} x(i) \quad (13.22)$$

The mean is found to be $E\{y(n)\} = nm_x$, and the covariance is

$$cov(y(n), y(n)) = E\{[y(n) - m_y][y(n) - m_y]\} = n[E\{x^2\} - m_x^2] = n\sigma_x^2$$

It interesting to note that the difference $x(n) = y(n) - y(n - 1)$ is purely random, and hence stationary.

13.4 Special random signals and probability density functions

White noise

A WSS discrete random sequence that satisfies the relation

$$f(x(0), x(1), \cdots) = f(x(0))f(x(1))f(x(2)) \cdots \quad (13.23)$$

is a pure random sequence whose elements $x(n)$ are statistically **independent** and **identically distributed** (iid). Therefore, the **zero mean iid** sequence has the following correlation function:

$$r_x(m-n) = E\{x(m)x(n)\} = \sigma_x^2 \delta(m-n) \tag{13.24}$$

where σ_x^2 is the variance of the signal and $\delta(m-n)$ is the discrete-time delta function. For $m \ne n$, $\delta(m-n) = 0$, and hence (13.24) becomes

$$r_x(k) = \begin{cases} \sigma_x^2 \delta(k) & k = 0 \\ 0 & k \ne 0 \end{cases} \tag{13.25}$$

For example, a random process consisting of a sequence of uncorrelated Gaussian rv's is a white noise process referred to as **white Gaussian noise** (WGN). MATLAB has a special function that will produce WGN. For example, writing **x=randn(1,500)** will produce a sequence (vector x) whose elements are normally distributed and variance is 1. We can use the MATLAB function **hist(x,20)** to produce the pdf of the time series $\{x(n)\}$. The function divides the range between the maximum and minimum values of the sequence in 20 sections. Then it plots the number of x's, whose values fall within each section, vs. the range of values of x's.

Gaussian processes

The pdf of a Gaussian rv $x(n)$ at time n is given by

$$f(x(n)) \triangleq N(m_n, \sigma_n) = \frac{1}{\sqrt{2\pi \sigma_n^2}} e^{-\frac{(x(n)-m_n)^2}{2\sigma_n^2}} \tag{13.26}$$

Algorithm to produce normalized Gaussian distribution

1. Generate two independent rv's u_1 and u_2 from uniform distribution (0, 1).
2. $x_1 = (-2\ln(u_1))^{1/2} \cos(2\pi u_2)$ [or $x_1 = (-2\ln(u_1))^{1/2} \sin(2\pi u_2)$].
3. Keep x_1 or x_2.

Book MATLAB function

```
function[x]=ssnormalpdf(m,s,N)
%function[x]=ssnormalpdf(m,s,N);
%s=standard deviation;m=mean value;
for i=1:N
r1=rand;
```

```
r2=rand;
  z(i)=sqrt(-2*log(r1))*cos(2*pi*r2);
end;
x=s*z+m;
```

Example 3.4.1: Find the joint pdf of a sequence of WGN with n elements, each one having a zero mean value and the same variance.

Solution: The joint pdf is

$$f(x(1), x(2), \cdots, x(n)) = f_1(x(1))f_2(x(2))\cdots f_n(x(n))$$

$$= \frac{1}{(2\pi)^{n/2} \sigma_x^n} \exp\left[-\frac{1}{2\sigma_x^2} \sum_{k=1}^{n} x^2(k)\right] \quad (13.27)$$

∎

A discrete-time random process $\{x(n)\}$ is said to be Gaussian if every finite collection of samples of $x(n)$ is jointly Gaussian. A Gaussian random process has the following properties: (a) it is completely defined by its mean vector and covariance matrix; (b) any linear operation on the time variables produces another Gaussian random process; (c) all higher moments can be expressed by the first and second moments of the distribution (mean, covariance); and d) white noise is necessarily generated by iid samples (independence implies uncorrelated rv's and vice versa).

Lognormal distribution

Let the rv x be $N(m, \sigma^2)$. Then $y = \exp(x)$ has the lognormal distribution with pdf

$$f(y) = \begin{cases} \dfrac{1}{\sqrt{2\pi}\sigma y} \exp\left[-\dfrac{(\ln y - m)}{2\sigma^2}\right] & 0 \leq y < \infty \\ 0 & \text{otherwise} \end{cases} \quad (13.28)$$

The values of sigma and mu must be small to form a lognormal-type distribution.

Algorithm to produce lognormal distribution

1. Generate z from $N(0, 1)$.
2. $x = \mu + \sigma z$ (x is $N(m, \sigma^2)$).
3. $y = \exp(x)$.
4. Keep y.

Book MATLAB function: [y]=sslognormalpdf(m,s,N)

```
function[y]=sslognormalpdf(m,s,N)
%function[y]=sslognormalpdf(m,s,N);
%m=mean value;s=standard deviation;N=number of samples;
%m and s is associated with the normal distribution and
%to find the mean and standard deviation of the
%lognormal distribution you must
%use the MATLAB functions mean(y) and std(y)
for i=1:N
    r1=rand;
    r2=rand;
    z(i)=sqrt(-2*log(r1))*cos(2*pi*r2);
end;
x=m+s*z;
y=exp(x);
```

Chi-square distribution

If $z_1, z_2, \cdots, z_n$ are $N(0, 1)$, then

$$y = \sum_{i=1}^{n} z_i^2 \qquad (13.29)$$

has the chi-square distribution with n degrees of freedom and it is denoted by $\chi^2(n)$.

13.5 Wiener–Kintchin relations

For a WSS process, the correlation function asymptotically goes to zero, and therefore, we can find its spectrum using the discrete Fourier transform (DFT). Hence, the **power spectrum** is given by

$$\boxed{S_x(e^{j\omega}) = \sum_{k=-\infty}^{\infty} r(k) e^{-j\omega k}} \qquad (13.30)$$

This function is periodic with period 2π ($\exp(-jk(\omega + 2\pi)) = \exp(-jk\omega)$). Given the power spectral density, the autocorrelation sequence is given by the relation

Chapter 13: Random variables, sequences, and power spectra densities

$$r_x(k) = \frac{1}{2\pi} \int_{-\pi}^{\pi} S_x(e^{j\omega}) e^{j\omega k} d\omega \qquad (13.31)$$

For real processes, $r_x(k) = r_x(-k)$ (symmetric function), and as a consequence, the power spectrum is an **even** function. Furthermore, the power spectrum of the WSS process is also nonnegative. These two assertions are given below in the form of mathematical relations:

$$S_x(e^{j\omega}) = S_x(e^{-j\omega}) = S_x^*(e^{j\omega})$$
$$S_x(e^{j\omega}) \geq 0 \qquad (13.32)$$

Example 13.5.1: Find the power spectral density of the sequence $x(n) = \sin(0.2 \times 2\pi n) + randn(1,64)$ with $n = 0, 1, 2, \cdots, 63$.

Solution: The following Book MATLAB m-file produces Figure 13.5.1.

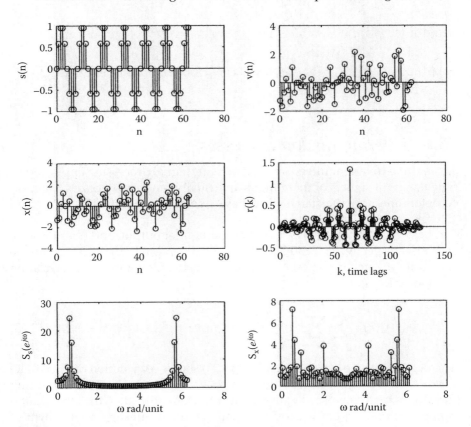

Figure 13.5.1 Illustration of Example 13.5.1.

Book MATLAB m-file: ex_13_5_1

```
%Book m-file:ex_13_5_1
n=0:63;
s=sin(0.1*2*pi*n);
v=randn(1,64);%Gaussian white noise;
x=s+v;
r=xcorr(x,'biased');%the biased autocorrelation is divided by N,
                   %here by 64;
fs=fft(s);
fr=fft(r,64);
w=0:2*pi/64:2*pi-(2*pi/64);
subplot(3,2,1);stem(n,s,'k');xlabel('n');ylabel('s(n)');
subplot(3,2,2);stem(n,v,'k');xlabel('n');ylabel('v(n)');
subplot(3,2,3);stem(n,x,'k');xlabel('n');ylabel('x(n)');
subplot(3,2,4);stem(n,r(1,32:63),'k');xlabel('k, time lags');
ylabel('r(k)');
subplot(3,2,5);stem(w,abs(fs),'k');xlabel('\omega rad/unit');
ylabel('S_s(e^{j\omega}');
subplot(3,2,6);stem(w,abs(fr),'k');xlabel('\omega rad/unit');
ylabel('S_x(e^{j\omega}');
```

∎

13.6 Filtering random processes

Linear time-invariant filters are used in many signal processing applications. Sine the input signals of these filters are usually random processes, we need to determine how the statistics of these signals are modified as a result of filtering.

Let $x(n)$, $y(n)$, and $h(n)$ be the filter input, filter output, and filter impulse response, respectively. It can be shown that if $x(n)$ is a WSS process, then the filter output autocorrelation $r_y(k)$ is related to the filter input autocorrelation $r_x(k)$ as follows:

$$r_y(k) = \sum_{l=-\infty}^{\infty}\sum_{m=-\infty}^{\infty} h(l)r_x(m-l+k)h(m) = r_{x(k)} * h(k) * h(-k) \qquad (13.33)$$

The right-hand side of the expression (13.33) shows convolution of three functions. We can take the convolution of two of the functions, and the resulting function is then convolved with the third function. Based on the convolution property discussed in the previous chapters, the results are independent of the order we operate on the functions.

Chapter 13: Random variables, sequences, and power spectra densities 563

From the z-transform table, we know that the z-transform of the convolution of two functions is equal to the product of their z-transform. Remembering the definition of the z-transform, we find the relationship (the order of summation does not change the results)

$$\mathcal{Z}\{h(-k)\} = \sum_{k=-\infty}^{\infty} h(-k)z^{-k} = \sum_{m=\infty}^{-\infty} h(m)(z^{-1})^{-m} = H(z^{-1}) \qquad (13.34)$$

since the summation is the same regardless of the direction in which we sum the series.

Therefore, the z-transform of (13.33) becomes

$$R_y(z) = \mathcal{Z}\{r_x(k) * h(k)\}\mathcal{Z}\{h(-k)\} = R_x(z)H(z)H(z^{-1}) \qquad (13.35)$$

If we set $z = e^{j\omega}$ in the definition of the z-transform of a function, we find the spectrum of the function. Having in mind the Wiener–Kintchin theorem, (13.35) becomes

$$\boxed{S_y(e^{j\omega}) = S_x(e^{j\omega})|H(e^{j\omega})|^2} \qquad (13.36)$$

Note: *The above equation indicates that the power spectrum of the output random sequence is equal to the power spectrum of the input sequence modified by the square of the absolute value of the spectrum of the filter transfer function.*

Example 13.6.1: A FIR filter is defined in the time domain by the difference equation $y(n) = x(n) + 0.5x(n-1)$. If the input signal is a white Gaussian noise with zero mean value, find the power spectrum of the output of the filter.

Solution: The z-transform of the difference equation is $Y(z) = (1 + 0.5z^{-1})X(z)$. Since the ratio of the output to input is the transfer function of the filter, the transformed equation gives $H(z) = Y(z)/X(z) = 1 + 0.5z^{-1}$. The absolute square value of the spectrum of the transfer function is given by

$$|H(e^{j\omega})|^2 = (1+0.5e^{j\omega})(1+0.5e^{-j\omega}) = 1+0.5(e^{-j\omega}+e^{j\omega})+0.25 = 1.25 + \cos\omega \qquad (13.37)$$

where the Euler identity $e^{\pm j\omega} = \cos\omega \pm j\sin\omega$ was used. Figure 13.6.1 shows the required results. Remember, the spectrum is valid in the range $0 \le \omega \le \pi$. We remind the reader that if the sequence were coming from a continuous function sampled at times T, the spectrum range would have been $0 \le \omega(\text{rad/s}) \le \pi/T$. The following Book MATLAB m-file produces Figure 13.6.1.

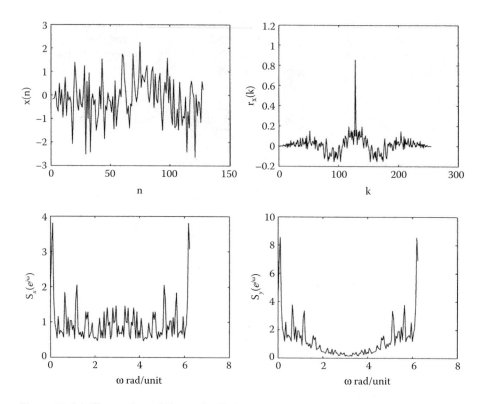

Figure 13.6.1 Illustration of Example 13.6.1.

Book MATLAB m-file: ex_13_6_1

```
%Book m-file:ex_13_6_1
x=randn(1,128);
rx=xcorr(x,'biased');
sx=fft(rx,128);
w=0:2*pi/128:2*pi-(2*pi)/128;
sy=abs(sx).*(1.25+cos(w));
subplot(2,2,1);plot(x,'k');xlabel('n');ylabel('x(n)');
subplot(2,2,2);plot(rx,'k');xlabel('n');ylabel('x(n)');
subplot(2,2,3);plot(w,abs(sx),'k');xlabel('\omega rad/unit');
ylabel('S_x^{j\omega}');
subplot(2,2,4);plot(w,abs(sy),'k');xlabel('\omega rad/unit');
ylabel('S_y^{j\omega}');
```

Observe that the filter is low-pass since it has attenuated the high frequencies (frequencies close to π). We have also plotted the sequences in a continuous format to show better the effects of filtering. ∎

Chapter 13: Random variables, sequences, and power spectra densities

Spectral factorization

The power spectral density $S_x(e^{j\omega})$ of a WSS process $\{x(n)\}$ is a real-valued, positive, and periodic function of ω. It can be shown that this function can be **factored** in the form

$$S_x(e^{j\omega}) = \sigma_v^2 Q(z)Q(z^{-1}) \tag{13.38}$$

where:

1. Any WSS process $\{x(n)\}$ may be realized as the output of a causal and stable filter $h(n)$ that is driven by white noise $v(n)$ having variance σ_v^2. This is known as the **innovations representation** of the process.
2. If $x(n)$ is filtered by the filter $1/H(z)$ (**whitening filter**), the output is a white noise $v(n)$ having variance σ_v^2. This process is known as the **innovations process** and is given by

$$[\sigma_v^2 H(z)H(z^{-1})]\left[\frac{1}{H(z)H(z^{-1})}\right] = \sigma_v^2 \tag{13.39}$$

where the first bracket is the input $S_x(z)$, the second represents the filter power spectrum, and the output is the variance of the noise.
3. Since $v(n)$ and $x(n)$ are related by inverse transformations, one process can be derived from the other. Therefore, both contain the same information.

Autoregressive process (AR)

A special and important type of the combined FIR and IIR filter, known also as the **autoregressive moving average** (ARMA) filter (system), is the **autoregressive filter** (AR). If we set $q = 0$ (see Figure 9.6.8), we obtain the relation

$$H(z) = \frac{b(0)}{1 + \sum_{k=1}^{p} a(k)z^{-k}} = \frac{b(0)}{A(z)} \tag{13.40}$$

Using (13.38), we find the power spectrum of an observed signal produced at the output of an AR filter to be equal to

$$S_x(e^{j\omega}) = \sigma_v^2 \frac{b(0)^2}{|A(e^{j\omega})|^2} \tag{13.41}$$

It can be shown that the following correlation relationship exists for the AR process:

$$r_x(k) + \sum_{m=1}^{p} a(m) r_x(k-m) = \sigma_v^2 b(0)^2 \delta(k) \qquad k \geq 0 \qquad (13.42)$$

The above equation is also written in a matrix form as follows:

$$\begin{bmatrix} r_x(0) & r_x(-1) & \cdots & r_x(-p) \\ r_x(1) & r_x(0) & \cdots & r_x(-p+1) \\ \vdots & \vdots & & \vdots \\ r_x(p) & r_x(p-1) & \cdots & r_x(0) \end{bmatrix} \begin{bmatrix} 1 \\ a(1) \\ \vdots \\ a(p) \end{bmatrix} = \sigma_v^2 b(0)^2 \begin{bmatrix} 1 \\ 0 \\ \vdots \\ 0 \end{bmatrix} \qquad (13.43)$$

For real sequences, the autocorrelation is symmetric, $r_x(k) = r_x(-k)$.

Example 13.6.2: Find the AR coefficients $a(1)$ to $a(5)$ if the autocorrelation of the observed signal $\{x(n)\}$ is given. Assume the noise variance to be equal to 1. Find the power spectrum of the process.

Solution: First we must produce a WSS process $\{x(n)\}$. This can be achieved by passing white noise through a linear time-invariant filter. In this example we use an AR filter with two coefficients, $x(n) - 0.9x(n-1) + 0.5x(n-2) = v(n)$. The variance of the white noise input to the filter $v(n)$ is 1, and its mean value is 0. The results are shown in Figure 13.6.2. The Book MATLAB m-file is also given below.

Book MATLAB m-file: ex_13_6_2

```
%Book m-file: ex_13_6_2
x(2)=0;x(1)=0;    %initial conditions to solve the difference
                  %equation;
for n=3:512
    x(n)=0.9*x(n-1)-0.5*x(n-2)+3.5*(rand-0.5);%input is WSS
                  %white noise;
end;              %this iteration produces the WSS process x(n);
x1=x(1,10:265);   %we skipped the first and last values to avoid
                  %edge effects;
rx1=sssamplebiasedautoc(x1,15); %we call the m-function here
                  %to obtain the autocorrelation
                  %function, the m-function must be located in a
                  %folder which
                  %is in the working path of MATLAB;
```

Chapter 13:   Random variables, sequences, and power spectra densities   567

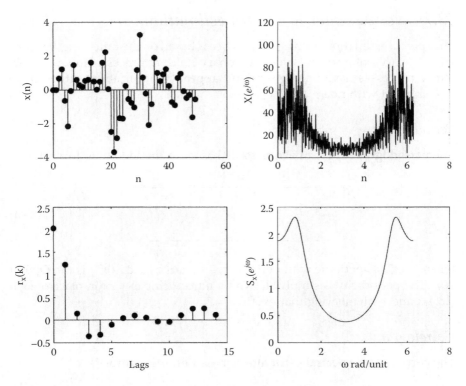

Figure 13.6.2 Illustration of Example 13.6.2.

```
R=toeplitz(rx1(1,1:6)); %we create the 6x6 autocorrelation
                        %matrix;
R1=R(2:6,2:6); %R1 is a sub-matrix of R by picking the 2nd to
               %6th row and the 2nd to 6th column;
a=-inv(R1)*R(2:6,1); %R(2:6,1)=the first column and rows 2 to
                     %6, a= a 5x1 column;
b2=R(1,:)*([1 a']'); %R(1,:)=a row vector of the 1st row of
                     %R and all 6 columns of the row; b2=b(0)2
H=abs(b2./(fft([1 a'],512))); %with the appropriate commands we
                              %plotted x(n), r(k) and
                              %Sx(exp(jω)_;
ftx=fft(x,512);
m=0:49;w=0:2*pi/512:2*pi-(2*pi/512);
subplot(2,2,1);stem(m,x(1,1:50),'k','filled');xlabel('n');
ylabel('x(n)');
subplot(2,2,2);plot(w,abs(ftx),'k');xlabel('n');ylabel...
('X(e^{j\omega})');
p=0:14;subplot(2,2,3);stem(p,rx1(1,1:15),'k','filled');
xlabel('Lags');ylabel('r_x(k)');
subplot(2,2,4);plot(w,H,'k');xlabel('\omega rad/unit');
ylabel('S_x(e^{j\omega})');
```
■

13.7 Nonparametric spectra estimation

The spectra estimation problem in practice is based on the finite length record $\{x(1), \ldots, x(N)\}$ of a second-order stationary random process. However, harmonic processes have line spectra and appear in applications either alone or combined with noise.

Periodogram

The **periodogram spectral estimator** is based on the following formula:

$$\hat{S}_p(e^{j\omega}) = \frac{1}{N}\left|\sum_{n=0}^{N-1} x(n)e^{-j\omega n}\right|^2 = \frac{1}{N}\left|X(e^{-j\omega})\right|^2 \qquad (13.44)$$

where $\hat{S}_p(e^{j\omega})$ is periodic with period $2\pi, -\pi \leq \omega \leq \pi$, and $X(e^{j\omega})$ is the DFT of $x(n)$. The periodicity is simply shown by introducing $\omega + 2\pi$ in place of ω in (13.44) and remembering that $\exp(j2\pi) = 1$.

Correlogram

The **correlogram spectral estimator** is based on the formula

$$\hat{S}_c(e^{j\omega}) = \sum_{m=-(N-1)}^{N-1} \hat{r}(m)e^{-j\omega m} \qquad (13.45)$$

where $\hat{r}(m)$ is the estimate of the biased autocorrelation (assumed mean value of $\{x(n)\}$ zero) given by (13.11). It can be shown that the correlogram spectral estimator, evaluated using the standard biased autocorrelation estimates, coincides with the periodogram spectral estimator. As in (13.44), the correlogram is a periodic function of ω with period 2π.

Computation of $\hat{S}_p(e^{j\omega})$ and $\hat{S}_c(e^{j\omega})$ using FFT

Since both functions are continuous functions of ω, we can sample the frequency as follows:

$$\omega_k = \frac{2\pi}{N}k \qquad k = 0, 1, 2, \cdots, N-1 \qquad (13.46)$$

This situation reduces (13.44) and (13.45) to finding the DFT at those frequencies:

Chapter 13: Random variables, sequences, and power spectra densities 569

$$X(e^{j\omega_k}) = \sum_{n=0}^{N-1} x(n) e^{-j\frac{2\pi}{N}nk} = \sum_{n=0}^{N-1} x(n) W^{nk} \quad W = e^{-j2\pi/N} \quad 0 \le k \le N-1 \quad (13.47)$$

and thus,

$$\hat{S}_p(e^{j\omega_k}) = \frac{1}{N} |X(e^{j\omega_k})|^2 \; ; \quad \hat{S}_c(e^{j\omega_k}) = \sum_{m=-(N-1)}^{N-1} \hat{r}(m) e^{-j\frac{2\pi}{N}km} \quad (13.48)$$

The most efficient way to find the DFT using fast Fourier transform (FFT) is to set $N = 2^r$ for some integer r. The following two Book MATLAB functions give the windowed periodogram and correlogram, respectively.

Book MATLAB function for windowed periodogram: [s,as,phs]=ssperiodogram(x,w,L)

```
%Book function:[s,as,phs]=ssperiodogram(x,w,L)
function[s,as,phs]=ssperiodogram(x,w,L)
%w=window(@name,length(x)),(name=hamming,kaiser,hann,rectwin,
%bartlett,tukeywin,blackman,gausswin,nattallwin,triang,black
%manharris);
%L=desired number of points (bins) of the spectrum;
%x=data in row form;s=complex form of the DFT;
xw=x.*w';
for m=1:L
    n=1:length(x);
    s(m)=sum(xw.*exp(-j*(m-1)*(2*pi/L)*n));
end;
as=((abs(s)).^2/length(x))/norm(w);
phs=(atan(imag(s)./real(s))/length(x))/norm(w);
wb=0:2*pi/L:2*pi-(2*pi/L);subplot(2,1,1);stem(wb,as,'k');
xlabel('Frequency bins, 2\pi/L');ylabel('Magnitude');
```

Book MATLAB function for windowed correlogram: [s,as,phs]=sscorrelogram(x,w,lg,L)

```
%Book MATLAB function:function[s,asc,psc]=sscorrelogram
%(x,w,lg,L);
function[s,as,ps]=sscorrelogram(x,w,lg,L)
%function[s]=aacorrelogram(x,w,lg,L);
```

```
%x=data with mean zero;w=window(@name,length(2*lg)), see
%ssperiodogram
%function and below this function);L=desired number of spectral
%points;
%lg=lag number<<N;rc=symmetric autocorrelation function;
r=sssamplebiasedautoc(x,lg);
rc=[fliplr(r(1,2:lg))  r  0];
rcw=rc.*w';
for m=1:L
   n=-lg+1:lg;
   s(m)=sum(rcw.*exp(-j*(m-1)*(2*pi/L)*n));
end;
as=(abs(s).^2)/norm(w);%amplitude spectrum;
ps=(atan(imag(s))/real(s))/norm(w);%phase spectrum;
```

To plot, for example, **as** or **ps**, we can use the command `plot(0:2*pi/L:2*pi-(2*pi/L),as)`.

General remarks on the periodogram

1. The variance of the periodogram does not tend to zero as $N \to \infty$. This indicates that the periodogram is an **inconsistent** estimator; that is, its distribution does not tend to cluster more closely around the true spectrum as N increases.
2. To reduce the variance, and thus produce a smoother spectral estimator, we must (a) average contiguous values of the periodogram or (b) average periodograms obtained from multiple data segments.
3. The effect of the side lobes of the windows on the estimated spectrum consists of transferring power from strong bands to less strong bands or bands with no power. This process is known as the **leakage** problem.

Blackman–Tukey (BT) method

Because the correlation function at its extreme lag values is not reliable due to the small overlapping of the correlation process, it is recommended to use lag values of about 30 to 40% of the total length of the data. The Blackman–Tukey estimator is a windowed correlogram and is given by

$$\hat{S}_{BT}(e^{j\omega}) = \sum_{m=-(L-1)}^{L-1} w(m)\hat{r}(m)e^{-j\omega m} \tag{13.49}$$

where $w(m)$ is the window with zero values for $|m| > L-1$ and $L \ll N$. The above equation can also be written in the form

Chapter 13: Random variables, sequences, and power spectra densities

$$\hat{S}_{BT}(e^{j\omega}) = \sum_{m=-\infty}^{\infty} w(m)\hat{r}(m)e^{-j\omega m}$$

(13.50)

$$= \hat{S}_c(e^{j\omega}) * W(e^{j\omega}) = \frac{1}{2\pi} \int_{-\pi}^{\pi} \hat{S}_c(e^{j\tau})W(e^{j(\omega-\tau)})d\tau$$

where we applied the discrete-time Fourier transform (DTFT) frequency convolution property (the DTFT of the multiplication of two functions is equal to the convolution of their Fourier transforms). Since windows have a dominant and relatively strong main lobe, the BT estimator corresponds to a "local" weighting average of the periodogram. Although the convolution smoothes the periodogram, it reduces resolution at the same time. It is expected that the smaller the L, the larger the reduction in variance and the lower the resolution. It turns out that the resolution of this spectral estimator is on the order of $1/L$, whereas its variance is on the order of L/N.

For convenience, we again give some of the most common windows below. For the Kaiser window, the parameter β trades the main lobe width for the side lobe leakage; $\beta = 0$ corresponds to a rectangular window and $\beta > 0$ produces a lower side lobe at the expense of a broader main lobe.

- Rectangle window

$$w(n) = 1 \qquad n = 0, 1, 2, \ldots, L-1$$

- Bartlett (triangle) window

$$w(n) = \begin{cases} \dfrac{n}{L/2} & n = 0, 1, \ldots, L/2 \\ \dfrac{L-n}{L/2} & n = \dfrac{L}{2}+1, \ldots, L-1 \end{cases}$$

- Hann window

$$w(n) = 0.5\left[1 - \cos\left(\frac{2n}{L}\pi\right)\right] \qquad n = 0, 1, \ldots, L-1$$

- Hamming window

$$w(n) = 0.54 - 0.46\cos\left(\frac{2\pi}{L}n\right) \qquad n = 0, 1, \ldots, L-1$$

- Blackman window

$$w(n) = 0.42 + 0.5\cos\left(\frac{2\pi}{L}\left(n-\frac{L}{2}\right)\right) + 0.08\cos\left(\frac{2\pi}{L}2\left(n-\frac{L}{2}\right)\right) \quad n = 1, 2, \ldots, L-1$$

- Kaiser window

$$w(n) = \frac{I_0\left[\beta\sqrt{1.0-\left(\frac{n}{L/2}\right)^2}\right]}{I_0(\beta)} \quad -(L-1) \le n \le L-1$$

$$I_0(x) = \sum_{k=0}^{\infty}\left[\frac{\left(\frac{x}{2}\right)^k}{k!}\right]^2 = \text{zero-order modified Bessel function}$$

$w(k) = 0$ for $|k| \ge L$, $w(k) = w(-k)$ and equations are valid for $0 \le k \le L-1$

Note: To use the window derived from MATLAB, we must write

```
w=window(@name,L);
name=the name of any of the following windows: bartlett,
    barthannwin, blackman, blackmanharris, bohmanwin, chebwin,
    gausswin, hanning, hann, kaiser, natullwin, hamming,
    rectwin, tukeywin, triang.
L=number window values
```

Bartlett method

Bartlett's method reduces the fluctuation of the periodogram by splitting up the available data of N observations into $K = N/L$ subsections of L observations each, and then averaging the spectral densities of all K periodograms (see Figure 13.7.1). The MATLAB function below provides the Bartlett periodogram.

**Book MATLAB function for Bartlett's method:
function[s,as,ps]=ssbartlettpsd(x,k,w,L)**

```
%Book MATLAB function for Bartlett spectra estimation
function[s,as,ps]=ssbartlettpsd(x,k,w,L)
%x=data;k=number of sections; w=window(@name,floor(length
%(x)/k));
```

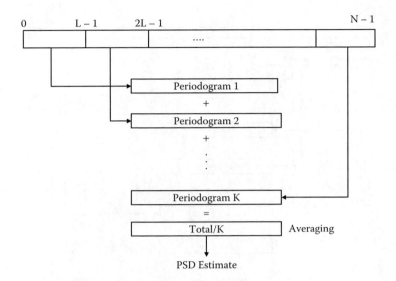

Figure 13.7.1 Bartlett method of spectra estimation.

```
%L=number of points desired in the FT domain;
%K=number of points in each section;
K=floor(length(x)/k);
s=0;
ns=1;
for m=1:k
   s=s+ssperiodogram(x(ns:ns+K-1),w,L);
   ns=ns+K;
end;
as=abs(s)/k;
ps=atan(imag(s)./real(s))/k;
```

Welch method

Welch proposed the following modifications to the Bartlett method: data segments are allowed to overlap, and each segment is windowed prior to computing the periodogram. Since, in most practical applications, only a single realization is available, we create smaller sections as follows:

$$x_i(n) = x(iD + n)w(n) \qquad 0 \le n \le M-1, \qquad 0 \le i \le K-1 \qquad (13.51)$$

where $w(n)$ is the window of length M, D is an offset distance, and K is the number of sections that the sequence $\{x(n)\}$ is divided into. Pictorially, the Welch method is shown in Figure 13.7.2.

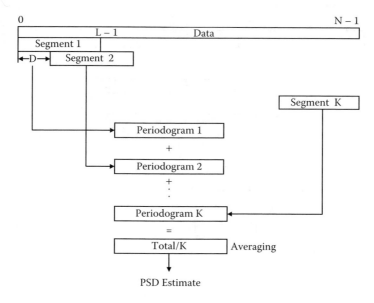

Figure 13.7.2 Welch method of spectra estimation.

The ith periodogram is given by

$$S_i(e^{j\omega}) = \frac{1}{L}\left|\sum_{n=0}^{L-1} x_i(e^{-j\omega n})\right|^2 \qquad (13.52)$$

and the average periodogram is given by

$$S(e^{j\omega}) = \frac{1}{K}\sum_{i=0}^{K-1} S_i(e^{j\omega}) \qquad (13.53)$$

If $D = L$, then the segments do not overlap and the result is equivalent to the Bartlett method, with the exception of the data being windowed.

Book MATLAB function for the Welch method:
function[s,as,ps,K]=sswelch(x,w,D,L)

```
function[s,as,ps,K]=sswelch(x,w,D,L)
%function[as,ps,s,K]=sswelch(x,w,D,L);
%M=length(w)=section length;
%L=number of samples desired in the frequency domain;
%w=window(@name,length of each segment=length(w));x=data;
```

Chapter 13: Random variables, sequences, and power spectra densities

```
%D=offset distance=fraction of length(w), mostly 50% of M;
%M<<N=length(x);
N=length(x);
M=length(w);
K=floor((N-M+D)/D);%K=number of processings;
s=0;
for i=1:K
    s=s+ssperiodogram(x(1,(i-1)*D+1:(i-1)*D+M),w,L);
end;
as=abs(s)/(M*K);%as=amplitude spectral density;
ps=atan(imag(s)./real(s))/(M*K);%ps=phase spectral density;
```

The MATLAB function is given as follows:

P=spectrum(x,m) %x=data; m=number of points of each section
%and must be a power of 2;the sections are
%windowed by
%a hanning window;P is a (m/2)x2 matrix whose
%first column is the
%power spectral density and the second is the
%95% confidence interval;

Modified Welch method

It is evident from Figure 13.7.2 that if the lengths of the sections are not long enough, frequencies close together cannot be differentiated. Therefore, we propose a procedure, defined as **symmetric method**, and its implementation is shown in Figure 13.7.3. Windowing of the segments can also be incorporated. This approach and the rest of the proposed schemes have the advantage of progressively incorporating longer and longer segments of the data, thus introducing better and better resolution. In addition, due to the averaging process, the variance decreases and smoother periodograms are obtained. Figure 3.7.4 shows another proposed method, which is defined as the **asymmetric method**. Figure 3.7.5 shows a suggested approach for better resolution and reduced variance. The procedure is based on the method of prediction and averaging and is defined as the **symmetric prediction method**. This procedure can be used in all the other forms, e.g., nonsymmetric. The above methods can also be used for spectral estimation if we substitute the word *periodogram* with the word *correlogram*.

Figure 13.7.6 shows data given by the equation

$$x(n) = \sin(0.4\pi n) + \sin(0.422\pi n) + randn(1,128) \qquad (13.54)$$

and 128 time units. Figure 13.7.6b shows the Welch method using the MATLAB function (P=spectrum(x,64)). Figure 13.7.6c shows the proposed

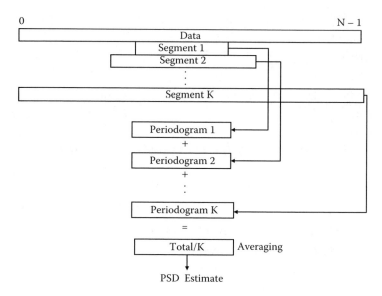

Figure 13.7.3 Modified symmetric Welch method.

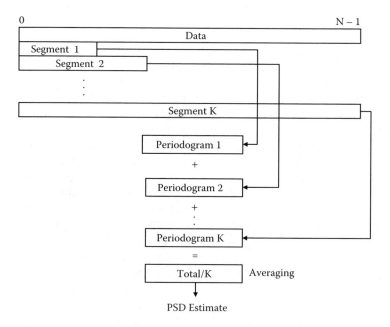

Figure 13.7.4 Modified asymmetric Welch method.

modified Welch method. We observe that we are able to differentiate a 0.022 radian difference in the spectrum. However, as expected, the variance was somewhat larger.

Chapter 13: Random variables, sequences, and power spectra densities 577

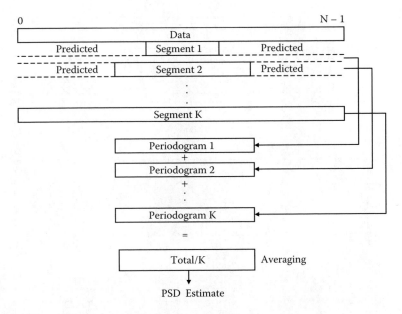

Figure 13.7.5 Modified with prediction Welch method.

The Blackman–Tukey periodogram with the Bartlett window

The PSD based on the Blackman–Tukey method is given by

$$\hat{S}_{BT}(e^{j\omega}) = \sum_{m=-L}^{L} w(m)\hat{r}(m)e^{j\omega m}$$

$$w(m) = \begin{cases} 1 - \dfrac{|m|}{L} & m = 0, \pm 1, \pm 2, \cdots, L \\ 0 & \text{otherwise} \end{cases}$$
(13.55)

Book MATLAB function for the Blackman–Tukey periodogram with triangle window

```
functon[s]=ssblackmantukeypsd(x,lg,L)
%function[s]=ssblackmantukeypsd(x,lg,L);
%the window used is the triangle (Bartlett) window;
%x=data;lg=lag number about 20-40% of length(x)=N;
%L=desired number of spectral points (bins);
[r]=sssamplebiasedautoc(x,lg);
n=-(lg-1):1:(lg-1);
w=1-(abs(n)/lg);
```

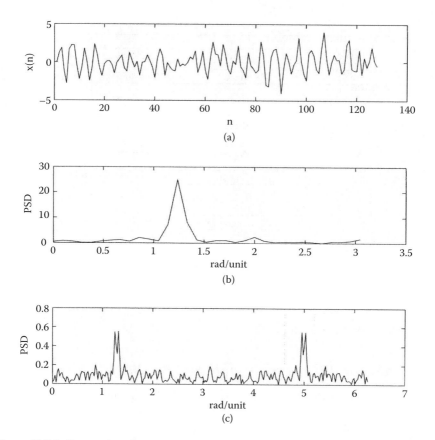

Figure 13.7.6 Comparison between the Welch method (b) and the proposed modified Welch method. (a) is the original random signal.

```
rc=[fliplr(r(1,2:lg))  r];
rcw=rc.*w;
s=fft(rcw,L);
```

Another approach to spectra estimation is the parametric method, which will not be developed in this text.

Important definitions and concepts

1. Discrete random signals
2. Probability density function
3. Cumulative density function
4. Multivariate distribution
5. Realization of random variables
6. Event
7. Stationary process

Chapter 13: Random variables, sequences, and power spectra densities 579

8. Ergotic process
9. Sample mean value
10. Sample variance
11. Mean value or expectation
12. Frequency interpretation of statistical characteristics
13. The time-average formula
14. Unbiased estimator
15. Correlation and cross-correlation
16. Autocorrelation
17. Variance and covariance
18. Independent rv's
19. Stationary, wide-sense stationary, and strict-sense stationary processes
20. Independent and orthogonal rv's
21. Autocorrelation matrix
22. Toeplitz matrix
23. White noise
24. Random walk
25. White Gaussian noise
26. Lognormal distribution
27. Chi-square distribution
28. Wiener–Kintchin relations
29. Power spectrum
30. Filtering of random processes
31. Spectral factorization
32. Nonparametric spectral estimation
33. Periodogram
34. Correlogram
35. Blackman–Tukey method
36. Bartlett method
37. Welch method
38. Modified Welch method

Chapter 13 Problems

Section 13.2

1. To add the rows of a matrix A we write sum(A,1). Produce four realizations using the script m-file (*realizations*) and produce the y1 = sum(x, 1)/4. Next, produce 1000 realizations (the script file must be modified) and then obtain the sum y2 = sum(x, 1)/1000. Explain any difference you observe between the y1 and y2 time series.
2. Find the sample mean of the first and third realizations shown in Figure 13.1.1. State your observation.
3. Modify the *ssmeanautocorensemble* m-file by substituting the line x=rand(M,N); with the line x=rand(M,N)-0.5; rand is a

MATLAB file producing uniform iid random samples. State your observation by changing the number of realizations.
4. Verify (13.12).

Section 13.3

1. Prove the basic properties of a wide-sense stationary (WSS) process (see (13.18)).
2. Modify the noise vector v in the m-file ex_13_2_1 as follows: v=0.1*randn(1,40). State your observation between Figure 13.3.1 and your result of the FT of rx. Also, using vector rx, create a correlation matrix R_x having dimensions 2 × 2, 5 × 5, and 19 × 10.
3. If the sequence {x(n)} is characterized by iid random variables, find the mean and variance of the signal y(n) = ax(n) + b, where a and b are constants.
4. The covariance function of WN (white noise) is given by (13.20). Using the MATLAB function *rand*, create a sample random sequence x=0.2*rand(1,20)+0.2 and find its sample covariance $c_x(k)$.
5. Create a random walk of 200 steps and plot y(n).
6. Find the mean, E{y(n)}, and the cov(y(n) – m_y,y(n) – m_y) of the random walk process y(n) (see (13.21)).

Section 13.4

1. Create the following vectors: x1 = randn(1, 50), x2 = randn(1, 500), and x3 = randn(1, 10,000). Next, plot different histograms for each sequence and observe their shapes. State your observations.
2. Using (13.26), fit the analytical Gaussian rv x to those produced in Problem 13.4.1.
3. Create a pdf f(x(1), x(2)) = f(x(1))f(x(2)), each one having zero mean and the same variance. Next, create a histogram and superimpose the analytical function given by (13.27).
4. Produce a simulated lognormal sequence and superimpose on its histogram its analytic representation given by (13.28).

Section 13.5

1. Repeat Example 13.5.1, but increase the noise by 1.5. That is, substitute v=randn(1,64) with v=1.5*rand(1,64). State your observation and conclusions.
2. Repeat Problem 13.5.1, but this time increase the length of the signal to 512. That is, substitute any number 64 in m-file Ex_15_3_1 with 512. State your observations and conclusions.

Section 13.6

1. Repeat Example 13.6.1 with a third-order FIR filter.
2. Repeat Example 13.6.1 with a third-order high-pass FIR filter.
3. Create a 3×3 similar to matrix (13.43). Solve the lower 2×2 matrix to find the a_i's. Next, substitute them in the first row to find $\sigma_v^2 b(0)^2$. Finally, find the AR spectrum density $S(e^{j\omega})$.

Section 13.7

1. Use the Book MATLAB functions for the periodogram and correlogram and study the following:
    a. $x(n) = \cos(0.2 \times 2\pi n) + v(n)$ $0 \le n \le 31$, $v(n)$ is noise.
    b. $x(n) = \cos(0.2 \times 2\pi n) + v(n)$ $0 \le n \le 255$, $v(n)$ is same noise as in a.
    c. $x(n) = \cos(0.2 \times 2\pi n) + \cos(0.25 \times 2\pi n) + v(n)$ $0 \le n \le 31$, $v(n)$ is noise.
    d. $x(n) = \cos(0.2 \times 2\pi n) + \cos(0.25 \times 2\pi n) + v(n)$ $0 \le n \le 255$, $v(n)$ same noise as in c).
    e. Repeat a) to d) above with increased (double) variance of the noise.
2. Compare the Blackman–Tukey method with that of Bartlett by repeating Problem 13.7.1.
3. Using the signal $x(n) = \sin(0.4 \times 2\pi n) + \sin(0.405 \times 2\pi n) + randn(1,N)$, do the following:
    a. Increase the length to $N = 64$, $N = 128$, and $N = 512$, and in each case compare the Welch method with the modified Welch method.
    b. Do the same, but now use `0.1*randn(1,N)`.

*chapter 14

Least square system design, Wiener filter, and the LMS filter

In this chapter we shall develop three basic and important principles used extensively in signal processing, communications, system identification, denoising signals, etc. The least squares principle, proposed first by Gauss when he was 18 years old, is widely applicable to the design of digital signal processing systems. We will first describe the use of the least square approach to modeling, interference canceling, as well as the cases involving prediction. Next, we will study the celebrated Wiener filter, which was developed during the Second World War, and finally, we will study another widely used filter known as the least mean squares (LMS) filter.

14.1 The least-squares technique

The principle of **least squares** will be used in this text for signals of one variable, although the concept applies equally well for signals with more than one variable. Furthermore, we shall study discrete signals, which can always be derived from continuous signals if the sampling frequency is sufficiently high enough so that aliasing does not take place.

Let us, for example, have a discrete signal that is represented in its vector form as follows: $\underline{x} = [x(1) \ x(2) \ \cdots x(N)]$. We would next like to approximate the function with a polynomial of the form

$$\hat{x}(nT) = c_1 + c_2(nT) + c_3(nT)^2 + \cdots + c_M(nT)^{M-1} \qquad T = \text{sampling time} \quad (14.1)$$

Since the approximating polynomial is a linear function of the unknown c_i's, we are dealing with **linear least squares approximation**. The difference between the exact and approximate function is the error. We can define a **cost function** J, which is equal to the sum of the difference of errors squared, the total square error. Hence, we write

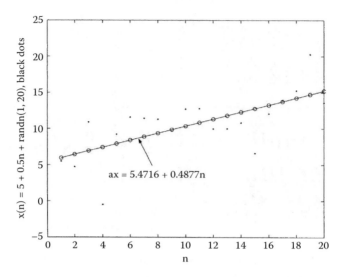

Figure 14.1.1 Least squares approximation of sampled data $T = 1$.

$$\boxed{J = \sum_{n=1}^{N} [x(n) - \hat{x}(n)]^2} \qquad (14.2)$$

Our next step is to minimize this error. To see how the above equation works, let us have the discrete function plotted in Figure 14.1.1 as black dots. It is required to find a straight line that minimizes the mean square error, the cost function. To do so, we assume an approximate function of the form $x = a + b(nT)$ $T = 1$, where a and b are unknown and must be determined based on the given data. Hence, we must minimize the cost function

$$J = \sum_{n=1}^{20} (x(n) - a - bn)^2$$

Next, we take the partial derivatives of J with respect to a and b and equate these derivatives equal to zero. Using the fundamental property of differentiation, the derivative of J with respect to the unknown a is

$$\frac{\partial J}{\partial a} = -2\sum_{n=1}^{20}(x(n) - a - bn) = 0 \quad \text{or} \quad a\sum_{n=1}^{20} 1 + b\sum_{n=1}^{20} n = \sum_{n=1}^{20} x(n)$$

$$\text{or} \quad 20a + b\frac{20(20+1)}{2} = 211.8391 \quad \text{or} \quad 20a + 210b = 211.8391$$

Chapter 14: Least square system design, Wiener filter, and the LMS filter   585

Similarly, the derivative of J with respect to the unknown b is

$$\frac{\partial J}{\partial b} = -2\sum_{n=1}^{20}(x(n)-a-bn)n = 0 \text{ or } a\sum_{n=1}^{20}n + b\sum_{n=1}^{20}n^2 = \sum_{n=1}^{20}nx(n)$$

or

$$a\frac{20(20+1)}{2} + b\frac{20(20+1)(40+1)}{6} = \sum_{n=1}^{20}nx(n)$$

or

$$210a + 2870b = 2548.6$$

With the help of MATLAB we found the following expressions:

$$\sum_{n=1}^{20} x(n) = sum(x)$$

$$\sum_{n=1}^{20} nx(n) = sum(x*n')$$

However, we could also use MATLAB for the other two summations, but instead, we used Appendix A at the end of the book. The final value is found with the help of MATLAB. However, we could have done this manually since the inverse of a 2×2 matrix is easy to perform. The final value is then found as follows:

$$\begin{bmatrix} a \\ b \end{bmatrix} = \begin{bmatrix} 20 & 210 \\ 210 & 2870 \end{bmatrix}^{-1} \begin{bmatrix} 211.8391 \\ 2548.6 \end{bmatrix} = \begin{bmatrix} 5.4716 \\ 0.4877 \end{bmatrix}$$

Introducing the values a and b found above in the equation of a straight line, we obtained the approximation to the exact line whose constant values are $a = 5$ and $b = 0.5$. The straight line indicated by x's, which is very close to the exact one, is the mean squares approximation to the data. The straight line found above gives the smallest sum of the squared errors from any other line that we may try to create. The error is the distance from the line to a scattered point along the same time. For example, at $n = 4$ the error is approximately equal to 8.

Example 14.1.1: Find the least squares approximation to the data shown in Figure 14.1.2 as black dots and superimpose on them the least squares error line.

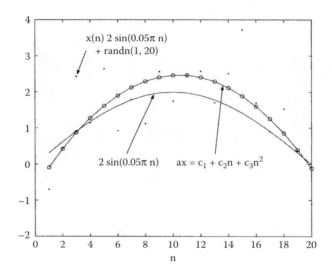

Figure 14.1.2 Least squares approximation for nonlinear data.

Solution: In Figure 14.1.1, it was obvious that the points have a linear trend, and as a consequence, we used a straight line. Here, however, we observe some nonlinear trend, and therefore, we will use a higher-order polynomial. We again assume sampling time $T = 1$. As previously, the cost function is

$$J = \sum_{n=1}^{20}[x(n)-\hat{x}(n)]^2 = \sum_{n=1}^{20}(x(n)-c_1-c_2n-c_3n^2)^2$$

The partial derivatives with respect to c_i's are:

$$\frac{\partial J}{\partial c_1} = -2\sum_{n=1}^{20}(x(n)-c_1-c_2n-c_3n^2) = 0$$

$$\text{or } \sum_{n=1}^{20}x(n)-c_1\sum_{n=1}^{20}1-c_2\sum_{n=1}^{20}n-c_3\sum_{n=1}^{20}n^2 = 0$$

$$\frac{\partial J}{\partial c_2} = -2\sum_{n=1}^{20}(x(n)-c_1-c_2n-c_3n^2)n = 0$$

$$\text{or } \sum_{n=1}^{20}x(n)n-c_1\sum_{n=1}^{20}n-c_2\sum_{n=1}^{20}n^2-c_3\sum_{n=1}^{20}n^3 = 0$$

$$\frac{\partial J}{\partial c_3} = -2\sum_{n=1}^{20}(x(n) - c_1 - c_2 n - c_3 n^2) n^2 = 0$$

$$\text{or} \sum_{n=1}^{20} x(n) n^2 - c_1 \sum_{n=1}^{20} n^2 - c_2 \sum_{n=1}^{20} n^3 - c_3 \sum_{n=1}^{20} n^4 = 0$$

Using the MATLAB help, we obtain the following system:

$$20 c_1 + 210 c_2 + 2870 c_3 = 30.4297$$

$$210 c_1 + 2870 c_2 + 44100 c_3 = 318.5917$$

$$2870 c_1 + 44100 c_2 + 687190 c_3 = 3847.1$$

The values of the unknown c_i's are found by using the expression

$$\begin{bmatrix} c_1 \\ c_2 \\ c_3 \end{bmatrix} = \begin{bmatrix} 20 & 210 & 2870 \\ 210 & 2870 & 44100 \\ 2870 & 44100 & 722666 \end{bmatrix}^{-1} \begin{bmatrix} 30.4297 \\ 318.5917 \\ 3847.1 \end{bmatrix} = \begin{bmatrix} -0.6580 \\ 0.5970 \\ -0.0285 \end{bmatrix}$$

Therefore, the estimate curve is given by: $\hat{x}(n) = -0.6580 + 0.5970 n - 0.0285 n^2$. Observe that although we used a sine function to create the scattered points, the estimate curve that gives the least squares error is different. ∎

Linear least squares

We proceed to generalize the linear least squares technique by a set of known functions $h(n)$ and write

$$J(\underline{c}) = \sum_{n=0}^{N-1} [x(n) - \hat{x}(n)]^2 = \sum_{n=0}^{N-1} [x(n) - \underline{c}^T \underline{h}(n)]^2$$

$$= \sum_{n=0}^{N-1} [x(n) - (c_1 h_1(n) + \cdots, c_M h_M(n))]^2 = (\underline{x} - \underline{H}\underline{c})^T (\underline{x} - \underline{H}\underline{c})$$

(14.3)

where the underline indicates a vector for small letters and matrices for capital letters. The exponent T means the transpose of a vector or matrix (see Appendix 1.1 of Chapter 1). We shall assume in this text that $N > M$ and full-rank M for $\underline{H}$ so that solutions exist.

Let us verify (14.3) by accepting a two-term expansion. Hence, we write

$$J = \sum_{n=0}^{1}(x(n) - c_1 h_1(n) - c_2 h_2(n))^2$$

$$= (x(0) - c_1 h_1(0) - c_2 h_2(0))^2 + (x(1) - c_1 h_1(1) - c_2 h_2(1))^2$$

The last expression of (14.3) for the above conditions is

$$J = \left(\begin{bmatrix} x(0) \\ x(1) \end{bmatrix} - \begin{bmatrix} h_1(0) & h_2(0) \\ h_1(1) & h_2(1) \end{bmatrix}\begin{bmatrix} c_1 \\ c_2 \end{bmatrix}\right)^T \left(\begin{bmatrix} x(0) \\ x(1) \end{bmatrix} - \begin{bmatrix} h_1(0) & h_2(0) \\ h_1(1) & h_2(1) \end{bmatrix}\begin{bmatrix} c_1 \\ c_2 \end{bmatrix}\right)$$

$$= \left([x(0) \ x(1)] - [c_1 \ c_2]\begin{bmatrix} h_1(0) & h_1(1) \\ h_2(0) & h_2(1) \end{bmatrix}\right)\begin{pmatrix} x(0) - h_1(0)c_1 - h_2(0)c_2 \\ x(1) - h_1(1)c_1 - h_2(1)c_2 \end{pmatrix}$$

$$= [x(0) - c_1 h_1(0) - c_2 h_2(0) \quad x(1) - c_1 h_1(1) - c_2 h_2(1)]\begin{pmatrix} x(0) - h_1(0)c_1 - h_2(0)c_2 \\ x(1) - h_1(1)c_1 - h_2(1)c_2 \end{pmatrix}$$

The last two expressions are identical. The reader should observe that the following matrix relationship was used:

$$(\underline{x} - \underline{H}\,\underline{c})^T = (\underline{x}^T - \underline{c}^T \underline{H}^T) \tag{14.4}$$

where the exponent T stands for transpose of a matrix.

To find the unknown c_i's, we differentiate the cost function of (14.3) with respect to each c_i and then set the developed equations equal to zero. Therefore, we have a system with c_i's, of which the unknowns are determined by solving the system. The following example will elucidate the procedure.

Example 14.1.2: Let a signal be a constant, $s(n) = A$, and let the received signal be given by $x(n) = 5 + randn(1,10)$. Find A.

Solution: According to the least squares approach, we can estimate 5 by minimizing the cost function. Taking the derivative of J with respect to A and setting the result equal to zero, we obtain

$$J(A) = \sum_{n=0}^{N-1}[A - x(n)]^2; \qquad \frac{\partial J(A)}{\partial A} = \sum_{n=0}^{N-1}[2A - 2x(n)] = 0 \text{ or}$$

$$A\sum_{n=0}^{N-1}1 = \sum_{n=0}^{N-1}x(n) \text{ or } A = \frac{1}{N}\sum_{n=0}^{N-1}x(n)$$

Using MATLAB help we obtain $\hat{A} = sum(x)/10 = 5.1022$. The $J(A)$ minimum is given by

$$J_{min} = \sum_{n=0}^{N-1}(x(n)-\hat{A})^2$$

and for the present case

$$J_{min} = \sum_{n=1}^{10}(x(n)-5.1022)^2 = 5.6484$$

∎

Example 14.1.3: Find the amplitude constants of the signal

$$s(n) = A\sin 0.1\pi n + B\sin 0.4\pi n \qquad n = 0, 1, \cdots, N-1$$

if the received signal is $x(n) = s(n) + randn(1, N)$ and their exact values are $A = 0.2$ and $B = 5.2$.

Solution: For this case, and for $n = 0, 1, 2,$ and 3 and $N = 4$, we obtain

$$J(A, B) = \sum_{n=0}^{3}(x(n)-h_1(n)A-h_2(n)B)^2$$

$$h_1(n) = \sin(0.1\pi n) \qquad h_2(n) = \sin(0.4\pi n)$$

If we differentiate first $J(A, B)$ with respect to A, and next with respect to B, and then equate the results to zero, we obtain

$$\begin{bmatrix} A \\ B \end{bmatrix} = \begin{bmatrix} \sum_{n=0}^{3}h_1^2(n) & \sum_{n=0}^{3}h_1(n)h_2(n) \\ \sum_{n=0}^{3}h_1(n)h_2(n) & \sum_{n=0}^{3}h_2^2(n) \end{bmatrix}^{-1} \begin{bmatrix} \sum_{n=0}^{3}x(n)h_1(n) \\ \sum_{n=0}^{3}x(n)h_2(n) \end{bmatrix} \quad (14.5)$$

With MATLAB help, the approximate values are $\hat{A} = 0.4934$ $\hat{B} = 5.0037$. The exact values are $A = 0.2$ $B = 5.2$. Further, using the above two sets of values we obtain $\underline{\hat{s}} = [0 \ 4.9113 \ 3.2311 \ -2.5419]$ and $\underline{s} = [0 \ 4.8171 \ 3.0565 \ -2.0771]$, respectively. The moment we define the vector $\underline{n}$, we find the vectors $\underline{x}, \underline{h}_1,$ and $\underline{h}_2$. The values of A and B are found from the column vector ab, which is given by the MATLAB expression

```
ab=inv([sum(h1.^2)  sum(h1.*h2);sum(h1.*h2)  sum(h2.^2)])*...
    [sum(x.*h1);sum(x.*h2)]
```
∎

14.2 The mean square error

In this chapter we develop a class of linear optimum discrete-time filters known as the **Wiener filters**. These filters are optimum in the sense of minimizing an appropriate function of the error, known as the **cost function**. The cost function that is commonly used in filter design optimization is the **mean square error** (MSE). Minimizing MSE involves only second-order statistics (correlations) and leads to a theory of linear filtering that is useful in many practical applications. This approach is common to all optimum filter designs. Figure 14.2.1 shows the block diagram presentation of the optimum filter problem.

The basic idea is to recover a desired signal $d(n)$ given a noisy observation $x(n) = d(n) + v(n)$, where both $d(n)$ and $v(n)$ are assumed to be wide-sense stationary (WSS) processes. Therefore, the problem can be stated as follows:

> Design a filter that produces an estimate $\hat{d}(n)$ using a **linear** combination of the data $x(n)$ such that the **mean square error** (MSE) function (cost function J)
>
> $$\boxed{J = E\{[d(n) - \hat{d}(n)]^2\} = E\{e^2(n)\}} \qquad (14.6)$$
>
> is minimized.

Depending on how the data $x(n)$ and the desired signal $d(n)$ are related, there are four basic problems that need solved: filtering, smoothing, prediction, and deconvolution.

Observe that in (14.6) the least squares formula appears again, but with the difference that we take the expectation (ensemble average) value of the square value of the error. To proceed with our development and create useful formulas, we will initially assume that we are able to repeat the experiments, produce the probability density function (pdf), and then find the mean square error using the formula (see also Chapter 13)

$$E\{e^2(n)\} = \int_{-\infty}^{\infty} e^2(n) f(e(n)) de(n)$$

However, when we find the final formulas, we will proceed to define the signals as ergotic and then find their means and variances using the approximate formulas developed in Chapter 13 (see (13.7) and (13.11)).

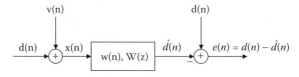

Figure 14.2.1 Block diagram representation of the optimum filtering problem.

The Wiener filter

Let the sample response (filter coefficients) of the desired filter be denoted by **w**. In this chapter we will use w_i's for filter coefficients since it is customary to use this symbol in books involving adaptive and optimum filtering of random signals. This filter will process the real-valued stationary process $\{x(n)\}$ to produce an estimate $\hat{d}(n)$ of the desired real-valued signal $d(n)$. Without loss of generality, we will assume, unless otherwise stated, that the processes $\{x(n)\}$, $\{d(n)\}$, etc., have zero mean values. Furthermore, assuming that the filter coefficients do not change with time, the output of the filter is equal to the convolution of the input and the filter coefficients. Hence, we obtain

$$\hat{d}(n) = w_n * x(n) = \sum_{m=0}^{M-1} w_m x(n-m) = \mathbf{w}^T \mathbf{x}(n) \quad (14.7)$$

where M is the number of filter coefficients and

$$\mathbf{w} = [w_0 \ w_2 \ \cdots \ w_{M-1}]^T, \ \mathbf{x}(n) = [x(n) \ x(n-1) \ \cdots \ x(n-M+1)]^T \quad (14.8)$$

In this chapter we will develop filtering problems using only finite impulse response (FIR) filters. This is due to the main property that FIR filters are stable.

The mean square error (see (14.6)) of the filtering problem is

$$\boxed{\begin{aligned} J(\mathbf{w}) &= \text{constant} = E\{e^2(n)\} = E\{[d(n) - \mathbf{w}^T \mathbf{x}(n)]^2\} \\ &= E\{[d(n) - \mathbf{w}^T \mathbf{x}(n)][d(n) - \mathbf{w}^T \mathbf{x}(n)]^T\} \\ &= E\{[d(n) - \mathbf{w}^T \mathbf{x}(n)][d(n) - \mathbf{x}^T(n)\mathbf{w}]\} \\ &= E\{d^2(n) - \mathbf{w}^T \mathbf{x}(n) d(n) - d(n)\mathbf{x}^T(n)\mathbf{w} + \mathbf{w}^T \mathbf{x}(n)\mathbf{x}^T(n)\mathbf{w}\} \\ &= E\{d^2(n)\} - 2\mathbf{w}^T E\{d(n)\mathbf{x}(n)\} + \mathbf{w}^T E\{\mathbf{x}(n)\mathbf{x}^T(n)\} \mathbf{w} \\ &= \sigma_d^2 - 2\mathbf{w}^T \mathbf{p}_{dx} + \mathbf{w}^T \mathbf{R}_x \mathbf{w} \end{aligned}} \quad (14.9)$$

where

$$\mathbf{w}^T \mathbf{x}(n) = \mathbf{x}^T(n)\mathbf{w} = \text{number}$$

$\sigma_d^2 =$ variance of the desired signal, $d(n)$

$$\mathbf{p}_{dx} = [p_{dx}(0) \ p_{dx}(1) \ \cdots \ p_{dx}(M-1)]^T = \text{cross-correlation vector} \quad (14.10)$$

$p_{dx}(0) \triangleq r_{dx}(0), \ p_{dx}(1) \triangleq r_{dx}(1), \ \cdots, p_{dx}(M-1) \triangleq r_{dx}(M-1)$

$$\mathbf{R}_x = E\left\{\begin{bmatrix} x(n) \\ x(n-1) \\ \vdots \\ x(n-M+1) \end{bmatrix}[x(n)\ x(n-1)\ \cdots\ x(n-M+1)]\right\} \quad (14.11)$$

$$= \begin{bmatrix} E\{x(n)x(n)\} & E\{x(n)x(n-1)\} & \cdots & E\{x(n)x(n-M+1)\} \\ E\{x(n-1)x(n)\} & E\{x(n-1)x(n-1)\} & \cdots & E\{x(n-1)x(n-M+1)\} \\ \vdots & & & \vdots \\ E\{x(n-M+1)x(n)\} & E\{x(n-M+1)x(n-1)\} & \cdots & E(x(n-M+1)x(n-M+1)\} \end{bmatrix}$$

$$= \begin{bmatrix} r_x(0) & r_x(1) & \cdots & r_x(M-1) \\ r_x(-1) & r_x(0) & \cdots & r_x(M-2) \\ \vdots & & & \vdots \\ r_x(-M+1) & r_x(-M+2) & \cdots & r_x(0) \end{bmatrix}$$

The above matrix is the correlation matrix of the input data to the filter, and it is symmetric because the random process is assumed to be stationary, and hence we have the equality $r_x(k) = r_x(-k)$. Since in practical cases we have only one realization, we will assume that the signal is ergodic. Therefore, we will use the sample biased autocorrelation coefficients given in (13.11).

Example 14.2.1: Let us assume that we have found the sample autocorrelation coefficients ($r_x(0) = 1.0$, $r_x(1) = 0$) from given data $x(n)$, which, in addition to noise, contain the desired signal. Furthermore, let the variance of the desired signal $\sigma_d^2 = 24.40$ and the cross-correlation vector be $\mathbf{p}_{dx} = [2\ 4.5]^T$. It is desired to find the surface defined by the mean square function $J(\mathbf{w})$.

Solution: Introducing the values given above in (14.9), we obtain

$$J(\mathbf{w}) = 24.40 - 2[w_0\ w_1]\begin{bmatrix} 2 \\ 4.5 \end{bmatrix} + [w_0\ w_1]\begin{bmatrix} 1 & 0 \\ 0 & 1 \end{bmatrix}\begin{bmatrix} w_0 \\ w_1 \end{bmatrix} \quad (14.12)$$

$$= 24.40 - 4w_0 - 9w_1 + w_0^2 + w_1^2$$

Note that the equation is quadratic with respect to filter coefficients, and it is true for any number of filter coefficients. This is because we used the mean square error approach for the minimization of the error. Figure 14.2.2

Chapter 14: Least square system design, Wiener filter, and the LMS filter

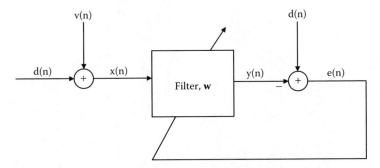

Figure 14.2.2 Block diagram representation of the Wiener filter.

shows the schematic representation of the Wiener filter. The data $x(n)$ are the sum of the desired signal and noise. From the data we find the correlation matrix and the cross-correlation between the desired signal and the data.

Note: *To find the optimum Wiener filter coefficients, the desired signal is needed.*

Figure 14.2.3 shows the **mean square error surface**. This surface is found by inserting different values of w_0 and w_1 in the function $J(\mathbf{w})$ and plotting its magnitude. The values of the coefficients that correspond to the bottom of the surface are the **optimum** Wiener coefficients. The vertical distance from the $w_0 - w_1$ plane to the bottom of the surface is known as the **minimum error**, J_{min}, and corresponds to the optimum Wiener coefficients. We observe that the minimum height of the surface corresponds to about $w_0 = 2$ and $w_1 = 4.5$, which are the optimum coefficients (we will learn how to find them in the next section). Figure 14.2.4 shows an adaptive FIR filter. The following Book MALAB m-file produces Figure 14.2.3.

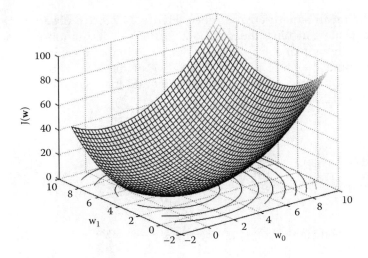

Figure 14.2.3 The mean square error surface.

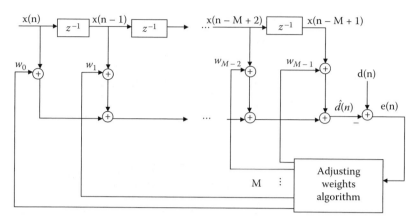

Figure 14.2.4 Schematic representation of an adaptive FIR filter.

Book MATLAB m-file: ex_14_2_1

```
%Book m-file:ex_14_2_1
x=-2:0.4:10;y=-2:0.4:10;
[w0,w1]=meshgrid(x,y);%MATLAB function;
J=24.40-4*w0-9*w1+w0.^2+w1.^2;
meshc(w0,w1,J);grid on;xlabel('w_0'),ylabel('w_1'),zlabel('J');
title('Mean-square error surface');
colormap([0 0 0]);%changes the color grading to black color;
```
■

The Wiener solution

From the mean square error surface, Figure 14.2.3, we observe that there exists a plane touching the parabolic surface at its minimum point and parallel to the **w**-plane. Furthermore, we observe that the surface is concave upwards, and therefore the first derivative of the MSE with respect to w_0 and w_1 must be zero at the minimum point and the second derivative must be positive. Hence, we write

$$\frac{\partial J(w_0,w_1)}{\partial w_0}=0 \qquad \frac{\partial J(w_0,w_1)}{\partial w_1}=0 \qquad a)$$

$$\frac{\partial^2(w_0,w_1)}{\partial^2 w_0}>0 \qquad \frac{\partial^2 J(w_0,w_1)}{\partial^2(w_1)}>0 \qquad b)$$

(14.13)

For a two-coefficient filter, (14.9) becomes

$$J(w_0,w_1)=w_0^2 r_x(0)+2w_0 w_1 r_x(1)+w_1^2 r_x(0)-2w_0 r_{dx}(0)-2w_1 r_{dx}(1)+\sigma_d^2 \quad (14.14)$$

Chapter 14: Least square system design, Wiener filter, and the LMS filter   595

Introducing (14.14) in part a of (14.13) produces the following set of equations:

$$2w_0^o r_x(0) + 2w_1^o r_x(1) - 2r_{dx}(0) = 0 \quad \text{a)}$$
$$2w_1^o r_x(0) + 2w_0^o r_x(1) - 2r_{dx}(1) = 0 \quad \text{b)} \quad (14.15)$$

The above system can be written in the following matrix form, called the **Wiener–Hopf** equation:

$$\boxed{\mathbf{R}_x \mathbf{w}^o = \mathbf{p}_{dx}} \quad (14.16)$$

where the superscript o indicates the optimum Wiener solution for the filter. Note that to find the correlation matrix R_x, we must know the second-order statistics (autocorrelation of the data $x(n)$). If, in addition, the matrix is invertible, the optimum filter is given by

$$\boxed{\mathbf{w}^o = \mathbf{R}_x^{-1} \mathbf{p}_{dx}} \quad (14.17)$$

For an M-order filter, R_x is an $M \times M$ matrix, $\mathbf{w}^o$ is an $M \times 1$ vector, and $\mathbf{p}$ is an $M \times 1$ vector.

If we differentiate $J(\mathbf{w})$ with respect to w_0^o and w_1^o twice, we find that it is equal to $2r_x(0)$. But $r_x(0) = E\{x(m)\,x(m)\} = \sigma_x^2 > 0$, and hence the surface is concave upwards. Therefore, the extreme is the minimum point of the surface. If we next introduce (14.17) in (14.9), we obtain the **minimum mean square error** (**MMSE**):

$$\boxed{J_{min} = \sigma_d^2 - \mathbf{p}_{xd}^T \mathbf{w}^o = \sigma_d^2 - \mathbf{p}_{dx}^T \mathbf{R}_x^{-1} \mathbf{p}_{xd}} \quad (14.18)$$

which indicates that the minimum point of the error surface is at a distance J_{min} above the $\mathbf{w}$-plane. The above equation shows that if no correlation exists between the desired signal and the data, the error is equal to the variance of the desired signal.

The problem we face is how to choose the length of the filter M. In the absence of **a priori** information, we compute the optimum coefficients, starting from a small reasonable number. As we increase the number, we check the MMSE, and if its value is small enough, e.g., MMSE < 0.01, we accept the corresponding number of the coefficients.

Example 14.2.2: We would like to find the optimum filter coefficients w_0 and w_1 of the Wiener filter, which approximates (models) the unknown system with coefficients $b_0 = 1$ and $b_1 = 0.38$ (see Figure 14.2.5).

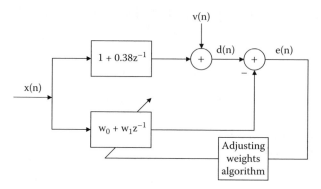

Figure 14.2.5 Illustration of Example 14.2.2.

Solution: The Book MATLAB program given below was used for the identification of the unknown system.

Book MATLAB m-file: ex_14_2_2

```
%Book m-file:ex_14_2_2
%example14_2_2 m-file;
v=0.5*(rand(1,20)-0.5);%v=noise vector(20 uniformly distributed
     %rv's with mean zero);
x=randn(1,20);%x=data vector entering the system and the
     %Wiener filter (20 normal
     %distributed rv's with mean zero;
sysout=filter([1.00 0.38],1,x);%sysout=system output with x
     %as input; filter(b,a,x) is a
     %MATLAB function, where b is the vector of the
     %coefficients of the ARMA numerator,
     %a is the vector of the coefficients of the ARMA
     %denominator;
dn=sysout+v;
rx=sssamplebiasedautoc(x,2);%book MATLAB function with lag=2;
Rx=toeplitz(rx);%toeplitz() is a MATLAB function that gives
     %the symmetric
     %autocorrelation matrix;
pdx=xcorr(x,dn,'biased');%xcorr() is a MATLAB function that
     %gives a symmetric
     %biased crosscorrelation;
p=pdx(1,19:20);
w=inv(Rx)*p';
dnc=sssamplebiasedautoc(dn,1);%gives the variance of dn;
jmin=dnc-p*w;
```

Chapter 14: Least square system design, Wiener filter, and the LMS filter

Some typical values found are $\mathbf{R}_x$ = [0.9979 0.1503;0.1503 0.9979], $\mathbf{p}$ = [0.5196 1.0674], $\mathbf{w}$ = [1.0142 0.3680], and J_{min} = 0.0179. ∎

Orthogonality condition

In order for the set of filter coefficients to minimize the cost function $J(\mathbf{w})$, it is necessary and sufficient that the derivatives of $J(\mathbf{w})$ with respect to w_k be equal to zero for k = 0, 1, 2, ..., M – 1,

$$\frac{\partial J}{\partial w_k} = \frac{\partial}{\partial w_k} E\{e(n)e(n)\} = 2E\left\{e(n)\frac{\partial e(n)}{\partial w_k}\right\} = 0 \qquad (14.19)$$

But

$$e(n) = d(n) - \sum_{m=0}^{M-1} w_m x(n-m) \qquad (14.20)$$

And hence it follows that

$$\frac{\partial e(n)}{\partial w_k} = -x(n-k) \qquad (14.21)$$

Therefore, (14.19) becomes

$$E\{e^o(n)x(n-k)\} = 0 \qquad k = 0, 1, 2, ..., M-1 \qquad (14.22)$$

where the superscript o denotes that the corresponding w_k's used to find the estimation error $e^o(n)$ are optimal. Figure 14.2.6 illustrates the orthogonality principle where the error $e^o(n)$ is orthogonal (perpendicular) to the data set $\{x(n)\}$ when the estimator employs the optimum set of filter coefficients.

14.3 Wiener filtering examples

This section illustrates the use of Wiener filtering in different engineering applications.

Example 14.3.1: Filtering: Filtering of noisy signals (noise reduction) is extremely important, and the method has been used in many applications, such as speech in a noisy environment, reception of data across a noisy channel, enhancement of images, etc.

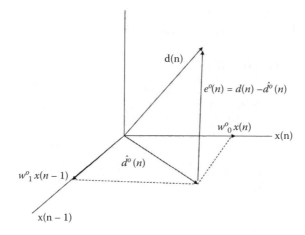

Figure 14.2.6 Pictorial illustration of the orthogonality principle.

Let the received signal be $x(n) = d(n) + v(n)$, where $v(n)$ is a noise signal with zero mean, variance σ_v^2, and it is uncorrelated with the desired signal, $d(n)$. We also assume that the signals are stationary (see Chapter 13). Hence,

$$p_{dx}(m) = E\{d(n)x(n-m)\} = E\{d(n)d(n-m)\} + E\{d(n)\}E\{v(n-m)\} \\ = E\{d^2(m)\} = r_d(m) \quad (14.23)$$

Similarly, we obtain

$$r_x(m) = E\{x(n)x(n-m)\} = r_d(m) + r_v(m) \quad (14.24)$$

where we used the assumption that $d(n)$ and $v(n)$ are uncorrelated and $v(n)$ has a zero mean value. Therefore, the Wiener–Hopf equation (14.16) becomes

$$(\mathbf{R}_d + \mathbf{R}_v)\mathbf{w}^o = \mathbf{p}_{dx} \quad (14.25)$$

The following Book MATLAB m-file is used to produce the results shown in Figure 14.3.1.

Book MATLAB m-file: ex_14_3_1

```
%Book MATLAB m-File:ex_14_3_1
n=0:511;
d=sin(.1*pi*n);%desired signal
v=0.5*randn(1,512);%white Gaussian noise;
x=d+v;%input signal to Wiener filter;
```

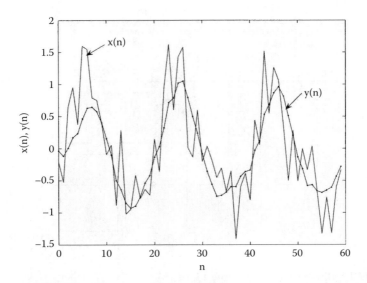

Figure 14.3.1 Illustration of Example 14.3.1.

```
rd=sssamplebiasedautoc(d,20);%rdx=rd=biased autoc. function of
       %the desired signal(see 14.3.1);
rv=sssamplebiasedautoc(v,20);%rv=biased autoc. function of the
       %noise;
R=toeplitz(rd(1,1:12))+toeplitz(rv(1,1:12));%see(14.3.3);
pdx=rd(1,1:12);
w=inv(R)*pdx';
y=filter(w',1,x);%output of the filter;
```

But

$$\sigma_x^2 = \sigma_d^2 + \sigma_v^2 ; \quad (\sigma_x^2 = r_x(0),\ \sigma_d^2 = r_d(0),\ \sigma_v^2 = r_v(0)) \qquad (14.26)$$

and hence, from the MATLAB function var(), we obtain vard = var(x) − var(v) = 0.4968 and J_{min} = 0.4968 − $\mathbf{p}_{dx}\mathbf{w}^o$ = 0.0320 . We can also use the Book MATLAB function **[w,jmin]=sswienerfirfilter(x,d,M)** to obtain the filter coefficients and the MMSE. ∎

Example 14.3.2: Filtering. It is desired to find a two-coefficient Wiener filter for the communication channel shown in Figure 14.3.2. Let $v_1(n)$ and $v_2(n)$ be white noises with zero mean, uncorrelated with each other and with $d(n)$, and have the following variances: $\sigma_1^2 = 0.31$, $\sigma_2^2 = 0.12$. The desired signal produced by the first filter shown in Figure 14.3.2 is

$$d(n) = -0.796 d(n-1) + v_1(n) \qquad (14.27)$$

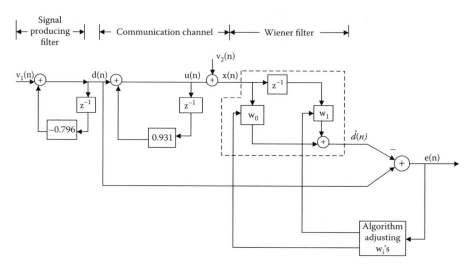

Figure 14.3.2 Illustration of Example 14.3.2.

Therefore, the autocorrelation function of the desired signal becomes

$$E\{d(n)d(n)\} = 0.796^2 E\{d^2(n-1)\} - 2 \times 0.796 E\{d(n-1)\}E\{v_1(n)\} \\ + E\{v_1^2(n)\} \tag{14.28}$$

or

$$\sigma_d^2 = 0.796^2 \sigma_d^2 + \sigma_1^2 \quad \text{or} \quad \sigma_d^2 = 0.31/(1 - 0.796^2) = 0.8461 \tag{14.29}$$

We must always have in mind that all the signals are assumed to be stationary. This means that the signal statistics (mean, variance, pdf, etc.) are independent of the time n.

From the second filter we obtain

$$d(n) = u(n) - 0.931 u(n-1) \tag{14.30}$$

Introducing (14.30) in (14.27) we obtain

$$u(n) - 0.135 u(n-1) - 0.7411 u(n-2) = v_1(n) \tag{14.31}$$

But $x(n) = u(n) + v_2(n)$, and hence the vector form of the set becomes $\mathbf{x}(n) = \mathbf{u}(n) + \mathbf{v}_2(n)$. Therefore, the autocorrelation matrix $\mathbf{R}_x$ becomes

$$E\{\mathbf{x}(n)\mathbf{x}^T(n)\} \triangleq \mathbf{R}_x = E\{[\mathbf{u}(n) + \mathbf{v}_2(n)][\mathbf{u}^T(n) + \mathbf{v}_2^T(n)]\} = \mathbf{R}_u + \mathbf{R}_{v_2} \tag{14.32}$$

Chapter 14: Least square system design, Wiener filter, and the LMS filter

where we used the assumption that $\mathbf{u}(n)$ and $\mathbf{v}_2(n)$ are uncorrelated zero mean random sequences (vectors), which implies that $E\{\mathbf{v}_2(n)\mathbf{u}^T(n)\} = E\{\mathbf{v}_2(n)\} E\{\mathbf{u}^T(n)\} = 0$.

We must remember that the expectation values of a vector and matrix are defined as follows:

$$E\{\mathbf{x}\} = E\begin{bmatrix} x(0) \\ x(1) \\ x(2) \end{bmatrix} = \begin{bmatrix} E\{x(0)\} \\ E\{x(1)\} \\ E\{x(2)\} \end{bmatrix}; \quad E\{\mathbf{x}\} = E\{[x(0)\ x(1)\ x(2)]\} = \quad (14.33)$$

$$= [E\{x(0)\}\ E\{x(1)\}\ E\{x(2)\}]; \quad E\{\mathbf{R}\} = E\begin{bmatrix} x(0) & x(1) \\ x(3) & x(4) \end{bmatrix} = \begin{bmatrix} E\{x(0)\} & E\{x(1)\} \\ E\{x(3)\} & E\{x(4)\} \end{bmatrix}$$

Next, we multiply (14.31) by $u(n - m)$ and then take the ensemble average of both sides, which results in the following expression:

$$r_u(m) - 0.135 r_u(m-1) - 0.7411 r_u(m-2) = r_{v_1 u}(m) = E\{v_1(n)u(n-m)\} \quad (14.34)$$

Setting $m = 1$ and next $m = 2$ in the above equation, we find the **Yule–Walker equations**:

$$\begin{bmatrix} r_u(0) & r_u(-1) \\ r_u(1) & r_u(0) \end{bmatrix}\begin{bmatrix} -0.1350 \\ -0.7411 \end{bmatrix} = \begin{bmatrix} -r_u(1) \\ -r_u(2) \end{bmatrix} = \text{Yule-Walker equations} \quad (14.35)$$

since $v_1(n)$ and $u(n - m)$ are uncorrelated. If we set $m = 0$ in (14.34), it becomes

$$r_u(0) - 0.135 r_u(-1) - 0.7411 r_u(-2) = E\{v_1(n)u(n)\} \quad (14.36)$$

If we next substitute the value of $u(n)$ from (14.31) in (14.36), and taking into consideration that v and u are independent rv's, we obtain the relation

$$\sigma_1^2 = r_u(0) - 0.135 r_u(1) - 0.7411 r_u(2) \quad (14.37)$$

where we used the symmetry property of the correlation function. From the first equation of (14.35) we obtain the relation

$$r_u(1) = \frac{0.135}{1 - 0.7411} r_u(0) = \frac{0.135}{0.2589} \sigma_u^2 \quad (14.38)$$

Substituting, next, the above equation in the second equation of the set (14.35), we find

$$r_u(2) = 0.135 \frac{0.135}{0.2589} \sigma_u^2 + 0.7411\sigma_u^2 \tag{14.39}$$

Hence, the last three equations give the variance of u:

$$\sigma_u^2 = \frac{\sigma_1^2}{0.3282} = \frac{0.31}{0.3282} = 0.9445 \tag{14.40}$$

Using (14.40), (14.38), and the value $\sigma_2^2 = 0.12$, we obtain the correlation matrix

$$\mathbf{R}_x = \mathbf{R}_u + \mathbf{R}_{v_2} = \begin{bmatrix} 0.9445 & 0.4925 \\ 0.4925 & 0.9445 \end{bmatrix} + \begin{bmatrix} 0.12 & 0 \\ 0 & 0.12 \end{bmatrix}$$

$$= \begin{bmatrix} 1.0645 & 0.4925 \\ 0.4925 & 1.0645 \end{bmatrix} \tag{14.41}$$

From Figure 14.3.2, and specifically from the communication channel, we find the relation

$$u(n) - 0.931u(n-1) = d(n) \tag{14.42}$$

Multiplying (14.42) by $u(n)$ and then by $u(n-1)$, and taking the ensemble averages of the results, we obtain the correlation vector

$$\mathbf{p}_{dx} = [0.4860 \quad -0.3868]^T \tag{14.43}$$

Minimum mean square error (MMSE)

Introducing the above results in (14.9), we obtain the mean square surface (cost function) as a function of the filter coefficients. Hence,

$$J(\mathbf{w}) = 0.8461 - 2[w_0 \quad w_1]\begin{bmatrix} 0.4860 \\ -0.3868 \end{bmatrix} + [w_0 \quad w_1]\begin{bmatrix} 1.0645 & 0.4925 \\ 0.4925 & 1.0645 \end{bmatrix} \tag{14.44}$$

$$= 0.8461 + 0.972w_0 - 0.7736w_1 + 1.0645w_0^2 + 1.0645w_1^2 + 0.985w_0w_1$$

The mean square surface and its contour plots are shown in Figure 14.3.3.

Chapter 14: Least square system design, Wiener filter, and the LMS filter   603

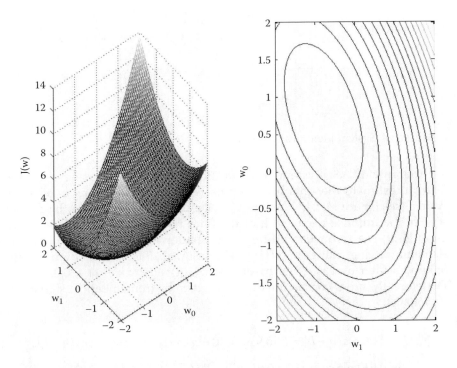

Figure 14.3.3 The MSE surface and its corresponding contour plots of Example 14.3.2.

Optimum filter (w^o)

The optimum filter coefficients are defined by (14.17), and in this case, the vector of optimum coefficients takes the following form:

$$\mathbf{w}^o = \mathbf{R}_x^{-1}\mathbf{p}_{dx} = \begin{bmatrix} 1.1953 & -0.5531 \\ -0.5531 & 1.1953 \end{bmatrix} \begin{bmatrix} 0.4860 \\ -0.3868 \end{bmatrix} = \begin{bmatrix} 0.7948 \\ -0.7311 \end{bmatrix} = \begin{bmatrix} w_0 \\ w_1 \end{bmatrix}$$

The MMSE is found using (14.18), which in this case gives the value

$$J_{\min} = \sigma_d^2 - \mathbf{p}_{dx}^T\mathbf{R}_x^{-1}\mathbf{p}_{dx} = 0.8461 - [0.4860 \ -0.3868]\begin{bmatrix} 1.1953 & -0.5531 \\ -0.5531 & 1.1953 \end{bmatrix}\begin{bmatrix} 0.4860 \\ -0.3868 \end{bmatrix}$$

$$= 0.1770$$

Observe that the center of the ellipse, minimum point, has the coordinates of the optimum vector coefficients 0.7948 and −0.7311.   ∎

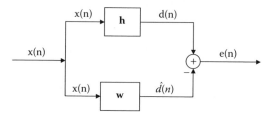

Figure 14.3.4 System identification setup.

Example 14.3.3: System identification. It is desired, using a Wiener filter, to estimate the unknown impulse response coefficients h_i of a FIR system (see Figure 14.3.4). The input $\{x(n)\}$ is a zero mean iid rv with variance σ_x^2. Let the impulse response **h** of the filter be $\mathbf{h} = [0.9\ 0.6\ 0.2]^T$. Since the input $\{x(n)\}$ is a zero mean iid rv, the correlation matrix $\mathbf{R}_x$ is diagonal with elements having values σ_x^2. The desired signal $d(n)$ is the output of the unknown filter and is given by $d(n) = 0.9x(n) + 0.6x(n-1) + 0.2x(n-2)$. Therefore, the cross-correlation output is given by (remember the signals are stationary):

$$p_{dx}(i) = E\{d(n)x(n-i)\} = E\{[0.9x(n) + 0.6x(n-1) + 0.2x(n-2)]x(n-i)\}$$
$$= 0.9E\{x(n)x(n-i)\} + 0.6E\{x(n-1)x(n-i)\} + 0.2E\{x(n-2)x(n-i)\}$$
$$= 0.9r_x(i) + 0.6r_x(i-1) + 0.2r_x(x-2)$$

Hence, we obtain ($r_x(m) = 0$ for $m \neq 0$ because the input data are independent to each other (white noise)): $p_{dx}(0) = 0.9\sigma_x^2$, $p_{dx}(1) = 0.6\sigma_x^2$.
The optimum filter is

$$\mathbf{w}^o = \mathbf{R}_x^{-1}\mathbf{p}_{dx} = (\sigma_x^2)^{-1}\begin{bmatrix}1 & 0 \\ 0 & 1\end{bmatrix}\begin{bmatrix}0.9 \\ 0.6\end{bmatrix} = (\sigma_x^2)^{-1}\begin{bmatrix}0.9 \\ 0.6\end{bmatrix}$$

and the MMSE is (assuming $\sigma_x^2 = 1$)

$$J_{min} = \sigma_d^2 - [0.9\ 0.6]\begin{bmatrix}1 & 0 \\ 0 & 1\end{bmatrix}\begin{bmatrix}0.9 \\ 0.6\end{bmatrix}.$$

But,

$$\sigma_d^2 = E\{d(n)d(n)\} = E\{[0.9x(n) + 0.6x(n-1) + 0.2x(n-2)]^2\}$$
$$= 0.81 + 0.36 + 0.04 = 1.21$$

and hence we find $J_{min} = 1.21 - (0.9^2 + 0.6^2) = 0.04$.

Book MATLAB function for system identification

```
function[w,jm]=sswienerfirfilter(x,d,M)
%function[w,jm]=aawienerfirfilter(x,d,M);
%x=data entering both the unknown filter(system) and the Wiener
%filter;
%d=the desired signal=output of the unknown system; length(d)=
%length(x);
%M=number of coefficients of the Wiener filter;
%w=Wiener filter coefficients;jm=minimum mean square error;
pdx=xcorr(d,x,'biased');
p=pdx(1,(length(pdx)+1)/2:((length(pdx)+1)/2)+M-1);
rx=sssamplebiasedautoc(x,M);
R=toeplitz(rx);
w=inv(R)*p';
jm=var(d)-p*w;%var() is a MATLAB function;
```

By setting, for example, the following MATLAB procedure, x=rand(1,256);d=filter([1 0.8 0.24],1,x); [w,jm]=sswienerfirfilter(x,d,6); we obtain: J_{min} = 0.0216 and **w** = [1.0002 0.7778 0.2399 0.0110 0.0074 0.0014]. Observe that the values after the third coefficient are close to zero. This indicates that the unknown system has only three coefficients. Observe that the derived coefficients are very close to the exact ones. ∎

Example 14.3.4: Noise canceling. In many practical applications there exists a need to cancel the noise added to a signal. For example, we are talking on a cell phone inside a car and the noise of the car, radio, etc., is added to the message we are trying to transmit. A similar circumstance appears when pilots in planes and helicopters try to communicate or tank drivers try to do the same. Figure 14.3.5 shows pictorially the noise contamination situations. Observe that the noise added to the signal and the other

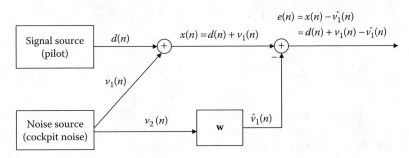

Figure 14.3.5 Noise canceling scheme.

components entering the Wiener filter emanate from the same source but follow different paths in the same environment. This indicates that there is some degree of correlation between these two noises. It is assumed that the noises have zero mean values. The output of the Wiener filter will approximate the noise added to the desired signal, and thus the error will be close to the desired signal. The Wiener filter in this case is

$$\mathbf{R}_{v_2}\mathbf{w}^o = \mathbf{p}_{v_1 v_2} \tag{14.45}$$

because the desired signal in this case is $\mathbf{v}_1$.
The individual components of the vector $\mathbf{p}_{v_1 v_2}$ are

$$p_{v_1 v_2}(m) = E\{v_1(n)v_2(n-m)\} = E\{(x(n) - d(n))v_2(n-m)\}$$
$$= E\{x(n)v_2(n-m)\} - E\{d(n)v_2(n-m)\} = p_{xv_2}(m) \tag{14.46}$$

Because $d(n)$ and $v_2(n)$ are uncorrelated,

$$E\{d(n)v_2(n-m)\} = E\{d(n)\}E\{v_2(n-m)\} = 0$$

Therefore, (14.45) becomes

$$\mathbf{R}_{v_2}\mathbf{w}^o = \mathbf{p}_{xv_2} \tag{14.47}$$

To demonstrate the effect of the Wiener filter, let $d(n) = 0.99^n \sin(0.1n\pi + 0.2\pi)$ be a decaying sine function, $v_1(n) = 0.8v_1(n-1) + v(n)$ and $v_2(n) = -0.95v_2(n-1) + v(n)$, where $v(n)$ is white noise with zero mean value and unit variance. The correlation matrix $\mathbf{R}_{v_2}$ and cross-correlation vector $\mathbf{p}_{xv_2}$ are found using the sample biased correlation equations

$$\hat{r}_{v_2 v_2}(k) = \frac{1}{N}\sum_{n=0}^{N-1} v_2(n)v_2(n-k) \qquad k = 0, 1, \cdots, K-1, K \ll N$$
$$\hat{p}_{xv_2}(k) = \frac{1}{N}\sum_{n=0}^{N-1} x(n)v_2(n-k) \qquad k = 0, 1, \cdots, K-1, K \ll N \tag{14.48}$$

Figure 14.3.6 shows simulation results for different-order filters using the Book MATLAB function given below (remember that you must have the Book MATLAB function in the MATLAB path).

Chapter 14: Least square system design, Wiener filter, and the LMS filter   607

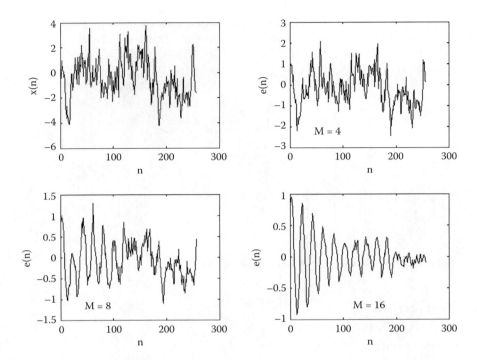

Figure 14.3.6 Noise canceling.

Book MATLAB function for noise canceling:
[d,w,xn]=sswienernoisecancelor(dn,a1,a2,v,M,N)

```
function[d,w,xn]=sswienernoisecancelor(dn,a1,a2,v,M,N)
%[d,w,xn]=sswienernoisecancelor(dn,a1,a2,v,M,N);dn=desired
%signal;
%a1=first order IIR coefficient,a2=first order IIR coefficient;
%v=noise;M=number of Wiener filter coefficients;N=number of
%sequence
%elements of dn(desired signal) and v(noise);d=output desired
%signal;
%w=Wiener filter coefficients;xn=corrupted signal;en=xn-v1=d;
v1(1)=0;v2(1)=0;
for n=2:N
    v1(n)=a1*v1(n-1)+v(n-1);
    v2(n)=a2*v2(n-1)+v(n-1);
end;
v2autoc=sssamplebiasedautoc(v2,M);
```

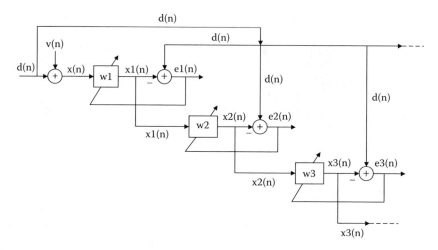

Figure 14.3.7 Self-correcting Wiener filter (SCWF).

```
xn=dn+v1;
Rv2=toeplitz(v2autoc);
p1=xcorr(xn,v2,'biased');
if M>N
   disp(['error:M must be less than N']);
end;
R=Rv2(1:M,1:M);
p=p1(1,(length(p1)+1)/2:(length(p1)+1)/2+M-1);
w=inv(R)*p';
yw=filter(w,1,v2);
d=xn-yw(:,1:N);
```
∎

We can also arrange the standard single Wiener filter in a series form, as shown in Figure 14.3.7. This configuration permits us to process the signal using filters with fewer coefficients, thus saving in computation. Because of the form, we can name it the **self-correcting Wiener filter (SCWF)**.

14.4 The least mean square (LMS) algorithm

In this chapter, we also present the celebrated least mean square (LMS) algorithm, developed by Widrow and Hoff in 1960. This algorithm is a member of stochastic gradient algorithms, and because of its robustness and low computational complexity, it has been used in a wide spectrum of applications.

Chapter 14: Least square system design, Wiener filter, and the LMS filter

The LMS algorithm has the following most important properties:

1. It can be used to solve the Wiener–Hopf equation without finding the matrix inversion. Furthermore, it does not require the availability of the autocorrelation matrix of the filter input and the cross-correlation between the filter input and its desired signal.
2. Its form and implementation are simple, yet it is capable of delivering high performance during the adaptation process.
3. Its iterative procedure involves (a) computing the output of a FIR filter produced by a set of tap inputs (filter coefficients), (b) generation of an estimated error by computing the output of the filter to a desired response, and (c) adjusting the tap weights (filter coefficients) based on the estimation error.
4. The correlation term needed to find the values of the coefficients at the $n + 1$ iteration contains the stochastic product $x(n)e(n)$ without the expectation operation that is present in the steepest descent method.
5. Since the expectation operation is not present, each coefficient goes through sharp variations (noise) during the iteration process. Therefore, instead of terminating at the Wiener solution, the LMS algorithm suffers random variation around the minimum point (optimum value) of the error-performance surface.
6. It includes a step-size parameter, μ, that must be selected properly to control stability and the convergence speed of the algorithm.
7. It is stable and robust for a variety of signal conditions.

The LMS algorithm

In many signal processing applications data are received one at a time, and because of this, the following question arises: Should we wait for all the data to be received first and then process (**batch** processing) them, or proceed to process as they arrive (**sequential** processing)? It turns out that sequential processing is desirable for two main reasons: we need less memory space and the results are given instantaneously.

Let, for example, that we have received N values of the signal $\{x(0), x(1), x(2), \ldots, x(N-1)\}$. The estimate mean value is given by (see also Chapter 13)

$$\hat{m}(N-1) = \frac{1}{N}\sum_{n=0}^{N-1} x(n) \qquad (14.49)$$

where the argument of $\hat{m}$ denotes the index of the most recent data point received. If we now observe the new data sample $x(N)$, then the mean value is given by

$$\hat{m}(N) = \frac{1}{N+1} \sum_{n=0}^{N} x(n) \qquad (14.50)$$

In computing this new estimate, we do not have to recompute the sum of the observations since the above equation can be cast in the form

$$\hat{m}(N) = \frac{1}{N+1} \left(\sum_{n=0}^{N-1} x(n) + x(N) \right) = \frac{N}{N+1} \hat{m}(N-1) + \frac{1}{N+1} x(N) \quad (14.51)$$

Note: *The new mean value is found by using the previous mean value plus the new sample.*

The sequential approach also lends itself to an interesting interpretation. Rearranging (14.51) we obtain

$$\hat{m}(N) = \hat{m}(N-1) + \frac{1}{N+1} [x(N) - \hat{m}(N-1)] \qquad (14.52)$$

Note: *The new estimate is equal to the old estimate plus a **correction** term. The correction term decreases with N. The term $[x(N) - \hat{m}(N-1)]$ can be thought of as the error in predicting x(N) by the previous samples.*

The LMS algorithm is similar to the above format, but with the difference that instead of numbers, vectors are involved. Hence, the sequential LMS algorithm is

$$\boxed{\begin{aligned} \mathbf{w}(n+1) &= \mathbf{w}(n) + 2\mu \mathbf{x}(n)[d(n) - \mathbf{x}^T(n)\mathbf{w}(n)] \\ &= \mathbf{w}(n) + 2\mu \mathbf{x}(n)[d(n) - \mathbf{w}^T(n)\mathbf{x}(n)] \\ &= \mathbf{w}(n) + 2\mu e(n)\mathbf{x}(n) \end{aligned}} \qquad (14.53)$$

where

$$y(n) = \mathbf{w}^T(n)\mathbf{x}(n) \quad \text{filter output} \qquad (14.54)$$

$$e(n) = d(n) - y(n) \quad \text{error} \qquad (14.55)$$

$$\mathbf{w}(n) = [w_0(n) \; w_1(n) \; \cdots \; w_{M-1}(n)]^T \quad \text{filter taps (coefficients) at time } n, \text{ an } M \times 1 \text{ vector} \qquad (14.56)$$

$$\mathbf{x}(n) = [x(n)\ x(n-1)\ x(n-2) \cdots x(n-M+1)]^T \quad \text{input data,} \quad (14.57)$$
$$\text{an } M \times 1 \text{ vector}$$

The algorithm defined by (14.53), (14.54), and (14.55) constitutes the LMS algorithm. The algorithm requires that at each time step $x(n)$, $d(n)$, and $\mathbf{w}(n)$ are known. The block diagram representation of the algorithm is given in Figure 14.4.1. The essential elements in adaptive signal processing are shown in Figure 14.4.2. As shown, the filter weights are adjusted at regular intervals, in accordance with an adaptive algorithm. The adaptive algorithm, as given above, usually uses, either explicitly or implicitly, the signals shown in Figure 14.4.2. These signals are the input signal, $x(n)$, the desired signal, $d(n)$, and the error signal, $e(n)$, which is the difference between $d(n)$ and $y(n)$, the filtered version of the input signal $x(n)$. The adaptive filtering algorithm continually varies the filter coefficients to reduce the mean square error toward its minimum value. Thus, the adaptive filter continually seeks to reduce the difference between the desired response, $d(n)$, and its own response, $y(n)$. It is this approach that the filter adjusts to its prescribed signal environment.

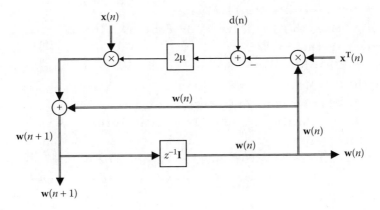

Figure 14.4.1 Block diagram representation of the LMS algorithm.

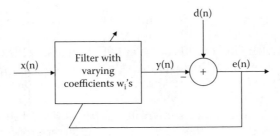

Figure 14.4.2 Schematic of adaptive signal processing.

Below is a step-by-step development of how the LMS algorithm works. The results were found using MATLAB help:

```
>>n=0:9;  x=[-0.4326  -1.3566  0.7131  1.0967  -0.1954  2.1909...
>>2.1402  0.7714  0.9151  0.4837];  mu=0.1;  dn=[0  0.3090...
>>0.5878  0.8090  0.9511  1.0000...
>>0.9511  0.8090  0.5878  0.3090];w0=[0  0]';
>>w1=w0+2*mu*x(1,1:2)'*(dn(1)-w0'*x(1,1:2)';%w1=[0  0]'
>>w2=w1+2*mu*x(1,2:3)'*(dn(2)-w1'*x(1,2:3)';%w2=[-0.0838
%0.0441];
>>w3=w2+2*mu*x(1,3:4)'*(dn(3)-w2'*x(1,3:4)';%w3=[0.0016
%0.1755]
etc
```

Book MATLAB function for LMS algorithm: [w,y,e,J]=sslms(x,dn,mu,M)

```
function[w,y,e,J]=sslms(x,dn,mu,M)
%function[w,y,e,J]=aalms(x,dn,mu,M);
%all quantities are real-valued;
%x=input data to the filter; dn=desired signal;
%M=order of the filter;
%mu=step-size factor; x and dn must be of the same length;
N=length(x);
y=zeros(1,N); %initialized output of the filter;
w=zeros(1,M); %initialized filter coefficient vector;
for n=M:N
   x1=x(n:-1:n-M+1); %for each n the vector x1 is
             %of length M with elements from x in reverse order;
   y(n)=w*x1';
   e(n)=dn(n)-y(n);
   w=w+2*mu*e(n)*x1;
end;
J=e.^2; %J is the learning curve of the adaptation;
```

Example 14.4.1: Find the 30-coefficient FIR filter using the Book MATLAB function for the LMS algorithm. Use the following: n=0:255; dn=sin(0.1*pi*n); x=dn+rand(1,256); mu=0.04; M=30;.

Solution: The results are shown in Figure 14.4.3. Observe that as the error, J, becomes smaller, the output of the filter approaches the desired signal. ∎

Chapter 14: Least square system design, Wiener filter, and the LMS filter

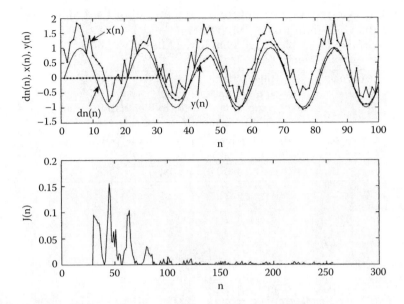

Figure 14.4.3 Illustration of Example 14.4.1.

The following Book MATLAB function provides the history of values for each filter coefficient.

Book MATLAB function providing the filter coefficient history: [w,y,e,J,w1]=sslms1(x,dn,mu,M)

```
function[w,y,e,J,w1]=sslms1(x,dn,mu,M)
%function[w,y,e,J,w1]=sslms1(x,dn,mu,M);
%this function provides also the changes of the filter coef-
%ficients
%versus iterations;
%all quantities are real-valued;
%x=input data to the filter; dn=desired signal;
%M=order of the filter;
%mu=step size; x and dn must be of the same length;
%each column of the matrix w1 contains the history of each
%filter coefficient;
N=length(x);
y=zeros(1,N);
w=zeros(1,M); %initialized filter coefficient vector;
for n=M:N
    x1=x(n:-1:n-M+1); %for each n the vector x1 of length M is
                      %produced
                      %with elements from x in reverse order;
```

Table 14.4.1 provides a comprehensive format of the LMS algorithm.

Table 14.4.1 LMS Algorithm for an Mth-Order FIR Adaptive Filter

Inputs:	M = filter length
	μ = step-size factor
	x(n) = input data vector to the adaptive filter
	w(0) = initialization filter vector = **0**
Outputs:	$y(n)$ = adaptive filter output = $\mathbf{w}^T(n)\mathbf{x}(n) \equiv \hat{d}(n)$
	$e(n) = d(n) - y(n)$ = error
	$\mathbf{w}(n+1) = \mathbf{w}(n) + 2\mu e(n)\mathbf{x}(n)$

```
  y(n)=w*x1';
  e(n)=dn(n)-y(n);
  w=w+2*mu*e(n)*x1;
  w1(n-M+1,:)=w(1,:);
end;
J=e.^2;%J is the learning curve of the adaptive process; each
       %column of the matrix w1
          %depicts the history of each filter coefficient;
```

14.5 Examples using the LMS algorithm

The following examples will elucidate the use of the LMS algorithm to different areas of engineering applications and will create an appreciation for the versatility of this important algorithm.

Example 14.5.1: Linear prediction. We can use an adaptive LMS filter as a predictor, as shown in Figure 14.5.1. The data $\{x(n)\}$ were created by passing a zero mean white noise $\{v(n)\}$ through an autoregressive (AR) process described by the difference equation $x(n) = 0.6010x(n-1) - 0.7225x(n-2) + v(n)$. The LMS filter is used to predict the values of the AR filter parameters 0.6010 and –0.7225. A two-coefficient LMS filter predicts $x(n)$ by

$$\hat{x}(n) = \sum_{i=0}^{1} w_i(n)x(n-1-i) \equiv y(n) \qquad (14.58)$$

Figure 14.5.2 gives w_0 and w_1 vs. the number of iterations for two different values of step size (μ = 0.02 and μ = 0.005). The adaptive filter is a two-coefficient filter. The noise is white and Gaussian distributed. The figure shows fluctuations in the values of coefficients as they converge to a neighborhood of their optimum value, 0.6010 and –0.7225, respectively. As the step-size μ becomes smaller, the fluctuations are not as large, but the convergence speed to the optimal values is slower.

Chapter 14: Least square system design, Wiener filter, and the LMS filter 615

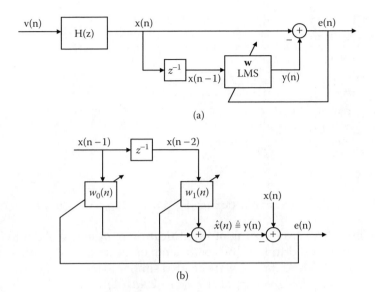

(a)

(b)

Figure 14.5.1 (a) Linear predictor LMS filter. (b) Two-coefficient adaptive with its adaptive weight-control mechanism.

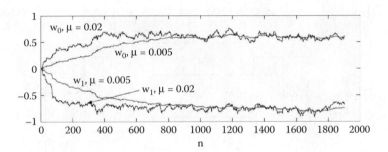

Figure 14.5.2 Convergence of two-element LMS adaptive filter used as a linear predictor.

Book one-step LMS predictor MATLAB function:
[w,y,e,J,w1]=sslmsonesteppredictor(x,mu,M)

```
function[w,y,e,J,w1]=sslmsonesteppredictor(x,mu,M)
%function[w,J,w1]=sslmsonesteppredictor(x,mu,M);
%x=data=signal plus noise;mu=step size factor;M=number of filter
%coefficients;w1 is a matrix and each column is the history
%of each
%filter coefficient versus time n;
N=length(x);
```

```
y=zeros(1,N);
w=zeros(1,M);
for n=M:N-1
   x1=x(n:-1:n-M+1);
   y(n)=w*x1';
   e(n)=x(n+1)-y(n);
   w=w+2*mu*e(n)*x1;
   w1(n-M+1,:)=w(1,:);
end;
J=e.^2;
   %J is the learning curve of the adaptive process;             ∎
```

Example 14.5.2: Modeling or identification of systems. Adaptive filtering can also be used to find the coefficients of an unknown filter, as shown in Figure 14.5.3. The data $x(n)$ were created similar to those in Example 14.5.1. The desired signal is given by $d(n) = x(n) - 2x(n-1) + 4x(n-2)$. If the output $y(n)$ is approximately equal to $d(n)$, it implies that the coefficients of the LMS filter are approximately equal to those of the unknown system. Figure 14.5.4 shows the ability of the LMS filter to identify the unknown system. After 500 iterations, the system is practically identified. For this example, we used $\mu = 0.01$ and $M = 4$. It is observed that the fourth coefficient is zero, as it

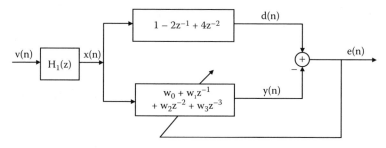

Figure 14.5.3 System identification setup.

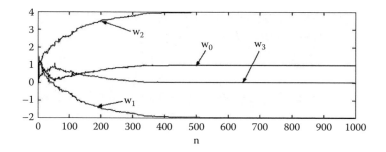

Figure 14.5.4 LMS adaptive filtering for system identification.

Chapter 14: Least square system design, Wiener filter, and the LMS filter

should be, since the system to be identified has only three coefficients and the rest are zero.

Note: *One important difference of adaptive filtering with Wiener filtering is that adaptive filtering does not need the desired signal to be known.* ∎

Example 14.5.3: Noise cancellation. A noise cancellation scheme is shown in Figure 14.5.5. We introduce in this example the following values: $H_1(z) = 1$ (or $h(n) = \delta(n)$), $v1(n)$ = white noise = $v(n)$, $L = 1$, $s(n) = \sin(0.2\pi n)$. Therefore, the input signal to the adaptive filter is $x(n) = s(n-1) + v(n-1)$ and the desired signal is $d(n) = s(n) + v(n)$. The Book LMS MATLAB algorithm **sslms** was used. Figure 14.5.6 shows the signal, the signal plus noise, and the outputs of the filter for three different sets of coefficients: $M = 4$, $M = 12$, and $M = 62$.

Book MATLAB m-file: ex_14_5_3

```
%Book m-file:ex_14_5_3
n=0:2000;mu=0.005;v=rand(1,2001)-0.5;
s=0.8*sin(0.1*pi*n);dn=s+v;
for m=1:1999
    x(m+1)=dn(m);
end;
subplot(2,2,1);plot(s(1,1:120),'k');xlabel('n');ylabel('s(n)');
subplot(2,2,2);plot(dn(1,1:120),'k');xlabel('n');ylabel...
    ('dn(n)');
[w,y,e,J,w1]=sslms1(x,dn,mu,4);
subplot(2,2,3);plot(y(1,4:124),'k');xlabel('n');ylabel...
    ('y(n), M=4');
[w,y,e,J,w1]=sslms1(x,dn,mu,32);
subplot(2,2,4);plot(y(1,32:152),'k');xlabel('n');ylabel...
    ('y(n), M=32');
```
∎

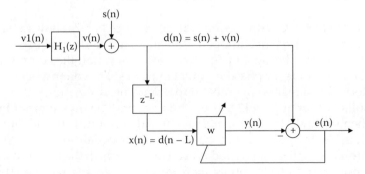

Figure 14.5.5 Adaptive LMS noise cancellation scheme.

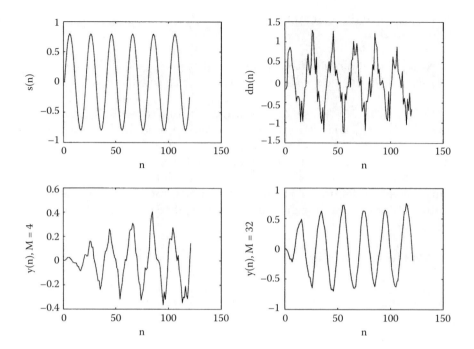

Figure 14.5.6 Noise cancellation using LMS filtering.

The transmission of signals through space, for example, change, to a large extent, due to noise introduced by the change of the index of refraction and scattering elements of the atmosphere. This phenomenon impedes, and sometimes completely destroys, the information to be transmitted. It is therefore desirable to find a transfer function that is equal to that of the atmosphere. Using the inverse of this transmission function, we will be able to neutralize to some extent the influence of the transmission channel, and thus receive the original signal undistorted or slightly so. In this example, we assume that the atmospheric channel changes slowly, and hence we can introduce the inverse filter periodically after we have specified the atmospheric channel by the adaptive filtering procedure.

Example 14.5.4: Channel equalization. Figure 14.5.7a shows a baseband data transmission system equipped with an adaptive channel equalizer and a training system. The signal $\{s(n)\}$ transmitted through the communication channel is amplitude- or phase-modulated pulses. The communication channel distorts the signal — the most important one is the pulse spreading — and results in overlapping of pulses, thus creating the **intersymbol interference** phenomenon. The noise $v(n)$ further deteriorates the fidelity of the signal. It is ideally required that the output of the equalizer is the signal $s(n)$. Therefore, an initialization period is used during which the transmitter sends a sequence of training symbols that are known to the receiver (**training mode**). This approach is satisfactory if the channel does not change its

Chapter 14: Least square system design, Wiener filter, and the LMS filter

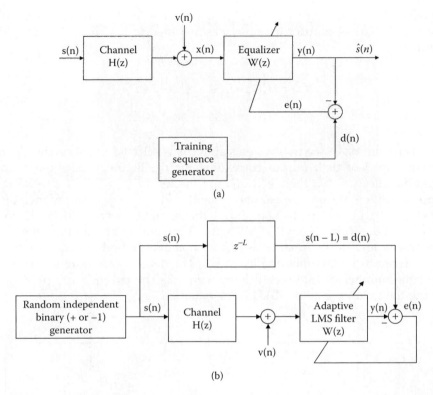

Figure 14.5.7 (a) Baseband data transmission system equipped with an adaptive channel equalizer and (b) a training system.

characteristics rapidly in time. However, for slow changes, the output from the channel can be treated as the desired signal for further adaptation of the equalizer, so that its variations can be followed (**decision-directed mode**).

If the equalization filter is the inverse of the channel filter, $W(z) = 1/H(z)$, the output will be that of the input to the channel, assuming, of course, that noise is small. To avoid singularities from the zero of the channel transfer function inside the unit circle, we select an equalizer such that $W(z)H(z) \cong z^{-L}$. This indicates that the output of the filter $W(z)$ is that of the input to the channel shifted by L units of time. Sometimes, more general filters of the form $Y(z)/H(z)$ are used, where $Y(z) \neq z^{-L}$. These systems are known as the partial-response signaling systems. In these cases, $Y(z)$ is selected such that the amplitude spectra are about equal over the range of frequencies of interest. The result of this choice is that $W(z)$ has a magnitude response of about 1, thereby minimizing the noise enhancement.

Figure 14.5.7b shows a channel equalization problem at the training stage. The channel noise $v(n)$ is assumed to be white Gaussian with variance σ_v^2. The equalizer is an M-tap (M-coefficient) FIR filter, and the desired output is assumed to be a delayed replica of the signal $s(n)$, $s(n - L)$. The signal $s(n)$

is white, has variance $\sigma_s^2 = 1$, has zero mean value, and is uncorrelated with $v(n)$. The channel transfer function was assumed to take the following FIR forms:

$$H(z) = H_1(z) = 0.34 + z^{-1} - 0.34z^{-2}$$
$$H(z) = H_2(z) = 0.34 + 0.8z^{-1} + 0.1z^{-2}$$
(14.59)

The solution of the two systems above was selected based on the eigenvalue spread of their output correlation matrix. Figure 14.5.8a shows the learning curve with a channel system of the form $H(z) = 0.34 + z^{-1} - 0.34z^{-2}$ and two different step-size parameters and $c = 0.5$. The variance of the noise was $\sigma_v^2 = 0.043$. Figure 14.5.8b shows the learning curve with a channel system of the form $H(z) = 0.34 + 0.8z^{-1} + 0.1z^{-2}$ and with the same step-size parameter and noise variance as in case a above. In both cases, the curves were reproduced 20 times and averaged. The delay L was assumed to be 3, and the number of filter coefficients were 12. The curves were produced using the following Book MATLAB function.

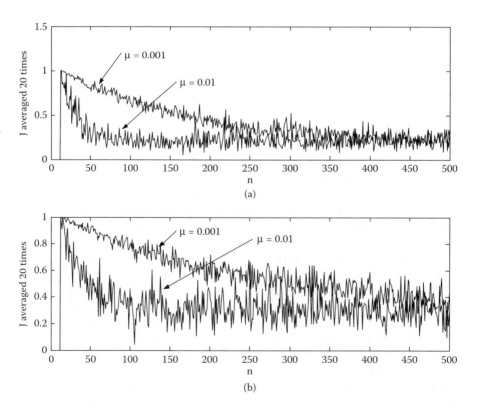

Figure 14.5.8 Learning curves of the channel system.

Book MATLAB function for channel equalization

```
function[Jav,wav,dn,e,x]=ssequalizer(av,M,L,h,N,mu,c)
%function[Jav,wav,dn,e,x]=aaequalizer(av,M,L,h,N,mu,c)
%this function solves the example depicted
%in Fig 14.5.7;av=number of times to average e(error or
%learning curve)and w(filter coefficient);N=length
%of signal s;L=shift of the signal s to become dn;h=assumed
%impulse response of the channel system;mu=step factor;
%M=number of adaptive filter coefficients;c=constant multiplier;
w1=[zeros(1,M)];
J=[zeros(1,N)];
for i=1:av
   for n=1:N
      v(n)=c*randn;
      s(n)=rand-.5;
      if s(n)<=0
         s(n)=-1;
      else
         s(n)=1;
      end;
   end;
      dn=[zeros(1,L) s(:,1:N-L)];
      ych=filter(h,1,s);
      x=ych(1,1:N)+v;
      [w,y,e]=sslms(x,dn,mu,M);
      w1=w1+w;
      J=J+e.^2;
   end;
Jav=J/av;
wav=w1/av;
```
∎

During the past 40 years, many more types of LMS filters were proposed. The main purpose of these proposed algorithms is to create a more robust and fast adaptive algorithm. One of the proposed algorithms that has often been used in practice is the **normalized** LMS (NLMS) algorithm:

$$\mathbf{w}(n+1) = \mathbf{w}(n) + \frac{\bar{\mu}}{\varepsilon + \mathbf{x}^T(n)\mathbf{x}(n)} e(n)\mathbf{x}(n) \qquad (14.60)$$

where $\bar{\mu}$ and ε are constants.

Book MATLAB normalized LMS filter algorithm

(Note: This algorithm can also be used with complex-valued signals.)

```
function[w,y,e,J,w1]=sscomplexnlms(x,dn,mubar,M,ep)
%function[w,y,e,J,w1]=sscomplexnlms(x,dn,mubar,M,c)
%x=input data to the filter;dn=desired signal;
%M=filter order;c=constant;mubar=step-size equivalent parameter;
%x and dn must be of the same length;J=learning curve;
N=length(x);
y=zeros(1,N);
w=zeros(1,M);%initialized filter coefficient vector;
for n=M:N
   x1=x(n:-1:n-M+1);%for each n vector x1 is of length
   %M with elements from x in reverse order;
   y(n)=conj(w)*x1';
   e(n)=dn(n)-y(n);
   w=w+(mubar/(ep+conj(x1)*x1'))*conj(e(n))*x1;
   w1(n-M+1,:)=w(1,:);
end;
J=e.^2;
%the columns of the matrix w1 depict the history of the
%filter coefficients;
```

Table 14.5.1 presents the NLMS algorithm.

Table 14.5.1 NLMS Algorithm

	Real-Valued Functions		Complex-Valued Functions
Input:			
	Initialization vector:	$\mathbf{w}(n) = \mathbf{0}$	
	Input vector:	$\mathbf{x}(n)$	
	Desired output:	$d(n)$	
	Step-size parameter:	$\bar{\mu}$	
	Constant:	ε	
	Filter length:	M	
Output:			
	Filter output:	$y(n)$	
	Coefficient vector:	$\mathbf{w}(n+1)$	
Procedure:			
1. $y(n) = \mathbf{w}^T(n)\mathbf{x} = \mathbf{w}(n)\mathbf{x}^T(n)$			1. $y(n) = \mathbf{w}^H(n)\mathbf{x}(n)$
2. $e(n) = d(n) - y(n)$			2. $e(n) = d(n) - \mathbf{w}^H(n)\mathbf{x}(n)$
3. $\mathbf{w}(n+1) = \mathbf{w}(n) + \dfrac{\bar{\mu}}{\varepsilon + \mathbf{x}^T(n)\mathbf{x}(n)} e(n)\mathbf{x}(n)$			3. $\mathbf{w}(n+1) = \mathbf{w}(n)$ $+ \dfrac{\bar{\mu}}{\varepsilon + \mathbf{x}^H(n)\mathbf{x}(n)} \mathbf{x}(n)e^*(n)$

Note: The superscript H stands for Hermitian or, equivalently, conjugate transpose.

Chapter 14: Least square system design, Wiener filter, and the LMS filter

Important definitions and concepts

1. Least squares technique
2. Linear least squares approximation
3. Mean square error
4. Wiener filter
5. Cost function
6. Correlation matrix
7. Variance of a signal
8. Mean square error surface
9. Optimum Wiener filter coefficients
10. Minimum error
11. Wiener solution
12. Orthogonality principle
13. Wiener filtering
14. Yule–Walker equations
15. System identification (modeling) using Wiener filtering
16. Noise canceling using Wiener filtering
17. Self-correcting Wiener filter
18. The least mean square (LMS) filter
19. Sequential adaptation
20. Batch processing
21. Applications using LMS filter: linear prediction, modeling, noise cancellation, channel equalization
22. Normalized LMS algorithm

Chapter 14 Problems

Section 14.1

1. Use the following equation to create 20 random points:

$$x = 5 + 0.5 * n + 5 * rand(1, 20)$$

   Accept $\hat{x}(n) = a + bn + cn^2$. Find a, b, and c and then superimpose the line $\hat{x}(n)$ on the random points.

2. Use the following equation to create 20 random points:

$$x = 1.2 + rand(1, 20)$$

   Accept $\hat{x}(n) = a + bn + cn^2$. Find a, b, and c and then superimpose the line $\hat{x}(n)$ on the random points.

3. Repeat Example 14.1.1, but now use $\hat{x}(n) = a + bn + cn^2 + dn^3$ and compare with the results of the example.

4. The cost function is

$$J(c) = \sum_{n=0}^{N-1}(x(n)-ch(n))^2.$$

Find the estimate value $\hat{c}$ and then find J_{min}.
5. Verify (14.5).
6. Repeat Example 14.1.3 for $N = 30$.

Section 14.2

1. Repeat Example 14.2.1 with the following data:

$$r_x(0) = 0.8, r_x(1) = 0.5, \sigma_d^2 = 15$$

$$\mathbf{p}_{dx} = [1.5 \; 5]^T$$

2. Introduce (14.15) in (14.13b) and show that the inequalities are true.
3. Use the values given and found in Example 14.2.1 to find J_{min}.
4. Prove (14.18).

Section 14.3

1. Verify (14.24).
2. Repeat Example 14.3.1 and substitute the fourth line of ex_14_3_1 with $v = 0.1 * randn(1,512)$.
3. Repeat Example 14.3.1 by introducing the following change to t_0 in the m-file:

   ```
   ex_14_3_1: v=1.5*randn(1,512); rd=sssamplebiasedautoc ...
       (d,80);
   rv=sssamplebiasedautoc(v,80); R=toeplitz(rd(1,1:60))+ ...
       toeplitz(rv(1,1:60));
   pdx=rd(1,1:60);
   ```

   State your observations and conclusions and compare to the results of Example 14.3.1.
4. Verify (14.31).
5. Verify (14.41).
6. Repeat Example 14.3.2 with the help of MATLAB to produce the different needed signals and the results, for example, $J(w)$, w^o, and J_{min}.

Chapter 14: Least square system design, Wiener filter, and the LMS filter

7. Repeat Example 14.3.3 using the inputs (a) x=2*randn(1,256), (b) n=0:255; x=sin(0.1*pi*n), and (c) n=0:255; x=sin(0.1*pi*n)+randn(1,256);. In all cases, use the following FIR filter: h = [0.95 0.423 0.110].
8. Repeat Example 14.3.4 with the difference by making the following substitution: $v2 = a2*v2(n-1) + v(n-1) + c*dn(n)$ for (a) $c = 0.05$, (b) $c = 0.1$, and (c) $c = 0.5$. State your observations and conclusions.
9. Find the outputs from a SCWF at the first and fourth states.

Section 14.4

1. Follow the development in the book and find the two-coefficient filter for four steps.
2. Repeat Example 14.4.1 for (a) $\mu = 0.005$, (b) $\mu = 0.01$, (c) $\mu = 0.5$. State your observations and conclusions.

Section 14.5

1. Repeat Example 14.5.1 with (a) x(n) = rand - 0.5 and (b) x(n) = 0.85x(n-1) + rand and different values of μ.
2. Repeat Example 14.5.1 with $\mu = 0.008$ and compare the signals $x(n)$ and $y(n)$.
3. Repeat Example 14.5.2 using the following unknown systems:
    a. $1 - 0.85z^{-1} - 0.6z^{-2}$
    b. $0.1 + 0.9z^{-1} + 0.23z^{-2}$
    c. $1 - 0.85z^{-2} + 0.55z^{-4}$
4. Repeat Example 14.5.2 by (a) using v = 1.5*rand(1,2001) - 0.5, (b) repeating (a) with mu = 0.01, and (c) repeating Example 14.5.2, but using a delay $L = 2$.
5. Use the NLMS algorithm in Examples 14.5.4, 14.5.3, 14.5.2, and 14.5.1.

appendix A

Mathematical formulas

A.1 Trigonometric identities

$$\cos(-a) = \cos a$$

$$\sin(-a) = -\sin a$$

$$\cos\left(a \pm \frac{\pi}{2}\right) = \mp \sin a$$

$$\sin\left(a \pm \frac{\pi}{2}\right) = \pm \cos a$$

$$\cos(a \pm \pi) = -\cos a$$

$$\sin(a \pm \pi) = -\sin a$$

$$\cos^2 a + \sin^2 a = 1$$

$$\cos^2 a - \sin^2 a = \cos 2a$$

$$\cos(a \pm b) = \cos a \cos b \mp \sin a \sin b$$

$$\sin(a \pm b) = \sin a \cos b \pm \cos a \sin b$$

$$\cos a \cos b = \frac{1}{2}[\cos(a-b) + \cos(a+b)]$$

$$\sin a \sin b = \frac{1}{2}[\cos(a-b) - \cos(a+b)]$$

$$\sin a \cos b = \frac{1}{2}[\sin(a-b) + \sin(a+b)]$$

$$c \cos a + d \sin a = \sqrt{a^2 - b^2} \cos[a - \tan^{-1}(d/c)]$$

$$\cos^2 a = \frac{1}{2}(1 + \cos 2a)$$

$$\cos^3 a = \frac{1}{4}(3 \cos a + \cos 3a)$$

$$\cos^4 a = \frac{1}{8}(3 + 4 \cos 2a + \cos 4a)$$

$$\sin^2 a = \frac{1}{2}(1 - \cos 2a)$$

$$\sin^3 a = \frac{1}{4}(3 \sin a - \sin 3a)$$

$$\sin^4 a = \frac{1}{8}(3 - 4 \cos 2a + \cos 4a)$$

$$e^{\pm ja} = \cos a \pm j \sin a$$

$$\cosh a = \cos ja$$

$$\cos a = \frac{1}{2}(e^{ja} + e^{-ja})$$

$$\sin a = \frac{j}{2}(e^{-ja} - e^{ja})$$

Appendix A: Mathematical formulas

$$\sinh a = -j \sin ja$$

$$\tanh a = -j \tan ja$$

A.2 Orthigonality

$$\sum_{n=0}^{N-1} \cos\frac{2\pi k}{N}n \cos\frac{2\pi l}{N}n = 0 \qquad 1 \leq k, l \leq N-1, k \neq l$$

$$\sum_{n=0}^{N-1} \sin\frac{2\pi k}{N}n \sin\frac{2\pi l}{N}n = 0 \qquad 1 \leq k, l \leq N-1, k \neq l$$

$$\sum_{n=0}^{N-1} \sin\frac{2\pi k}{N}n \cos\frac{2\pi l}{N}n = 0 \qquad 1 \leq k, l \leq N-1, k \neq l$$

$$\sum_{n=0}^{N-1} \cos^2\frac{2\pi k}{N}n = \begin{cases} N/2 & 1 \leq k, l \leq N-1, k \neq N/2 \\ N & k = 0, N/2 \end{cases}$$

$$\sum_{n=0}^{N-1} \sin^2\frac{2\pi k}{N}n = \begin{cases} N/2 & 1 \leq k, l \leq N-1, k \neq N/2 \\ 0 & k = 0, N/2 \end{cases}$$

The above formulas are correct if all k, l, and n are replaced by $k \bmod N$ and $l \bmod N$.

A.3 Summation of trigonometric forms

$$\sum_{n=0}^{N-1} \cos\frac{2\pi k}{N}n = \begin{cases} 0 & 1 \leq k \leq N-1 \\ N & k = 0, N \end{cases}$$

$$\sum_{n=0}^{N-1} \sin\frac{2\pi k}{N}n = \begin{cases} 0 & 1 \leq k \leq N-1 \\ N & k = 0, N \end{cases}$$

(k, l and n are integers)

A.4 Summation formulas

Finite summation formulas

$$\sum_{k=0}^{n} a^k = \frac{1-a^{n+1}}{1-a}, \quad a \neq 1$$

$$\sum_{k=1}^{n} ka^k = \frac{a(1-(n+1)a^n + na^{n+1})}{(1-a)^2} \quad a \neq 1$$

$$\sum_{k=1}^{n} k^2 a^k = \frac{a[(1+a)-(n+1)^2 a^n + (2n^2+2n-1)a^{n+1} - n^2 a^{n+2}]}{(1-a)^3} \quad a \neq 1$$

$$\sum_{k=1}^{n} k = \frac{n(n+1)}{2}$$

$$\sum_{k=1}^{n} k^2 = \frac{n(n+1)(2n+1)}{6}$$

$$\sum_{k=1}^{n} k^3 = \frac{n^2(n+1)^2}{4}$$

$$\sum_{k=0}^{2n-1} (2k+1) = n^2$$

Infinite summation formulas

$$\sum_{k=0}^{\infty} a^k = \frac{1}{1-a} \quad |a| < 1$$

$$\sum_{k=0}^{\infty} ka^k = \frac{a}{(1-a)^2} \quad |a| < 1$$

$$\sum_{k=0}^{\infty} k^2 a^k = \frac{a^2 + a}{(1-a)^3} \qquad |a| < 1$$

A.5 Series expansions

$$e^a = 1 + a + \frac{a^2}{2!} + \frac{a^3}{3!} + \cdots$$

$$\ln(1+a) = a - \frac{a^2}{2} + \frac{a^3}{3} - \frac{a^4}{4} + \cdots \qquad |a| < 1$$

$$\sin a = a - \frac{a^3}{3!} + \frac{a^5}{5!} - \frac{a^7}{7!} + \cdots$$

$$\cos a = 1 - \frac{a^2}{2!} + \frac{a^4}{4!} - \frac{a^6}{6!} + \cdots$$

$$\tan a = a + \frac{a^3}{3} + \frac{2a^5}{15} + \frac{17a^7}{315} + \cdots \qquad |a| < \frac{\pi}{2}$$

$$\sinh a = a + \frac{a^3}{3!} + \frac{a^5}{5!} + \frac{a^7}{7!} + \cdots$$

$$\cosh a = 1 + \frac{a^2}{2!} + \frac{a^4}{4!} + \frac{a^6}{6!} + \cdots$$

$$\tanh a = a - \frac{a^3}{3} + \frac{2a^5}{15} - \frac{17a^7}{315} + \cdots \qquad |a| < \frac{\pi}{2}$$

$$(1+a)^n = 1 + na + \frac{n(n-1)}{2!} a^2 + \frac{n(n-1)(n-2)}{3!} a^3 + \cdots \qquad |a| < 1$$

A.6 Logarithms

$$\log_b N = \log_a N \log_b a = \frac{\log_a N}{\log_a b}$$

A.7 Some definite integrals

$$\int_0^\infty x^2 e^{-ax} dx = \frac{2}{a^3}$$

$$\int_0^\infty x^n e^{-ax} dx = \frac{n!}{a^{n+1}} \qquad a > 0$$

$$\int_0^\infty e^{-a^2 x^2} dx = \frac{\sqrt{\pi}}{2a} \qquad a > 0$$

$$\int_0^\infty x e^{-a^2 x^2} dx = \frac{1}{2a^2} \qquad a > 0$$

$$\int_0^\infty \frac{e^{-ax}}{x} \sin mx\, dx = \tan^{-1} \frac{m}{a} \qquad a > 0$$

$$\int_0^\infty \frac{\sin mx}{x} dx = \frac{\pi}{2}$$

appendix B

Suggestions and explanations for MATLAB use

Before using the text, it is suggested that readers who do not have a lot of experience with MATLAB review this appendix and try to execute the presented material in MATLAB.

B.1 Creating a directory

It was found by the author that it is less confusing if for a particular project we create our own directory where our developed MATLAB m-files are stored. However, any time we need any of these files, we must include the directory in the MATLAB path. Let us assume that we have the following directory path: c:\ap\sig-syt\ssmatlab. We can use the following two approaches:

```
>>cd 'c:\ap\sig-syst\ssmatlab'
```

or

```
>>path(path,'c:\ap\sig-syst\ssmatlab')%remember to introduce
              %the path any time you start new
              %MATLAB operations; the symbol % is necessary
              %for the MATLAB to ignore the explanations;
```

The MATLAB files are included in the 'ssmatlab' directory.

B.2 Help

If we know the name of a MATLAB function and would like to know how to use it, we write the following command in the command window:

```
>>help sin
```

or

```
>>help exp
```
etc.

If we want to look for a keyword, we write:
```
>>lookfor filter
```

B.3 Save and load

When we are in the command window and have created many variables and, for example, would like to save two of them in a particular directory and in a particular file, we proceed as follows:

```
>>cd 'c:\ap\matlabdata'
>>save data1 x dt %it saves in the matlabdata directory the
                 %file data1 having
                 %the two variables x and dt;
```

Let us assume now that we want to bring these two variables in the working space to use them. First, we change directory, as we did above, and then we write in the command window:

```
>>load data1
```

Then, the two variables will appear in the working space ready to be used.

B.4 MATLAB as calculator

```
>>pi^pi-10;
>>cos(pi/4);
>>ans*ans; %the result will be (√2/2) × (√2/2) = 1/2 because
%the first output
%is eliminated, only the last output is kept in the form of ans;
```

B.5 Variable names

```
>>x=[1 2 3 4 5];
>>dt=0.1;
>>cos(pi*dt);%since no assignment takes place there is no
            %variable;
```

B.6 Complex numbers

```
>>z=3+j*4;%note the multiplication sign;
>>zs=z*z;%or z^2 will give you the same results;
>>rz=real(z);iz=imag(z):%will give rz=3, and iz=4;
>>az=angle(z); abz=abs(z);%will give az=0.9273 rad, and abz=5;
>>x=exp(-z)+4;%x=3.9675+j0.0377;
```

B.7 Array indexing

```
>>x=2:1:6;    %x is an array of the numbers {2, 3, 4, 5, 6};
>>y=2:-1:-2:  %y is an array of the numbers {2, 1, 0, -1, -2};
>>z=[1 3 y];  %z is an array of the numbers {1, 3, 2, 1, 0, -1, -2};
              %note the required space between array numbers;
>>xt2=2*x;    %xt2 is an array of numbers of x each one multiplied
              %by 2;
>>xty=x.*y;   %xty is an array of numbers which are the result of
              %multiplication of corresponding elements, that is
              %{4, 3, 0, -5, -12};
```

B.8 Extracting and inserting numbers in arrays

```
>>x=2:1:6;
>>y=[x zeros(1,3)];%y is an array of the numbers {2, 3, 4, 5,
              %6, 0, 0, 0};
>>z=y(1,3:7); %1 stands for row 1 which y is and 3:7 instructs
              %to keep columns
              %3 through 7 the result is the array {4, 5, 6,
              %0, 0};
lx=length(x);%lx is the number equal to the number of columns
              %of the row
              %vector x, that is lx=5;
x(1,2:4)=4.5*(1:3);%this assignment substitutes the elements
              %of x at column
              %positions 2,3 and 4 with the numbers 4.5*
              %[1 2 3]=4.5, 9,
              %and 13.5, note the columns of 2:4 and
              %1:3 are the same;
x(1,2:2:length(x))=pi;%substitutes the columns 2 and 4 of x with
              %the value of pi, hence the array is
              %{2, 3.1416, 4, 3.1416, 6};
```

B.9 Vectorization

```
>>n=0:0.2:1;
>>s=sin(0.2*pi*n);%the result of these two commands gives the
              %signal s
              %(sine function) at times (values of n)
              %0, 0.2, 0.4, 0.6, 0.4, 1;
```

This approach is preferable since MATLAB executes faster with the vectorization approach rather than the loop approach, which is

```
>>s=[];%initializes all values of vector s to zero;
>>for n=0:5%note that the index must be integer;
>>s(n+1)=sin(0.2*pi*n*0.2);%since we want values of s every
                %0.2 seconds
                %we must multiply n by 0.2; note also that
                %for n=0 the variable becomes s(1) and this
                %because the array in MATLAB always starts
                %counting columns from 1;
>>end
```

The results are identical with those of the previous approach.

B.10 Matrices

If a and b are matrices such that a is 2×3 and b is 3×3, then c = a*b is a 2×3 matrix.

```
>>a=[1 2; 4 6]; %a is a 2x2 matrix [1 2; 4 6];
>>b=a';%b is a transposed 2x2 matrix of a and is [1 4; 2 6];
>>da=det(a);%da is a number equal to the determinant of a, da=-2;
>>c=a(:);%c is a vector which is made up of the columns of a,
        %c=[1 4 2 6];
>>ia=inv(a); ia is a matrix which is the inverse of a;
>>sa1=sum(a,1);%sa1 is a row vector made up of the sum of the
               %rows, sa1=[5 8];
>>sa2=sum(a,2);%sa2 is a column vector made up by the sum of
               %the columns, sa2=[3 10]';
```

B.11 Produce a periodic function

```
>>x=[1 2 3 4];
>>xm=x'*ones(1,5);%xm is 4×5 matrix and each of its column is x';
>>xp=xm(:)';%xp is a row vector, xp=[x x x x x];
```

B.12 Script files

Script files are m-files that when we introduce their names in the command window we receive the results. However, we must have the directory that

Appendix B: Suggestions and explanations for MATLAB use

includes the file in the MATLAB search directories. You can modify the file any desired way and get new results. Suppose that any time we ask for the file pexp.m the magnitude and angle of the exponential function $e^{j\omega}$ are plotted. To accomplish this, we first go to the command window and open a new m-file. At the window we type the file as shown below. As soon as we finish typing, we click on *Save as* and save the file in, say, :c:\ap\ssmatlab. If we want to see the results, at the command window we just write *pexp* and hit the enter key.

Script file pexp.m

```
>>w=0:pi/500:pi-pi/500;%they are 500 at pi/500 apart;
>>x=exp(j*w);ax=abs(x);anx=angle(x);
>>subplot(2,1,1);plot(w,ax,'k')%'k' means plot line in black;
>>xlabel('\omega rad/s');ylabel('Magnitude');
>>subplot(2,1,2);plot(w,anx,'k');
>>xlabel('\omega rad/s');ylabel('Angle');
```

If we have the function

$$\frac{2e^{j\omega}}{e^{j\omega} - 0.5}$$

and want to plot the results as above, we substitute in the script file the function x with the function

```
x=2*exp(j*w)./(exp(j*w)-0.5);
```

In the above MATLAB expression note the dot before the slash. This instructs MATLAB to operate at each value of w separately, and thus give results at each frequency point.

B.13 Functions

We will present here an example of how to write functions. The reader should also study the functions that are presented throughout the book. In Fourier series, for example, we have to plot functions of the form

$$s(t) = \sum_{n=0}^{N} A_n \cos n\omega_0 t$$

and we want to plot this sum of cosines with each one having a different amplitude and frequency. Let $A = [1\ 0.6\ 0.4\ 0.1]$, $\omega_0 = 2$, and $0 \le t \le 4$. We approach this solution by vectorizing the summation. The MATLAB function is of the form

```
function[s]=sumofcos(A,N,w0,rangeoft)
n=0:N-1;
s=A*cos(w0*n'*rangeoft)
```
%when we want to use this function at the command window to
%find s we write for example:
```
>>A=[1 0.6 0.4 0.1];N=4;w0=2;rangeoft=0:0.05:6;
>>[s]=sumofcos(A,N,w0,rangeoft); %at the enter key click the
        %vector s is one of the variables in the command window
        %and it can be plotted at the wishes of the reader; we
        %must secure that the directory in which sumofcos function
        %exists is in the MATLAB path; after you type the function
        %in the editing window you save as.. in the directory,
        %for example, c:\ap\ssmatlab and filename: sumofcos.m
```

It is recommended that the reader set small numbers for N ($N = 4$) and the range of t (0:0.2:1) and produce first the matrix cos($w0*n'*t$) and then see the result $A*\cos(w0*n'*t)$.

B.14 Subplots

If, for example, we would like to plot four separate plots in a page, we write:

```
>>subplot(2,2,1);plot(x,y)%plots on the first half row a plot
                    %of x versus y;
>>subplot(2,2,2);plot(x1,y1); %plots on the second half of the
                    %first row;
>>subplot(2,2,3);plot(x2,y2); %plots the first half of the
                    %second row;
>>subplot(2,2,4);plot(x3,y3); %plots the second half of the
                    %second row;
                    %if we are working, say, in
                    %subplot(2,2,4) and we want
                    %to add labels in subplot(2,2,2)
                    we write;
>>subplot(2,2,2);
>>xlabel('Frequency');ylabel('Magnitude in dB');
```

B.15 Figures

If we have plotted something in figure(1) and we want to see another plot, without erasing the first plot, we write:

```
>>figure(2);
>>plot(x,y);%it will leave the plot in figure(1) but will now
        %plot in figure(2) and both
```

Appendix B: Suggestions and explanations for MATLAB use

```
                %will be available; if we want, next, to add
                %something to figure(1), we
                %write:
>>figure(1);
>>              %and continue on figure(1);
```

B.16 Changing the scales of the axes of a figure

To change scale we write:

```
>>axis([minimum x maximum x minimum y maximum y]);%for example,
                %we write: axis([1 166 -60 0]);
```

B.17 Writing Greek letters

Say, for example, we have a figure in which we want to place the following words: ω in rad/s, Ω in rad per unit. In the command window we write:

```
>>gtext('\omega in rad/s, \Omega in rad per unit');
                %at the return, a cross hair will appear in the
                %figure and we can set the cross
                %point anywhere we want; at the click of the mouse
                %the above sentence
                %will appear; observe the slash direction and the
                %greek letter designation
                %starting with small letter and a capital letter;
                %the same approach we use for
                %labeling, for example, xlabel('\lambda and \Pi')
                %will create the following
                %label along the x axis: λ and Π;
```

The following table gives some additional information:

≈ \approx	. \bullet	o \circ	÷ \div	↓ \downarrow
ε \epsilon	≡ \equiv	∃ \exists	∀ \forall	≥ \geq
∞ \infty	∫ \int	← \leftarrow	≤ \leq	≠ \neq
∂ \partial	± \pm	⊂ \subset		

For the Greek letters we have delta, gamma, lambda, omega, phi, pi, psi, sigma, theta, upsilon, xi, eta, iota, kappa, mu, nu, omega, rho, tau, and zeta. All these need the slash before them, and if we want capital letters, the first letter must be capital.

B.18 Subscripts and superscripts

If we want to write the following expression in a figure, $t_{i,j}^{2\pi x}$, in the command window we write:

```
>>gtext('t_{i,j}^{2\pix}');
```

B.19 Lines in plots

```
>>plot(x,y,'.'); %will plot blue dots at the {x,y} points;
>>plot(x,y,'.r') %will plot red points; if we substitute the r
                 %with k, we get black dots; g
                 %letter will give green.
>>plot(x,y,'g');%will plot green lines;
>>plot(x,y,'xr');%will plot red lines with x's;
>>plot(x,y,'-r',x,y,'xk');%plots green continuous line and at
                          %the {x,y} points puts black x's;
```

The following information is valuable:

Color: 'y', 'r', 'g', 'b', 'k'
Line style: '-' for solid, '- -' for dashed, ':' for dotted, '–·' for dash-dot
Markers: '+', 'o', '*', 'x', 's' for square, 'd' for diamond

Index

A

abs, 442, 445
Aliasing, 247, 251
Amplitude, 1
    modulation, 14, 204, 205
        pulse, 223
    response, 479
    sinusoidal wave, 4
    transfer function, 227
Analog filter(s). *See also* Filter(s)
    design, 477–501
        using MATLAB functions, 500–501
        ideal frequency response characteristics, 478
Analog signal, 250
    construction from sampled values, 252–253
Analog systems
    digital simulation, 118–131
        problems, 144–146
Analog-to-digital converter, 2
Angle, 442, 445
Angular velocity, 50
Asymmetric method, 575
Autocorrelation, 219, 550
    matrix, 554–557
    unbiased sample, 551
Autoregressive filter, 565
Autoregressive process, 565–567
Averages, 548–553
    mean value, 548–549
    problems, 579–580

B

Bandwidth, 4, 205, 224, 496
    finite, 239
    limitation, 4
    low-pass filter, 249
    time, 225
    transition, 519
    widening, 291
Banking, 144
Bartlett method, 572–573
Bartlett window, 17, 266, 513, 571
    Blackman-Tukey periodogram and, 577–578
Basis functions, 18
    orthonormal, 20
Basis set, 18, 19, 23
Batch processing, 623
Bilinear transformation, 532–538, 543–544
    problems, 543–544
Bioengineering, 325–327
Biomedical engineering, 102
Blackman-Tukey method, 570–572
Blackman-Tukey periodogram, Bartlett window and, 577–578
Blackman window, 266, 572
Block diagram(s), 42, 68
    elementary rules, 69
    operator, 69
    pick-off point, 68, 69
    representation, 68–71, 72–74
        problems, 103–104
    summation point, 69
Bode plot(s), 377–382
    of constants, 377–378
    differentiator, 378–379
    problems, 396–399
    for real pole, 379–382
Butterfly, 306
Butterworth filter(s), 479–486
    design, 500
    left-hand-side poles of fourth-order, 481
    phase and amplitude characteristics of second-order, 484

polynomials, 482
problems, 501–502

C

Carrier frequency, 204
Cascade stabilization, 365–366
Casoratian determinant, 451
Causality, 43, 137, 505
Central ordinate, 219
Channel, defined, 231
Channel equalization, 618–621
Chebyshev filter, 489–492
    files, 501
    low-pass, 486–493, 502
    problems, 502
Chebyshev polynomials, 486, 487
Chemical engineering, 128–131
Chi-square distribution, 560
Comb function, 13
Communication channel, 3
Communication system, 2
    distortion, 3
    message, 3
    receiving signal, 3
Complex conjugate, 181
Complex functions, 157–161
    problems, 181–182
Complex signals, 6
Compressed, 293
Compression, 293
Computer system, 57–60
Conductance, 47
Continuous functions
    features of periodic, 169–174
    Fourier series, 161–169
Continuous signal, sampling, 239–256.
    See also Sampling
Continuous-time signal(s)
    convolution, 71, 74–83
    correlation, 84–86
    problems, 105–106
Controller, 371
Convolution, 56, 173–174, 261
    associative property, 80–83
    of continuous-time signals, 71, 74–83
        problems, 105–106
    of discrete-time signals, 135–142
        problems, 148–149
    frequency, 217–218, 275–277
    property, 417–423
Correlation, 84–86, 261, 300, 549–552
    of sequences, 149

Correlation matrix, 592, 593, 595, 602, 604, 606
    output, 620
Correlogram spectral estimator, 568
Cost function, 583, 590
Covariance, 553
Critically damped response, 371
Critically damped second-order system, 452
Cross-correlation, 549
Cumulative density function, 546

D

D'Alembert's principle, 325
Damper element, 49–50, 52
Decimation, 284–291
    in time procedure, 304–307
Deconvolution, 83
Delay element, 112
Delta functions, 10, 190
Delta sampling, 248
Demodulation, 206
Dependent variable, 53
Detection schemes, 206
Determinant, 34
Deterministic systems, 43
DFT. *See* Discrete Fourier transform (DFT)
Difference equation(s), 113
    first order, 114
    higher order, 450–458
        problems, 471–473
Differential equation(s)
    Euler's approximation, 152–155
    higher order, 450–458
    homogeneous, 54
    nonhomogeneous, 55
    standard solution techniques, 56–63
        for computer system, 57–60
        for economics, 62–63
        for electro-optics, 60–62
    systems, 350–351
Digital signal(s), 2
Digital signal processing
    multirate, 284–295
Digital simulation
    analog systems, 118–131
        problems, 144–146
    of higher-order differential equations, 131–135
        problems, 146–147
Digital systems, higher-order, frequency response in, 443–450
Dirac delta function, 10
Dirichlet conditions, 162, 190
Discrete energy spectral density, 303

Discrete Fourier transform (DFT), 269–271
  inverse, 269
  properties, 271–284
    frequency, 275–277
    Parseval's theorem, 277–278
    proofs, 300–304
    shifting property, 272–274
    time convolution, 274–275
Discrete frequency, 7
Discrete random signal, 545, 548
Discrete source, 111
Discrete system(s)
  block diagram representation (*See* Block diagram(s))
  equations and, 111–118
    problems, 142–144
  first-order, frequency response of, 438–443
  linear time-invariant, 268
  phase shift in, 443
Discrete-time Fourier transform (DTFT), 257–267
  approximating, 257–260
  of finite time sequences, 262–267
  properties, 260–262
    problems, 296–298
    proofs of, 298–300
  windowing, 266–267
Discrete-time stochastic process, 546
Discrete-time systems, 43
  signals, 135–142, 148–149
Discrete-time transforms, 257–307
Distortion, 3
Distortionless filter, 228–229
Distributed parameters, 42
Dot product, 18
Double-sided suppressed carrier modulation, 206
Down sampler, 284
Down sampling, 284–286, 285
  block diagram representation, 285
  frequency domain of signals, 286–291
  MATLAB function for, 285–286
  signals, frequency domain of, 286–291
DTFT. *See* Discrete-time Fourier transform (DTFT)

E

Economics, differential equation for, 62–63
Edge detection, 106
Edge detector, 106
Electro-optics, 60–62
Electromechanical system, 323–325

Energy relation, 170
Energy-storing element, 53
Engineering economics, 117–118
Ensemble of realization, 547
Environmental engineering, 102–103, 349
Equilibrium price, 63
Ergodic process, 548, 549
Ergotic process, 551, 579
Error signal, 70, 325, 372, 611
Euler equations, 6
Euler method, 153–155
Event, 547
Expectation, 548–549
Exponential order, 310

F

Fast Fourier transform (FFT), 189, 257, 304–307, 568–569
Feedback stabilization, 367–368
Feedback system, 69, 365–377
  error signal, 372
  output sensitivity function, 369
  rejection of disturbance using, 369
  sensitivity in, 368–369
FFT. *See* Fast Fourier transform (FFT)
Filter(s). *See also* Finite impulse response (FIR) filters
  autoregressive, 565
  Butterworth, 479–486
  Chebyshev, 489–492
    files, 501
    low-pass, 486–493, 502
    problems, 502
  defined, 477
  design using MATLAB functions, 500–501
  digital, frequency transformations for, 538–542, 544
  distortionless, 228–229
  frequency transformations, 494–500
  general aspects of, 477–479
  ideal high-pass, 230
  ideal low-pass, 229–230
  linear time-invariant, 226–230
  matched, 83–84
  maximally flat, 480
  normalized, 483
  optimum, 603–608
  phase characteristics, 494
  prescribed specifications using Kaiser window, 519–522
    problems, 524
  problems, 501–505
  types, 501

whitening, 565
Wiener, 590, 591–594
   examples, 597–608
   problems, 624–625
   self-correcting, 608
Filtering, 4
  matched, 83–84
  properties, 357
  random processes, 562–567
  problems, 581
Final value property, 417
Finite duration impulse response, 115–116
Finite impulse response (FIR) filters, 436, 437, 505–524
  antisymmetric, 506
  band-pass, 522
  band-stop, 523
  causal, 506
  design, 522–523
  high-pass, 522
  linear phase, 506
  low-pass, 522
    symmetric, 508–509
  MATLAB design, 522–523
  properties, 505–509
    causality, 505
    frequency, 505–506
    phase consideration, 506–507
    problems, 523
    scaling transfer function, 508
  stability, 505
  symmetric, 506, 508–509
  using Fourier series approach, 509–512
    problems, 523
  using windows, 513–519, 523
    main lobe, 513
    problems, 524
    ripple ratio, 513
Finite impulse response (FIR) system, 115–116
Finite signals, 172–173
FIR filters. *See* Finite impulse response (FIR) filters
First order discrete system, 438–443
  problems, 467–468
First-order hold, 252
First order system(s), 52–63
  problems, 100–103
Fold-over frequency, 7
Fourier series, 162–169
  in complex exponential form, 162–165
  transform pair, 163
  in trigonometric form, 165–169
Fourier series of continuous functions, 161–169
  problems, 182–185

Fourier transform(s)
  direct and inverse, 190–196
    problems, 231–232
  pairs, 220–224, 237–238
  properties, 196–220
    central ordinate, 200–201
    continuous-time, 218–219
    derivatives, 208–213
    frequency convolution, 217–218
    frequency shifting, 202
    linearity, 196
    modulation, 202–207
    problems, 232–236
    scaling, 200
    symmetry, 196–198
    time convolution, 216
    time shifting, 199
  real functions, 194–196
Frequency convolution, 217–218, 275–277
Frequency differentiation, 219
Frequency division multiplexing, 206
Frequency interpretation, 549, 550
Frequency response
  discrete system, 438–443
  first order discrete system, 438–443, 467–468
  function, 439
  higher-order digital systems, 443–450
  LTI systems, 352–360
Frequency shifting, 219, 261, 299, 301
Frequency transformations
  for digital filters, 538–542
    problems, 544
  low-pass-to-band-pass, 495–498
  low-pass-to-band-stop, 498–500
  low-pass-to-high-pass, 495
  low-pass-to-low-pass, 494–495
  MATLAB functions for, 500–501
  problems, 503
Full-wave rectified signal, 183

G

Gaussian processes, 558–559
Generalized functions, 190
Gibbs' phenomenon, 164, 225–226
Gram–Schmidt orthogonalization process, 31

H

Half-wave rectified signal, 183
Hamming window, 17, 266, 513, 523, 571
Hanning window, 17, 266, 523, 572
Heat transfer, 101

Index

Higher-order digital systems, 443–446
Hurwitz polynomials, 484

I

Ideal high-pass filter, 230
Ideal low-pass filter, 229–230
Impulse function, 10
Impulse-invariant design method, 525–532
    problems, 543
Impulse response, 74, 86–96, 337
    problems, 107–108
    symmetrical, 507
Independent and identical distribution, 550, 553, 559, 580
Independent random variables, 553, 558
Indexed random variables, 545, 548
Inductor, 45–47
Infinite impulse response (IIR) filters, 525–544
    design, 525–543
        bilinear transformation, 532–538, 543–544
        frequency transformation for digital filters, 538–542, 544
        impulse-invariant method, 525–532
        recursive vs. non-recursive, 542–543
Infinite impulse response (IIR) system, 114, 436, 437
Infinite order, 510
Information source, 2
Initial condition, 53, 54
Initial state, 55
Initial time, 53
Initial value problems, 152
Initial value property, 417
Innovations process, 565
Innovations representation, 565
Input, 41
Integer, 285
Integral, 313–314
Integration constant(s), 152
    evaluation of, 63–68
        circuit behavior L and C, 65
        conservation of charge, 64
        conservation of flux linages, 64–65
        general switching, 65–68
        problems, 103–104
        switching of sources, 64
Interpolation, 284, 291–295
    block diagram of, 292
    by factor U, 291–292
    signals, 292–295
Interpolation function, 244
Intersymbol interference, 618

K

Kaiser window, 513, 519–522

L

Lags, 550
Laplace transform, 309–399
    inverse, 328–336
        problems, 390
    one-sided, 309–312
        problems, 383–384
    problem solving with, 336–351
        impulse response, 337
        problems relating to, 390–394
        step response, 338
        superposition, 338
        time invariance, 338
    properties, 312–316
        complex frequency shift, 398
        final value, 316, 399
        frequency shift, 315
        initial value, 316, 399
        integral, 313–314, 397–398
        linearity, 397
        multiplication by exponential, 314, 398
        multiplication by t, 314, 398
        problems, 384–385
        proofs, 397–399
        scaling, 315, 398
        time convolution, 315, 399
        time derivative, 313, 397
        time shifting, 314–315, 398
Leakage, 283, 570
Least mean square
    algorithm, 609
        examples using, 614–622
        problems, 625
    applications, 614–622
    normalized, 621
Least squares technique, 583–589
    problems, 623–624
Left-shifting property, 415–416
Legendre function, 22, 31
Linear and rotational mechanical systems, 388
Linear continuous-time systems, 41–109
    problems, 97–109
Linear interpolation, 252
Linear least square(s), 587–589
Linear least square approximation, 583
Linear mechanical system, 389
Linear prediction, 614–616
    modeling, 616–617

Linear system(s), 43
  continuous-time, 41–109
    problems, 97–109
  feedback, 365–377
    cascade stabilization of systems, 365–366
    parallel composition, 366–367
    problems, 396
    proportional integral differential controllers, 375–376
    rejection of disturbance using, 369
    sensitivity in, 368–369
    stabilization, 367–368
    step response, 370–375
  with periodic inputs, 174–180
    problems, 186–187
Linear time-invariant filters, 226–230
Linear time-variant (LTI) discrete system(s), 268
Linear time-variant (LTI) system(s), 21, 43
  frequency response, 352–360
    problems, 394–395
  pole location and stability of, 361–364
  transfer functions, 316–328
    problems, 386–389
Linearity, 137, 218, 260, 298
Lognormal distribution, 559–560
LTI systems. *See* Linear time-variant (LTI) system(s)
Lumped parameters, 42

M

Main lobe, 513
Mass element, 48
Matched filters, 83–84
Mathematical modeling, 7, 41
MATLAB, manipulation of matrices in, 35–36
MATLAB function residue, 329–332
Matrix(ices)
  addition and substractions of, 36
  adjoint, 34
  correlation, 592, 593, 595, 602, 604, 606, 620
  determinant, 34
  in elementary algebra, 33–35
  manipulations in MATLAB, 35–36
  multiplication, 34
  Toeplitz, 555, 567
  transposition of, 36
Maximally flat filter, 480
Mean square error, 23, 590–597
  minimum, 595
  surface, 593
Mean value, 548–549

Mechanical system, 321–323
Message, 3
Method of undetermined coefficients, 54, 150, 454–458
Method of variation of parameters, 149–152
Microphone, 386
Minimum error, 593
Minimum mean square error, 595, 602
Modulation, 2, 14, 202, 219, 261, 299
  amplitude, 14, 204, 205
    pulse, 223
  double-sided suppressed carrier, 206
  frequency, 29, 204
  index, 205
  need for, 204
  properties, 203
  single-sideband, 206
  sinusoidal, 205
  tone, 205
Monte Carlo method, 98
Multiplication
  matrices, 34
Multiplication by n property, 416
Multirate digital signal processing, 284–295
Multivariate distributions, 546

N

Noise cancellation, 605, 617–618
Nonanticipatory systems, 43
Noncausal transfer function, 510
Nonlinear systems, 43
  linearization of, 132
Nonparametric spectral estimation, 568
  Bartlett method, 572–573
  Blackman–Tukey method, 570–572
  Blackman–Tukey periodogram with Bartlett, 577–578
  correlogram, 568
  modified Welch method, 575–577
  periodogram, 568, 570
  problems, 581
  Welch method, 573–575
Nonrecursive system, 115
Normalized filter, 483
Nyquist frequency, 239, 244, 247, 248
Nyquist interval, 246

O

Operator, 47
Optimum filter, 603–608
Optimum Wiener coefficients, 593
Orthogonal variable, 553
Orthogonality condition, 597

Index

Orthonormal function, 20
Output response, 41
Overdamped response, 371
Overdamped second-order system, 452

P

Parallel composition, 366–367
Parseval formula, 169–170, 261
Parseval's theorem, 213, 219, 261, 277, 300, 303
Particular solution, 57
Percent overshoot, 370
Periodic continuous functions, 4
    features, 169–174
    problems, 18–186
Periodic sequence property, 416
Periodogram(s), 570, 574, 575
    nonparametric spectra estimation, 568
Periodogram spectral estimate, 303
Periodogram spectral estimator, 569
Phase characteristics, 494
Phase transfer function, 227
Physical system, 41
Plant, 371
Power spectrum, 560–561, 563, 565, 566
Probability density function, 545
Projection theorem, 31
Proper fractions, 423
Proper function, 333
Proportional controllers, 371–377
Proportional integral differential controllers, 375–377
pth order IIR filter, 438
Pulse amplitude modulation, 223
Pure sinusoid function, 250

Q

qth order FIR filter, 438
Quantitative description, 2

R

Radius of gyration, 51
Random processes, 562–567
Random sequels, 557–560
Random sequence, 546
Random signals, 545–548
    special, 557–560
        problems, 580
Random variables, 545, 548, 553, 558
    realization of, 555
Realization, 547, 592
    defined, 546, 547
    ith, 549
    of random variables, 555
    of stochastic process, 548
Receiving signal, 3
Rectangle window, 17, 226, 263, 266, 283, 513, 571
Rectangular pulse function, 8
Recursive filters, Infinite impulse response (IIR) filters
Replacement rate, 349
Resistance, 47
Resistor, 47–48
Ripple factor, 181, 183, 184, 495, 543
Ripple ratio, 513
Rise time, 370
Root locus, 372

S

Sample mean value, 548, 549
Sample variance, 548
Sampled function, 13–14, 243, 246, 251, 259, 269, 402
Sampling, 239–256
    delta, 248
    down, 284–286, 285
        block diagram representation, 285
        frequency domain of signals, 286–291
        MATLAB function for, 285–286
        signals, frequency domain of, 286–291
    frequency, 6, 240, 506, 571, 583
    fundamentals, 239–244
    interval, 239
    problems, 254–256
    rate, 240
    theorem, 244–253
    time, 6, 125
        changing, 130
    up, 291–295
        frequency domain characterization of signals, 292–295
    values, 239
Scalar multiplier, 112
Scaling, 200, 315, 398
    digital transfer function, 507–508
    time, 15–16, 219, 415
Second-order systems, 344
Separate roots, 329
Sequence(s)
    correlation of, 149
    down-sampled, 286–291
    DTFT of finite time, 262–267
    up-sampled, 291–295

Sequential processing, 623
Shape, 1
Shifting property, 272–274, 413–415
Signal(s)
  amplitude, 1
  applications involving, 1
  bandwidth, 4, 205, 224, 496 (*See also* Bandwidth)
  coding, 2
  complex, 5
  conditioning and manipulation, 14–17
    modulation, 14
    shifting and flipping, 14
    time scaling, 15–17
    windowing of scaling, 17
  continuous-time, 71, 74–86, 105–106
  convolution (*See* Convolution)
  digital, 2
  discrete, 1
    nonperiodic special, 10–13
      arbitrary sampled function, 13
      comb function, 13
      delta function, 10–12
  modulation, 2
  nonperiodic continuous, 7–10
    rectangular pulse function, 8
    sinc function, 9–10
    unit step, 7–8
  nonperiodic special discrete, 10–13
    arbitrary sampled function, 13
    comb function, 13
    delta function, 10–12
  periodic discrete time, 6–7
  quantitative description, 2
  receiving, 3, 5
  representation, 18–25
  shape, 1
  simple time, 3–13
  time duration, 1
  transducer, 1
  variance, 548, 591, 592, 595
  windowing, 17, 266–267 (*See also* Windowing)
Simple continuous system(s)
  electrical elements of, 43–48
    capacitor, 43–45
    inductor, 45
    resistor, 46–48
  mechanical rotational elements of, 50–52
    damper, 52
    inertial element, 50–51
    spring, 51–52
  mechanical translation elements of, 48–50
    damper, 49–50
    ideal mass element, 48
    spring, 48–49
  modeling of, 43–52
    problems, 97–100
Sinc function, 9–10, 225, 226, 244
Sine, 3, 6
Single-sideband modulation, 206
Sinusoidal wave(s)
  amplitude, 4
  phase spectra, 4
Smoothing system, 144
Spectral factorization, 565
Spectrum function, 190
Spring element, 48–49
Stability, 137
State, 54
Stationary process(es), 548, 554–557
  problems, 580
  purely random process, 557
  random walk, 557
  wide-sense, 554, 580
Steady-state response, 372
Steady-state solution, 56. 176
Step response, 370–375
Stochastic process, discrete-time, 546
Stochastic systems, 43
Survival function, 349
Symmetric functions, 170–172
  even, 172
  odd, 172
Symmetric method, 575
Symmetric prediction method, 575
Symmetry, 219
System(s)
  across variable, 44, 48
  causal, 43, 80, 138, 140, 209
    response, 74
  deterministic, 43
  discrete-time, 43
  first order
    solutions, 52–63
      problems, 100–103
  identification, 616–617
  impulse response, 74, 86–96, 337
    problems, 107–108
    symmetrical, 507
  linear, 43
  linear time invariant, 43
  nonanticipatory, 43
  nonlinear, 43
  physical, 41
  properties, 41–43

simple continuous
  electrical elements of, 43–48
    capacitor, 43–45
    inductor, 45
    resistor, 46–48
  mechanical rotational elements of, 50–52
    damper, 52
    inertial element, 50–51
    spring, 51–52
  mechanical translation elements of, 48–50
    damper, 49–50
    ideal mass element, 48
    spring, 48–49
  modeling of, 43–52
    problems, 97–100
stochastic, 43
time-varying, 43
System identification, 604–605
System transfer function, 432

T

Terminal properties, 42
Three-dimensional vector space, 19
Through variable, 44, 48
Time-average formula, 549
Time constant, 55
Time convolution, 216–217, 219, 274–275, 298–299, 301–303
Time derivative, 313
Time differentiation, 219
Time duration, 1
Time invariance, 338
Time multiplex, 239
Time multiplication, 299
Time reversal, 219, 260, 298, 304
Time scaling, 15–16, 219
  property, 415
Time series, 545
Time shifting, 218, 260, 298
Time-varying systems, 43
Toeplitz matrix, 555, 567
Torque, 50
Tracking error, 372
Training mode, 618
Training sequence, 366
Transducer, 1, 174, 545
  rotational electromechanical, 323, 324
Transfer function, 82, 177, 179, 209, 215, 228, 431–438

amplitude, 227
causal, 511
of distortionless filter, 228
feedback, 70
forward, 70
Fourier transform and, 216
higher-order, 437–438
inverse Fourier transform of, 226
loop, 70
of LTI systems, 316–328
noncausal, 510
open-loop, 70
phase, 227
problems, 466–467
scaling the digital, 507–508
system, 432
value, 211
z-transform, 431–438
Transient solution, 56, 100
Transition bandwith, 519
Transversal response, 115–116
Transversal system, 115
Triangle window, 17, 513, 571

U

Unbiased estimator, 549
Unbiased sample autocorrelation, 551
Uncorrelated random variables, 553
Underdamped response, 371
Underdamped second-order system, 452
Undetermined coefficients, 54, 150, 454–458
Up-sampled signals, 292–295
Up sampler, 284
  block diagram representation, 292
Up sampling, 291–295
  frequency domain characterization of signals, 292–295

V

Variable(s)
  dependent, 53
  independent random, 553, 558
  indexed random, 545, 548
  orthogonal, 553
  random, 545, 548
  through, 44, 48
  uncorrelated random, 553
Variance, 548, 591, 592, 595
Variation of parameters, 129
Vibration isolator, 102

W

Walsh functions, 32
Warping, 534
Weakly stationary process, 554
Welch method, 573–575
    modified, 575–577
White Gaussian noise, 558, 563
White noise, 550, 557–558
Whitening filter, 565
Wide-sense stationary process, 554, 580
Wiener coefficients, 593
Wiener filter(s), 590, 591–594
    examples, 597–608
    problems, 624–625
    self-correcting, 608
Wiener–Hopf equation, 595
Wiener–Kintchin relations, 560–562
    even function, 561
    power spectrum, 560
    problems, 580
Wiener solution, 594–597
Windowing, 17, 266–267
    Bartlett, 17, 266, 513, 571, 577–578
    Blackman, 266, 572
    Hamming, 17, 266, 513, 523, 571
    Hanning, 17, 266, 523, 572
    Kaiser, 513, 519–522
    rectangle, 17, 226, 263, 266, 283, 513, 571
    triangle, 17, 513, 571
Wroskian determinant, 451

Y

Yule-Walker equation, 601

Z

Z-transform
    convergence, 405–412
        problems, 460–461
    essential features, 401–405
    inverse, 423–431
        problems, 464–466
    Laplace transform and, 402
    pairs, 459–460
    problems, 459–460
    properties, 412–423
        convolution, 417–423
        final value, 417
        initial value, 417
        inverse, 423–431
        left shifting, 415–416
        linearity, 413
        multiplication by n, 417
        pairs, 423
        periodic sequence, 416
        problems, 461–464
        proofs, 473–476
        shifting property, 413–415
        time scaling, 415
        transfer function, 431–438
            higher-order transfer, 437
    region of convergence, 405
    region of divergence, 405
    solution of first-order difference equations, 447–450
    problems, 468–471
Zero-input response, 54, 55, 80
Zero-input solution, 337, 450
Zero mean sequence, 558, 604
Zero-order hold, 252
Zero-state response, 80
Zero-state solution, 337, 450
Zplane, 481